1994

Methods in Neurosciences

Volume 22

Neurobiology of Steroids

Methods in Neurosciences

Editor-in-Chief

P. Michael Conn

Methods in Neurosciences

Volume 22
Neurobiology of Steroids

Edited by

E. Ronald de Kloet
Win Sutanto

Division of Medical Pharmacology
Center for Bio-Pharmaceutical Sciences
Leiden/Amsterdam Center for Drug Research
University of Leiden
Sylvius Laboratories
2300 RA Leiden
The Netherlands

ACADEMIC PRESS
San Diego   New York   Boston   London   Sydney   Tokyo   Toronto

Front cover photograph: Light micrograph of the rat MBH which shows the result of a double immunostaining experiment for NPT and ß-endorphin using the NiDAB/DAB double immunostaining technique. Immunoreactivity for NPY was labeled by the dark-blue to black NiDAB reaction, while ß-endorphin-immunoreactive neurons were labeled with the brown DAB chromogen. For more details, see Chapter 25, Figure 6. Photograph courtesy of Drs. Csaba Leranth, Frederick Naftolin, Marya Shanabrough, and Tamas L. Horvath, Yale University School of Medicine, New Haven, Connecticut.

Copyright © 1994 by ACADEMIC PRESS, INC.

All Rights Reserved.
No part of this publication may be reproduced or transmitted in any form or by any means, electronic or mechanical, including photocopy, recording, or any information storage and retrieval system, without permission in writing from the publisher.

Academic Press, Inc.
A Division of Harcourt Brace & Company
525 B Street, Suite 1900, San Diego, California 92101-4495

United Kingdom Edition published by
Academic Press Limited
24-28 Oval Road, London NW1 7DX

International Standard Serial Number: 1043-9471

International Standard Book Number: 0-12-185292-X

PRINTED IN THE UNITED STATES OF AMERICA
94 95 96 97 98 99 EB 9 8 7 6 5 4 3 2 1

Table of Contents

Section III Molecular Effects of Steroids

Section IV   Cellular Effects of Steroids

Section V   Steroid Effects on Integrated Systems

Perspectives

Contributors to Volume 22

Article numbers are in parentheses following the names of contributors. Affiliations listed are current.

LUIGI F. AGNATI (9, 23), Department of Human Physiology, University of Modena, 41100 Modena, Italy

GUNNAR AKNER (9), Department of Medical Nutrition, Karolinska Institute, Novum F60, Huddinge University Hospital, S-141 86 Huddinge, Sweden

EFRAIN C. AZMITIA (22), Department of Biology, New York University, New York, New York 10003

ETIENNE-EMILE BAULIEU (3), INSERM U 33, Lab. Hormones, 94276 Le Kremlin-Bicêtre Cedex, France

D. BELELLI (27), Department of Pharmacology and Clinical Pharmacology, University of Dundee, Ninewells Hospital and Medical School, Dundee DD1 9SY, Scotland

NICHOLAS BODOR (2), Center for Drug Discovery, University of Florida, Gainesville, Florida 32610

DEBBIE BOWLBY (8), Department of Obstetrics and Gynecology, University of Toronto, Toronto, Ontario, Canada M5G 2C4

THEODORE J. BROWN (8), Department of Obstetrics and Gynecology, The Toronto Hospital Research Institute, University of Toronto, Toronto, Ontario, Canada M5G 2C4

J. PETER H. BURBACH (17), Department of Medical Pharmacology, Rudolf Magnus Institute for Neurosciences, Utrecht University, 3508 TA Utrecht, The Netherlands

GERSON CHADI (23), Department of Neuroscience, Karolinska Institute, S-171 77 Stockholm, Sweden

GEORGE P. CHROUSOS (14), Developmental Endocrinology Branch, National Institute of Child Health and Human Development, National Institutes of Health, Bethesda, Maryland 20892

MARTHA CHURCHILL BOHN (28), Department of Neurobiology and Anatomy, University of Rochester Medical Center, Rochester, New York 14642

ANTONIO CINTRA (9, 23), Department of Neuroscience, Karolinska Institute, S-171 77 Stockholm, Sweden

RAFAEL COVEÑAS (9), Depto de Biología Celular y Patología, 37007 Salamanca, Spain

JOKE J. COX (17), Department of Medical Pharmacology, Rudolf Magnus Institute for Neurosciences, Utrecht University, 3508 TA Utrecht, The Netherlands

KLAUS DAMM (16), Pharmacological Research, Dr. Karl Thomae Gmb H, 88397 Biberach an der Riss, Germany

MERCEDES DE LEÓN (9), Depto de Biología Celular y Patología 37007 Salamanca, Spain

LINDA A. DOKAS (20), Departments of Neurology, Biochemistry and Molecular Biology, Medical College of Ohio, Toledo, Ohio 43699

JAMES H. EBERWINE (19), Department of Pharmacology, University of Pennsylvania Medical School, Philadelphia, Pennsylvania 19104

CALEB E. FINCH (18), Department of Biological Sciences, Andrus Gerontology Center, University of Southern California, Los Angeles, California 90089

KJELL FUXE (9, 23), Department of Neuroscience, Karolinska Institute, S-171 77 Stockholm, Sweden

ELISE P. GÓMEZ-SÁNCHEZ (30), Department of Internal Medicine, University of Missouri, and Harry S. Truman Veteran Affairs Hospital, Columbia, Missouri 65201

KELVIN W. GEE (13), Department of Pharmacology, College of Medicine, University of California at Irvine, Irvine, California 92717

ELIZABETH GOULD (24), Laboratory of Neuroendocrinology, The Rockefeller University, New York, New York 10021

JAN-ÅKE GUSTAFSSON (4, 9, 23), Department of Medical Nutrition, Karolinska Institute, Novum F60, Huddinge University Hospital, S-141 86 Huddinge, Sweden

WILJAN HENDRIKS (17), Department of Cell Biology and Histology, University of Nijmegen, Trigon, 6500 HB Nijmegen, The Netherlands

JAMES P. HERMAN (12), Department of Anatomy and Neurobiology, University of Kentucky Medical Center, Lexington, Kentucky 40536

C. HILL-VENNING (27), Department of Pharmacology and Clinical Pharmacology, University of Dundee, Ninewells Hospital and Medical School, Dundee DD1 9SY, Scotland

RICHARD B. HOCHBERG (8), Department of Obstetrics and Gynecology, Yale University School of Medicine, New Haven, Connecticut 06510

TAMAS L. HORVATH (25), Department of Obstetrics and Gynecology, Yale University School of Medicine, New Haven, Connecticut 06510

XIAO PING HOU (22), Department of Biology, New York University, New York, New York 10003

MARIAN JOËLS (26), Department of Experimental Zoology, University of Amsterdam, 1098 SM Amsterdam, The Netherlands

IMRE KALLÓ (11), Department of Anatomy, Albert Szent-Györgyi Medical University, H-6720 Szeged, Hungary

MICHAEL KARL (14), Developmental Endocrinology Branch, National Institute of Child Health and Human Development, National Institutes of Health, Bethesda, Maryland 20892

ZYGMUNT KROZOWSKI (5), Molecular Hypertension Laboratory, Baker Institute for Medical Research, Prahran, Melbourne, 3181 Victoria, Australia

SEUNG P. KWAK (12), Mental Health Research Institute, University of Michigan, Ann Arbor, Michigan 48109

J. J. LAMBERT (27), Department of Pharmacology and Clinical Pharmacology, University of Dundee, Ninewells Hospital and Medical School, Dundee DD1 9SY, Scotland

CSABA LERANTH (25), Department of Obstetrics and Gynecology, and Section of Neurobiology, Yale University School of Medicine, New Haven, Connecticut 06510

ZSOLT LIPOSITS (11), Department of Anatomy, Albert Szent-Györgyi Medical University, H-6720 Szeged, Hungary

SOFIA LOPES DA SILVA (17), Department of Medical Pharmacology, Rudolf Magnus Institute for Neurosciences, Utrecht University, 3508 TA Utrecht, The Netherlands

NEIL J. MacLUSKY (8), Department of Obstetrics and Gynecology, The Toronto Hospital Research Institute, University of Toronto, Toronto, Ontario, Canada M5G 2C4

JEFFREY N. MASTERS (18), Ohio State Biotechnology Center, The Ohio State University, Columbus, Ohio 43210

MARGARET M. MCCARTHY (21), Department of Physiology, University of Maryland, School of Medicine, Baltimore, Maryland 21201

LINDA D. MCCAULEY (13), Department of Pharmacology, College of Medicine, University of California at Irvine, Irvine, California 92717

BRUCE S. MCEWEN (32), Laboratory of Neuroendocrinology, The Rockefeller University, New York, New York 10021

THOMAS F. MURRAY (7), College of Pharmacy, Oregon State University, Corvallis, Oregon 97331

FREDERICK NAFTOLIN (25), Department of Obstetrics and Gynecology, Yale University School of Medicine, New Haven, Connecticut 06510

SURESH M. NAIR (19), Department of Pharmacology, University of Pennsylvania Medical School, Philadelphia, Pennsylvania 19104

NANCY R. NICHOLS (18), Andrus Gerontology Center, University of Southern California, Los Angeles, California 90089

MILES ORCHINIK (7), Laboratory of Neuroendocrinology, The Rockefeller University, New York, New York 10021

WILLIAM M. PARDRIDGE (1), Department of Medicine, University of California at Los Angeles, Los Angeles, California 90024

J. A. PETERS (27), Department of Pharmacology and Clinical Pharmacology, University of Dundee, Ninewells Hospital and Medical School, Dundee DD1 9SY, Scotland

DONALD W. PFAFF (15), Laboratory of Neurobiology and Behavior, The Rockefeller University, New York, New York 10021

PAUL ROBEL (3), INSERM U 33, Lab. Hormones, 94276 Le Kremlin-Bicêtre Cedex, France

BERND SCHÖBITZ (31), Department of Neuroendocrinology, Max Planck Institute of Psychiatry, Clinical Institute, 80804 Munich, Germany

HEINRICH M. SCHULTE (14), Institute for Hormone and Fertility Research, University of Hamburg, 22529 Hamburg, Germany

MARYA SHANABROUGH (25), Department of Obstetrics and Gynecology, Yale University School of Medicine, New Haven, Connecticut 06510

MELLY SILVANA OITZL (29), Division of Medical Pharmacology, Leiden/

Amsterdam Center for Drug Research, University of Leiden, 2300 RA Leiden, The Netherlands

JAMES W. SIMPKINS (2), Center for the Neurobiology of Aging, University of Florida, Gainesville, Florida 32610

THOMAS F. SZURAN (6), Department of Animal Science, Swiss Federal Institute of Technology, CH-8092 Zürich, Switzerland

ERIKA P. VAN BINNENDIJK (10), E. C. Slater Institute, University of Amsterdam, 1018 TV Amsterdam, The Netherlands

ROEL VAN DRIEL (10), E. C. Slater Institute, University of Amsterdam, 1018 TV Amsterdam, The Netherlands

AERNOUT D. VAN HAARST (6), Department of Medical Pharmacology, Leiden/Amsterdam Center for Drug Research, University of Leiden, 2300 RA Leiden, The Netherlands

BAS VAN STEENSEL (10), E. C. Slater Institute, University of Amsterdam, 1018 TV Amsterdam, The Netherlands

MARGARET WARNER (4), Department of Medical Nutrition, Karolinska Institute, Huddinge University Hospital, Novum F60, S-141 86 Huddinge, Sweden

STANLEY J. WATSON (12), Mental Health Research Institute, University of Michigan, Ann Arbor, Michigan 48109

ANN-CHARLOTTE WIKSTRÖM (9), Department of Medical Nutrition, Karolinska Institute, Novum F60, Huddinge University Hospital, S-141 86 Huddinge, Sweden

CATHERINE S. WOOLLEY (24), Department of Neurological Surgery, University of Washington, Seattle, Washington 98195

ADRIAN WYSS (4), Department of Medical Nutrition, Karolinska Institute, Huddinge University Hospital, Novum F60, S-141 86 Huddinge, Sweden

SHIGETAKA YOSHIDA (4), Department of Medical Nutrition, Karolinska Institute, Huddinge University Hospital, Novum F60, 141 86 Huddinge, Sweden

HE YUAN (8), Department of Obstetrics and Gynecology, The Toronto Hospital Research Institute, University of Toronto, Toronto, Ontario, Canada M5G 2C4

YUAN-SHAN ZHU (15), Laboratory of Neurobiology and Behavior, The Rockefeller University, New York, New York 10021

Preface

Steroid hormone action in the brain has been actively explored for over a hundred years. Breakthroughs in techniques and methodologies were invariably followed by the elucidation of novel mechanisms underlying steroid control of brain function. Now, in the 1990s, there has been a rapid advance in sophisticated molecular and cellular techniques that are extremely important for the understanding of the genomic mechanism of action of steroid hormones. Another recent advance concerns the view that the brain too is able to generate neurosteroids independent of a peripheral source.

Steroid hormones are unique compounds in that they are active at the interface of peripheral endocrine events and neural mechanisms. Thus steroid hormone effects present an important peripheral signaling system to alter brain function. These effects, which are profound and long lasting, are observed in early life, after recovery from damage by stress and toxic events, in reproduction and adaptation, and during aging when homeostasis is more readily challenged.

This book is a compendium of state-of-the-art techniques in the field of steroid hormone research from molecule to behavior. Students and researchers venturing into the study of the neurobiology of steroids will find it timely and highly stimulating. The presentation of some of the chapters allows the reader to follow the techniques in a step-by-step manner. Several chapters, without giving the fine detail of methodology, present conceptual and in-depth evaluations. By combining these approaches in one volume, we hoped to achieve a balance between detailed description of novel as well as classical techniques and a critical review of the subject.

E. RONALD DE KLOET
WIN SUTANTO

Methods in Neurosciences

Section I

Steroid Kinetics and Metabolism

[1] Steroid Hormone Transport through Blood–Brain Barrier: Methods and Concepts

William M. Pardridge

Introduction

The blood–brain barrier (BBB) is composed of the brain capillary endothelial wall. The capillary endothelium in the brains of all vertebrates is endowed with epithelial-like, high-resistance tight junctions that create an absence of either paracellular (i.e., open junctions) or transcellular (i.e., pinocytosis) movement from blood to the brain interstitial fluid (1). Accordingly, steroid hormones may undergo transport through the BBB via only one of two mechanisms: lipid mediation or carrier mediation. There is no convincing evidence that steroid hormones undergo carrier-mediated transport through the BBB, because the BBB transport of these molecules is not saturable (2). Rather, steroid hormones gain access to brain interstitial fluid from blood via lipid-mediated transport through the BBB based on the lipid solubility of the compound (2). However, because virtually all steroid hormones are bound to plasma proteins, including albumin and specific globulins, plasma protein-mediated transport plays a dominant role in determining the extent to which steroid hormones undergo transport through the BBB *in vivo* (3).

The binding of steroid hormones to albumin and specific globulins raises the issue as to whether it is the free or protein-bound moiety in plasma that is the principal component available for uptake by brain. Plasma proteins, per se, do not undergo significant exit from the plasma compartment in brain. Therefore, the availability to brain of circulating plasma protein-bound steroid hormones involves a mechanism of enhanced dissociation of the ligand from the protein at the endothelial plasma membrane surface (4). The rate of hormone dissociation from the plasma protein binding site *in vivo* within the brain microcirculation exceeds the rate of hormone dissociation *in vitro* as determined by such methods as equilibrium dialysis or ultrafiltration. The enhanced dissociation of hormones from plasma protein binding sites *in vivo* is believed to be mediated via conformational changes about the ligand-binding sites that are induced by adsorption of the plasma protein to the endothelial surface and glycocalyx. Therefore, the complexities of steroid hormone transport through the BBB *in vivo* arise from the mechanisms underlying plasma protein-mediated transport of steroid hormones through the BBB. Whereas the methods for studying plasma protein-mediated transport of steroid hormones through the BBB are relatively straightforward, the data analysis requires a firm understanding of the concepts underlying the fusion of two entirely separate disciplines, that is, mass-action binding equilibria and *in vivo* capillary physiology. Therefore, this chapter emphasizes both the methods and concepts of steroid hormone transport through the BBB.

Blood–Brain Barrier Transport: Methodological Approaches

In Vivo Methodologies

Cerebrospinal Fluid Steroid Hormone Measurement

The two extracellular fluid compartments in brain are the brain interstitial fluid (ISF) and the cerebrospinal fluid (CSF), the latter occupying the four cerebral ventricles and the subarachnoid space. These two extracellular fluid compartments are segregated by two anatomically and functionally distinct membrane systems, that is, the BBB, which segregates the blood and ISF, and the choroid plexus or blood–CSF barrier, which segregates blood and CSF (Fig. 1). The CSF, like salivary gland fluid, is often regarded as an ultrafiltrate of plasma. Therefore, it is assumed that the measurement of steroid hormone concentrations in the CSF (5), or salivary fluid (6), reflects the concentration in plasma of steroid hormone that is available for transport into the organ, that is, the bioavailable fraction. Moreover, it is often assumed that this bioavailable fraction is equivalent to the portion of plasma hormone that is free (dialyzable) *in vitro*. For example, in one study, the CSF/plasma ratio of a series of steroid hormones was compared to the percentage free (dialyzable) fraction of plasma hormone, and found to be approximately equivalent (5). On this basis it was concluded that the concentration of steroid hormone entry into CSF, which is also assumed to be in equilibrium with ISF, was a function of the plasma free hormone and not the lipid solubility of the individual steroid hormone. (As discussed in Section III, BBB transport of steroid hormones is a function of both lipid solubility and plasma protein binding.) On the basis of these measurements of steroid hormone distribution to CSF, it was concluded that the reason that dihydrotestosterone (DHT) does not suppress luteinizing hormone (LH) secretion is because DHT transport across the BBB is minimal owing to avid plasma protein binding (7). However, DHT does undergo transport through the BBB (8). The failure of DHT to suppress LH is likely to be because this androgen is nonaromatizable in brain.

The interpretation of steroid hormone distribution in the CSF should consider the following general principles: (a) solute distribution in the CSF

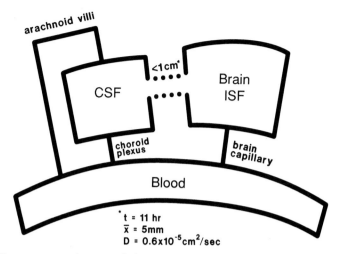

FIG. 1 The two central extracellular compartments of brain are the cerebrospinal fluid (CSF) and the brain interstitial fluid (ISF). The CSF is segregated from the blood by the choroid plexus or blood–CSF barrier and the ISF is segregated from blood by the brain capillary or BBB. Although the two extracellular fluid compartments are separated by a distance of less than 1 cm, there is a functional barrier preventing complete equilibration between these two fluid compartments. This functional barrier arises from the fact that CSF is constantly formed at the choroid plexus and drained into the peripheral bloodstream at the arachnoid villi, and this CSF volume is completely cleared in the human brain approximately every 4 to 5 hr. The bulk flow of CSF through the brain is rapid compared with the relatively slow rate of solute diffusion within brain. For example, small molecules such as a steroid hormone, with a diffusion coefficient of 0.6×10^{-5} cm^2/sec, diffuse 5 mm in approximately 11 hr, assuming no enzymatic metabolism or tissue sequestration of the compound. Therefore, the relative rate of the solute diffusion into brain parenchyma from the CSF is slow compared to the rate of solute efflux from the CSF compartment back to blood. [From Pardridge (24 and 67).]

reflects primarily transport across the choroid plexus epithelium, that is, the blood–CSF barrier, and not transport across the brain capillary endothelium, that is, the BBB (Fig. 1); and (b) solute distribution into CSF from plasma is a function not only of the concentration of bioavailable steroid hormone in plasma, but is also inversely related to the metabolic degradation of the steroid hormone within the choroid plexus epithelium. That is, CSF is not an ultrafiltrate of plasma, but rather is the secretory product of the choroid plexus epithelium. Therefore, the near equivalence of the CSF/plasma ratio of steroid hormone and the percentage free (dialyzable) fraction of plasma hormone indicates either (a) steroid hormone transport across the choroid plexus epithelium is restricted to the free (dialyzable) fraction within choroid plexus capillaries, or (b) the bioavailable steroid hormone in choroid plexus includes portions of the albumin or globulin-bound hormone, but the pool size of exchangeable steroid hormone in choroid plexus epithelium is contracted relative to the pool size of plasma-exchangeable hormone. This contraction of cellular exchangeable hormone pool size is due to active metabolism of the steroid hormone by choroid plexus epithelium. For example, the concentration of steroid hormone in salivary gland fluid is much less than the concentration of plasma-exchangeable hormone (which includes the free plus the albumin-bound hormone fraction in plasma), owing to the rapid degradation of steroid hormone by salivary gland epithelium (8).

Intracerebral Dialysis Fibers

The concentration of exchangeable steroid hormone in brain ISF is potentially amenable to direct *in vivo* measurement using intracerebral dialysis fibers. This methodology allows for direct sampling of brain ISF. When the concentrations of solute in ISF and CSF have been simultaneously measured, previous studies have shown that there is poor equilibration of solute between these two compartments (9). That is, the measurement of solute distribution in CSF should not be used as an indirect measurement of solute distribution in brain ISF. Rather, the latter parameter should be directly measured with dialysis fibers. The two most important experimental variables underlying the use of dialysis fibers in quantitation of solute distribution in the brain ISF are (a) the *in vivo* recovery, and (b) the effects of local brain injury caused by implantation of the dialysis fiber or cannula on the local tissue physiology. With respect to dialysis fiber recovery, this parameter generally ranges from 1 to 10%, and is inversely related to the flow rate (e.g., 0.1–10 μl/min) through the dialysis fiber or cannula. The recovery is also directly related to the length of the dialyzable surface of the cannula or probe. Although the *in vitro* recovery may be as high as 44% when 5 cm of dialysis fiber is used (10), the *in vitro* recovery may be lower with the use of small dialysis fiber probes, which may have a dialysis surface length of 1–10 mm. It was originally assumed that the dialysis fiber recovery, which reflects the incomplete equilibration of solute across the dialysis fiber wall, could be measured *in vitro* in a "beaker experiment." However, subsequent studies have shown that the dialysis fiber recovery *in vivo* in brain tissue is considerably lower than the recovery recorded *in vitro* (10–13). The importance of measuring the *in vivo* recovery is greatest when attempting to quantitate the actual concentration of cellular exchangeable hormone in brain. In one study, the cellular exchangeable fraction of diazepam, which is regulated by the same factors that control cerebral exchangeable steroid hormone levels, was identical to the free (dialyzable) fraction of plasma diazepam (14). However, this study assumed that the *in vitro* recovery was identical to that found *in vivo*. Subsequent investigations have shown that the *in vivo* recovery is much lower than the *in vitro* recovery (10–13). Therefore, the use of the *in vitro* recovery fraction will lead to the underestimation of the concentration of cellular exchangeable hormone in brain. For example, in a study of cocaine, a lipophilic amine bound by plasma proteins, the concentration of brain cellular exchangeable cocaine was threefold higher than the free fraction of cocaine in plasma (15). These studies are pertinent to steroid hormone transport because both cocaine and diazepam, like some steroid hormones, are bound avidly by both albumin and a specific globulin called α_1-acid glycoprotein (8). In addition to the lack of equivalence between *in vivo* and *in vitro* recovery with dialysis fibers, a second problem is the introduction of local brain injury with the implantation of the fiber (16, 17). Although the BBB may be relatively intact following implantation of the catheter (13), other studies have shown that the dialysis fibers disturb the local neurochemistry of brain such that an acute brain injury model is created by implantation of the fiber or probe (16, 17).

Measurements of Net Metabolic Clearance or Extraction of Steroid Hormones by Brain

The metabolic clearance rate (MCR) of steroid hormones by brain is computed from the product of the rate of cerebral blood flow (F) times the net extraction (E). The net extraction is defined as

$$E = (C_v - C_a)/C_a$$

where C_v and C_a equal the concentrations of steroid hormone in the cerebral venous and arterial compartments, respectively, as determined by arterial–venous difference measurements. The measurement of arterial–venous differences across the brain should be performed carefully, using techniques that minimize extracerebral contamination of the cerebral venous effluent (18). If internal jugular venous blood is removed by aspiration under pressure, then the direction of blood flow within the head and neck may be reversed such that the venous drainage of extracerebral structures in the head contributes to the internal jugular venous effluent. Thus, it is important to collect the internal jugular venous effluent by gravity to minimize extracerebral contamination.

The measurement of net extraction of steroid hormones by the brain has led to some anomalous findings. For example, the net extraction of progesterone by brain is reported to be in the range of 25–60% (19). In the steady state, the net extraction of a substance by brain reflects the metabolic conversion of the circulating compound into a metabolite. However, the metabolism of circulating progesterone by brain is relatively slow compared to the rapid rates of net extraction of the steroid hormone by brain from blood (20, 21). A similar anomalously high net extraction of estrone across the dog hind limb has been reported, and this high net extraction has been attributed in part to binding of steroid hormone to albumin and globulin pools in the lymphatic circulation, which provides for a net drainage of solute and plasma constituents from the organ back to the systemic circulation (22). However, this mechanism is difficult to propose in the case of brain because the brain lacks a lymphatic system. The brain does have a measurable bulk flow of fluid through the parenchymal spaces, but this rate is very low, on the order of 0.1 μl/min/g in the rat brain (23). This rate of flow movement is <5% of the flow rate of CSF (24), and would seem insufficient as a drainage mechanism for cerebral progesterone. Therefore, the high cerebral net extraction of progesterone, in the absence of rapid rates of metabolic conversion of this steroid hormone by brain *in vivo*, is an anomalous finding and at present has no explanation.

Unidirectional Brain Extraction

The unidirectional transport clearance rate (TCR) is a function of cerebral blood flow (F) and the unidirectional extraction of steroid hormone by brain. The latter parameter is measured with radioisotopes and either constant infusion or single-injection methodologies and reflects the one-way movement of steroid hormone from blood to brain (called influx) or from brain to blood (called efflux). The unidirectional extraction (E) is defined by the Kety–Renkin–Crone (KRC) equation of capillary physiology, that is,

$$E = 1 - e^{-fPS/F}$$

where PS equals the permeability–surface area product and f equals the fraction of plasma-exchangeable hormone, also called the bioavailable hormone fraction. For a single plasma protein-binding system, the exchangeable fraction is further defined as (25)

$$f = \frac{K_D^a/n}{A_F + K_D^a/n}$$

where A_F equals the unbound plasma protein concentration, for example, the albumin concentration, K_D^a equals the apparent dissociation constant governing the plasma protein/ligand-binding reaction *in vivo* within the brain microcirculation, and n equals the number of ligand-binding sites on the protein. This derivation of the KRC equation, modified for solute that undergoes plasma protein binding, assumes that the plasma protein-binding reaction is in the steady state *in vivo,* and this assumption requires only that the rate of ligand dissociation and/or association with the plasma protein proceeds at a rate in excess of the rate of ligand permeation through the BBB membrane (25). That is, k_1 or $k_2 A_F \gg k_3$, where k_1 equals the rate constant of ligand dissociation from the plasma protein (sec^{-1}), k_2 equals the rate constant of ligand plasma protein association ($M^{-1} sec^{-1}$), and k_3 equals the membrane permeation rate constant (sec^{-1}). If the capillary transit time is denoted by t (in seconds), then the $k_3 t$ product is identical to the *PS/F* ratio, as defined above in the KRC formulation (25).

The unidirectional extraction of steroid hormones across the BBB, the *PS* product, the rate of cerebral blood flow, and the *in vivo* K_D^a in the cerebral microcirculation may all be conveniently measured with the carotid arterial single injection technique developed by Oldendorf, also called the brain uptake index (BUI) method (26). With this method, an approximately 200-μl bolus of buffered Ringer's solution containing ^{3}H-labeled steroid hormone and [^{14}C]butanol, a highly diffusible internal reference, and various concentrations of plasma protein or serum, is rapidly injected (less than 0.5 sec) into the common carotid artery of an anesthetized rat or rabbit, or a conscious animal that has an external carotid artery catheter preinserted (27). Owing to the rapid injection, mixing of the bolus with the circulating rat plasma is less than 5% (28). The bolus passes through the head nearly completely within the first 2 sec after injection and the animal is decapitated at 5 to 15 sec after carotid arterial injection. Owing to the large ratio of extravascular to vascular volume in brain, there is minimal efflux of the radiolabeled substance during the 5 to 15 sec after injection. Therefore, the measurement of steroid hormone extraction at 5 to 15 sec after injection represents unidirectional extraction depicted in the KRC formulation shown above. This extraction represents actual extravascular distribution in brain and not simply binding of steroid hormone to the capillary endothelium. Thaw-mount autoradiography has been used to demonstrate that the distribution of [^{3}H]estradiol at 15 sec after a single common carotid injection is uniform throughout the brain as compared to the focal distribution of serum albumin within the capillary lumen (29).

The extraction of steroid hormone by brain may be measured in the presence of varying concentrations of a specific protein added to the carotid arterial injection solution (2, 3). As shown in Fig. 2, the unidirectional extraction of [^{3}H]testosterone by rat brain is measured versus the concentration of bovine serum albumin in the carotid arterial injection solution. If only the fraction of testosterone that is free (dialyzable) *in vitro* is available for transport through the BBB *in vivo,* then the extraction/albumin concentration

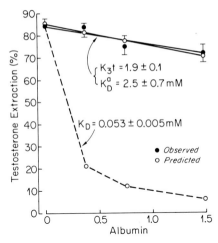

FIG. 2   The unidirectional extraction of [³H]testosterone by rat brain is plotted versus the concentration of arterial bovine albumin. The experimentally observed values are given by the solid circles (mean ± SE, n = three to six animals per point). The extraction values predicted on the basis of fitting the experimental data to the KRC equation are shown by the open circles, and the curve fitting gives the two parameters, K_3t and K_D^a. The K_3t product is identical to the PS/F ratio depicted in the KRC equation. The dashed line represents the extraction values predicted by substituting into the KRC equation the individual albumin concentrations, the K_3t product, and the *in vitro* albumin–testosterone dissociation constant, $K_D = 53 \pm 1 \, \mu M$. Therefore, the dashed curve gives the expected inhibition of testosterone transport caused by hormone binding to albumin if testosterone was not available for transport into brain from the circulating albumin bound pool. [From Pardridge (8) by permission of Oxford University Press.]

curve should conform to the dashed line in Fig. 2. However, at all concentrations of albumin, the observed testosterone extraction is greatly in excess of that predicted form *in vitro* measurements of albumin binding of testosterone, and this discrepancy is explicable within the context of the model of enhanced dissociation (25). That is, owing to conformational changes about the testosterone-binding site on albumin, there is a markedly increased rate of testosterone dissociation from this binding site *in vivo* in the brain microcirculation. This is reflected by the 50-fold greater K_D^a value *in vivo* in the brain microcirculation as compared to the *in vitro* K_D value (Table I). The testosterone extraction values shown in Fig. 2 were computed from the corresponding BUI values. The conversion of BUI values into unidirectional extraction values requires the additional estimation of the rate of unidirectional extraction of the [¹⁴C]butanol reference (E_R) and the rate of cerebral blood flow (F). These parameters may be conveniently measured using other highly

TABLE I   Comparison of Bovine Albumin or Human α_1-Acid Glycoprotein Dissociation Constant *in Vivo* within Brain Capillary (K_D^a) with Corresponding *in Vitro* Dissociation Constant (K_D)

Plasma protein	Ligand	K_D (μM) (*in vitro*)	K_D^a (μM) (*in vivo*)
Bovine albumin	Testosterone	53 ± 1	2520 ± 710
	Tryptophan	130 ± 30	1670 ± 110
	Corticosterone	260 ± 10	1330 ± 90
	Dihyrotestosterone	53 ± 6	830 ± 140
	Estradiol	23 ± 1	710 ± 100
	Propranolol	290 ± 30	220 ± 40
	Bupivacaine	141 ± 10	211 ± 107
	$T_3{}^a$	4.7 ± 0.1	46 ± 4
Human α_1-acid glycoprotein	Propranolol	3.3 ± 0.1	19 ± 4
	Bupivacaine	6.5 ± 0.5	17 ± 4

[a] T_3, Triiodothyronine. From Pardridge (8).

diffusible isotopes, such as ^{3}H-labeled water, [^{3}H]diazepam, or [^{125}I]iodoamphetamine, as described previously (30).

In Vitro Measurements of Transport: Isolated Cells

The transport of steroid hormones across the BBB may be measured *in vitro* using either isolated brain capillaries or cultures of brain capillary endothelium grown on filters that are placed in side-by-side diffusion chambers. In the case of isolated capillary experiments, the actual transport through the BBB is not measured; rather, cellular uptake by brain capillary endothelial cells is measured. The steroid hormone transport across the plasma membrane of isolated cells has been measured in a variety of tissues and in some cases has led to the model that plasma membrane transporters are involved in movement of steroid hormones through plasma membranes (31, 32). However, the movement of steroid hormones through the plasma membrane of isolated or cultured cells has generally been found to be due to free diffusion (33–35). If isolated cells are used to measure plasma membrane transport of steroid hormones, it is essential that initial rates of transport be obtained. Initial rates of transport reflect actual movement across the plasma membrane as opposed to net rates of transport, which are dominated by intracellular mechanisms accounting for sequestration of radiolabeled steroid hormone within the cell, such as cytoplasmic binding, nuclear receptor binding, or cytosolic metabolism. The confusion arises from the fact that the rate of uptake of the radiolabeled steroid hormone by the isolated or cultured cell may increase with the length of the incubation period, despite the fact that net rates of uptake are being recorded. The increasing uptake of steroid hormone by the cell relative to incubation period is interpreted by some investigations as evidence that initial rates are being determined. In fact, initial rates of transport are probably reached within a few seconds of incubation of isolated cells with the radiolabeled steroid hormones. A convenient way of determining whether initial rates are being actually measured is to compute the cell/medium ratio of radiolabeled steroid hormone in units of microliters per milligram of protein. This cell/medium ratio may also be regarded as volume of distribution (V_D) and net rates of transport are being recorded when the V_D value exceeds the intracellular water space, which in most cultured cells is on the order of 3–5 μl/mg protein (36). When the V_D value exceeds the water space, then the radiolabeled steroid hormone is being sequestered by the isolated cell owing to cytoplasmic metabolic events, not owing to plasma membrane transport. This sequestration may result in increasing tissue uptake with time that is saturable and is erroneously interpreted as saturation of plasma membrane transport.

Steroid hormone transport may also be studied *in vitro* by measuring solute flux across a monolayer of brain capillary endothelial cells grown in tissue culture on a filter disk and placed in side-by-side diffusion chambers. However, the use of this model system is optimal only if the "*in vitro* BBB" approximates the permeability of the BBB *in vivo*. Unfortunately, this equivalence has not been experimentally verified (37). The rate of solute clearance across the endothelial monolayer is equivalent to the *PS* product as defined above in the KRC formulation. The *in vitro PS* product may be converted to an *in vitro* P_e value, that is, the permeability coefficient (units are centimeters per second) based on the endothelial surface area across the filter disk. The *in vitro* P_e value may then be compared to the *in vivo* P_e value (37). The latter is computed from measurements of the *in vivo PS* product as defined above in the KRC formulation, and estimates of the BBB surface area *in vivo*, $S = 100$ cm^2/g, that is, $P_e = PS/S$. The *in vitro* P_e value for a series of compounds that are transported through the BBB by lipid mediation was found to be, on average, 150-fold higher than the *in vivo* P_e values (37). The high *in vitro* P_e value reflects the leakiness of the "*in vitro* BBB" and a very large transcellular pathway of solute movement across the *in vitro* endothelial monolayer as compared to minimal fluid-phase movement across the BBB *in vivo*.

Major Determinants of Blood–Brain Barrier Transport of Steroid Hormones

Blood–Brain Barrier Permeability–Surface Area (PS) Product

As depicted in the KRC equation above, the three principal parameters underlying the transport of steroid hormones through the BBB are the *PS* product, the rate of cerebral blood flow (*F*), and the *in vivo* plasma protein K_D^a value. The *PS* product may be measured in either the absence or the presence of varying concentrations of steroid hormone-binding plasma proteins. As defined by the KRC equation above, the *PS* product is

$$PS = F \ln(1 - E)$$
$$E = (\text{BUI})(E_R)$$

Therefore, the *PS* product is logarithmically related to the BUI value. As shown in Fig. 3, the BUI value for five different steroid hormones varies over two log orders of magnitude (2). These BUI measurements were recorded following carotid arterial injection of Ringer's solution containing ^{3}H-labeled steroid hormone in the presence of an insignificant concentration (0.1%) of bovine albumin. Thus, these BUI measurements reflect the *PS* product and are not influenced by plasma protein binding.

Figure 3A shows the structures of the five different steroid hormones and the polar functional groups involved in hydrogen bonding with solvent water are emphasized in each structure. There are two hydrogen bonds formed with each hydroxyl moiety and one hydrogen bond formed with each ketone or aldehyde moiety (38). The BBB permeability to steroid hormones falls a log order of magnitude with the addition of each hydroxyl group (2). The BUI value is not significantly different in comparing progesterone, testosterone, and estradiol, despite the fact that these steroid hormones form, respectively, two, three, and four hydrogen bonds with water, because of the logarithmic relationship between BUI value and *PS* product. The higher the total number of hydrogen bonds formed between the steroid hormone and the solvent water (where the total hydrogen bond number is denoted as *N* in Fig. 3), the lower the lipid solubility of the steroid hormone. The latter may be measured by estimating the distribution of steroid hormone in two phases, such as Ringer's solution versus 1-octanol (2). These data demonstrate that lipid solubility plays an important role in regulating BBB transport

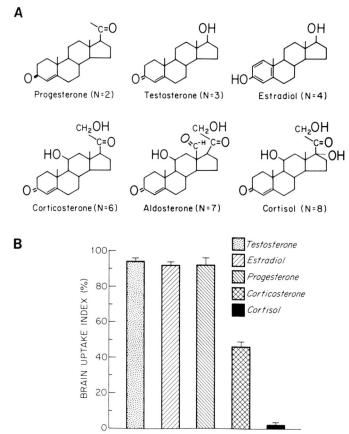

FIG. 3   (A) Structures of steroid hormones with emphasis on the polar functional groups that form hydrogen bonds with water. The hydrogen bond number (*N*) is given in parentheses and is equal to the total number of hydrogen bonds formed between solvent water and the individual steroid molecule. (B) The brain uptake index of five different ³H-labeled steroid hormones is shown as mean ± SE (*n* = three to five rats per point). [From Pardridge and Mietus (2) and Pardridge (3).] [(A) Reproduced from the Journal of Clinical Investigation, 1979 by copyright permission of The American Society for Clinical Investigation, (B) is copyright © 1981, The Endocrine Society.]

of steroid hormones. However, the hydrogen bond number is conveniently computed by simply inspecting the structure of the given steroid hormone, and the determination of the hydrogen bond number is a powerful predictor of BBB transport. This prediction may be made in the absence of any measurements of lipid solubility of steroid hormones.

Cerebral Blood Flow (F)

The KRC equation predicts that the unidirectional extraction (E) of steroid hormones across the BBB is inversely related to the rate of cerebral blood flow (F). The TCR, where TCR = EF, is essentially independent of flow rate when $E < 15\%$, and the TCR value is equivalent to the PS product. Thus, the clearance of steroid hormones may be altered by modulations that affect cerebral blood flow, such as phenobarbital (39), when $E > 15\%$. Barbiturate anesthesia results in a reduction of cerebral blood flow (30), and can be expected to cause a decrease in the TCR of steroid hormones by brain if the extraction values are in the range of 15–85%, which is the blood flow-dependent region of TCR values (30).

In Vivo Plasma Protein Dissociation Constant K_D^a

The experiments in Fig. 2 demonstrate how the K_D governing the albumin–testosterone binding reaction *in vivo* in the living cerebral microcirculation may be measured using the BUI technique. This approach shows that the K_D value *in vivo* for most, but not all, ligands is much greater than the K_D value found *in vitro* (Table I). This *in vivo*/*in vitro* discrepancy is attributed to enhanced dissociation reactions caused by conformational changes about the albumin ligand-binding site *in vivo* (4). Possible mechanisms underlying this enhanced dissociation are discussed below.

Some investigators have advanced an alternative model to explain the discrepancy between *in vivo* and *in vitro* K_D values shown in Fig. 2 or Table I. The model that is an alternative explanation to the enhanced dissociation model is called the dissociation limited model and posits that the rate of ligand dissociation from the plasma protein-binding site is slow relative to either ligand reassociation with the plasma protein or ligand permeation through the biological membrane (40–42). In fact, the dissociation limited model originally formulated is flawed and is actually a misnomer because both the rate of ligand dissociation and the rate of ligand reassociation with the plasma protein must be slow compared to the rate of ligand permeation through the BBB membrane (25). To invalidate the enhanced dissociation model, advocates of the dissociation limited model are forced to propose rates of hormone/albumin association that are log orders lower than the rates of ligand association with proteins that are recorded experimentally, wherein k_2 generally ranges from 10^6 to 10^8 M^{-1}sec^{-1} (43–45). For example, at an albumin concentration of 0.5 mM, the rate of hormone reassociation with albumin ranges form 500 to 50,000 sec^{-1}. These values are log orders greater than the rate constant (k_3) of steroid hormone permeation through the BBB membrane, that is, k_3 values range from 0.1 to 2 sec^{-1} (8). The model of enhanced hormone dissociation from albumin has also been criticized because if the albumin-bound pool is a major source of hormones for tissues, a markedly reduced rate of hormone delivery to tissues would be expected in special situations where serum albumin is extremely low, such as in the Nagase analbuminemic rat (NAR) (46). However, other studies have shown that plasma protein-mediated transport in the NAR occurs owing to plasma protein-mediated transport via the γ-globulin fraction (47).

Evidence in support of the enhanced dissociation model may be summarized as follows. First, there is no evidence in support of the alternative model, that is, dissociation limitation, because this model is forced to propose rate constants for ligand association that are log orders too low and rate constants for ligand permeation that are log orders too high, as compared to the experimentally observed values (48). Second, conformational changes about albumin hormone-binding sites have been experimentally confirmed (49), and other studies have shown that the interaction of albumin even with glass surfaces may induce conformational changes in the protein (50). Third, studies from X-ray diffraction experiments have shown that a slight modification of the binding interaction between a ligand and a protein, such that a single hydrogen bond between the protein and the ligand is removed, may result in log order increases in the rate of ligand dissociation from the protein and log order increases in the K_D value governing the protein/ligand-binding reaction (51).

The mechanism underlying the conformational change about plasma protein-binding sites that occurs within the microcirculation *in vivo* is not known. The binding of albumin to cellular surfaces was originally proposed to be receptor mediated (52, 53). However, the earlier data indicating the presence of an albumin receptor on cellular surfaces were subsequently not confirmed (54–56), although other studies have shown that albumin nonspecifically absorbs to the endothelial glycocalyx of the microcirculation (57). Indeed, this absorption of albumin to the endothelial glycocalyx is a critical factor maintaining normal capillary permeability in tissues (58). The important role played by adsorption of plasma albumin to the endothelial surface should be considered when evaluating perfusion experiments measuring organ transport of steroid or thyroid hormones, wherein the organ is preperfused for 30 min with buffer containing no plasma protein (42, 46).

The nonspecific binding of albumin to the brain endothelial glycocalyx (57) may cause conformational changes on the albumin molecule (49, 50), that are albumin binding site specific (4). In addition, local changes in membrane pH (59), membrane phospholipid (60), or small diffusible molecules may account for the conformational change (61). Because not all albumin-bound molecules undergo enhanced association within the brain microcirculation (Table I), the conformational changes on the albumin molecule are specific to certain domains on the albumin molecule (4). Albumin is composed of six primary ligand-binding domains that bind specific groups of ligands

(62). Finally, the enhanced dissociation of ligands from albumin-binding sites in the brain microcirculation is not restricted to the compounds listed in Table I, but has been observed for a variety of different compounds (24).

Steady State Model of Steroid Hormone Distribution in Brain

The biological effects of steroid hormones in brain are believed to be triggered by the interaction of the steroid hormone with a specific nuclear receptor, and the biological effect is believed to be proportional to the steroid occupancy of the nuclear receptor. Assuming the receptor interacts with the free intracellular steroid hormone, then the nuclear receptor occupancy, and thus the biological effect of the hormone, is proportional to the concentration of cellular exchangeable hormone in brain, which is depicted in Fig. 4A as the L_M pool (63). The pool of cytoplasmic bound steroid hormone in brain is depicted in Fig. 4A as the PL pool. The rates of association and dissociation of the cellular exchangeable hormone with cytoplasmic binding systems are denoted as K_5 and K_6, respectively, in Fig. 4A. The rates of hormone dissociation and reassociation with the albumin-bound pool (AL) in the capillary plasma compartment are denoted as K_7 and K_8, respectively, and the rates of hormone dissociation and reassociation with the globulin bound pool (GL) in capillary plasma are denoted as K_1 and K_2 in Fig. 4A. The rates of hormmone influx and efflux across the BBB are denoted as K_3 and K_4 in Fig. 4A. No rate constants are assigned for hormone permeation across brain cell membranes. Because the surface area of brain cell membranes is log orders greater than the surface area of the BBB membrane, the rate-limiting transport step in hormone movement from the plasma-exchangeable pool to the cellular exchangeable pool is at the BBB (63).

The BUI technique has been used in previous studies to estimate the rates of testosterone interaction with the albumin- and globulin-bound pools in plasma, as well as testosterone transport through the BBB and testosterone interaction with brain cytoplasmic binding systems (2, 20, 64). These combined parameters have been analyzed within the context of a steady state model of testosterone distribution in brain (63), to yield the predicted pool sizes of testosterone in brain depicted in Fig. 4B. The analytical solution of the equations underlying the steady state model have been reported previously (63). This model has also been published in the form of a microcomputer program in BASIC for interactive testing of hormone transport theory (65). The values of plasma protein concentrations and the various rate constants have been published previously (63). The simulation analysis predicts that the concentration of cellular exchangeable testosterone in brain is more than 10-fold greater than the concentration of free (dialyzable) testosterone

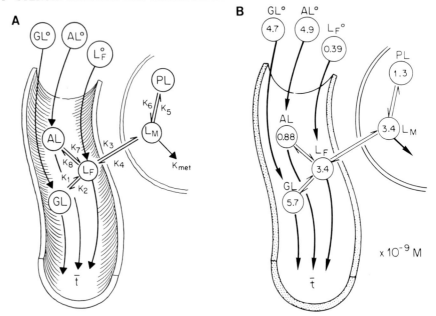

FIG. 4   (A) Steady state model of testosterone transport through the brain capillary wall and into brain cells. Pools of the globulin-bound, albumin-bound, and free ligand in the systemic circulation are denoted as $GL°$, $AL°$, and $L_F°$, and pools of globulin-bound, albumin-bound, and plasma-exchangeable hormone in the brain capillary are denoted as GL, AL, and L_F, respectively. Pools of free and cytoplasmic bound steroid hormone in brain cells are given by L_M and PL, respectively; $\bar{t}$ is the mean capillary transit time in brain. (B) Predicted steady state concentrations of testosterone in the various pools of the brain capillary and in brain cells. The pool sizes represent the basal state, which is simulation 1 from Ref. 8. The concentration of free cytosolic testosterone in brain cells is predicted to approximate the concentration of albumin-bound hormone in the circulation, and to be more than 10-fold greater than the concentration of free hormone measured by equilibrium dialysis ($L_F°$). [From Pardridge (8) and Pardridge and Landau (63) by permission of Oxford University Press and The American Physiological Society.]

in plasma measured *in vitro* but is approximately equal to the free plus albumin-bound fraction of plasma testosterone. This finding is important because the concentration of steroid hormone that induces 50% nuclear receptor occupancy is approximately equal to the albumin-bound concentration of steroid hormone in plasma, but is at least 10-fold greater than the concentration of free, dialyzable steroid hormone (8). Another important principle elucidated by this analysis is that the concentration of cellular exchangeable hormone (L_M) is solely a function of membrane transport rates (K_3 and K_4), the concentration of capillary plasma-exchangeable hormone

(L_F), and the rate of hormone metabolism as denoted by K_{met} in Fig. 4A (63). The concentration of cellular exchangeable steroid hormone has not been measured in brain or any other tissue *in vivo,* because this would require the invention of a steroid-specific electrode that could be placed directly into the cell cytoplasm. However, the concentration of cellular exchangeable hormone may be measured indirectly, because this parameter is proportional to the concentration of plasma-exchangeable hormone, and the latter is directly measurable with experimental techniques such as the BUI method. A corollary of this analysis is that the concentration of cellular exchangeable hormone *in vivo* is independent of cytoplasmic binding proteins, but rather is a function of the concentration of capillary-exchangeable hormone, membrane transport, and cytoplasmic metabolism. Although binding of ligand by cytoplasmic proteins does not influence the concentration of cellular exchangeable hormone *in vivo,* this cytoplasmic binding plays a dominant influence in controlling cellular exchangeable hormone *in vitro.* The latter is measured when aliquots of tissue cytosol are utilized to estimate the concentration of cytosolic exchangeable hormone *in vitro* with techniques such as equilibrium dialysis. Previous studies have actually used this *in vitro* approach to measure purportedly the concentration of cellular exchangeable hormone (66). However, this approach measures the concentration of cellular exchangeable hormone in the absence of the dominant function that occurs *in vivo,* that is, the continuous flow of plasma-exchangeable hormone (65). The situation *in vivo* as depicted in Fig. 4A is a partly flow–partly compartmental model, as opposed to the flow-independent compartmental model that exists *in vitro* when dialyzable (exchangeable) hormone is measured with tissue cytosol preparations (66).

Concluding Remarks

In summary, the two most important factors determining the transport of steroid hormones through the BBB *in vivo* are the K_D^a value that governs the plasma protein/steroid hormone-binding reaction *in vivo* within the brain microcirculation, and the BBB *PS* product, which reflects the permeability of the brain capillary endothelial membrane for the steroid hormone, and which is inversely related to the hydrogen bond number on the steroid hormone structure (Fig. 3). Finally, the pool size of cellular exchangeable steroid hormone in brain has never been experimentally measured, but may be estimated from measurements of plasma-exchangeable hormone as depicted in Fig. 4. Plasma-exchangeable hormone may be measured with the BUI technique, and the estimated transport parameters may be further analyzed with the steady state model described in Fig. 4, using an interactive approach for the testing of hormone transport theory (63, 65).

Acknowledgment

We thank Sherri J. Chien for skillfully preparing the manuscript.

References

1. M. W. Brightman, *Exp. Eye Res.* **25,** 1 (1977).
2. W. M. Pardridge and L. J. Mietus, *J. Clin. Invest.* **64,** 145 (1979).
3. W. M. Pardridge, *Endocr. Rev.* **2,** 103 (1981).
4. W. M. Pardridge, *Am. J. Physiol.* **252,** E157 (1987).
5. S. P. Marynick, G. B. Smith, M. H. Ebert, and D. L. Loriaux, *Endocrinology (Baltimore)* **101,** 562 (1977).
6. D. Riad-Fahmy, G. F. Read, and K. Griffiths, *Endocr. Rev.* **3,** 367 (1982).
7. S. P. Marynick, W. W. Havens, II, M. H. Ebert, and D. L. Loriaux, *Endocrinology (Baltimore)* **99,** 400 (1976).
8. W. M. Pardridge, *Oxford Rev. Reprod. Biol.* **10,** 237 (1988).
9. J. Lerma, A. S. Herranz, O. Herreras, V. Abraira, and R. Martin del Rio, *Brain Res.* **384,** 145 (1986).
10. P. Lönnroth, P.-A. Jannson, and U. Smith, *Am. J. Physiol.* **253,** E228 (1987).
11. J. K. Hsiao, B. A. Ball, P. F. Morrison, I. N. Mefford, and P. M. Bungay, *J. Neurochem.* **54,** 1449 (1990).
12. K. H. Dykstra, J. K. Hsiao, P. F. Morrison, P. M. Bungay, I. N. Mefford, M. M. Scully, and R. L. Dedrick, *J. Neurochem.* **58,** 931 (1992).
13. T. Terasaki, Y. Deguchi, Y. Kasama, W. M. Pardridge, and A. Tsuji, *Int. J. Pharm.* **81,** 143 (1992).
14. R. K. Dubey, C. B. McAllister, M. Inoue, and G. R. Wilkinson, *J. Clin. Invest.* **84,** 1155 (1989).
15. Y. L. Hurd, J. Kehr, and U. Ungerstedt, *J. Neurochem.* **51,** 1314 (1988).
16. R. D. O'Neill, J.-L. Gonzalez-Mora, M. G. Boutelle, D. E. Ormonde, J. P. Lowry, A. Duff, B. Fumero, M. Fillenz, and M. Mas, *J. Neurochem.* **57,** 22 (1991).
17. J. A. Yergey and M. P. Heyes, *J. Cereb. Blood Flow Metab.* **10,** 143 (1990).
18. M. M. Hertz and T. G. Bolwig, *Brain Res.* **107,** 333 (1976).
19. B. Little, R. B. Billiar, S. S. Rahman, W. A. Johnson, Y. Takaoka, and R. J. White, *Am. J. Obstet. Gynecol.* **123,** 527 (1975).
20. W. M. Pardridge, T. L. Moeller, L. J. Mietus, and W. H. Oldendorf, *Am. J. Physiol.* **239,** E96 (1980.
21. R. B. Billiar, Y. Takaoka, P. S. Reddy, D. L. Hess, C. Longcope, and B. Little, *Endocrinology (Baltimore)* **108,** 1643 (1981).
22. D. C. Collins, E. L. Bradley, III, P. I. Musey, and J. R. K. Preedy, *Steroid* **30,** 455 (1977).
23. H. F. Cserr, D. N. Cooper, P. K. Suri, and C. S. Patlak, *Am. J. Physiol.* **240,** F319 (1981).

24. W. M. Pardridge, "Peptide Drug Delivery." Raven, New York, 1991.
25. W. M. Pardridge and E. Landaw, *J. Clin. Invest.* **74,** 745 (1984).
26. W. H. Oldendorf, *Brain Res.* **24,** 372 (1970).
27. L. D. Braun, L. P. Miller, W. M. Pardridge, and W. H. Oldendorf, *J. Neurochem.* **44,** 911 (1985).
28. W. M. Pardridge, E. M. Landaw, L. P. Miller, L. D. Braun, and W. H. Oldendorf, *J. Cereb. Blood Flow Metab.* **5,** 576 (1985).
29. W. M. Pardridge, *in* "Protein Binding and Drug Transport," 20th Symposium Medicum Hoechst (J. P. Tillement and E. Lindenlaub, eds.), pp. 277–292. Schattauer, Stuttgart, 1986.
30. W. M. Pardridge and G. Fierer, *J. Cereb. Blood Flow Metab.* **5,** 275 (1985).
31. E. Milgrom, M. Atger, and E.-E. Baulieu, *Biochim. Biophys. Acta* **320,** 267 (1973).
32. R. J. Pietras and C. M. Szego, *Nature (London)* **265,** 69 (1977).
33. R. E. Müller and H. H. Wotiz, *Endocrinology (Baltimore)* **105,** 1107 (1979).
34. E. P. Giorgi and W. D. Stein, *Endocrinology (Baltimore)* **108,** 688 (1981).
35. K. Kilvik, K. Furu, E. Haug, and K. M. Gautvik, *Endocrinology (Baltimore)* **117,** 967 (1985).
36. W. M. Pardridge and D. Casanello-Ertl, *Am. J. Physiol.* **263,** E234 (1979).
37. W. M. Pardridge, D. Triguero, J. Yang, and P. A. Cancilla, *J. Pharmacol. Exp. Ther.* **253,** 884 (1990).
38. E. M. Wright and N. Bindslev, *J. Membr. Biol.* **29,** 289 (1976).
39. G. Bidder, *Endocrinology (Baltimore)* **83,** 1353 (1968).
40. R. A. Weisiger, *Proc. Natl. Acad. Sci. U.S.A.* **82,** 1563 (1985).
41. R. P. Ekins and P. R. Edwards, *Am. J. Physiol.* **255,** E403 (1988).
42. C. M. Mendel, R. A. Weisiger, and R. R. Cavalieri, *Endocrinology (Baltimore)* **123,** 1817 (1988).
43. M. Rowland, D. Leitch, G. Fleming, and B. Smith, *J. Pharmacokinet. Biopharm.* **12,** 129 (1984).
44. T. W. Smith and K. M. Skubitz, *Biochemistry* **14,** 1496 (1975).
45. J. Janin and C. Chothia, *J. Biol. Chem.* **265,** 16027 (1990).
46. C. M. Mendel, R. R. Cavalieri, L. A. Gavin, T. Pettersson, and M. Inoue, *J. Clin. Invest.* **83,** 143 (1989).
47. S. C. Tsao, Y. Sugiyama, K. Shinmura, Y. Sawada, S. Nagase, T. Iga, and M. Hanano, *Drug Metab. Dispos.* **16,** 482 (1988).
48. W. M. Pardridge and E. M. Landaw, *Am. J. Physiol.* **258,** E396 (1990).
49. T. Horie, T. Mizuma, S. Kasai, and S. Awazu, *Am. J. Physiol.* **254,** G465 (1988).
50. R. G. Reed and C. M. Burrington, *J. Biol. Chem.* **264,** 9867 (1989).
51. P. A. Bartlett and C. K. Marlowe, *Science* **235,** 569 (1987).
52. R. Weisiger, J. Gollan, and R. Ockner, *Science* **211,** 1048 (1981).
53. E. L. Forker and B. A. Luxon, *J. Clin. Invest.* **72,** 1764 (1983).
54. Y. R. Stollman, U. Gärtner, L. Theilmann, N. Ohmi, and A. W. Wolkoff, *J. Clin. Invest.* **72,** 718 (1983).
55. A. B. Fleischer, W. O. Shurmantine, F. L. Thompson, E. L. Forker, and B. A. Luxon, *J. Lab. Clin. Med.* **105,** 185 (1985).
56. M. Simionescu, N. Ghinea, A. Fixman, M. Lasser, L. Kukes, N. Simionescu, and G. E. Palade, *J. Submicrosc. Cytol. Pathol.* **20,** 243 (1988).
57. W. M. Pardridge, J. Eisenberg, and W. T. Cefalu, *Am. J. Physiol.* **249,** E264 (1985).

58. C. C. Michel, *J. Physiol.* (*London*) **404,** 1 (1988).
59. S. Urien, F. Brée, B. Testa, and J.-P. Tillement, *Biochem. J.* **280,** 277 (1991).
60. T. Endo, M. Eilers, and G. Schatz, *J. Biol. Chem.* **264,** 2951 (1989).
61. T. H. Lin, Y. Sugiyama, Y. Sawada, T. Iga, and M. Hanano, *Biochem. Pharmacol.* **37,** 2957 (1988).
62. X. M. He and D. C. Carter, *Nature* (*London*) **358,** 209 (1992).
63. W. M. Pardridge and E. M. Landaw, *Am. J. Physiol.* **249,** E534 (1985).
64. W. M. Pardridge, L. J. Mietus, A. M. Frumar, B. Davidson, and H. L. Judd, *Am. J. Physiol.* **239,** E103 (1980).
65. W. M. Pardridge and E. M. Landaw, *Endocrinology* (*Baltimore*) **120,** 1059 (1987).
66. J. H. Oppenheimer and H. L. Schwartz, *J. Clin. Invest.* **75,** 147 (1985).
67. W. M. Pardridge *in* "Pathophysiology of the Blood Brain Barrier: Long Term Consequences of Barrier Dysfunction for the Brain" (B. B. Johansson, ed.), pp. 217–225. Elsevier, Amsterdam.

[2] Enhanced Delivery of Steroids to the Brain Using a Redox-Based Chemical Delivery System

James W. Simpkins and Nicholas Bodor

Introduction

The delivery of steroids to the brain is of major importance. Steroids modulate a variety of brain-mediated behavioral, thermic, antinociceptive, and autonomic functions (1) in addition to their well-known role in the feedback regulation of the anterior pituitary gland and hence of secretion of the adrenal cortex and the gonads. As such, the modification of the hormone environment of the brain can be used therapeutically to modify behavior and endocrine abnormalities. The observations that both ovarian and adrenal steroids can affect neuronal survival (2) and memory and cognition (3) provide evidence that the endocrine environment of the brain is essential to its normal function throughout life.

Steroids are highly lipophilic and therefore readily distribute to the brain following their secretion from the adrenal glands and the gonads or following peripheral administration. This central nervous system (CNS) distribution of steroids is, however, accompanied by the distribution of the steroid to virtually every tissue in the body. The therapeutic modification of brain function through the peripheral administration of steroids necessarily results in the treatment of the entire body and causes their associated peripheral side effects. To achieve the selective distribution of steroids to the CNS, we undertook a program of research to discover means for the brain-enhanced delivery of steroids.

Redox Methods for Brain-Enhanced Delivery of Steroids

The redox approach for the site-specific delivery of drugs to the brain is based on the unique architecture of the blood–brain barrier (BBB) that excludes transport of substances that lack specific transport carriers or that

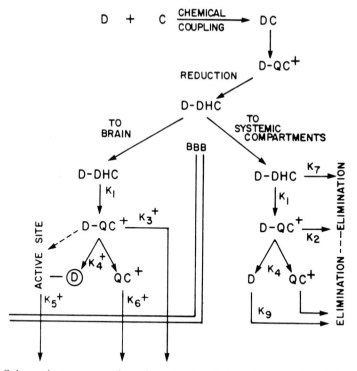

FIG. 1   Schematic representation of the *in vivo* distribution of a drug (D) attached to the dihydropyridine carrier (C). The drug is coupled by two reaction steps to the charged, pyridinium form of the carrier molecule (D–Q$^+$). The drug–carrier complex (D–CDS) is the lipophilic form of the delivery system administered to the animal. After intravenous administration of D–CDS, it penetrates the BBB to partition into the brain and is distributed to peripheral tissues. In both the central and peripheral compartments, oxidation of D–CDS to D–Q$^+$ occurs. In the periphery, the oxidation to the charged D–Q$^+$ results in a complex that is readily eliminated by kidney or biliary processes; in the brain the formation of the charged moiety serves to "lock-in" the complex, forming a local brain depot of the drug in an as-yet inactive form. Hydrolysis of the D–Q$^+$ to release the drug and the carrier results in the sustained release of the active drug. The small, inert carrier molecule, trigonelline, released during the hydrolysis is easily eliminated from the brain.

are hydrophilic (4). The redox approach is depcited in Fig. 1 and utilizes a targetor moiety that is able to undergo enzymatic and nonenzymatic oxidation and reduction to produce a charged or neutral site alternatively on the drug–targetor complex. The complex is administered in the reduced (neutral) form, which can readily cross the BBB. In the brain, the targetor is oxidized

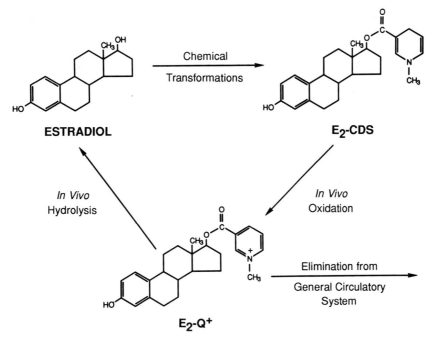

ESTRADIOL   E₂-CDS

E₂-Q⁺

FIG. 2   A schematic representation of the brain-enhanced delivery of estradiol using the redox approach. The estradiol–chemical delivery system (E₂–CDS) is E₂ attached by a covalent bond to the lipoidal, dihydropyridine targetor. Oxidation of the delivery system to E₂–Q⁺ serves to trap the E₂ inside the brain in a form that is inactive. Hydrolysis of this charged, depot form of the delivery system results in the sustained release of E₂ in the brain.

to the charged form, thus preventing it from recrossing the BBB. The portion of the administered dose that remains in the periphery is also oxidized, but this serves to enhance its rate of clearance by virtue of its increased hydrophilicity. This theoretical distribution of the redox-based, brain-enhanced chemical delivery system (CDS) has provided the rationale for our efforts to synthesize and evaluate candidate compounds for the delivery of steroids to the brain.

The delivery of estradiol (E₂) to the brain using the CDS approach is depicted in Fig. 2. The most widely studied redox-based system is based on the dihydropyridine–pyridinium salt redox reaction. In this system, the lipoidal dihydropyridine moiety is covalently attached to E₂ by an ester bond at the C-17 position. For E₂, this addition does not remarkably increase its lipophilicity and like E₂ this delivery system can readily pass the BBB. In the brain and in the periphery, the dihydropyridine moiety is oxidized to the charged pyridinium ion by the ubiquitous NADP–NADPH redox system. This oxidation is rapid ($t_{1/2}$ of 29 min in brain tissue) and results in a 44,000-fold decrease in the lipoid solubility of the complex, thereby preventing its egress from the brain by way of the BBB (5). Hence, a local brain depot of E_2–Q^+ is produced whose clearance is dependent on two processes. First, the enzymatic cleavage of the targetor releases E_2, which can redistribute down its concentration gradient across the BBB to the periphery. Second, E_2–Q^+ can be cleared by bulk flow of cerebrospinal fluid (CSF). Both of these processes are documented to be comparatively slow processes.

Synthesis of Brain-Enhanced Estradiol–Chemical Delivery System

The preparation of the C-17 dihydropyridine derivative of E_2 was undertaken as 17-substituted esters of E_2 are known not to interact with estrogen receptors (6). Estradiol was reacted with nicatinoyl chloride hydrochloride in pyridine to give the 3,17-bisnicotinate. This compound was subjected to methanolic potassium bicarbonate, which results in selective hydrolysis of the phenolic nicotinate. The resulting secondary ester was quaternized with methyl iodide to give the 17-trigonellinate (E_2–Q^+) and then reduced using sodium dithionite to give the 17-(1,4-dihydrotrigonellinate) (E_2–CDS). Elemental analysis was used to demonstrate the products of each synthetic step.

The delivery system approach requires that the administered dihydrotrigonellinate (the CDS) be lipophilic in order to cross the BBB and that the corresponding pyridinium salt be highly polar to allow CNS retention and rapid peripheral elimination. The log P values of the octanol : water partition coefficient for E_2, E_2–CDS, and E_2–Q^+ were determined using standard methods (5). The log P was 4.50 for E_2–CDS, 3.76 for E_2, and -0.144 for E_2–Q^+. These data indicate that the quaternary salt (E_2–Q^+) is 8000-times more hydrophilic than the parent E_2 and 44,000 times more hydrophilic than the E_2–CDS. Hence the synthesized E_2–CDS and its metabolic products possess the necessary physicochemical properties to function as a brain-enhanced chemical delivery system.

Distribution of Estradiol–Chemical Delivery System

In Vitro Studies

The chemical delivery system concept requires that the dihydropyridine–drug conjugate rapidly convert to the charged pyridinium salt and that the "lock-in" form of the delivery system slowly hydrolyze, releasing the

TABLE I   Stability of E_2–CDS in Various
Biological Media

Medium	Half-life (min)	Correlation coefficient (r)
Plasma	156.6	0.98
Liver homogenate	29.9	0.99
Brain homogenate	29.2	0.99

drug (i.e., E_2). Table I shows the half-lives of E_2–CDS in three biological media: the brain, liver, and plasma. In both liver and brain tissue, E_2–CDS is converted to the corresponding E_2–Q^+ much faster than in plasma. This is consistent with a membrane-bound enzyme acting as the oxidative catalyst. A candidate enzyme is NADH transhydrogenase, which mediates the redox reaction of NADH, a molecule similar in structure to the active portion of the CDS.

The second metabolic step in the brain-enhanced delivery of E_2 is the hydrolysis of E_2–Q^+, releasing E_2 and forming trigonelline, which readily passes out of the brain. We assayed the production of E_2, following the *in vitro* administration of the E_2–Q^+ to plasma or tissue homogenates (7). At 2 hr following dosing with E_2–Q^+, 1.2, 2.3, and 20% of the E_2–Q^+ dose was converted to E_2 in the brain, liver, and plasma, respectively. This slow hydrolysis of E_2–Q^+ would suggest that *in vivo* E_2 should be slowly produced from the brain depot of E_2–Q^+. Given the expected rapid elimination of peripheral E_2–Q^+, the slow hydrolysis of E_2–Q^+ in the brain should result in sustained and marked increases in the brain-to-plasma ratio of E_2.

In Vivo Studies

The acid test for the CDS concept is the selective deposition and retention of E_2 in the brain following peripheral administration of E_2–CDS. Our initial studies on the pharmacokinetics of the E_2–CDS involved the administration of high doses of the drug (60 mg/kg) to conscious, restrained rats. The CDS was given intravenously so as to allow the initial distribution of the CDS to all tissues and thereby to assess the capacity of the CDS to concentrate E_2 in the brain. The E_2–CDS rapidly disappeared from the blood, as would be expected given the rapid oxidation of E_2–CDS in plasma and the large volume of distribution of this lipophile. E_2–Q^+ appeared in all tissues examined soon after administration of E_2–CDS. Evaluation of the terminal portion of the

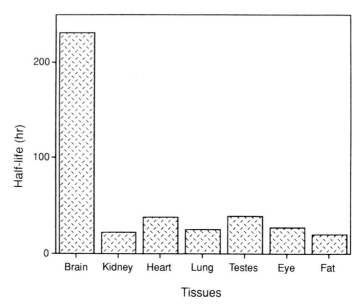

FIG. 3   The half-life of elimination of E_2–Q^+ formed in various tissues after iv administration of E_2–CDS. E_2–CDS was administered at a dose of 5 mg/kg and samples were obtained for up to 48 hr postdosing. [From Mullerman *et al.* (8), with permission.]

tissue concentration vs time curve revealed that the half-lives for elimination of E_2–Q^+ was 46 min in liver, 5.5 hr in lung, 8 hr in kidney, and nearly 1 day in brain (5). The dose of drug used in these early studies was extremely high owing to the insensitivity of the analytical technique used in these early studies was extremely high owing to the insensitivity of the analytical technique used to quantitate E_2–Q^+.

This problem was partially solved through the use of a precolumn-enriching (high-performance liquid chromatography (HPLC) technique (8). This method employed the injection of a large volume (1800 μl) of the tissue solvent onto a precolumn that allowed the absorption and concentration of the compounds of interest. The precolumn is then washed and E_2, E_2–CDS, and E_2–Q^+ are eluted from the precolumn onto an analytical column. Detection of the substances is achieved using a variable-wavelength ultraviolet (UV) detector. This approach improved assay sensitivity for each of the delivery system compounds to the level of 20 to 50 ng/ml plasma or g tissue, which allowed an evaluation of the clearance of E_2–CDS and E_2–Q^+ following a much lower dose (5 mg/kg) of E_2–CDS administered. Animals were sampled for up to 48 hr postinjection of the delivery system. As shown in Fig. 3, at this dose, the half-life of E_2–Q^+ was five times longer in the brain than in the kidney, heart, lung, testes, eye, or adipose tissue (8).

To evaluate the formation of E_2 from the locked-in E_2–Q^+, we developed an assay system based on the solid-phase separation of the various forms of the delivery system and their detection with sensitive radioimmunoassay (RIA) procedures (9). Inasmuch as RIAs for E_2 have sensitivities in the low picogram range, this assay method allowed us to evaluate low doses of E_2–CDS. Tissue was extracted with 100% methanol, which resulted in recovery of E_2 from brain and plasma samples of 71 to 89%. Estradiol was then separated by C_{18} column chromatography, eluted, and dissolved in assay buffer. Additional samples were subjected to NaOH-driven hydrolysis to liberate the E_2 bound to the E_2–Q^+. This E_2 was isolated on a C_{18} column, eluted, and dissolved in assay buffer for analysis by RIA. The column extraction efficiency was 92% for E_2 and the hydrolysis of E_2–Q^+ was 66% in brain tissue and 30% in plasma, with the coefficient of variation of these estimates being small and less than 5% for both tissues. As such, we had a procedure for the simultaneous and sensitive measurement of E_2 and E_2–Q^+ in a variety of tissues.

Ovariectomized rats were treated with a single intravenous (iv) dose of E_2–CDS (10, 100, or 1000 μg/kg) or its vehicle, dimethyl sulfoxide (DMSO), and tissue samples were obtained 1 to 28 days thereafter. E_2–CDS caused a dose-dependent increase in concentrations of both E_2–Q^+ and its metabolite, E_2, in both the brain (Fig. 4) and the hypothalamus (data not shown). In the brain, concentrations of E_2–Q^+ were 5- to 10-fold higher than E_2 levels, as would be expected on the basis of the predicted slow rate of cleavage of the steroid from the locked-in form of the delivery system. The clearance of both E_2–Q^+ and E_2 from brain was nearly identical and showed a half-life of about 8 days. By contrast both E_2–Q^+ and E_2 were rapidly cleared from all peripheral tissue examined, including plasma, kidney, lung, heart, liver, fat, and uterus (10).

For the sake of comparison, we administered E_2 as a single iv dose equimolar to the highest E_2–CDS dose tested (1 mg/kg). Brain levels of E_2 achieved 1 day postinjection were 82-fold lower than those observed after administration of the E_2–CDS (0.33 ng/g for E_2 vs 27 ng/g for E_2–CDS), indicating the extent of enhancement of brain delivery of the steroid when administered as the delivery system.

Pharmacology

To determine if the long half-lives of E_2–Q^+ and E_2 in brain were reflected in a corresponding increase in the duration of action of the delivery system, we assessed the effects of E_2–CDS on secretion of luteinizing hormone

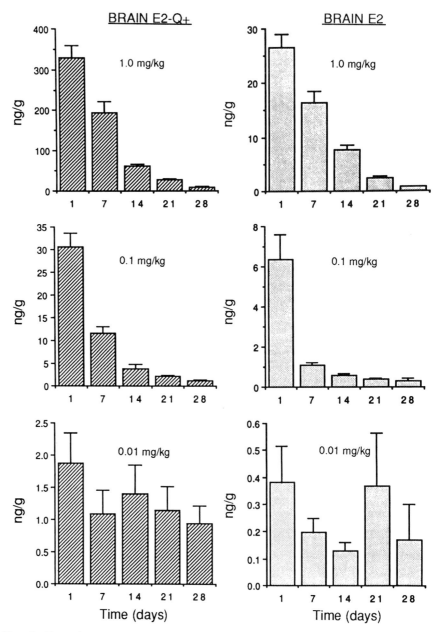

FIG. 4   Dose-dependent effects of E_2–CDS on brain concentrations of E_2–Q^+ (left-hand column) and E_2 (right-hand column) in ovariectomized rats. Rats, ovariectomized for 2 weeks, were administered at a single iv dose of E_2–CDS dissolved in

(LH) and on the induction of masculine sexual behavior, both of which are mediated in large part by the effects of E_2 in the brain.

In an initial study, castrated male rats were administered a single iv dose of E_2–CDS (3 mg/kg), an equimolar dose of E_2, or the DMSO vehicle for both steroids. Both E_2 and E_2–CDS reduced serum LH concentrations equivalently from 4 to 48 hr postinjection (Fig. 5). However, from 4 to 12 days, the E_2-treated rats showed a progressive increase in LH to levels observed in the DMSO-treated controls. By contrast, LH levels in animals treated with E_2–CDS remained suppressed by 82, 88, and 90% when compared to DMSO-treated rats at 4, 8, and 12 days after treatment, respectively. Serum E_2 concentrations were elevated through day 2 in both E_2- and E_2–CDS-treated rats. By day 4, however, E_2 levels in both groups had returned to those seen in the DMSO-treated rats (11). These data indicate that the preferential delivery of E_2 to the brain, using E_2–CDS, results in the chronic suppression of LH secretion, despite low plasma levels of the steroid. Apparently the long half-life of E_2–Q^+ and the resulting slow release of E_2 in the brain results in a prolonged biological response to the delivery system.

To evaluate the duration of the effect of E_2–CDS on LH secretion, we repeated this study, but sampled animals at 12, 18, and 24 days following a single injection of E_2–CDS or the vehicle DMSO (12). E_2–CDS suppressed LH levels relative to DMSO controls by 88, 86, and 66% at 12, 18, and 24 days, respectively (Table II).

In an additional study, we assessed the effects of E_2–CDS on LH secretion in ovariectomized rats (13). Two weeks after ovariectomy, rats were treated with E_2–CDS at doses of 10, 100, or 1000 μg/kg or the vehicle for the drugs, hydroxypropyl-β-cyclodextrin (HPCD). Animals were killed by decapitation 1, 7, 14, 21, and 28 days later. E_2–CDS caused a dose-dependent suppression in LH secretion at each time point evaluated (Fig. 6). The maximal suppression of serum LH was observed at 7 days, at which time E_2–CDS caused a reduction in LH of 21, 46, and 86% at the 10-, 100-, and 1000-μg/kg doses, respectively. By contrast, E_2 at a dose equimolar to the highest dose of E_2–CDS tested was comparatively ineffective in suppressing LH secretion. In view of our observation that E_2–CDS markedly increases the response of the anterior pituitary gland to luteinizing hormone-releasing hormone

---

HPCD. The 1-, 0.1-, and 0.01-mg/kg doses are represented in the upper, middle, and lower panels, respectively. Brain concentrations of E_2–Q^+ and E_2 were determined from the same sample following the separation of the two E_2–CDS metabolites. The animals used in this study were also used for assay of serum LH levels (see Fig. 6). [From Rahimy et al. (10), with permission.]

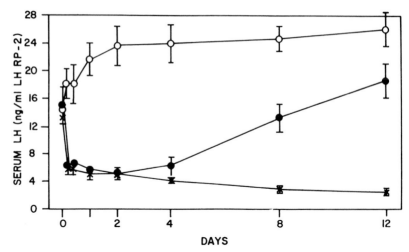

FIG. 5   Effects of a single dose of E_2–CDS ($\times$), an equimolar dose of E_2 ($\bullet$), or the vehicle, DMSO ($\bigcirc$), on serum LH in orchidectomized rats. Values depicted are means $\pm$ SEM. [From Simpkins *et al.* (11), with permission.]

(LHRH), the suppression of LHRH secretion by E_2–CDS must be even greater than is reflected through our evaluation of LH release. This conclusion is consistent with the observation of Sarkar *et al.* (14), who reported that E_2–CDS reduced LHRH secretion into the hypophyseal portal vessels at both 1 and 16 day postinjection.

Finally, we assessed the effects of E_2–CDS on masculine sexual behavior

TABLE II   Concentrations of Serum Luteinizing Hormone and Estradiol at 12 to 24 Days after Treatment with a Single Intravenous Dose of E_2–CDS in Orchidectomized Rats

Treatment group	Days posttreatment	Serum LH (ng/ml)	Serum E_2 (ng/ml)
DMSO	12	6.8 ± 0.8	26 ± 4
E_2–CDS	12	0.8 ± 0.3[a]	26 ± 4
DMSO	18	14 ± 3	<20[b]
E_2–CDS	18	1.7 ± 0.7[a]	<20
DMSO	24	8.6 ± 1.7	24 ± 3
E_2–CDS	24	2.9 ± 1.1[a]	24 ± 3

[a] $p < 0.05$ vs the DMSO group at the same time.
[b] Samples were below the limits of detection of the assay used.

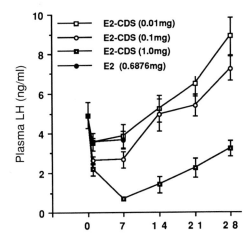

FIG. 6   Dose- and time-dependent effects of E_2–CDS on plasma concentrations of LH in ovariectomized rats. Animals received a single iv injection of E_2–CDS at time 0 and were sampled by decapitation at the times indicated. The doses of E_2–CDS administered were 0.01, 0.1, or 1 mg/kg for E_2–CDS. Estradiol was administered at a dose of 0.7 mg/kg, equimolar to the 1-mg/kg dose of E_2–CDS. Values depicted are means ± SEM. (□) E_2–CDS (0.01 mg); (○) E_2–CDS (0.1 mg); (■) E_2–CDS (1.0 mg); (●) E_2 (0.6876 mg). [From Rahimy *et al.* (13), with permission.]

in male rats to confirm the brain-mediated action of E_2–CDS, as it is recognized that E_2 stimulates copulatory behavior in male rats. We compared the copulatory response of castrated male rats to a single injection of E_2–CDS (3 mg/kg), an equimolar dose of E_2–valorate and the vehicle for both drugs, DMSO. E_2-valorate (E_2–V) was chosen for these studies because, like E_2–CDS, it is a C-17 substituted estrogen. As shown in Fig. 7, castration for 4 weeks resulted in a decline in performance in all three groups. Treatment with E_2–V improved performance transiently, as measured by percent responders (Fig. 7) or other measures of sexual activity; such as latency to the first mount, intromission number and latency and ejaculatory behavior (15). By contrast, E_2–CDS stimulated male sexual behavior through 35 days postinjection (Fig. 7). These data provide clear evidence that E_2–CDS has a long biological half-life in the brain, as is predicted on the basis of the design of the delivery system and our pharmacokinetic evaluations.

Concluding Remarks

In summary, the E_2–CDS described in this chapter has the physicochemical, kinetic, and pharmacodynamic properties necessary for the delivery of steroids to the brain and for their sustained local release. As estrogens are

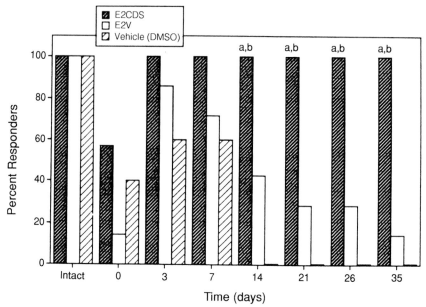

FIG. 7   Effects of E_2–CDS and E_2–V on masculine sexual behavior in orchidectomized rats. Intact rats were tested every 5 days until four successive, consistent behavioral patterns were observed, then all were orchidectomized. At 28 days after orchidectomy, rats were again tested and were randomly assigned to one of three treatment groups, which received E_2–CDS, E_2–V, or DMSO. Behavioral tests were repeated through 35 days postinjection. (a) Different from DMSO group; (b) different from E_2–V group. [From Anderson *et al.* (15), with kind permission from Pergamon Press Ltd., Headington Hill Hall, Oxford OX3 OBW, U.K. © 1987.]

candidates for the therapy of a variety of brain-mediated diseases, including the treatment of the menopausal syndrome, alleviation of depression, and the improvement of memory and cognition, the development of a delivery system for the enhanced and selective release of estrogens in the brain is needed. The E_2–CDS is one such delivery system.

Acknowledgments

This work was supported in part by Grants NIH AG10485 and HD 22540.

References

1. L. A. Berglund, H. Derendorf, and J. W. Simpkins, *Endocrinology* (*Baltimore*) **122,** 200 (1988).
2. P. W. Landfield, R. K. Baskin, and T. A. Pitler, *Science* **214,** 581 (1981).

3. M. Singh, E. M. Meyer, W. J. Millard, and J. W. Simpkins, *Brain Res.* **644,** 305 (1994).

4. N. Bodor and J. W. Simpkins, *Science* **221,** 65 (1983).

5. N. Bodor, J. McCornick, and M. E. Brewster, *Int. J. Pharmacol.* **35,** 47 (1987).

6. L. Janocko, J. Larner, and R. K. Hochberg, *Endocrinology (Baltimore)* **114,** 1180 (1984).

7. M. E. Brewster, K. S. Estes, and N. Bodor, *J. Med. Chem.* **31,** 244 (1987).

8. G. Mullerman, H. Derendorf, M. E. Brewster, K. S. Estes, and N. Bodor, *Pharm. Res.* **5,** 172 (1988).

9. M. H. Rahimy, N. Bodor, and J. W. Simpkins, *J. Steroid Biochem.* **33,** 179 (1989).

10. M. H. Rahimy, J. W. Simpkins, and N. Bodor, *Pharm. Res.* **7,** 1061 (1990).

11. J. W. Simpkins, J. McCormack, K. S. Estes, M. E. Brewster, E. Shek, and N. Bodor, *J. Med. Chem.* **29,** 1809 (1986).

12. K. S. Estes, M. E. Brewster, J. W. Simpkins, and N. Bodor, *Life Sci.* **40,** 1327 (1987).

13. M. H. Rahimy, J. W. Simpkins, and N. Bodor, *Pharm. Res.* **7,** 1107 (1990).

14. D. K. Sarkar, S. J. Friedman, S. S. C. Yen, and S. A. Frautschy, *Neuroendocrinology* **50,** 204 (1989).

15. W. R. Anderson, J. W. Simpkins, M. E. Brewster, and N. Bodor, *Pharmacol. Biochem. Behav.* **27,** 265 (1987).

[3]   Neurosteroids: Biosynthesis and Function

Paul Robel and Etienne-Emile Baulieu

The relationships between steroid hormones and brain function have mostly been envisioned within the framework of endocrine mechanisms, as responses elicited by secretory products of steroidogenic endocrine glands, borne by the bloodstream, and exerting action on the brain.

In fact, the brain is a target organ for steroid hormones. Intracellular receptors involved in the regulation of specific gene transcription have been identified in neuroendocrine structures, with each class of receptor having a unique distribution pattern in the complex anatomy of the brain (1,2). Mechanisms involving steroid receptors account for most steroid-induced feedback and many behavioral effects, for the regulation of the synthesis of several neurotransmitters, hormone-metabolizing enzymes, and hormone and neuromediator receptors, and also for the organizational effects on neural circuitry that occur during development and persist to adulthood.

However, it is now well established that local target tissue metabolism is an important factor in the mechanism of action of sex steroid hormones. Not only may such metabolism be involved in the regulation of intracellular hormone levels, but it may also provide an essential contribution to the cellular response. The brain is a site of extensive steroid metabolism. Aromatization and 5α-reduction represent major routes of androgen metabolism (3–5). The importance of these two pathways lies in the fact that they give rise to metabolites with considerable biological activity and thus are involved in the mechanism by which circulating androgens influence neuroendocrine function and behavior.

Progesterone (PROG) is also a substrate of 5α-reductase, and is converted to several metabolites, particularly 5α-dihydroprogesterone (5α-DH PROG) and 3α-hydroxy-5α-pregnan-20-one (allopregnanolone, $3\alpha,5\alpha$-TH PROG), which exert progesterone-like effects on neuroendocrine functions such as gonadotropin regulation and sexual behavior (6).

The characterization of pregnenolone (PREG) and dehydroepiandrosterone (DHEA) in the rat brain, as nonconjugated steroids and their sulfate (S) and fatty acid (L) esters, at higher concentrations in brain than in blood, has led to reconsideration of steroid–brain interrelationships (reviewed in Ref. 7).

The accumulation of DHEA, PREG, and their conjugates in the brain appeared to be independent of adrenal and gonadal sources, as shown by the persistence of these steroids in the brain for up to 1 month after gland

Methods in Neurosciences, Volume 22

ablation or pharmacological suppression. This contrasted with testosterone and corticosterone, the concentrations of which readily decline to zero after removal of the corresponding endocrine glands. This observation led to the discovery of a steroid biosynthetic pathway in the central nervous system (CNS).

The term *neurosteroids* was applied in 1981 to steroids synthesized in the brain, either *de novo* from cholesterol or by *in situ* metabolism of blood-borne precursors (8). These steroids have already been chemically characterized in other tissues, but what is peculiar is their site of synthesis, and hence their eventual involvement in autocrine and paracrine processes that might ultimately regulate brain functioning. This chapter provides a current view of neurosteroid biosynthesis in the brain, and of their possible mechanisms of action and physiological roles. The peripheral nervous system has not been included, although PREG is also synthesized and metabolized in Schwann cells and may produce active metabolites (9, 10). The endocrine effects of peripheral steroid hormones on neuronal activity have been extensively reviewed (2) and remain outside the scope of this chapter.

Neurosteroids: A Brief History

Several years elapsed between the discovery of DHEA and PREG accumulation in the rat brain, as well as in that of several other mammalian species, including the human, and the conclusive demonstration of *de novo* steroid biosynthesis. The first step in steroid synthesis is the conversion of cholesterol (CHOL) to PREG. Cytochrome P-450scc (for side-chain cleavage) is found in the mitochondria of steroidogenic endocrine cells as part of a ternary complex with adrenodoxin reductase and adrenodoxin (the cholesterol desmolase complex). Bovine and rat P-450scc enzymes have been purified, and specific antisera have been generated. The corresponding immunoglobulins have been used to set up an immunohistochemical technique for the detection of cytochrome P-450scc in rat tissues and human brain (11, 12). Specific immune staining was detected in the white matter throughout the brain. The immunohistochemical results fulfilled all the criteria of specificity. This was consistent with the detection of an antigen with the expected molecular size of P-450scc on Western blots (13). It was nevertheless mandatory to obtain the biochemical demonstration of side-chain cleavage activity. Mitochondria from oligodendrocytes were shown to contain this enzymatic activity, converting [^{3}H]CHOL to [^{3}H]PREG. Because the oligodendrocytes are the glial cells that synthesize myelin, these results confirm the immunologic localization of P-450scc. The presence of PREG throughout the brain can be explained by the generalized distribution of oligodendrocytes in the CNS (14).

The combination of immunohistochemical and biochemical evidence thus allowed the conclusion that brain cells can perform steroid biosynthesis from CHOL, thus justifying the term "neurosteroids" proposed in 1981 and adopted since then in many publications.

The biosynthesis of PREG from sterol precursors was confirmed by the incubation of newborn rat glial cell cultures in the presence of [^{3}H]mevalonate [^{3}H](MVA). These cells undergo a process of differentiation in culture resembling that *in vivo*, which can be followed by the measurement of 2,3'-cyclic-nucleotide-3'-phosphodiesterase (CNPase) activity. After day 15 of culture, CNPase activity and the biosynthesis of PREG increased in parallel, reaching their highest level at day 21 (15). At that point, the cells immunolabeled with antibodies against galactocerebroside, another marker of oligodendrocyte differentiation, were also positive for P-450scc. Cells incubated with [^{3}H]MVA in the presence of aminoglutethimide (AG), a potent inhibitor of P-450scc, accumulated [^{3}H]CHOL in their cytoplasm. On release of AG blockade, [^{3}H]CHOL was readily converted to [^{3}H]PREG (16), thus confirming the role of P-450scc in PREG biosynthesis.

Neurosteroid Metabolism in Brain

Both PREG and DHEA are found in part as their sulfate esters, the concentration of DHEAS exceeding that of DHEA. The conversion in the brain of 3β-hydroxy-Δ^5-steroids to their S esters is likely (Fig. 1), but is not documented, although preliminary evidence has been obtained for a low sulfotransferase activity (K. Rajkowski, unpublished observations). In contrast, the major conjugation forms of PREG and DHEA are their L esters. The acyltransferase responsible for their formation is enriched in the microsomal fraction (17). Its activity is highest at the time of myelin formation. It is likely that several isoforms exist, because CHOL- and corticosterone-esterifying activities, distinct from 3β-hydroxy-Δ^5-steroid acyltransferase, have been reported in the rat brain. [^{3}H]PREG can be converted to [^{3}H]PROG, and DHEA to androst-4-ene-3,17-dione, by a 3β-hydroxy-Δ^5-steroid dehydrogenase isomerase (3β-HSD) enzyme that is inhibited by specific steroid inhibitors such as trilostane. Four isoforms of 3β-HSD are known in the rat, types I and II are both expressed in adrenals, gonads, and adipose tissue, whereas type III is specific to the liver and type IV is expressed in skin and placenta (18, 19, 20). The brain isoform(s) has not yet been completely characterized.

The corresponding 20α-DH derivatives can be formed from either PREG or PROG. Progesterone is mainly converted to 5α-reduced metabolites. Two isoforms of the rat enzyme involved, 5α-reductase, have been cloned. RNAs encoding the type 2 isozyme are more abundant than type 1 mRNA in most male reproductive tissues, whereas the type 1 isozyme predominates in

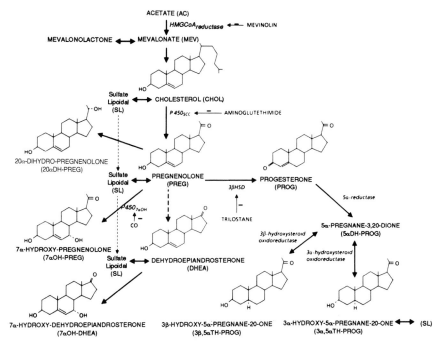

FIG. 1   Neurosteroid biosynthesis and metabolism in the rat brain. Dotted arrows indicate metabolic conversions not yet formally demonstrated.

peripheral tissues, including the brain (21). The 5α-reduced metabolite of PROG, 5α-DH PROG, is in turn converted to 3α- and 3β-hydroxy-5α-pregnan-20-ones. The corresponding hydroxysteroid oxidoreductases in the brain have not yet been cloned.

Finally, both PREG and DHEA give rise to large amounts of polar metabolites, the respective 7α-hydroxylated compounds (22, 23) of unknown biological significance.

Steroid-Metabolizing Enzymes in Rat Brain Cell Types

The cholesterol/desmolase complex appears localized almost exclusively in oligodendrocytes (Table I). The P-$450_{17\alpha}$ enzyme (a 17α-hydroxylase with C_{17-20}-desmolase activity) that would be responsible for the conversion of PREG to DHEA has not yet been found in the brain, and the origin of DHEA, DHEAS, and DHEAL accumulated there remains unexplained. P-

TABLE I   Enzymes of Steroid Metabolism in Rat Brain Cells[a]

Enzyme	Substrate → product	Newborn mixed glial[b]	Fetal astrocytes	Neurons
P-450scc	CHOL → PREG	+	nd	nd
P-450$_{17\alpha}$	PREG → DHEA			
	PROG → ADIONE	nd	nd	nd
20α-HOR	PREG → 20α-DH PREG			
	PROG → 20α-DH PROG	+	+	+
P-450$_{7\alpha}$	PREG → 7α-OH PREG			
	DHEA → 7α-OH DHEA	+	+	nd
3β-HSD	PREG → PROG			
	DHEA → ADIONE	+	+	(+)
5α-Reductase	PROG → DH PROG	+	+	+
3ξ-HOR	DH PROG → TH PROG	3α ≫ 3β	3β > 3α	3β ≫ 3α

[a] P450, cytochrome P-450; 20α-HOR, 20α-hydroxysteroid oxidoreductase; 3β-HSD, 3β-hydroxy-Δ5-steroid dehydrogenase-isomerase; 3ξ-HOR, 3α- or 3β-hydroxysteroid oxidoreductase; PREG, pregnenolone; PROG, progesterone; DHEA, dehydroepiandrosterone; ADIONE, androst-4-ene-3,17-dione, DH PROG; 5α-pregnane-3,20-dione; TH PROG, 3α- or 3β-hydroxy-5α-pregnan-20-one. (+), detected at low cell density; nd, not detected.
[b] Consist predominantly of oligodendrocytes.

450scc activity is notable in all types of glial cells, but negligible in neurons. Low 3β-HSD activity is present in glial cells and neurons.

Therefore, the lack of key enzymes indicates that neurons are unable to convert CHOL to PROG, whereas 5α-reductase and 3α-hydroxysteroid oxidoreductase activities are present in both glial cells and neurons (24). Whereas glial cells from newborn rats mainly hydroxylate 5α-DH PROG in the 3α position, astrocytes from fetal rats produce more of the 3β-hydroxy isomer and neurons seem to form mainly the 3β-hydroxy isomer.

Regulation of Neurosteroid Metabolism

Little is known about the mechanisms regulating neurosteroid formation. Both cAMP and glucocorticosteroids have been shown to enhance PREG formation from [³H]MVA in mixed glial cell cultures, but these effects might be related to an acceleration of cell differentiation *in vitro* (15). Cholesterol side-chain cleavage activity is regulated in the C6 glioma cell line by a mitochondrial benzodiazepine receptor, which serves to increase intramitochondrial cholesterol transport, thereby increasing the substrate availability to P-450scc as previously described for adrenals and gonads. It would seem likely that steroidogenesis in the brain is under some type of control by second messengers, like those in classic steroidogenic organs, but the trophic factors, their source, and the physiological stimuli are not known (25).

The metabolism of PREG and DHEA by astroglial cells (and probably also by neurons) is regulated by cell density: 3β-HSD activity is strongly inhibited at high cell density (see Ref. 26, and Y. Akwa, unpublished observations).

Mechanisms of Neurosteroid Action: Genomic Effects

Considering that PROG can be synthesized by glial cells, the demonstration of an estrogen-inducible PROG receptor in cultured oligodendrocytes of male and female rats suggests a classic intracellular mechanism of action in an autocrine/paracrine manner (27).

The effects of PROG and estradiol (E_2) were tested on mixed glial cell primary cultures. Cell growth was inhibited by PROG and stimulated by E_2 (28). Both hormones induced dramatic morphologic changes in both oligodendrocytes and astrocytes, in the sense of a more differentiated phenotype. This was accompanied by an increased synthesis of myelin basic protein in oligodendrocytes and of glial fibrillary acidic protein in astrocytes.

Nongenomic, Membrane Receptor-Mediated Activities of Neurosteroids

The anesthetic and sedative properties of steroids such as PROG were reported by Hans Selye as early as 1941 (29). It was initially assumed that they acted by partitioning into membrane lipids and altering neuronal function indirectly, as a consequence of a membrane-disordering effect similar to that reported for CHOL, barbiturates, and inhalation anesthetics (see Ref. 30 for a review). Indeed, PREG and DHEA are structurally related to CHOL. The hydrophobic fatty esters of PREG and DHEA are also good candidates, as are those of CHOL, for integration into plasma membranes, and are recovered mainly in myelin after subcellular fractionation of brain homogenates (7). The S esters of neurosteroids are amphipathic molecules, and they also could become integral components of cell membranes. Indeed, several steroidal compounds, including PREGS and DHEAS, have been delivered by iontophoresis and/or pressure, and produced an excitatory response from some neurons in the septopreoptic area of the guinea pig brain (see Ref. 8 for a review). Responses displayed a short latency in onset and fast decay, suggesting an action at the membrane level.

However, several observations contradicted the hypothesis of a simple

TABLE II   Ligand Specificity of Synaptosomal
Binding Component[a]

Steroid sulfates[b]	Isomers	Relative competition ratio
DHEA		100
PREG		43
17-OH PREG		15
ADIOL-3-mono		44
Cholesterol		0
TH PROG	$5\alpha, 3\alpha$-	468
	$5\alpha, 3\beta$	95
Testosterone		0
Androsterone	$5\alpha, 3\alpha$-	26
	$5\alpha, 3\beta$-	125
	$5\beta, 3\alpha$-	7
	$5\beta, 3\beta$-	45
Estradiol-3-mono-		30
Estradiol-17-mono-		45
Estradiol-3,17-di-		83

[a] Noncompetitors: DHEA, $5\alpha,3\alpha$-TH PROG up to 25 μM; GABA, muscimol, flunitrazepam, bicuculline, picrotoxin, mebubarbital up to 1 mM.
[b] ADIOL (androst-5-ene-$3\beta,17\beta$-diol) and estradiol can be esterified at either position 3 or 17 (3-mono- or 17-mono-) or at both (3,17-di-).

membrane-disordering mechanism of steroid action. Early structure–activity studies showed that the reduced metabolites of PROG [and of deoxycorticosterone (DOC)] were markedly more potent sedative–anesthetic agents than their less polar precursor. Also, unlike the 3α-hydroxy isomers, the 3β-hydroxy isomers of similar polarity were shown to be inactive. Furthermore, a synaptosomal component of proteinaceous nature, displaying a saturable binding affinity for DHEAS and PREGS, has been reported (see Ref. 31 for a review). [³H]DHEAS also binds to purified synaptosomal membranes, and the results are best fitted by a model involving a single class of saturable binding sites (32). The relative competition ratios of several steroid sulfates have been reported, in the order $3\alpha,5\alpha$-TH PROGS > DHEAS > PREGS (Table II). The nature of the binding component is discussed below.

Neurosteroids and GABA_A Receptors

The γ-aminobutyric acid (GABA) receptor type A (GABA$_A$-R) is an oligomeric protein complex that, when activated by agonists, produces an increase in neuronal membrane conductance to Cl⁻ ions, resulting in membrane hy-

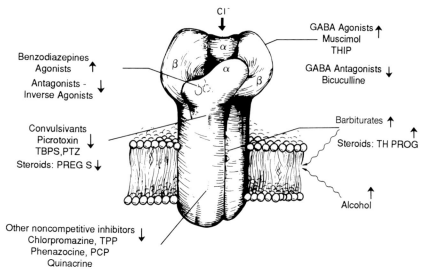

Cl⁻

GABA Agonists ↑
Muscimol
THIP

GABA Antagonists ↓
Bicuculline

Benzodiazepines
Agonists ↑

Antagonists -
Inverse Agonists ↓

Convulsivants
Picrotoxin ↓
TBPS,PTZ

Steroids: PREG S ↓

Barbiturates ↑

Steroids: TH PROG ↑

Alcohol ↑

Other noncompetitive inhibitors ↓
Chlorpromazine, TPP
Phenazocine, PCP
Quinacrine

FIG. 2 A simplified model of the GABA$_A$ receptor. This scheme does not represent either the exact assembly of subunits or the stoichiometry of ligand recognition sites associated with subunits. The continuous lines indicate the categories of sites on the major subunits α and β. The γ subunit (not represented) is necessary for the binding of benzodiazepines. The wavy lines indicate a possible modification of membrane structure. Vertical arrows indicate the stimulation or potentiation (↑) and the inhibition (↓) of GABAergic neurotransmission by the ligands listed. GABA, γ-Aminobutyric acid; THIP, 4,5,6,7-tetrahydroisoxazolo[5,4-c]pyridin-3-ol; TBPS, *tert*-butylbicyclophosphorothionate; PTZ, pentylene tetrazole; TPP, tetraphenylphosphonium; PCP, phencyclidine. [After Schwartz (51).]

perpolarization and reduced neuronal excitability. A number of centrally active drugs, including convulsants, anticonvulsants, anesthetics, and anxiolytics, bind to distinct but interacting domains of this receptor complex, to modulate Cl⁻ conductance (Fig. 2) (33).

There are both inhibitory and excitatory steroids that, respectively, excite or inhibit GABA activity, mediated by GABA$_A$ Rs in nerve cell membranes.

The inhibitory steroid metabolites, such as $3\alpha,5\alpha$-TH PROG and $3\alpha,5\alpha$-TH DOC ($3\alpha,21$-dihydroxy-5α-pregnan-20-one, tetrahydrodeoxycorticosterone), both mimic and enhance the effects of GABA. These steroids potentiate both benzodiazepine and muscimol binding, whereas they inhibit the binding of the convulsant tert-butylbicyclophosphorothionate (TBPS). Pharmacological evidence indicates that steroid interaction site(s) is (are) distinct from those of both barbiturates and benzodiazepines.

The most active inhibitory molecules operate in the low nanomolar range. The 3α-hydroxyl group is an absolute structural requirement. A planar conformation of the A/B ring junction (5α-H) is preferred; however, 5β-H derivatives still display significant activity.

PREGS and DHEAS are prototypic naturally excitatory neurosteroids (31). At low micromolar concentrations, they antagonize GABA$_A$ receptor-mediated ^{36}Cl$^-$ uptake into synaptoneurosomes and Cl$^-$ conductance in cultured neurons. PREGS bimodally modulates [^{3}H]muscimol binding to synaptosomal membranes, slightly potentiates benzodiazepine binding, and inhibits the binding of the convulsant [^{35}S]TBPS to the GABA$_A$ receptor chloride channel. Conversely, PREGS is displaced from its synaptosomal binding site by barbiturates at millimolar concentrations. Moreover, despite mutual competition in these membranes DHEAS and PREGS behave differently in their capability to displace [^{35}S]TBPS and 1-[^{3}H]-phenyl-4-*tert*-butyl-2,6,7-trioxabicyclo[2.2.2]octane ([^{3}H]TBOB) and to enhance [^{3}H]benzodiazepine binding. Furthermore, unlike the sulfate esters of 3β-hydroxy-Δ^5-steroids, those of 3α-hydroxysteroids behave as their unconjugated counterparts, potentiating GABA-induced Cl$^-$ transport in synaptoneurosomes (34). Therefore, distinct sites for neurosteroids, mediating distinct allosteric modes of interaction, seem to exist on the GABA$_A$-R or in its membrane vicinity. Nevertheless, the complexity of GABA$_A$-Rs must be kept in mind when interpreting these results, and transfection experiments with defined wild-type or mutated GABA$_A$-R subunits should provide more direct evidence for their sites of interaction with neurosteroids.

GABA$_A$ Receptors: In Vivo

The behavioral effects of the GABA-antagonistic neurosteroid PREGS contrast with the hypnotic actions of 3α,5α-TH PROG. When injected intracerebroventricularly, PREGS (8 μg/10 μl) shortens the sleep time in rats under pentobarbital hypnosis. Large amounts of PREGS, injected intraperitoneally, have the same effect, with a dose–effect relationship (7, 31).

Several reports have indicated a modulatory role of steroids in models of aggressiveness. DHEA inhibits the aggressive behavior of castrated male mice against lactating female intruders (35). To eliminate the possibility that the activity of DHEA was related to its conversion to metabolites with documented androgenic or estrogenic potency, behavioral experiments were repeated with the DHEA analog 3β-methylandrost-5-en-17-one (CH$_3$-DHEA) which, although not demonstrably estrogenic or androgenic in rodents, inhibited the aggressive behavior of castrated mice, dose relatedly, at least as efficiently as DHEA. Both molecules produced a marked and significant decrease of PREGS concentrations in the brain of treated castrated mice (36). It is tempting to speculate that DHEA and CH$_3$-DHEA, by decreasing

PREGS levels in brain, might increase the GABAergic tone, which has repeatedly been involved in the control of aggressiveness.

The memory-enhancing effects of DHEAS and of PREGS in male mice have been documented. When administered intracerebroventricularly after training, they showed memory-enhancing effects in foot-shock avoidance training. PREGS was the most potent, showing effects at 3.5 fmol/mouse (37). Infusion of PREGS (12 fmol) into the nucleus basalis magnocellularis (NBM) of the rat after the acquisition trial enhanced memory performance in a two-trial memory task (38). Conversely, TH PROG (6 fmol) disrupted performance when injected before the acquisition trial. A role for neurosteroids in memory processes subserved by the NBM is of interest in view of the implication of this structure in neurodegenerative processes leading to memory loss.

Other Membrane Effects of Neurosteroids

In contrast to GABA-activated Cl^- currents, glycine-activated Cl^- currents are unaffected by $3\alpha,5\alpha$-TH PROG in cultured chick spinal cord neurons. PROG inhibits glycine responses at high concentrations [50% effective concentration (EC_{50}), 16 μM]. PROG also modulates a neuronal nicotinic acetylcholine receptor reconstituted in *Xenopus* oocytes (EC_{50}, 2.9 μM) (39) (see Table III). A PROG-3-(*O*-carboxymethyl)oxime–BSA (bovine serum albumin) conjugate also inhibited the currents evoked by acetylcholine, indicating that the steroid interacts with a site located on the extracellular part of the membrane. Radioiodinated BSA conjugated to PROG had previously been shown to bind to nerve cell membranes with an affinity in the nanomolar range.

Neurosteroid Physiology

If the neuromodulatory activity of neurosteroids is not merely pharmacological, then their concentrations in brain, at least in defined biological situations, should be in the range of "neuroactive" concentrations.

Radioimmunoassays (RIAs) have been developed for measuring PREG, DHEA, their S and L esters, PROG, 5α-DH PROG, and $3\alpha,5\alpha$-TH PROG in plasma and tissues, including brain (7, 28).

Neither PREG nor DHEA disappear from the brain after removal of adrenals and gonads. Moreover, they undergo prominent circadian variations,

TABLE III   Neurosteroid Effects on Brain Cell Surface Events

Effect	Steroid	Effective concentration (nM)	Ref.
Binding to synaptosomal membranes	PROG	10^1	Towle and Sze (43)
	PROG-3–BSA	3×10^1	Ke and Ramirez (44)
	DHEAS	3×10^3	Demirgören et al. (45)
	PREGS	2×10^4	Majewska et al. (46)
	TH PROGS	8×10^2	Sancho et al. (32)
Allosteric modulation of drug–membrane interactions			
	GABA$_A$ receptor		
Positive	TH PROG	10^1	Paul and Purdy (28)
	TH PROGS	10^2	El Etr et al. (34)
Negative	PREGS	10^3–10^5	Majewska (31)
	DHEAS		
	NMDA receptor[a]		
Positive	PREGS	10^2–10^3	Wu et al. (47)
	Glycine receptor		
Negative	PROG	10^2	Wu et al. (48)
	PREGS		
	Nicotinic receptor		
Negative	PROG	10^2	Valera et al. (49)
	PROG-3–BSA		
	Oxytocin receptor		
Positive	PROG	0.1	Schumacher et al. (50)

[a] Glutamate receptor of the N-methyl D-aspartate type.

with an acrophase at the beginning of the dark period, preceding the corticosterone acrophase.

PROG also persists, in the low nanomolar range, in the brains of male and female rats after the removal of adrenals and gonads (40), consistent with the operation of a biosynthetic pathway from CHOL, observed in glial cells in culture.

The brain concentrations of PROG and $3\alpha,5\alpha$-TH PROG are markedly different according to gender. In particular, whereas the amounts of TH PROG are negligible in intact males, they are in the 10–100 nM range in cyclic and pregnant females, concentrations that are sufficient to exert clear-cut potentiation of GABAergic neurotransmission (40).

The ovary seems to serve a dual role here: it secretes PROG, which is converted to TH PROG in the brain, concurrently with endogenous PROG and it secretes estradiol, which may increase 3α-hydroxysteroid oxidoreductase activity in certain rat brain areas (41).

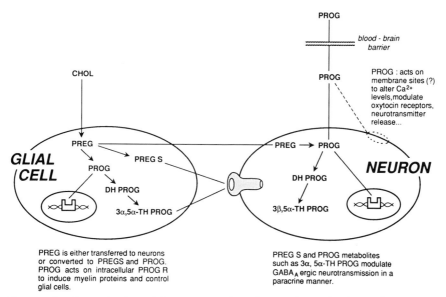

FIG. 3   A schematic view of the nongenomic and genomic mechanisms of neurosteroid action in neural tissue. The cross-talk between glial cells and neurons implies the provision of steroid precursors and the modulation of GABA$_A$ receptors by steroid metabolites of glial origins.

The concentrations of neurosteroids in brain also depend on stress conditions (30) and on behavioral situations such as heterosexual exposure (35).

Concluding Remarks

The effects of steroids on CNS neurotransmitter functions involve both genomic and nongenomic actions (8, 42). The GABA$_A$ receptor complex is a target for naturally occurring steroids, as shown by subcellular, cellular, and *in vivo* experiments. Because several steroids are accumulated in brain, independently (at least in part) of the steroidogenic gland contribution, and their presence can be related to steroid biosynthetic pathways in brain, their designation as "neurosteroids" is justified. Their concentrations in the brain appear to be controlled by mechanisms proper to that organ, although a modulatory role of peripheral steroid hormones, for example, ovarian E$_2$ secretion, is likely. The levels of neurosteroids reached in brain are compatible with their playing a physiological neuromodulatory role in situations such as the estrous cycle, pregnancy, and stress, and with influencing sexual behavior, mood, memory, and developmental and aging processes.

Further work is required to establish firmly the physiological significance of neurosteroids and to define their sites of interaction in molecular terms. The data at hand strongly suggest paracrine/autocrine functions of neurosteroids (Fig. 3).

Acknowledgments

We thank Krzysztof Rajkowski for careful revision of the manuscript, and Corinne Legris, Françoise Boussac, and Jean-Claude Lambert for editorial assistance.

References

1. K. Fuxe, J. A. Gustafsson, and L. Wetterberg, "Steroid Hormone Regulation of the Brain." Pergamon, Oxford, 1981.
2. B. McEwen, *Trends Endocrinol. Metab.* **2,** 62 (1991).
3. F. Naftolin, K. J. Ryan, I. J. Davies, V. V. Reddy, F. Flores, Z. Retro, M. Kahn, R. Y. White, Y. Takaoka, and L. Wolin, *Recent Progr. Horm. Res.* **31,** 295 (1975).
4. I. Celotti, R. Massa, and L. Martini, *in* "Endocrinology" (L. J. De Groot, G. F. Cahill, E. Steinberger, Jr., and A. I. Winegrad, eds.), Vol. 1, p. 41. Grune & Stratton, New York, 1979.
5. N. J. McLusky, A. Philip, C. Harlburt, and F. Naftolin, *in* "Metabolism of Hormonal Steroids in the Neuroendocrine Structures" (F. Celotti and F. Naftolin, eds.), p. 105. Raven, New York, 1984.
6. Y. G. Cheng and H. J. Karavolas, *J. Biol. Chem.* **250,** 7997 (1975).
7. P. Robel, Y. Akwa, C. Corpéchot, Z. Y. Hu, I. Jung-Testas, K. Kabbadj, C. Le Goascogne, R. Morfin, C. Vourc'h, J. Young, and E. E. Baulieu, *in* "Brain Endocrinology" (M. Motta, ed.), p. 105. Raven, New York, 1991.
8. E. E. Baulieu, *in* "Steroid Hormone Regulation of the Brain" (K. Fuxe, J. A. Gustafsson, and L. Wetterberg, eds.), p. 3. Pergamon, Oxford, 1981.
9. R. Morfin, J. Young, C. Corpéchot, B. Egestad, J. Sjövall, and E. E. Baulieu, *Proc. Natl. Acad. Sci. U.S.A.* **89,** 6790 (1992).
10. Y. Akwa, M. Schumacher, I. Jung-Testas, and E. E. Baulieu, *C. R. Acad. Sci.* **316,** 410 (1993).
11. C. Le Goascogne, P. Robel, M. Gouézou, N. Sananès, E. E. Baulieu, and M. Waterman, *Science* **237,** 1212 (1987).
12. C. Le Goascogne, M. Gouézou, P. Robel, G. Defaye, E. Chambaz, M. R. Waterman, and E. E. Baulieu, *J. Neuroendocrinol.* **1,** 153 (1989).

13. M. Warner, P. Tollet, M. Strömstedt, K. Carlström, and J. A. Gustafsson, *J. Endocrinol.* **122,** 341 (1989).

14. Z. Y. Hu, I. Jung-Testas, P. Robel, and E. E. Baulieu, *Proc. Natl. Acad. Sci. U.S.A.* **84,** 8215 (1987).

15. I. Jung-Testas, Z. Y. Hu, E. E. Baulieu, and P. Robel, *Endocrinology (Baltimore)* **125,** 2083 (1989).

16. Z. Y. Hu, I. Jung-Testas, P. Robel, and E. E. Baulieu, *Biochem. Biophys. Res. Commun.* **161,** 917 (1989).

17. C. Vourc'h, B. Eychenne, D. H. Jo, J. Raulin, D. Lapous, E. E. Baulieu, and P. Robel, *Steroids* **57,** 210 (1992).

18. G. Pelletier, E. Dupont, J. Simard, V. Luu-The, A. Bélanger, and F. Labrie, *J. Steroid Biochem. Mol. Biol.* **43,** 451 (1992).

19. J. I. Mason, *Trends Endocrinol. Metab.* **4,** 199 (1993).

20. J. Simard, J. Couet, F. Durocher, Y. Labrie, R. Sanchez, N. Breton, C. Turgeon, and F. Labrie, *J. Biol. Chem.* **168,** 19659 (1993).

21. K. Normington and D. W. Russel, *J. Biol. Chem.* **267,** 19548 (1992).

22. M. Warner, M. Strömstedt, L. Möller, and J. A. Gustafsson, *Endocrinology (Baltimore)* **124,** 2699 (1989).

23. Y. Akwa, R. F. Morfin, P. Robel, and E. E. Baulieu, *Biochem. J.* **288,** 959 (1992).

24. K. Kabbadj, M. El Etr, E. E. Baulieu, and P. Robel, *Glia* **7,** 170 (1993).

25. P. Guarneri, V. Papadopoulos, B. Pan, and E. Costa, *Proc. Natl. Acad. Sci. U.S.A.* **89,** 5118 (1992).

26. Y. Akwa, N. Sananès, M. Gouézou, P. Robel, E. E. Baulieu, and C. Le Goascogne, *J. Cell Biol.* **121,** 135 (1993).

27. I. Jung-Testas, J. M. Renoir, J. M. Gasc, and E. E. Baulieu, *Exp. Cell Res.* **193,** 12 (1991).

28. I. Jung-Testas, J. M. Renoir, G. L. Greene, and E. E. Baulieu, *J. Steroid Biochem. Mol. Biol.* **41,** 621 (1992).

29. H. Selye, *Proc. Soc. Exp. Biol. Med.* **46,** 116 (1941).

30. S. M. Paul and R. H. Purdy, *FASEB J.* **6,** 2311 (1992).

31. M. D. Majewska, *Progr. Neurobiol.* **38,** 379 (1992).

32. M. J. Sancho, B. Eychenne, J. Young, C. Corpéchot, and P. Robel, *Annu. Meet. Endocr. Soc., 73rd, Washington, D.C.* Abstr. 1039 (1991).

33. R. W. Olsen and A. J. Tobin, *FASEB J.* **4,** 1469 (1990).

34. M. El Etr, C. Corpéchot, J. Young, Y. Akwa, P. Robel, and E. E. Baulieu, *Annu. Meet. Eur. Neurosci. Assoc. 15th, Munich, Eur. J. Neurosci. Suppl. No. 5,* Abstr. 2215 (1992).

35. E. E. Baulieu, P. Robel, O. Vatier, M. Haug, C. Le Goascogne, and E. Bourreau, *in* "Receptor Interactions" (K. Fuxe and L. F. Agnati, eds.), Vol. 48, p. 89. Macmillan, New York, 1987.

36. J. Young, C. Corpéchot, M. Haug, S. Gobaille, E. E. Baulieu, and P. Robel, *Biochem. Biophys. Res. Commun.* **174,** 892 (1991).

37. J. F. Flood, J. E. Morley, and E. Roberts, *Proc. Natl. Acad. Sci. U.S.A.* **89,** 1567 (1992).

38. W. Mayo, F. Dellu, P. Robel, J. Cherkaoui, M. Le Moal, E. E. Baulieu, and H. Simon, *Brain Res.* **607,** 324 (1993).
39. P. Robel and E. E. Baulieu, *Trends Endocrinol. Metab.* **5,** 1 (1994).
40. C. Corpéchot, J. Young, M. Calvel, C. Wehrey, J. N. Veltz, G. Touyer, M. Mouren, V. V. K. Prasad, C. Banner, J. Sjövall, E. E. Baulieu, and P. Robel, *Endocrinology* (*Baltimore*) **133,** 1003–1009 (1993).
41. Y. G. Cheng and H. J. Karavolas, *Endocrinology* (*Baltimore*) **93,** 1157 (1973).
42. B. McEwen, *Trends Pharmacol. Sci.* **12,** 141 (1991).
43. A. C. Towle and P. Y. Sze, *J. Steroid Bichem.* **18,** 135 (1983).
44. F. C. Ke and V. D. Ramirez, *J. Neurochem.* **54,** 467 (1990).
45. S. Demirgören, M. D. Majewska, C. E. Spivak, and E. D. London, *Neuroscience* **45,** 127 (1991).
46. M. D. Majewska, S. Demirgören, and E. D. London, *Eur. J. Pharmacol.* **189,** 307 (1990).
47. F. S. Wu, T. T. Gibbs, and D. H. Farb, *Mol. Pharmacol.* **40,** 333 (1991).
48. F. S. Wu, T. T. Gibbs, and D. H. Farb, *Mol. Pharmacol.* **37,** 597 (1990).
49. S. Valera, M. Ballivet, and D. Bertrand, *Proc. Natl. Acad. Sci. U.S.A.* **89,** 9949 (1992).
50. M. Schumacher, H. Coirini, D. W. Pfaff, and B. S. McEwen, *Science* **250,** 691 (1990).
51. R. D. Schwartz, *Biochem. Pharmacol.* **17,** 3369 (1988).

[4] Cytochrome P450 Enzymes in Brain

Margaret Warner, Adrian Wyss, Shigetaka Yoshida, and Jan-Åke Gustafsson

Introduction

The idea that cytochromes P450 in the central nervous system (CNS) could be playing important roles in the endocrinology, pharmacology, and toxicology of the brain is not new. The presence of aromatase in the brain and its role in sexual imprinting of the brain has been known for over two decades (1) and since the first report of the spectral detection of P450 in brain microsomes (2) there have been numerous attempts to quantitate and characterize the forms of P450 in the brain (3–11). Because of the large number of forms of the enzyme (12), their varied substrate and product specificities, their tissue distribution, and the role of individual P450s in cellular physiology and toxicology, a careful characterization of the forms in the brain is warranted. Some controversy has arisen over the quantity of P450 in the brain and over which forms are constitutive and which are inducible. Most of the uncertainty concerns forms of P450 involved in the activation of xenobiotics to carcinogens and cytotoxins. The levels of these enzymes in the brain and their role in CNS toxicity, particularly in degenerative diseases of the brain, is an important issue that needs to be clarified. In this chapter we summarize the available data and try to analyze the conflicting results emerging from various laboratories by evaluating the methods used for measuring the quantity, distribution (cellular and subcellular), catalytic activity, immunological characteristics, mRNA levels, regulation, and functions of brain P450.

Quantitation and Subcellular Distribution of Brain P450

In tissues where its concentration is high, P450 can be measured by its CO difference spectrum as originally described by Omura and Sato (13). In brain homogenates or subcellular fractions, other pigments such as mitochondrial cytochrome oxidase and hemoglobin are present in higher concentrations than P450 and this makes it all but impossible to quantitate P450 accurately from spectra. In the first publication of the spectral evidence for the presence of P450 in brain microsomes (2) a small deflection at 450 nm was evident but no attempt was made to quantitate the P450. Since then several labora-

Methods in Neurosciences, Volume 22
Copyright © 1994 by Academic Press, Inc. All rights of reproduction in any form reserved.

51

TABLE I   Spectral Quantitation of Brain Microsomal P450

P450 content[a]	Spectra shown	Possible problems	Ref.
Not calculated	+		Cohn *et al.*, 1977 (2)
35	+	Difficult baseline	Marietta *et al.*, 1979 (3)
17	−		Guengerich and Mason, 1979 (4)
34	+		Shiverick and Notelovitz, 1979 (5)
19–37	+	Difficult baseline	Holtzman and Desautel, 1980 (6)
10	+	—	Nabeshima *et al.*, 1981 (7)
3–5	+	—	Qato and Maines, 1985 (8)
17	+	Difficult baseline and other chromophores	Walther *et al.*, 1986 (9)
88	−		Ravindranath *et al.*, 1990 (10)
100	−		Anandatheerhavarada *et al.*, 1991 (11)

[a] In picomoles per milligram of microsomal protein.

tories have calculated the concentration of brain P450 from spectra of both homogenates (3) and microsomal fractions (4–11). The general finding is that the level of P450 in brain microsomes is low, 5–20 pmol/mg microsomal protein, which is 0.5–2.0% of the hepatic microsomal level. In one laboratory (10, 11) values as high as 88–100 pmol/mg microsomal protein have been reported (Table I).

The extent to which this low and variable level reflects a low cellular level, a poor recovery of microsomes, or an inaccuracy in spectral quantitation in brain microsomes has been widely discussed and attempts have been made in different laboratories to resolve these issues. In publications in which the CO difference spectra are presented it is possible to evaluate whether some of the problems are due to baseline errors or interference with the P450 spectrum by other chromophores. However, in many studies no spectra are shown. We, as well as others, have difficulty in accurately measuring P450 in brain microsomes. There have been suggestions for modifications of the assay (9, 14, 15) but even with these modifications we are not able to quantitate brain P450 reliably.

Anandatheerhavarada *et al.* (10, 11) find quite high levels of P450 in brain microsomes and this they accredit to their method of preparation of brain microsomes (V. Ravindranath, personal communication). In these studies brains are first perfused and then homogenized in a buffer composed of 20% (v/v) glycerol, 100 mM Tris (pH 7.4), 1.15% (w/v) KCl, 1 mM ethylenediaminetetraacetic acid (EDTA), and 0.1 mM phenylmethylsulfonyl fluoride (PMSF) and microsomes are prepared by standard differential centrifugation. The only difference between this method and the protocols used in other laboratories is the replacement of 20% glycerol by 0.25 M sucrose. Because the level of cytochrome-c reductase, a microsomal marker, is the same in brain microsomes prepared with a sucrose buffer (17) as that in the Ananda- theerhavarada *et al.* study, it seems that it is the recovery of P450 that is higher and not the recovery of microsomes. We have also prepared brain microsomes in glycerol buffer, but do not find high levels of P450. The recovery of P450 from the brains of five male Wistar rats (200 g body weight) was 44 pmol/g tissue or 10 pmol/mg microsomal protein.

One approach to quantitation of brain P450 that avoids the issues of purity of subcellular fractions and the difficulties of spectral quantitation is extrac- tion of P450 from brain homogenates (18). With p-chloroamphetamine-cou- pled Sepharose as a hydrophobic matrix it is possible to wash away most chromophores and selectively retain P450 from solubilized brain homoge- nates. The P450 is then eluted from the matrix and easily quantitated by its CO difference spectrum according to Omura and Sato (13). The yield of P450 is 30–300 pmol/g wet weight of tissue, depending on the brain region. Most of this P450 is reducible with NADPH–cytochrome-P450 reductase and is therefore microsomal. One of the main reservations about the use of this approach is the question of recovery of P450 or selective losses of individual forms during isolation. Because it is not possible to quantitate P450 accurately in the starting material (brain homogenates) a yield cannot be calculated. However, on the basis of the P450 measurable after p-chloroamphetamine- coupled Sepharose chromatography, the P450 content of brain microsomes is 3–30 pmol/mg microsomal protein, which is close to the values obtained from many laboratories where P450 has been measured in brain microsomes. The P450 prepared in this way has been characterized on the basis of catalytic activity, immunological reactivity, and N-terminal amino acid sequence.

Detection of Specific Forms of P450 in Brain by Catalytic Activities

Xenobiotic Metabolism

Brain microsomes and homogenates have been examined for catalytic activi- ties characteristic of the major hepatic, adrenal, and gonadal forms of P450. In general it can be concluded that when they are measurable, these activities are low compared to the other organs. In addition it is evident that the activities are lower than would be predicted on the basis of the P450 content of the brain and this indicates that the major forms of brain P450 do not utilize the same substrates as the P450s characterized in the other tissues (Table II).

As with the P450 content of brain microsomes, Anandatheerhavarada *et*

TABLE II Catalytic Activities in Brain Microsomes of Untreated Rats

Enzyme	Catalytic activity (nmol/mg microsomal protein/min)		P450 involved	Ref.
	Brain	Liver		
Aminopyrine N-demethylase	0.026	2.93		Marietta et al., 1979 (3)
	173	107		Anandatheerhavarada et al., 1990 (19)
Meperidine N-demethylase	0.0023	6.8		Marietta et al., 1979 (3)
Hexobarbital hydroxylase	0.0056	21.3		
Benzo[a]pyrene hydroxylase	0.004	0.22		Guengerich and Mason, 1979[a] (4)
	0.0035	0.039		Marietta et al., 1979 (3)
	0.01	1.78	1A1/2	Anandatheerhavarada et al., 1990 (19)
Naphthalene hydroxylase	0.246	×200		Mesnil et al., 1988 (20)
Ethoxyresorufin	0.0036			Walther et al., 1987 (16)
Arylhydrocarbon hydroxylase	0.0016			Das et al., 1982 (21)
Morphine N-demethylase	128	74	2B1/2	Anandatheerhavarada et al., 1990 (19)
Ethoxycoumarin O-deethylase	0.001		2B1/2	Naslund et al., 1988 (17)
	0.0017	1.2		Guengerich et al., 1979[a] (4)
	0.0088			Qato and Maines, 1985 (8)
	0.17	3.1		Rouet et al.,1981 (22)
	1.5	6.5	2B1/2	Anandatheerhavarada et al., 1990 (19)
	0.0016	0.353		Srivastava and Seth, 1983 (23)
Aniline hydroxylase	1.4	36	2E1	Anandatheerhavarada et al., 1993 (24)
p-Nitrophenol hydroxylase	0.07	0.78	2E1	
N-Nitrosodimethylamine N-demethylase	0.8	0.64	2E1	
Pentoxyresorufin O-dealkylase	0.012		2B1/2	
Codein N-demethylase	0.00023	0.170		Kodaira and Spector, 1989 (26)
Aromatase	0.0006			Roselli et al., 1984[a] (27)
5α-Androstane-3β,17β-diol hydroxylase	0.20	—		Warner et al., 1989 (28)
Dehydro epiandrosterone 7α-hydroxylase	0.322			Akwa et al., 1992 (29)
Bufuralol hydroxylase	0.00084	1.29	2D	Fonne-Pfister, 1987 (30)
Estrogen 2/4-hydroxylase	0.0002	2.95		Brown et al., 1985 (31)
Testosterone hydroxylations (6α, 15β, 15α, 6β, 16β, 1β, 2β)	0.00016– 0.002323		2B1/2	Jayyosi et al., 1992 (32)

[a] Activity was expressed per milligram of tissue in the paper by Guengerich et al. (4) and has been converted using the value of 10 mg of microsomal protein per gram of tissue.

al. (11, 19) find catalytic activities typical of P450 2B1, 1A1, and 2E1 that are 10 to 700-fold higher than those reported in other laboratories.

The better yield of P450 (fivefold higher than other laboratories) cannot fully explain these high catalytic activities. In a more recent study (25) the catalytic activity of P450 2B is much lower than in the previous studies. It is not clear whether these large differences are due to the different substrates used or to some other factors. In the case of P450 2E1 (24) the three catalytic activities attributed to this enzyme, aniline hydroxylase, *p*-nitrophenol hydroxylase, and *N*-nitrosodimethylamine *N*-demethylase have brain levels that are 1/25, 1/10, and 1.3 times the liver levels. It is, therefore, unlikely that all three activities are due to P450 2E1. The authors' explanation (V. Ravindranath, personal communication) is that perhaps the enzyme in the brain is not identical to P450 2E1. Some clarification of these results is required before conclusions can be made about whether there are indeed high levels of P450 2B and 2E1 in the brain.

Neurosteroid Metabolism

The discovery of high levels of pregnenolone and dehydroepiandrosterone in the brain and the evidence that these metabolites are synthesized within the brain (33) led to the term *neurosteroids*. Today a neurosteroid refers to steroids that act on, but are not necessarily synthesized in, the CNS. Although catalytic activity of P450scc (the cholesterol side-chain cleavage enzyme) is below the level of detection in the brain, the enzyme was detected in the brain by immunohistochemical techniques (34) and its mRNA could be detected by PCR (polymerase chain reaction) amplification (35). The immunohistochemical staining revealed widespread distribution in the white matter and shows that a low level of the enzyme is present in cells throughout the brain. Whether these levels in the brain are of physiological relevance in the synthesis of pregnenolone is not known.

Brain P450 is involved in another aspect of neurosteroid metabolism. 5α-Reduced metabolites of corticosteroids and progesterone are anesthetics and sedatives through their interaction with γ-aminobutyric acid (GABA) receptors (36). The level of these potent steroids in the brain is regulated by a specific form of brain P450 (37). 3α-Hydroxy-5α-pregnan-20-one (3α,5α-THP) is not a substrate for brain P450, but 3β-hydroxy-5α-pregnan-20-one (3β,5α-THP) is efficiently hydroxylated at the 6α and 7α positions. These metabolites are not formed to any detectable extent in the liver or kidney, but are formed in the prostate, pituitary, and brain. The enzyme responsible for this activity has been isolated from the prostate and brain and its substrate specificity characterized (37, 38). The best substrates are 5α-reduced, 3β-hydroxysteroids. Dihydrotestosterone and dehydroepiandrosterone are poor substrates and testosterone and androstenedione are not substrates. To test the hypothesis that the hydroxylation of 3β,5α-THP represents a pathway for regulation of the level of 3α,5α-THP in the brain, the effect of inhibition of the hydroxylation of 3β,5α-THP on the duration of 3α,5α-THP-induced anesthesia was examined. The nonanesthetic steroid, 5α-androstane-$3\beta,17\beta$-diol (3β-Adiol) was used as a competitive inhibitor of the metabolism of 3β, 5α-THP. The duration of anesthesia on intravenous (iv) administration of $3\alpha,5\alpha$-THP was increased by 33% when 3β-Adiol was coadministered, but 3β-Adiol in itself was not an anesthetic.

It can be concluded that, in the CNS, in analogy with the elimination of androgen from the prostate, it is the 3β- and not the $3\alpha,5\alpha$-THP that represents the major pathway for the formation of more polar metabolites and thus the elimination from the brain. A specific form of P450 catalyzes the hydroxylation of $3\beta,5\alpha$-THP and this catalytic activity plays an important role in regulation of the levels of the neuroactive steroid, $3\alpha,5\alpha$-THP.

Arachidonic Acid Metabolism

Brain P450s are responsible for both the synthesis and degradation of active metabolites of arachidonic acid. Arachidonic acid epoxides, formed through the action of P450, are thought to be the intracellular signals involved in the release of somatostatin from the hypothalamus (39, 40) and in the regulation of vascular tone (41). The active metabolite 5,6-epoxyeicosatrienoic acid has been measured in the brain and its synthesis from arachidonic acid can be blocked by P450 inhibitors (42). Although the P450s in the brain that are responsible for these activities have not yet been characterized, arachidonic acid epoxygenases in the liver are members of the 2C family (43).

Members of the 4A family are present in the brains of untreated rats (44). These P450s have as their substrates fatty acids and prostaglandins (45). Their preferred positions of hydroxylation are at the ω and ω-1 positions. These hydroxylations are thought to be important in the elimination of prostaglandins, but they may also be involved in the inactivation of other arachidonic acid metabolites. The cellular localization of 2C and 4A P450s in the brain will, no doubt, help to elucidate the roles of these P450s in pathways of arachidonic acid metabolism in the brain.

Immunological Detection of Brain P450

On Western blots 50 ng or 1 pmol of P450 gives a reliable and easily detected signal. With the microsomal content of brain P450 between 5 and 20 pmol/mg microsomal protein, it is necessary to load 50–100 μg of protein per lane in order to detect the most abundant forms of P450. It is not so surprising, therefore, that convincing signals on Western blots are not usually obtained with 20–30 μg of brain microsomes. Jayyosi *et al.* (32) loaded 50 μg of microsomal protein and reported a "small amount of immunoreactive protein related to P450 2E1" and a "lightly stained band" comigrating with P450 2B. Ding *et al.* (46) loaded 90 μg of microsomal protein from rabbit brains and could see no evidence of P450 2E1. In 1987 Sugita *et al.* (47) partially purified P450 from brain microsomes of untreated rats. There was no detectable P450 1A or 2B in the preparation. We also partially purified P450 from the brains of untreated rats and showed that P450 1A and 2B represented less than 0.1% of this P450 (18). More recently we have probed blots containing 100 pmol of brain P450 with antibodies against P450 2A, 3A, 2E1, and 4A. P450 4A was the only one detectable (44). It can be concluded from the above-mentioned studies that the concentration of P450 1A, 2B, and 2E1 in the brain is low (less than 20 pmol/mg microsomal protein). In some studies strong signals with antibodies against P450 1A, 2B, and 2E1 in 30 and 90 μg of brain microsomal protein have been reported (19). The intensity of signals on Western blots for P450 1A and 2B in brain microsomal protein indicates that the concentration of each of these forms of P450 is 1.9 nmol/mg brain microsomal protein and this is 20-fold higher than the total P450 content measured in the same study. These results clearly differ from those in other laboratories and require some clarification.

In the case of immunohistochemical studies in the brain with antibodies against P450 1A, 2B, and 2E1, several laboratories have found widespread specific staining throughout the brains of untreated rats (19, 48–51). In all cases specific staining means that the signals are abolished by preabsorption of the antibody with the purified antigens and no staining is detected with preimmune serum. These controls still leave open the possibility that the antibodies can react with similar but unrelated epitopes in the brain. With antibodies against P450 2C, more limited localization in the brain was found (52, 53). Two members of this family, 2C11 and 2C12, were localized in the hypothalamic–preoptic area. These forms, which are sexually differentiated in the liver (54), also exhibited sexual differentiation in the brain with 2C11 in the male and 2C12 in the female brain. These P450s were localized primarily in somatostatin- and oxytocin-containing neurons. Because of the difficulties in obtaining Western blots with brain microsomes and the low catalytic activities corresponding to these forms of P450 in the brain, the issue of whether the immunohistochemical signals in fact do represent P450 is not yet resolved. One possible method of confirming the presence of P450 in brain cells is purification and characterization of the enzymes; another is localization of the mRNA by *in situ* hybridization.

Brain P450 Purification and Characterization

Because of the low content of P450 few laboratories have attempted a purification of brain P450 from control animals. Sugita *et al.* used 650 rat brains and managed to obtain a P450 preparation with a specific content of 0.18 nmol/mg protein (47). This P450 was active in the 6α-, 6β-, 2-, 4-, 16-, and 15α-hydroxylation of estradiol, but there was no detectable P450 1A or 2B in the preparation. We have also partially purified brain P450 from untreated rats. We routinely use 20 rats and obtain 2–3 nmol of P450 with a specific content of 1–3 nmol/mg protein. The major P450 in these preparations is active in the 6α-, 7α-, and 7β-hydroxylations of 5α-reduced 3β-hydroxysteroids (18, 28, 44). The major hepatic forms of P450, 1A, 2A, 2B, 2C, 2E1, and 3A, either could not be detected or gave weak signals on Western blots with this P450. Bergh and Strobel (55) used 90 rats to partially purify brain P450 from β-naphthoflavone treated rats. They obtained 200 pmol of P450 with a specific content of 0.60 nmol/mg protein. On reconstitution of catalytic activity, the turnover numbers with ethoxyresorufin and ethoxycoumarin, two typical P450 1A and 2B substrates, were approximately 0.4, indicating that these two forms of P450 were minor components of the total brain P450. Anandatheerhavarada *et al.* (56) have purified to homogeneity P450 2B from the brains of phenobarbital-treated rats. They started with 100 rats and produced 9.4 nmol of a homogeneous preparation with a specific content of 12.7 nmol/mg protein. On reconstitution of catalytic activity the turnover numbers with the substrates aminopyrine, morphine, and ethoxycoumarin were 80, 38, and 10, respectively. This was all done with two chromatography steps. Perhaps the reason for the ease of purification in this study is the high level of 2B P450 in the brain microsomes of these rats. The Western blots show a concentration of approximately 1.3 nmol/mg microsomal protein. This is 65-fold higher than the content of 2B in the control rat liver.

Detection of P450 mRNAS in Brain

All our attempts to clone the cDNAs of brain P450s by screening of cDNA libraries have failed. Two approaches were used: (1) screening with full-length cDNAs of liver P450s under conditions of reduced stringency, and (2) the use as a hybridization probe of a cocktail of oligonucleotides (50-mers) encompassing the heme-binding regions of the known P450 families. Several attempts were made with both commercial and homemade cDNA libraries but no P450 cDNAs were found (57).

To date no hepatic forms of P450 mRNA have been detected in the brain by *in situ* hybridization or by Northern blotting. High levels of P450 2C6 were found in female rat brains on dot blots (58), but this could not be confirmed with Northern blots (59). With PCR, P450 2C6, 2C11, 2C12, 2C13, 2C23, 2E1 (59), 4A2, 4A8 (60), scc, and cholesterol 27-hydroxylase (34) have been amplified from cDNA reversed transcribed from brain mRNA. None of these mRNAs were detectable on Northern blots.

Lephart *et al.* detected brain aromatase mRNA on Northern blots (61) and Harada and Yamada have developed a quantitative PCR assay for detection of aromatase mRNA. With this technique, it has been possible to quantitate this rare message and study its developmental regulation (62). One interesting observation of this study is the lack of correlation between the mRNA and protein levels in the brain. Between gestational days 17 and 19 both the mRNA and the protein increase in parallel. After day 19 the mRNA remains elevated but there is a marked decrease in the protein. Future use of such quantitative methods with other P450s will provide some insight as to how the brain regulates the expression of these proteins.

Induction of Brain P450

With standard protocols used for the induction of liver P450s by phenobarbital (PB), 80 mg/kg/day for 5–10 days (19), or after a prolonged treatment protocol, 0.025% in drinking water for 70 days followed by intraperitoneal administration 80 mg/kg for 10 days (55), a twofold induction of catalytic activities characteristic of P450 2B has been observed (Table III). Other chronic treatments that induce P450 2B in the brain are nicotine (25) (1.76 mg/kg subcutaneously twice daily for 10 days), and testosterone (11) (daily administration of 10 mg/kg for 1 month followed by 5 mg/kg for 40 days). The regulation by testosterone is particularly interesting because this enzyme was twofold higher in the male than in the female rat brain and the difference was abolished by chronic treatment of female rats with testosterone (11). In most studies the catalytic activities are low and even after induction they are less than 5% of that in a control liver. However, the level of P450 2B in the control rat brain, measured in different laboratories, can vary by three orders of magnitude. In the mouse, chronic treatment with phenytoin (63), 30 mg/kg/day for 6 days followed by 60 mg/kg/day for 20–90 days induced P450 2B catalytic activity in the cerebellum to levels similar to that in the induced liver. The reasons for these large variations in P450 2B levels in the brain and the mechanisms involved in its induction have not been investigated.

In most studies the increase in brain of 1A P450 catalytic activities on treatment of rats with polycyclic aromatic hydrocarbons is less than twofold. Ethanol has been shown to induce brain P450 both after chronic consumption and after a single dose. The reported chronic effects of ethanol on brain

TABLE III   Induction of Rat Brain P450

Compound	Catalytic activity[a]						Ref.
	Control brain	PAH treated	PB treated	Ethanol	Nicotine	Mn	
P450 content[b]	0.010		0.010				Nabeshima et al. (7)
	0.017	0.024	0.028				Guengerich and Mason (4)
	0.09	0.09	0.17	0.13	0.18		Anandatheerthavarada et al. (24)
Pentoxyresorufin	0.0013				0.0020		Anandatheerthavarada et al. (25)
Aniline	1.4			3.4			Anandatheerthavarada et al. (24)
Ethoxycoumarin	1.5	1.17	3.5				Anandatheerthavarada et al. (19)
	0.0022	0.0042	0.0030				Shrivastava and Seth (23)
	0.0003	0.0005					Walther et al. (9)
	0.00088					0.00172	Qato and Maines (8)
	0.0016	0.0024					Näslund et al. (17)
	0.0017	0.054	nd				Guengerich and Mason (4)
	0.0042	0.0062					Rouetet et al. (22)
Benzopyrene	0.00041	0.00087	0.0008				Guengerich and Mason (4)
	0.00001					0.00003	Qato and Maines (8)

[a] Catalytic activities are expressed as nanomoles of product formed per minute per milligram of microsomal protein. PAH, p-Aminohippuric acid/ PB, phenobarbital; nd, not determined.
[b] Catalytic activity in nanomoles per milligram of protein.

P450 require some clarification. Ethanol (in drinking water 2% for 3 days followed by 5% for 1 week and 10% for 30 days) was reported to induce P450 2E1 in the brain (24), but there are several problems with this study that make it difficult to judge which isozyme was induced and what was the extent of the induction. The level of the aniline hydroxylase activity in the brain in this study was 1/25 of that in the liver, that is, 1.5 nmol/mg protein/min. Because aniline hydroxylase activity in the liver is due to P450 2E1 (64), if the activity in the brain were also due to P450 2E1, the signal for this protein in the brain microsomes, on Western blots, should be 25-fold lower than that in the liver, but it is not. It is closer to one-half of the liver level. In addition, the catalytic activities in both the liver and brain are much higher than those found in other laboratories. In the laboratory of Ingelman-Sundberg, where P450 2E1 has been studied extensively (48, 65), the catalytic activity of P450 2E1 in the brain is 0.001 nmol/min/mg microsomal protein (M. Ingelman-Sundberg, personal communication). There is also a discrepancy between the fold induction of catalytic activity (2.4-fold) and the increase in the signal on Western blots (manyfold) on ethanol treatment of rats. The authors' explanation (personal communication) is that perhaps there is a significant amount of catalytically inactive 2E1 in the brains of ethanol-treated rats. An equally likely explanation for these data is that the antibody is recognizing another antigen in the brain. Whether this antigen is a form of P450 similar to 2E1 or some unrelated protein remains to be determined but under the circumstances it cannot be concluded that P450 2E1 was induced in the brain by chronic ethanol treatment.

Acute effects of ethanol on brain P450 have also been observed (44). A single dose of ethanol (1 ml/kg intraperitoneally) increased the P450 content of the brain three- to fivefold and resulted in detectable levels of six distinct forms of P450 in the brain. These were all microsomal forms and were identified by Western blots and N-terminal sequencing of the individual proteins. In contrast to the chronic ethanol induction study mentioned above, P450 2E1 was not a major form of the enzyme throughout the brain. It was induced in the olfactory lobes and could easily be detected in 20 pmol of P450 prepared from the olfactory lobes but was a minor component in total brain P450. The forms identified by N-terminal sequencing were 2C11, 4A3, 4A8, 2E1, and 2D1/5. In the liver P450 2E1 is induced by ethanol but this is a posttranscriptional effect. The mechanism by which ethanol induces several forms of brain P450 is under investigation.

There is also evidence for the hormonal regulation of brain P450. During pregnancy and lactation and on long-term treatment of rats with dihydrotestosterone there is a large induction (up to 20-fold) in the content of the olfactory lobes and hypothalamic–preoptic area (66). There is no evidence for the induction of P450 1A, 2B, 2C, 2E1, or 3β-diol hydroxylase under these conditions but P450 4A and 2D are detectable on Western blots and by N-terminal sequencing (M. Warner, unpublished observations). Most of this hormonally induced P450 remains to be identified.

Human Brain P450

Ravindranath et al. (65) measured several P450 catalytic activities in six individual human brains. With the exception of aniline hydroxylase, these activities were all as high or higher than that in the human liver or kidneys and higher than in rat liver microsomes. These workers have purified four forms of P450, labeled A, B, C, and D, from a human brain (66). They were not typical P450s because they all had molecular masses around 60 kDa. The corresponding human liver enzymes have molecular masses between 50 and 57 kDa. The catalytic activities and immunological reactivities of these forms are shown in Table IV. The P450 content of the human brain was 100 pmol/mg microsomal protein, which is 33% of that measured in the human liver (67). The idea that the human brain is as efficient as the rat liver in the metabolism of xenobiotics and could have 33% of the P450 content of the liver is contrary to all that has been previously known about the

TABLE IV   Immunological Reactivities and Catalytic Activities
of Human Brain P450[a]

Enzyme	Catalytic activity[a]				
	Microsomes	Form A	Form B	Form C	Form D
Aminopyrine N-demethylase	40	13	45		80
Morphine N-demethylase	30	12			
7-Ethoxycoumarin O-deethylase	4			50	
Aniline hydroxylase	1.0	27			
Nifedipine hydroxylase			46		

[a] Catalytic activities are expressed as nanomoles per minute per milligram of microsomal protein in the
microsomes and as turnover numbers (moles per mole of P450 per minute) with the purified P450s.
[b] Immunological reactivity: form A, 2B; form B, 3A; form C, 1A; form D, 2B.

metabolism and pharmacokinetics of xenobiotics and these studies need to
be confirmed.

The mRNAs for two P450s have been detected in human brains. These
are P450 2E1 (69) and P450 2D6 (70). Although catalytic activities corresponding to these enzymes have not been found in the human brain, their potential
presence in the brain is of toxicological significance. P450 2E1 is of interest
because it catalyzes the formation of the neurotoxin acetaldehyde from ethanol and because of the increase in active oxygen, lipid peroxidation, and
destruction of membranes that have been shown to occur with this isozyme
of P450 (63).

P450 2D6 is involved in the metabolism of many CNS-active drugs including neuroleptics and narcotic analgesics. Its presence in the brain has implications for the pharmacology and toxicology of these drugs. In the Caucasian
population 10% of the population have a genetic defect in the 2D6 gene and
a reduced metabolism of many pharmaceutical agents (72). P450 2D6 also
catalyzes the hydroxylation and presumably the inactivation of 1-methyl-
4-phenyl-1,2,3,6-tetrahydropyridine (MPTP) and tetrahydroisoquinoline (73,
74), two agents that can cause Parkinson's disease. If Parkinson's disease
is caused by some endogenous or environmentally acquired neurotoxin, then
P450 could protect against these toxins by metabolically inactivating them.

Epidemiologists have searched for a correlation between poor metabolizers
of debrisoquine and the incidence of Parkinson's disease. Evidence for overrepresentation of slow metabolizers in Parkinson's patients was found in one
(75) study, but not in another (76). These epidemiological studies do not
invalidate the hypothesis that P450 could be involved in the protection of
the brain against neurotoxins. Perhaps the relationship between Parkinson's
disease and P450 will be clearer when the forms of P450 in the human brain

have been characterized and their role in metabolism of neurotoxins evaluated.

General Conclusions

Many questions still remain about the quantity of the various known forms of P450 in the brain, particularly with respect to their role in CNS pharmacology and toxicology. Those studies that show high levels of P450 in the brain need to be confirmed because they challenge the concept that the brain is not an organ of quantitative importance in drug metabolism in the body. In addition to this, new and interesting functions of P450 in brain physiology are emerging as the evidence accumulates for a role of brain P450 in regulating GABA receptor function and in the synthesis and degradation of intracellular messenger molecules in the form of arachidonic acid metabolites.

Acknowledgments

This work was supported by grants from the Work Health and Environment Fund and from the Swedish Medical Research Council (Nos. 13X-06807 and B92-03P-08561-04A).

References

1. Naftolin, F., Ryan, K. J., Reddy, V. V., Flores, F., Petro, Z., Kuhn, M., White, R. J., Takaoka, Y., and Wolin, L. (1975). *Recent Progr. Horm. Res.* **31,** 295–319.
2. Cohn, J. A., Alvares, A. P., and Kappas, A. (1977). *J. Exp. Med.* **145,** 1607–1611.
3. Marietta, M. P., Vessell, E. S., Hartman, R. D., Weisz, J., and Dvorchik, B. H. (1979). *J. Pharmacol. Exp. Ther.* **208,** 271–279.
4. Guengerich, F. P., and Mason, P. S. (1979). *Mol. Pharmacol.* **15,** 154–164.
5. Shiverick, K. T., and Notelovitz, M. (1983). *Biochem. Pharmacol.* **32,** 101–106.
6. Holtzman, D., and Desautel, M. (1980). *J. Neurochem.* **34,** 1535–1537.
7. Nabeshima, T., Fontenot, J., and Ho, I. K. (1981). *Biochem. Pharmacol.* **30,** 1142–1144.
8. Qato, M. K., and Maines, M. D. (1985). *Biochem. Biophys. Res. Commun.* **128,** 18–24.
9. Walther, B., Gheresi-Egea, J. M., Minn, A., and Siest, G. (1986). *Brain Res.* **375,** 338–344.
10. Ravindranath, V., Anandatheerhavarada, H. K., and Shankar, S. K. (1990). *Biochem. Pharmacol.* **39,** 1013–1018.

11. Anandatheerhavarada, H. K., and Ravindranath, V. (1991). *Neurosci. Lett.* **125,** 238–240.

12. Nebert, D. W., Nelson, D. R., Coon, M. J., Estabrook, R. W., Feyereisen, R., Fujii-Kuriyama, Y., Gonzales, F. J., Guengerich, F. P., Gunsalus, I. C., Johnson, E. F., Loper, J. C., Sato, R., Waterman, M. R., and Waxman, D. J. (1991). *DNA Cell Biol.* **10,** 1–14.

13. Omura, T., and Sato, R. (1964). *J. Biol. Chem.* **239,** 2370–2378.

14. Matsubara, T., Koike, M., Touchi, A., Tochino, Y., and Sugeno, K. (1976). *Anal. Biochem.* **76,** 596–603.

15. Johannessen, K. A. M., and DePierre, J. (1978). *Anal. Biochem.* **86,** 725–732.

16. Walther, B., Ghersi-Egea, J.-F., Jayyosi, Z., Minn, A., and Siest, G. (1987). *Neurosci. Lett.* **76,** 58–62.

17. Näslund, B., Glauman, H., Warner, M., Gustafsson, J.-Å., and Hansson, T. (1987). *Mol. Pharmacol.* **33,** 31–37.

18. Warner, M., Köhler, C., Hansson, T., and Gustafsson, J.-Å. (1988). *J. Neurochem.* **50,** 1057–1065.

19. Anandatheerhavarada, H. K., Shankar, S. K., and Ravindranath, V. (1990). *Brain Res.* **536,** 339–343.

20. Mensil, M., Testa, B., and Jenner, P. (1988). *Xenobiotica* **18,** 1097–1106.

21. Das, M., Seth, P. K., Dixit, R., and Mukhtar, H. (1982). *Arch. Biochem. Biophys.* **217,** 205–215.

22. Rouet, P., Alexandrov, K., Markovits, P., Frayssinet, C., and Dansette, P. M. (1981). *Carcinogenesis* **2,** 919–926.

23. Srivastava, S. P., and Seth, P. K. (1983). *Biochem. Pharmacol.* **32,** 3657–3660.

24. Anandatheerhavarada, H. K., Shankar, S. K., Bhamre, S., Boyd, M. R., Song, B.-J., and Ravindranath, V. (1993). *Brain Res.* **601,** 279–285.

25. Anandatheerhavarada, H. K., Williams, J. F., and Wecker, L. (1993). *J. Neurochem.* **60,** 1941–1944.

26. Kodaira, H., and Spector, S. (1988). *Proc. Natl. Acad. Sci. U.S.A.* **85,** 1267–1271.

27. Roselli, C. E., Ellinwood, E. E., and Resko, J. A. (1984). *Endocrinology (Baltimore)* **114,** 192–200.

28. Warner, M., Strömstedt, M., Möller, L., and Gustafsson, J.-Å. (1989). *Endocrinology (Baltimore)* **124,** 2699–2706.

29. Akwa, Y., Morfin, R. F., Robel, P., and Baulieu, E.-E. (1992). *Biochem. J.* **288,** 959–964.

30. Fonne-Pfister, R., Bargetzl, M. J., and Meyer, U. (1987). *Biochem. Biophys. Res. Commun.* **148,** 1144–1150.

31. Brown, C. G., White, N., and Jefcoate, S. L. (1985). *J. Endocrinol.* **107,** 191–196.

32. Jayyosi, Z. A., Cooper, K. O. A., and Thomas, P. E. (1992). *Arch. Biochem. Biophys.* **298,** 265–270.

33. Corpechot, C., Robel, P., Axelson, M., Sjovall, J., and Baulieu, E.-E. (1981). *Proc. Natl. Acad. Sci. U.S.A.* **78,** 4704–4707.

34. Le Gascogne, C., Robel, P., Gouezow, M., Snanes, N., Baulieu, E.-E., and Waterman, M. (1987). *Science* **237,** 1212–1214.

35. Ahlgren, R., Warner, M., and Gustafsson, J.-Å. (1990). *In* "Drug Metabolizing Enzymes: Genetics, Regulation and Toxicology" (M. Ingelman-Sundberg, J.-Å. Gustafsson, and S. Orrenius, eds.), p. 161. Stockholm.

36. Majewska, M. D., Harrison, N. L., Schwartz, R. D., Barker, J. L., and Paul, S. M. (1986). *Science* **232**, 1004–1007.
37. Strömstedt, M., Warner, M., Banner, C., Macdonald, P. C., and Gustafsson, J.-Å. (1993). *Mol. Pharmacol.* **44**, 1077–1083.
38. Sundin, M., Warner, M., Haaparanta, T., and Gustafsson, J.-Å. (1987). *J. Biol. Chem.* **262**, 12293–12297.
39. Snyder, G. D., Capdevila, J., Chacos, N., Manna, S., and Flack, J. R. (1983). *Proc. Natl. Acad. Sci. U.S.A.* **80**, 3504–3507.
40. Junier, M.-P., Dray, F., Blair, I., Capdevilla, J., Dishman, E., Flack, J. R., and Ojeda, S. R. (1990). *Endocrinology (Baltimore)* **126**, 1534–1540.
41. Murphy, R. C., Falck, J. R., Lumni, S., Yadagari, P., Zirrolli, J. A., Balazy, M., Masferrer, J. L., Abraham, N. G., and Schwartzman, M. L. (1988). *J. Biol. Chem.* **263**, 17197–17202.
42. Capdevilla, J., Chacos, N., Flack, J. R., Manna, S., Negro-Vilar, A., and Ojeda, S. R. (1983). *Endocrinology (Baltimore)* **126**, 1534–1540.
43. Laethem, R. M., and Koop, D. R. (1992). *Mol. Pharmacol.* **42**, 958–963.
44. Warner, M., and Gustafsson, J.-Å. (1994). *Proc. Natl. Acad. Sci. U.S.A.* **94**, 1019–1023.
45. Capdevila, J. H., Falck, J. R., and Estabrook, R. W. (1992). *FASEB J.* **6**, 731–736.
46. Ding, X., Koop, D. R., Crump, B. L., and Coon, M. J. (1986). *Mol. Pharmacol.* **30**, 370–378.
47. Sugita, O., Miyairi, S., Sassa, S., and Kappas, A. (1987). *Biochem. Biophys. Res. Commun.* **147**, 1245–1250.
48. Hansson, T., Tindberg, N., Ingelman-Sundberg, M., and Köhler, C. (1990). *Neuroscience* **34**, 451–463.
49. Cammer, W., Downing, M., Clarke, W., and Schenkman, J. B. (1991). *J. Histochem. Cytochem.* **39**, 1089–1094.
50. Kapitulnik, J., Gelboin, H. V., Guengerich, F. P., and Jacobowitz, D. M. (1987). *Neuroscience* **20**, 829–833.
51. Köhler, C., Eriksson, L. G., Hansson, T., Warner, M., and Gustafsson, J.-Å. (1988). *Neurosci. Lett.* **20**, 829–832.
52. Hagihara, K., Shiosaka, S., Lee, Y., Kato, J., Hatano, O., Takakusu, A., Emi, Y., Omura, T., and Toyama, M. (1990). *Brain Res.* **515**, 69–78.
53. Lin, L.-P., Lee, Y., Tohyama, M., and Shiosaka, S. (1991). *Neuroendocrinology* **54**, 127–135.
54. Gustafsson, J.-Å., Mode, A., Norstedt, G., and Skett, P. (1983). *Annu. Rev. Physiol.* **45**, 51–60.
55. Bergh, A. F., and Strobel, H. W. (1992). *J. Neurochem.* **59**, 575–581.
56. Anandatheerhavarada, H. K., Boyd, M. R., and Ravindranath, V. (1992). *Biochem. J.* **288**, 483–488.
57. Strömstedt, M., Zaphiropoulos, P. G., and Gustafsson, J.-Å., in "Methods in Enzymology, Vol. 206" (M. R. Waterman and E. F. Johnson, eds.), pp. 640–648. Academic Press, San Diego, 1991.
58. Kimura, H., Sogawa, K., Saki, Y., and Fujii-Kuriyama, Y. (1989). *J. Biol. Chem.* **264**, 2338–2342.
59. Zaphiropoulos, P., and Wood, T. (1993). *Biochem. Biophys. Res. Commun.* **193**, 1006–1013.

60. Strömstedt, M., Warner, M., and Gustafsson, J.-Å. (1993). In press.
61. Lephart, E. D., Simpson, E. R., McPhaul, M. J., Kilgore, M. W., Wilson, J. D., and Ojeda, S. R. (1992). *Mol. Brain Res.* **16**, 187–192.
62. Harada, N., and Yamada, K. (1992). *Endocrinology (Baltimore)* **131**, 2306–2312.
63. Volk, B., Amelizad, Z., Anagnostopoulos, J., Knoth, R., and Oesch, F. (1988). *Neurosci. Lett.* **84**, 219–224.
64. Faunae, Y., and Imoaka, S. (1993). *Handb. Exp. Pharmacol.* **105**, 221–238.
65. Tindberg, N., and Ingelman-Sundberg, M. (1989). *Biochemistry* **28**, 4499–4504.
66. Warner, M., Tollet, P., Strömstedt, M., Carlström, K., and Gustafsson, J.-Å. (1989). *J. Endocrinol.* **122**, 341–349.
67. Ravindranath, V., Anandatheerthavarada, H. K., and Shankar, S. K. (1989). *Brain Res.* **496**, 331–335.
68. Bhamre, S., Anandatheerthavarada, H. K., Shankar, S. K., Boyd, M. R., and Ravindranath, V. (1993). *Arch. Biochem. Biophys.* **301**, 251–255.
69. Gut, J., Catin, T., Dauer, P., Kronbach, T., Zanger, U., and Meyer, U. A. (1986). *J. Biol. Chem.* **261**, 11734–11743.
70. Adams, M. D., Dubnick, M., Kerlavage, A. R., Moreno, R., Kelley, J. M., Utterback, T. R., Nagle, J. W., Fields, C., and Venter, J. C. (1992). *Nature (London)* **355**, 632–634.
71. Tyndale, R. F., Sunahara, R., Inaba, T., Kalow, W., Gonzalez, F. J., and Niznik, H. B. (1991). *Mol. Pharmacol.* **40**, 63–68.
72. Gonzales, F. J., and Meyer, U. A. (1991). *Clin. Pharmacol. Ther.* **50**, 233–238.
73. Jimenez-Jimenez, F. J., Tabernero, C., Mena, M. A., Garcia de Yebenes, J., Garcia de Yebenes, M. J., Casarejos, M. J., Pardo, B., Garcia-Agundez, J. A., Benitez, J., and Martinez, A. (1991). *J. Neurochem.* **57**, 81–87.
74. Ohla, S., Tachikawa, O., Makino, Y., Tasaki, Y., and Hirobe, M. (1990). *Life Sci.* **46**, 599–605.
75. Armstrong, M., Daly, A. K., Cholerton, S., Bateman, D. N., and Idle, J. R. (1992). *Lancet* **339**, 1017–1018.
76. Kallio, J., Marttila, R. J., Rinne, U. K., Sonninen, V., and Syvalahti, E. (1991). *Acta Neurol. Scand.* **83**, 194–197.

[5] Methods for Estimating 11β-Hydroxysteroid Dehydrogenase Activity

Zygmunt Krozowski

Introduction

The enzyme 11β-hydroxysteroid dehydrogenase (11β-HSD) has risen to prominence in the field of steroid endocrinology, where it has been postulated to play a critical role in reducing the level of active glucocorticoid in mineralocorticoid target cells (Fig. 1). Given the equivalent affinity of glucocorticoids and mineralocorticoids for the mineralocorticoid receptor (1), and the much higher circulating levels of glucocorticoids, it is thought that this enzymatic mechanism keeps glucocorticoids out of mineralocorticoid receptors (2, 3). Furthermore, there is also evidence that 11β-HSD may also modulate glucocorticoid activity by degrading glucocorticoids in glucocorticoid target tissues.

Genetic defects involving 11β-HSD give rise to the syndrome of apparent mineralocorticoid excess, a condition characterized by increased sodium retention, hypokalemia, and severe hypertension (4, 5). The 11β-HSD enzyme activity can also be compromised by the ingestion of licorice, the active constituent of which is glycyrrhetinic acid. This compound has been shown to inhibit enzyme activity directly by competing for the substrate-binding site. Inhibition of 11β-HSD in the kidney is thought to produce overstimulation of the mineralocorticoid receptor by the elevated intracellular levels of glucocorticoids, leading to sodium retention and potassium excretion. In addition, the enzyme has also been identified in the brain, where the administration of inhibitors has also been shown to lead to increases in blood pressure (6).

The original purification and cloning studies of Monder and White (6a, 6b) isolated an NADP-dependent enzyme from rat liver (EC 1.1.1.146) with micromolar affinity for cortisol (F) and a slightly higher affinity for corticosterone (B). However, immunohistochemical studies showed that the enzyme does not colocalize with the mineralocorticoid receptor in the kidney (7, 8). Furthermore, the low affinity of this enzyme for glucocorticoids suggests that this may not be the species responsible for endowing mineralocorticoid specificity on mineralocorticoid target tissues. The appropriate enzyme should colocalize with the mineralocorticoid receptor in the distal convoluted tubules and would need to have nanomolar affinity for glucocorticoids if it is to reduce the concentration of these steroids to the levels needed to keep them out of mineralocorticoid receptors. Although a low-affinity NAD-

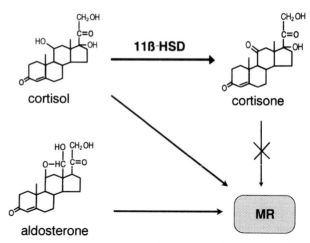

FIG. 1   Protection of the mineralocorticoid receptor from occupation by cortisol. The action of 11β-HSD converts cortisol to cortisone, a steroid with 1/300 the affinity of the parent compound for the mineralocorticoid receptor. Aldosterone is not metabolized because the C-11 oxygen atom forms a hemiacetal group with C-19.

dependent enzyme was originally identified in the distal nephron (9), it is now clear that a high-affinity, NAD-dependent dehydrogenase exists in distal tubular cells, and it is probable that it is this enzyme that endows mineralocorticoid specificity (10).

There is now considerable evidence in the literature of different isoforms of 11β-HSD. Because of this plurality we have suggested that the NADP-dependent form be designated 11β-HSD1, and the low-affinity NAD-dependent form as 11β-HSD2 (9). Furthermore, 11β-HSD1 can be further divided into A and B congeners, the latter form being an N-terminal truncated variant of the A form (11). In addition 11β-hydroxysteroid dehydrogenase from collecting duct (11-OHSD/CD), the high-affinity NAD-dependent 11β-HSD, has also been identified in immunodissected renal distal tubule cells (10), but this enzyme appears to be different from 11β-HSD2. Further 11β-HSD isoforms may also be present, because in some tissues NADP-dependent dehydrogenase activity can be demonstrated where no antigen is detected, using antibodies directed against the 11β-HSD1 protein (12).

Methods of Detecting 11β-Hydroxysteroid Dehydrogenase Activity

Measurement of 11-Keto Steroid Product

11β-Hydroxysteroid dehydrogenase activity is best detected by estimating the production of the oxidized 11β-steroid. The reaction can be performed with whole cells in culture, whole tissue homogenates, or microsomes. When using whole homogenate or microsomes it is possible to add exogenous cofactor to drive the reaction. It is important to remember that the absence of detectable dehydrogenase activity does not mean a complete absence of the protein. The enzyme may exist in a latent form and may be activated under various conditions such as the addition of detergents (13). Furthermore, it is not always clear which direction the enzyme is acting in *in vivo*. Thus, whereas 11β-HSD1 was purified from the liver by assaying dehydrogenase activity, it is well known that the major reaction in hepatic cells is the conversion of cortisone to F. Thus, *in vitro,* it is sometimes useful to assess the enzymatic reaction in both directions. The advantage of using microsomes is that most soluble endogenous cofactor has been removed during the preparation of the pellet and it is usually possible to drive the reaction in the reverse direction by the addition of the appropriate cofactor and steroid.

Method

Either fresh tissue or samples stored frozen at $-70°C$ can be used. Dehydrogenase activity can also be assayed in whole cells in culture.

1. To assay 11β-HSD activity in tissues use a total tissue homogenate. Homogenize tissues at 4°C in 4 vol of buffer A. Buffer A consists of 0.25 M sucrose, 1 mM phenylmethylsulfonyl fluoride (PMSF) in 10 mM phosphate buffer, pH 7.0.

To quantitate accurately the enzyme activity of any preparation it is important to use sufficient sample to obtain a direct linear correlation between enzyme activity and protein levels. It is thus necessary to construct a plot of the amount of sample versus enzyme activity and determine the linear portion of the curve.

2. Add 50 μl of the predetermined amount of protein to a 1-ml solution containing 10^{-8} M unlabeled corticosterone, 10^5 counts per minute (cpm) [^{3}H]corticosterone, and 1 mM NADP or NAD.

3. Incubate for 2 hr at 37°C.

4. Stop the reaction by addition of 3 vol of ethyl acetate. Vortex vigorously.

5. Centrifuge at room temperature for 5 min at 2000 g to sediment fat particles onto the aqueous phase.

6. Transfer the ethyl acetate layer to a fresh tube and dry under a stream of air.

7. Reconstitute the dried steroid in 100 μl of ethanol. Add 5 μl of 4 $\times$ 10^{-3} M 11-dehydrocorticosterone to aid visualization of the product under ultraviolet (UV) light.

8. Spot 10 μl on thin-layer chromatography (TLC) plates. It is convenient to use DC-Plastikfolien kieselgel 60 F254 plates (Merck, Darmstadt, Germany) of 0.2-mm thickness. Radioactive spots can readily be cut out with the aid of a pair of scissors.

9. Equilibrate the TLC tank with 92% (v/v) chloroform : 8% (v/v) ethanol.
10. Run TLC plates in the above-described organic solvent mixture.
11. Dry plates and mark spots after visualizing bands under UV light.

Tritiated steroids can be quantitated either by liquid scintillation counting or by a bioimaging instrument, using detection plates that have not been coated with Teflon. Enzyme activity is normally expressed as nanomoles of steroid oxidized per minute per milligram of protein, although percent conversion of substrate is also commonly used.

When using whole cells in culture simply add 10^{-8} M unlabeled corticosterone and 10^5 cpm of [^{3}H]corticosterone per milliliter of culture medium. Incubate for at least 1 hr at 37°C. 11-Dehydro[^{3}H]corticosterone is secreted into the medium and can be extracted and measured using the above protocol starting at step 4.

The activity of 11β-HSD appears to be somewhat dependent on the immediate environment surrounding the enzyme. Purification of 11β-HSD has shown that the reductase activity can be separated from the oxidase function of the enzyme and the judicious use of detergents has sometimes exposed latent activity (14). The cloning and expression of 11β-HSD1 have shown that despite the absence of reductive activity in the purified liver enzyme, the corresponding cDNA does produce a protein possessing reductase activity when expressed in Chinese hamster ovary (CHO) cells. In contrast, when the same cDNA is transfected into a toad bladder mucosal cell line only reductase activity is evident (15). Similar apparently unidirectional 11β-HSD enzymes have been described in heart and vascular tissue (16), and in MDCK (Maden–Darby canine kidney) and LLC-PK1 cells (a pig kidney epithelial cell line) (17). In addition, lung epithelia and fibroblast cells perform oxidation and reduction of glucocorticoids, respectively (18). Because all enzyme reactions are theoretically reversible, these observations suggest that accessory factors are required for net reductase activity or that there is perhaps a posttranslational modification of the enzyme that generates two congeners, one possessing predominantly dehydrogenase and the other mainly oxidoreductase activity.

When assaying specifically for either activity it may be desirable to choose conditions that optimize each reaction in cell-free preparations. The oxidase reaction should then be performed in the presence of NAD or NADP and the oxidoreductase reaction in the presence of NADH or NADPH. Maximal activities should be obtained near the pH optima for each reaction; the pH optimum of the reductase is pH 5.5–6.5 whereas for the dehydrogenase the pH optimum of the reaction is pH 8.5–9.5. Occasionally exogenous dehydrogenases can be added to generate the required cofactor. Addition of glucose-6-phosphate dehydrogenase and NADP can be used to generate NADPH where it is suspected that the availability of cofactor is limiting because of utilization by other reactions (19).

Not all 11β-hydroxysteroids will act as substrates for all isoforms of 11β-HSD. It has been shown that whereas an 11β-HSD converts both F and B to their respective 11-keto metabolites in the kidney, the 11β-HSD from the brain and heart appears to be incapable of utilizing F, despite the detection of 11β-HSD1 mRNA in both these tissues (20). Whereas B is always preferred over F as the substrate, the degree to which this occurs also appears to be tissue specific. This phenomenon may reflect the existence of different isoforms of the enzyme (21).

The isolation of various isoforms of 11β-HSD is currently being pursued in a number of laboratories. In order to measure 11β-hydroxysteroid dehydrogenase activity accurately it is important to consider which isoform of the enzyme is being assayed. The rat hepatic NADP-dependent 11β-HSD1 dehydrogenase enzyme has a micromolar affinity for B, catalyzes both oxidation and reduction, and does not display end-product inhibition, whereas the NAD-dependent 11-OHSD/CD isoform identified in immunopurified renal collecting duct cells has been shown to have an affinity of about 25 nM, to perform the reductase reaction poorly, and to be strongly inhibited by 11-dehydrocorticosterone. In addition, there is also evidence for a low-affinity renal NAD-dependent enzyme, 11β-HSD2. Thus, in order to distinguish between the 11-OHSD/CD and 11β-HSD2 enzymes some type of kinetic analysis is required.

An accepted procedure for determining the affinity of an enzyme is to construct double-reciprocal plots, where the reciprocal of the initial rate of formation of the product is plotted against the reciprocal of the concentration of the substrate. This approach is best used when there is a single isoform of the enzyme present. For example, in the kidney, where there appear to be several isoforms, the unequivocal demonstration of the high-affinity enzyme was made possible only by the purification of the appropriate cell types before analysis of enzyme activity (10).

Tetrazolium–Formazan System

Tetrazolium salts have long been used in the histochemical detection of dehydrogenase activity. In addition to 11β-HSD, other oxidoreductases such as pyruvate oxidase, isocitrate dehydrogenase, α-ketoglutarate oxidase, succinate dehydrogenase, malate dehydrogenase, and furfuryl alcohol dehydrogenase have also been detected on tissue sections (22). The salivary gland, kidney, and adrenal cortex have been shown to possess high levels of 11β-HSD activity histochemically, although the results obtained are clearly spe-

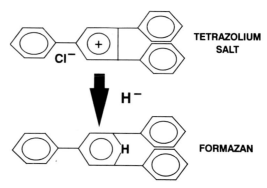

FIG. 2 Schematic representation of the chemical events occurring during formation of the insoluble formazan compound.

cies specific (23). In the salivary gland 11β-HSD displays marked substrate specificity utilizing 11β-hydroxyandrostenedione and cortisol, but not 11β-hydroxyestrone or 11β-hydroxyprogesterone (22). In the rat kidney, in response to the oxidation of 11β-hydroxyandrostenedione by 11β-HSD2, formazan deposition can be selectively inhibited utilizing carbenoxolone sodium, a derivative of glycyrrhetinic acid (9).

The chemical basis for the histochemical detection is the production of an insoluble blue formazan compound from the soluble, colorless tetrazolium salt (Fig. 2). The tetrazolium participates in a redox reaction acting as a hydride ion acceptor or as an oxidant of the reduced cofactor. There is a requirement for a flavoprotein to act as an intermediary in this reaction, but exogenously added phenazine methosulfate can also be substituted (24).

Method

1. Freeze fresh tissue in OCT (Miles Inc., Elkhart, IN) mounting fluid. Section at 6 μm on a cryostat maintained at $-20°C$.

2. Thaw the sections onto glass slides coated with 1% (w/v) gelatin and allow to dry for 10 min at room temperature.

3. Perform the histochemical reaction by incubating tissue sections at 37°C for 20 min under a large drop of phosphate-buffered saline (PBS) containing 1% (w/v) magnesium chloride, NAD or NADP (1 mM), nitro blue tetrazolium (2,2'-dinitrophenyl-5,5'-diphenyl-3,3'-(3,3'-dimethoxy-4,4'-diphenylene) tetrazolium chloride; 0.5 mg/ml), and 11β-hydroxysteroid (1 mM).

4. Wash sections briefly in PBS.

5. Fix tissue section with phosphate-buffered paraformaldehyde and counterstain with hematoxylin.

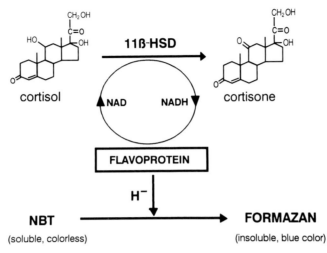

FIG. 3   Schematic representation of the events thought to occur during oxidation of cortisol and the production of formazan. NBT, Nitro blue tetrazolium.

In whole cells tetrazolium reduction is very much dependent on the level of flavin proteins, including nonheme iron-binding proteins and ubiquinone (Fig. 3). Furthermore, the reducibility of tetrazolium is not constant and can be affected by experimental procedure and the existence of other factors competing for the reduced cofactor. This may be the reason why formazan deposition does not parallel the immunostaining pattern obtained using antibody directed against 11β-HSD1 in the rat kidney (9). Other factors can also influence the results obtained. There is an increase in tetrazolium reducibility in the presence of detergents, while some salts are sensitive and others are insensitive to oxygen. Control experiments should be performed to monitor the effects of these variables in any particular system (25).

Western Blot Analysis

The 11β-HSD1 protein can be detected on Western blots with antibodies raised against the purified rat hepatic antigen. Monder and Lakshmi (26) have generated a number of antibodies of different specificities. The 56-125 and 56-126 antibodies produce different staining patterns for kidney, liver, heart, brain, and stomach. In the brain there is consistent detection of a 26-kDa 11β-HSD epitope that does not display dehydrogenase activity (27), reminiscent of the inactive 26-kDa 11β-HSD1B protein cloned from the kidney (11, 28). Protease digestion experiments carried out with the kidney enzyme suggest that the two bands detected in this tissue may be structurally or conformationally different.

Western blots have also been used to analyze the 11β-HSD enzyme when expressed in transfected cells. Although Cos-7 cells produce significant amounts of endogenous 11β-hydroxysteroid dehydrogenase activity, it is possible to estimate the much larger amounts of the transfected enzyme when using low numbers of cells. The transfection of Cos-7 cells with plasmids harboring 11β-HSD1A and 11β-HSD1B cDNAs has been used to study the enzymatic activity of the truncated form of the enzyme (28). T-antigen expression in these cells allows the replication of constructs derived from the pCDNA1 vector (Invitrogen, San Diego, CA) and the subsequent expression of large amounts of enzyme protein. Using this type of approach it has been possible to demonstrate that whereas the 11β-HSD1B protein is highly expressed in Cos-7 cells transfected with the pHSD1B construct the truncated protein does not appear to be endowed with either the 11β-hydroxysteroid dehydrogenase or reductase activities (28).

Method

1. Cos-7 cells are cultured in RPMI-1640 medium containing 10% (v/v) fetal calf serum and plated at a density of 1×10^5 cells/60-mm well.

2. Transfection is performed 3 days after plating by applying 4 μg of plasmid construct per plate in phosphate-buffered saline, containing DEAE-dextran (500 μg/ml), for 30 min at 37°C, followed by addition of 4 ml of RPMI medium containing 80 μM chloroquine for 2.5 hr at 37°C.

3. After a 2.5-min treatment with dimethyl sulfoxide (DMSO) the cells are returned to RPMI–10% fetal calf serum for 1 hr at 37°C and the medium changed to OPTI-MEM I reduced serum medium (1×) (Life Technologies, Inc., Grand Island, NY).

4. After 48 hr the cells are harvested by trypsinization, homogenized using a Dounce (Wheaton, Millville, NJ) homogenizer, and frozen in aliquots at −70°C at a protein concentration of 3 mg/ml.

5. Load 20 μg of transfected Cos-7 cell homogenate onto a 12% (w/v) sodium dodecyl sulfate (SDS)-polyacrylamide gel, perform electrophoresis, and transfer to nitrocellulose by standard procedures (29).

6. Block the nitrocellulose filters with 1% (w/v) bovine serum albumin–0.05% (v/v) Tween 20 in PBS for 1 hr at room temperature, and incubate at 4°C overnight with antibody 56-125 diluted 1/100 with blocking solution.

7. Wash three times with PBS–0.05% Tween 20 and incubate with iodinated protein A (0.5 μCi/ml) in blocking buffer for 1 hr at room temperature.

8. Wash the filter twice with PBS–0.05% Tween and once with PBS–0.01% SDS, dry, and expose to Kodak (Rochester, NY) X-Omat AR film with an intensifying screen.

Under these conditions we have observed 5% conversion of [³H]B per hour per 100 μg of homogenate protein, whereas the same cells transfected with the 11β-HSD1A construct display about 40% conversion (28). Both the exogenous 11β-HSD1A and 11β-HSD1B proteins are readily detected on Western blots. The endogenous simian 11β-HSD protein is either below the limits of detection or is not recognized by the 56-125 antibody.

Concluding Remarks

Not all possible means of detecting 11β-hydroxysteroid dehydrogenase activity have been described in this chapter; only those approaches that have been used in the author's laboratory have been presented. Immunohistochemistry and *in situ* hybridization are two other techniques that have also been successfully employed (16, 27). However, the ultimate criterion, that of an enzymatic activity producing an 11-keto compound from an 11β-hydroxysteroid substrate, must always be true. In this sense all other techniques provide only indirect evidence of such an enzyme activity and should be confirmed by measurement of the metabolized product.

References

1. Z. S. Krozowski and J. W. Funder, *Proc. Natl. Acad. Sci. U.S.A.* **80,** 6056 (1983).
2. J. W. Funder, P. T. Pearce, R. Smith, and A. I. Smith, *Science* **242,** 583 (1988).
3. C. R. Edwards, P. M. Stewart, D. Burt, L. Brett, M. A. McIntyre, W. S. Sutanto, E. R. deKloet, and C. Monder, *Lancet* **ii,** 986 (1988).
4. R. Tedde, A. Pala, A. Melis, and S. Ulick, *J. Endocrinol. Invest.* **15,** 471 (1992).
5. S. Ulick, S. L. Levine, P. Gunczler, G. Zanconato, L. C. Ramirez, W. Rauh, A. Rosler, L. H. Bradlow, and M. I. New, *Clin. Endocrinol. Metab.* **49,** 757 (1979).
6. E. P. Gomezsanchez and C. E. Gomezsanchez, *Am. J. Physiol.* **263,** E1125 (1992).
6a. V. Lakshmi and C. Monder, *Endocrinology* **123,** 2390 (1988).
6b. A. K. Agarwal, C. Monder, B. Eckstein, and P. C. White, *J. Biol. Chem.* **264,** 18939 (1989).
7. S. E. Rundle, J. W. Funder, V. Lakshmi, and C. Monder, *Endocrinology (Baltimore)* **125,** 1700 (1989).
8. Z. S. Krozowski, S. E. Rundle, C. Wallace, M. J. Castell, J. H. Shen, J. Dowling, J. W. Funder, and A. I. Smith, *Endocrinology (Baltimore)* **125,** 192 (1989).
9. W. R. Mercer and Z. S. Krozowski, *Endocrinology (Baltimore)* **130,** 540 (1992).
10. E. Rusvai and A. Narayfejestoth, *J. Biol. Chem.* **268,** 10717 (1993).
11. Z. Krozowski, V. Obeyesekere, R. Smith, and W. Mercer, *J. Biol. Chem.* **267,** 2569 (1992).
12. C. Monder, *J. Steroid Biochem. Mol. Biol.* **40,** 533 (1991).
13. C. Monder and V. Lakshmi, *J. Steroid Biochem.* **32,** 77 (1989).
14. V. Lakshmi and C. Monder, *Endocrinology (Baltimore)* **116,** 552 (1985).

15. H. Duperrex, S. Kenouch, H. P. Gaeggeler, J. R. Seckl, C. Edwards, N. Farman, and B. C. Rossier, *Endocrinology (Baltimore)* **132,** 612 (1993).

16. B. R. Walker, J. L. Yau, L. P. Brett, J. R. Seckl, C. Monder, B. C. Williams, and C. R. Edwards, *Endocrinology (Baltimore)* **129,** 3305 (1991).

17. C. Korbmacher, W. Schulz, M. Konig, H. Siebe, I. Lichtenstein, and K. Hierholzer, *Biochim. Biophys. Acta* **1010,** 311 (1989).

18. M. Abramovitz, R. Carriero, and B. E. Murphy, *J. Steroid Biochem.* **21,** 677 (1984).

19. A. K. Agarwal, L. M. Tusie, C. Monder, and P. C. White, *Mol. Endocrinol.* **4,** 1827 (1990).

20. R. E. Smith, W. R. Mercer, P. H. Provencher, V. Obeyesekere, and Z. S. Krozowski, *Clin. Exp. Pharmacol. Physiol.* **19,** 365 (1992).

21. C. Monder and V. Lakshmi, *J. Steroid Biochem.* **32,** 77 (1989).

22. M. M. Ferguson, *Histochemie* **9,** 269 (1967).

23. M. M. Ferguson, J. B. Glen, and D. K. Mason, *J. Endocrinol.* **47,** 511 (1970).

24. H. J. Lippold, *Histochemistry* **76,** 381 (1982).

25. H. Anderson and P. E. Hoyer, *Histochemistry* **38,** 71 (1974).

26. C. Monder and V. Lakshmi, *Endocrinology (Baltimore)* **126,** 2435 (1990).

27. R. R. Sakai, V. Lakshmi, C. Monder, and B. S. Mcewen, *J. Neuroendocrinol.* **4,** 101 (1992).

28. W. Mercer, V. Obeyesekere, R. Smith, and Z. Krozowski, *Mol. Cell. Endocrinol.* **92,** 247 (1993).

29. H. Towbin, T. Staehlin, and J. Gordon, *Proc. Natl. Acad. Sci. U.S.A.* **79,** 5157 (1979).

Section II

Steroid Receptors

[6] Steroid Hormone Binding to Intracellular Receptors: *In Vitro* and *in Vivo* Studies

Aernout D. van Haarst and Thomas F. Szuran

Introduction

For over 25 years tritium-labeled steroids have been used in binding studies of steroid hormones in the brain. These studies have revealed important information on the localization and properties of intracellular receptors for these circulating hormones. Initial studies have exploited the fact that steroid hormones are lipophilic and readily enter the brain. Thus, tracer amounts (about 1 μg/100-g rat) of tritium-labeled steroids were infused in rats, which had their endocrine glands extirpated for depletion of endogenous hormone. The receptor sites were then available for the radiolabeled steroid, which is retained by the receptors and slowly accumulates in the cell nuclear compartment, as was shown by biochemical and autoradiographical techniques. Surprisingly, these studies revealed a highly discrete distribution of the steroid receptors in the nervous tissue. Thus, the mineralocorticoid receptors (MRs) and glucocorticoid receptors (GRs) were found in abundance in the limbic system and in areas with a regulatory function in the endocrine and autonomous components of the stress response system (see Ref. 1 for review). Receptors for the sex steroids prevailed in the hypothalamus.

In vitro studies allowed a careful and routine analysis of the binding properties of the receptors. The procedure for labeling receptor sites in tissue sections and subsequent autoradiography are described by MacLusky *et al.* ([8] in this volume). The present chapter focuses on the application of radiolabeled steroids in cell nuclear uptake studies *in vivo* and in tissue slices, and on the *in vitro* binding to soluble receptors in cytosol (the cytosol receptor assay). It appears that these binding studies provide information complementary to the more recently developed immunocytochemical and hybridization procedures. The latter procedures provide insight concerning the abundance of receptor transcripts and the cellular localization of immunoreactive receptor protein. The binding experiments allow assessment of pharmacological parameters such as binding capacity and affinity constants. In this chapter we illustrate the radioligand-binding studies, using corticosteroid receptors as examples.

Cell nuclear uptake studies were performed with [3H]corticosterone, which is the naturally occurring glucocorticoid of the rat.

Methods in Neurosciences, Volume 22

Corticosterone is less suitable for use in cytosol receptor-binding experiments, because it does not allow discrimination between binding to MRs and GRs. In addition, binding of corticosterone to transcortin (corticosteroid-binding globulin) may interfere with the estimation of the receptor parameters. Therefore, ^{3}H-labeled RU 28362 is used routinely in the cytosol binding assay of GRs. For MRs, on the other hand, we routinely use tritiated aldosterone in the presence of excess unlabeled RU 28362 to prevent aldosterone from binding to GRs.

Cell Nuclear Retention of Corticosteroids

Because the action of steroid hormones takes place via modulation of DNA transcription, it is of interest to determine retention of steroids by cell nuclei. Not only may such an approach reveal the target areas of a certain steroid, it will in addition give an indication of the efficacy of a steroid under varying circumstances. The efficacy depends on the binding kinetics of the steroid to the receptor, dissociation of heat-shock proteins, binding of the steroid–receptor complex to DNA, and the dissociation rates of both steroid–receptor complex and DNA binding. Naturally, the availability of both active steroid (or metabolites) and cytoplasmic receptors determines the magnitude of the signal as well.

The retention of steroids by the cell nucleus may be studied in intact animals, in order to reveal, for example, physiological variations in nuclear steroid levels. The steroids are extracted from isolated cell nuclei and determined by a radioimmunoassay procedure. Such a method is described by Yongue and Roy (2) for endogenous corticosteroids. In this chapter we describe an assay employing (tissues of) adrenalectomized subjects to study nuclear uptake and retention of radiolabeled corticosteroids (see also Ref. 3). Advantages of the latter approach are that the use of radioligands allows routine quantitation of steroid retention and that the experimental conditions can be more stringently controlled.

The procedures described in this chapter are focused on hippocampus and pituitary, but may be similarly applicable to other tissues and other steroids.

In Vivo Cell Nuclear Uptake and Retention

Animals

Rats are adrenalectomized, preferably in the morning, when corticosterone levels are low. Rats are maintained on standard laboratory chow, tap water, and a saline solution [0.9% (w/v) NaCl] *ad libitum*. To allow clearance of all endogenous corticosteroids from the cell nuclei and from the blood, the animals are not used for the nuclear uptake procedure until three days after adrenalectomy.

Administration of Radiolabeled Steroid

Rats are injected subcutaneously (sc) ($t = 0$ min) with either 0.5–0.75 mg of unlabeled steroid in ethanol–saline per 100 g of body weight for determination of nonspecific binding or with the ethanol–saline solution. If steroids are administered in order to study their competitive activity, lower doses of steroid are more appropriate. The steroids are dissolved in 25% (v/v) ethanol–saline. The injection volume is 0.2 ml. After 30 min ($t = 30$ min) 50 μCi of the tritiated steroid per 100 g of body weight is infused into the tail vein. The radiolabeled steroid solutions are prepared in 5% (v/v) ethanol–saline. The infusion volume is 0.2 ml.

At $t = 90$ min rats are decapitated and the tissues under investigation (e.g., hippocampus) are dissected immediately and frozen on dry ice. Trunk blood is collected for measurement of plasma radioligand levels.

Isolation and Extraction of Cell Nuclei

The procedure described here is adapted from McEwen and Zigmond (3).

Solutions

S1 is a solution in 1 mM phosphate buffer (pH 6.5), containing 0.32 M sucrose, 3 mM MgCl$_2$, and 0.25% (v/v) Triton X-100. Buffer S2 is identical to S1, except that S2 does not contain Triton X-100. S3 is also a solution in 1 mM phosphate buffer (pH 6.5), containing, however, 2.39 M sucrose and 3 mM MgCl$_2$. To make S3 it is necessary to heat the buffer while stirring.

From three rats we pool tissue (in our case hippocampus or pituitary) to be homogenized at 4°C in 2 ml of the Triton X-100-containing buffer S1. Tissue is homogenized using a Potter–Elvehjem tissue homogenizer (10 strokes, ~700 rpm) and an aliquot of 50 μl of the homogenate is taken for estimation of the protein concentration.

The homogenate is centrifuged for 10 min at 4°C at 600 g and the supernatant is removed carefully. The pellet is resuspended in 2 ml of buffer S2 and centrifuged again for 10 min at 4°C at 600 g. The pellet is now resuspended in 0.72 ml of S2 and subsequently 3.78 ml of S3 is added. The mixture is transferred into ultracentrifuge tubes and centrifuged in a swing-out rotor [e.g., a Beckman (Fullerton, CA) SW 50.1 rotor] at 60,000 g for 60 min at 4°C.

The superficial pellicle is first removed (e.g., by using a pipette tip). The supernatant is decanted. It is preferable to wipe the tube wall with a cotton swab. The pellet, consisting of cell nuclei, is resuspended in 0.25 ml of

0.01 M citric acid solution. The nuclear suspension is transferred into extraction tubes and extracted by mixing thoroughly with 4 ml of Ultima Gold scintillation liquid. Cell nuclei are recollected by centrifugation for 5 min at 600 g. The supernatant, containing the extracted tritiated steroids, is decanted carefully into minivials for scintillation counting. The extraction step is repeated and the extracts are counted. The nuclear pellet is now washed with 5 ml of ethanol and recovered by centrifugation for 5 min at 600 g. The pellet is dissolved in 0.25 ml of 0.3 M KOH at 36°C in order to determine protein content, following the method of Lowry *et al.* (4). Nuclear uptake can thus be expressed per milligram of cell nuclear protein.

Discussion

In Fig. 1 representative data are shown from a nuclear uptake study, carried out in our laboratory, in which [³H]corticosterone is used. To illustrate the specificity of the process, Fig. 1 also shows the effect of pretreatment with a variety of unlabeled steroids on [³H]corticosterone cell nuclear uptake. Remarkably, both aldosterone and corticosterone, but not dexamethasone, a synthetic glucocorticoid, effectively block the nuclear uptake of the [³H]corticosterone, suggesting that the radioactivity recovered from the cell nuclei represents only radioligand bound to MRs (5). With the tracer amounts used, the plasma level of [³H]corticosterone achieved will be low and, as a consequence, radioligand retained by the nuclei will predominantly be bound to MRs rather than GRs. Interestingly, *in vitro* dexamethasone also binds with appreciable affinity to MRs. Yet, *in vivo* the steroid does not compete with the retention of the [³H]corticosterone, and the uptake of [³H]dexamethasone in hippocampal cell nuclei is low (6). Other factors apparently determine the high uptake of [³H]corticosterone (see also Section II,B).

The *in vivo* uptake studies provide useful information on the mechanism of corticosteroid binding and retention by the cell nuclei.

In Vitro Cell Nuclear Uptake

Animals

Animals are housed and adrenalectomized as described in Section II,A.

Incubation Procedure
Solutions

For artificial cerebrospinal fluid (ACSF) several similar formulations exist. The ACSF we use contains 134 mM NaCl, 5 mM KCl, 1.25 mM KH$_2$PO$_4$, 2 mM MgSO$_4$, 16 mM NaHCO$_3$, and 10 mM glucose. After saturation of the

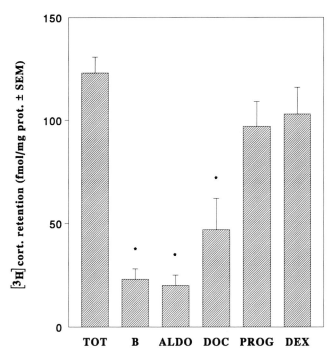

FIG. 1   *In vivo* cell nuclear uptake in the hippocampus. Hippocampal cell nuclear retention of [^{3}H]corticosterone was determined *in vivo* as described in Section II,A. Thirty minutes prior to the infusion of 50 μCi of radioligand, rats were injected with saline (TOT) or with unlabeled steroid at a dose of 30 μg/100 g of body weight. Unlabeled steroid was either corticosterone (B), aldosterone (ALDO), 11-deoxycorticosterone (DOC), progesterone (PROG), or dexamethasone (DEX). *, $p < 0.05$ vs saline.

ACSF with 95% O_2/5% CO_2 at 25°C, the pH is adjusted to 7.4 with concentrated HCl. Shortly before use, $CaCl_2$ is added to a final concentration of 1 mM.

Adrenalectomized animals are sacrificed by decapitation and the hippocampi are quickly dissected on ice. The hippocampi are pooled and sliced immediately at a thickness of 300 μm using a MacIlwain tissue chopper. The hippocampi of two rats are pooled and preincubated in 3.5 ml of ACSF, saturated with 95% O_2/5% CO_2, at 25°C. After 30 min, 0.5 ml of radioligand-containing ACSF is added, followed by incubation for 60 min at 25°C. The final label concentration varies between 1 and 20 nM. For determination of nonspecific nuclear uptake, unlabeled corticosterone is added to the preincubation medium to reach a final 500-fold excess in the incubation. Similarly, competitors may be added to the preincubation medium. Both preincubation and incubation are performed in a shaking water bath, while 95% O_2/5% CO_2 is blowing over the surface of the ACSF.

The incubation is stopped by adding 6 ml of ice-cold ACSF to each tube and putting the tubes on ice. The contents are transferred to centrifugation tubes and spun for 5 min at 600 g, while kept cool (4°C). Then the tissue slices are washed twice with 10 ml of ACSF and collected by repeating the centrifugation step and decanting the ACSF. Finally, to the pellet of tissue slices 2 ml of buffer S1 is added for homogenization.

Isolation and Extraction of Cell Nuclei

Homogenization of the tissue slices and the subsequent steps for isolation and extraction of cell nuclei are performed as described for *in vivo* cell nuclear uptake (Section II,A,3).

Discussion

The advantage of this procedure is that tissue slices can be incubated at fixed radioligand concentrations and under controlled *in vitro* conditions.

In Fig. 2 an example of an *in vitro* nuclear uptake assay is represented. Although the levels of the [3H]corticosterone concentration in the incubation medium should be high enough to ensure binding to GRs, the selective glucocorticoid RU 28362 reduces the amount of corticosterone retained by the cell nuclei only slightly. Aldosterone, on the other hand, can prevent specific retention of corticosterone effectively. This observation reinforces the findings from *in vivo* nuclear uptake studies, as described above, in which dexamethasone showed low uptake and could not affect nuclear retention of corticosterone (6). It therefore seems likely, in agreement with previous observations (7), that the corticosterone that is bound to GRs and retained by the cell nuclei may be readily lost during the isolation procedure of these nuclei; this may either be due to a faster dissociation rate for corticosterone from GRs as compared to MRs, or to a weaker interaction between the corticosterone–receptor complex and nuclear components in the case of GRs. Retention of GRs can be measured, however, using 3H-labeled RU 28362 (7).

In conclusion, [3H]corticosterone can be used as a tool for measuring cell nuclear uptake of MR-bound steroid.

Binding Assay for Cytosolic Corticosteroid Receptors

Animals and Tissues

Rats (200 g) are adrenalectomized, under ether anesthesia, in the morning when corticosterone levels are low. Adrenalectomized animals are main-

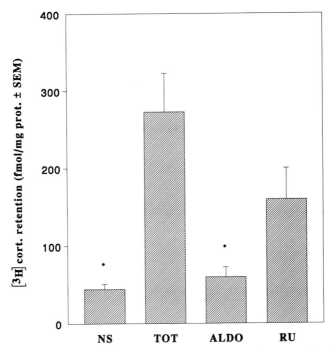

FIG. 2   *In vitro* cell nuclear uptake in the hippocampus. Hippocampal cell nuclear retention of [³H]corticosterone was measured as described in Section II,B, at an [³H]corticosterone concentration of 20 nM. TOT, Total; NS, nonspecific nuclear uptake; ALDO and RU, retention levels in the presence of a 100-fold excess of unlabeled aldosterone and RU 28362, respectively, as competitors. *, $p < 0.05$ vs total.

tained on a standard diet and 0.9% saline and tap water *ad libitum*. One day after adrenalectomy, animals are decapitated and the hippocampus is dissected from each and immediately frozen on dry ice. Hippocampal tissue may be stored for at least several weeks at −80°C until use.

Cytosol Preparation and Incubation

Solutions

For homogenization and incubation a buffer is used that contains 5 mM Tris, 1 mM ethylenediaminetetraacetic acid (EDTA), 10 mM sodium molybdate, 5% (v/v) glycerin, and 1 mM 2-mercaptoethanol (TEMG buffer, pH 7.4). A Tris-EDTA buffer similar to TEMG, but lacking the glycerin (TEM buffer), is used for elution and washing, respectively, in the Sephadex LH-20 and polyethyleneimine (PEI) methods (see below). For both buffers, 2-mercaptoethanol is added on the day of use. 2-Mercaptoethanol protects sulfhydryl groups (SH) against oxidation and thus helps to maintain the receptor in an activated (i.e., steroid-binding) state. Molybdate is also added to stabilize the receptors; although it is known to act as a phosphatase inhibitor, its exact mechanism of action is unclear (see Ref. 8 for review; see also Ref. 9). The addition of glycerin serves as an adjustment of the osmolarity of the buffer. In some cases loss of receptors may be due to protease activity and addition of protease inhibitors may be required.

Hippocampal tissue is homogenized, at 4°C in 1.1–1.5 ml of TEMG buffer per pair of hippocampal lobes, with a Potter–Elvehjem tissue homogenizer (10 strokes up and down, 1000 rpm). (In our procedure hippocampal tissue from two rats is sufficient for determination of MR and GR parameters.) Subsequently, the homogenate is centrifuged at 100,000 g in a fixed-angle rotor in a Beckman ultracentrifuge for 30 min at 4°C. The cytosol (i.e., supernatant) is collected and kept on ice.

Incubation is started by adding 100 μl of cytosol, except for blanks, to tubes containing 50 μl of TEMG buffer with radioligand concentrations varying from 0.1 to 15 nM. In our assays seven concentration points are incubated in duplicate. After mixing, all tubes are incubated at 4°C. After 18–24 hr the samples are subjected to a procedure for the separation of bound and free radioligand (see Section III,C).

For the Sephadex method, the blanks contain 100 μl of a bovine serum albumin (BSA) solution in TEMG buffer instead of cytosol. In the blanks for the PEI method 100 μl of a BSA solution in TEMG buffer is added instead of the cytosol. Cytosolic protein concentrations are measured following the procedure of Lowry et al. (4) with BSA as the standard. The final protein concentration in the incubate (except Sephadex blanks) will usually be ~2.7 mg/ml. The radioligands used in our assays are [³H]aldosterone and tritiated RU 28362 to measure MRs and GRs, respectively. When [³H]aldosterone is used, a 100-fold excess of unlabeled RU 28362 is added to prevent aldosterone from binding to GRs.

Separation of Receptor-Bound and Free Ligand

Sephadex Method

An established method for separating receptor-bound from free ligand in cytosol samples is the use of Sephadex LH-20 columns. Application of the incubation samples on such columns followed by elution with buffer leads to a retarded elution of the free steroid, because it is retained by the Sephadex particles. Thus, receptor–steroid complexes are readily eluted, whereas unbound steroid is removed only after prolonged elution.

Preparation of Elution Columns

Sephadex LH-20 is dissolved and stirred in water (LH-20 : H_2O, 2 : 1, v/v). Small glass beads are put into Pasteur pipettes (one each) to prevent the Sephadex from running through, while allowing the water to leak out. The pipettes are placed in an upright position. The Sephadex can be added simply by using plastic syringes (for convenience the syringe can be equipped with a pipette tip cut to a size wide enough to allow the Sephadex to run through it). The pipettes are filled up to about 2 cm from the top with dense Sephadex. No air bubbles should be left in the column, because these may severely affect the elution process. Before and during the elution step the columns need to be cooled thoroughly (2–4°C).

After use, the Sephadex can be collected and reused after multiple washing steps with ethanol and distilled water.

Elution Procedure

From each incubation sample 100-μl aliquots are applied on top of a column. As soon as the sample has run through the Sephadex, the bound fraction is eluted by adding TEM buffer with a total volume of 1.1 ml. From the eluent 600 μl is then transferred to a glass vial and counted after the addition of 10 ml of Ultima Gold scintillation liquid. Aliquots are also taken from the eluent for determination of the protein concentration. For evaluation of the binding, the bound fraction is usually expressed per milligram of protein in the eluent to correct for variations in eluent volume. However, a considerable amount of protein is retained by the Sephadex column, resulting in a relatively higher binding capacity when compared to bound fraction expressed per milligram of cytosolic protein. This is also discussed in the next section.

Polyethyleneimine Method

The use of glass fiber (GF) filters for separation of receptor-bound and free ligand is common in membrane receptor-binding studies (10). However, in studies dealing with binding to soluble proteins, for example, cytosolic receptors, GF filters are not generally applied. In fact, retention of soluble proteins by GF filters requires pretreatment of these filters with a reagent such as polyethyleneimine (PEI). In binding assays for cytoplasmic receptors this principle is used to separate free ligand from ligand bound to the receptor; the latter will be retained by the filter, whereas the former will be eluted. Such a procedure has been described to be effective for measuring binding to estrogen receptors (11), but has not been reported for corticosteroid receptors. Here we evaluate the use of GF filters as an appropriate tool in performing binding assays for MRs and GRs.

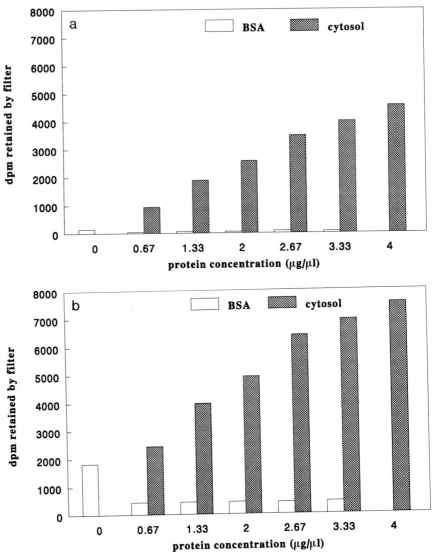

FIG. 3 PEI method: filter retention of radioligand at various protein concentrations. Dilutions of cytosolic protein (crossed bars) in a range of 0.67 to 4 mg/ml and blanks of BSA (open bars) in a range of 0 to 3.33 mg/ml were incubated with tritiated (a) aldosterone (10 nM), (b) RU 28362 (10 nM), or (c) corticosterone (10 nM) in a total volume of 150 μl of TEMG buffer. Separation was performed as described in Section III,C,2.

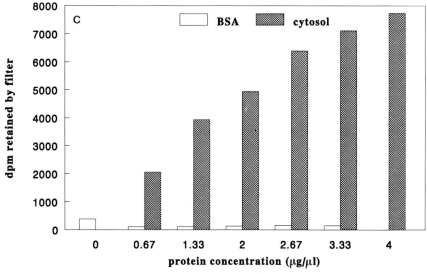

FIG. 3   (*Continued*)

Filtration Procedure

A 0.3% (v/v) solution of PEI in water is made by diluting a 10% (v/v) stock solution. Sheets of GF filter (GF/C; Whatman, Clifton, NJ) cut to size are immersed in the 0.3% PEI solution for 60 to 90 min at 4°C. The washing buffer is a Tris-EDTA buffer containing 0.008% (v/v) 2-mercaptoethanol (TEM buffer, pH 7.4) and is cooled on ice. Before the filtration is started, 30-μl aliquots are pipetted from all incubation tubes into scintillation vials and are counted for calculation of label concentration.

The filtration is performed using a 24-well Brandel cell harvester. When the PEI-coated GF/C filters are placed in the cell harvester with the vacuum on, 3 ml of TEM buffer is added to the remaining 120 μl of the incubates and the whole amount is immediately sucked up over the filters. As a rule, the filters are washed directly twice, each time with 3 ml of TEM buffer. After the vacuum is released, circular parts of the filter, which correspond to the bound fractions of the samples, are easily removed. These filter parts are then counted with 3 ml of scintillation liquid.

Although some ligands, especially RU 28362 in our case, may display some binding to the filter, we noticed that the presence of protein in the incubate, for example, cytosolic protein or BSA, substantially reduced such binding. Presumably, low amounts of protein (such as the concentrations employed in a normal binding assay) provide some kind of coating of the

PEI-treated filter that prevents high backgrounds arising from binding of free radioligand to this filter.

Because the cytosolic protein concentration should not affect the relative bound fraction of (radio)ligand, we tested the retention of ligand with varying concentrations of cytosolic protein and BSA. Figure 3 demonstrates that the protein concentration of cytosol prepared following the procedure described above is within a range where the bound fraction is directly proportional to the cytosolic protein content of the incubate.

Furthermore, the washing volume (or time) is critical for the efficiency of the separation step, because too-feeble washing results in nonspecific binding that is too high, relatively, whereas too-stringent washing leads to loss of ligand bound specifically to proteins retained by the filter and will thus attenuate the values of the specific binding. In Fig. 4 the retention by the filters of (non)specifically bound radioligand is shown for different washing volumes. It appears that brief washing is sufficient for removal of most of the remnants of (free) ligand, while leaving the receptor-bound fraction unaffected.

Analysis

The radioactivity measured in 30-μl samples of the incubates is converted into molar concentration, using the declared specific activity. These values represent the total concentration of ligand (t). The radioactivity measured either in the 600-μl samples from the eluents (Sephadex method) or on the GF filters is similarly converted into molar units, and is expressed per milligram of cytosolic protein in the eluent or in the incubate, respectively; these values represent the fraction of bound ligand (b).

Binding parameters can be estimated by means of different evaluation methods. In fact, the currently used Scatchard analysis (12) is less suitable for these purposes (13–16). First, the linearly transformed binding data are subjected to a high scatter for both low and high concentrations of the ligand (17). Second, the actual model does not account for phenomena such as cooperativity between and heterogeneity of receptor sites (18).

In our procedure, analysis is based on a two-term model [Eq. (1)], using the routine InPlot 4.0 (GraphPad-Software, Inc., San Diego, CA).

$$b = \frac{B_{max} * t}{K_d + t} + C * t \tag{1}$$

where b is the bound fraction, t is the total concentration of ligand, B_{max} is the binding capacity, K_d is the dissociation constant, and C is the constant for nonspecific binding. The first term describes the binding to a single binding

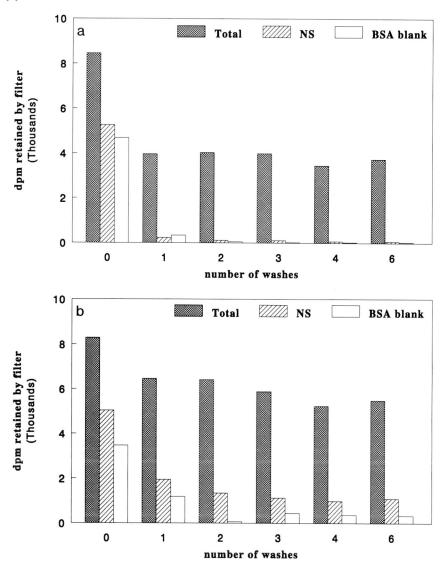

FIG. 4   PEI method: filter retention of radioligand with various washing volumes. Incubations were performed at label concentrations of around 10 nM as described in Section III,B, using either tritiated (a) aldosterone (10 nM), (b) RU 28362 (10 nM), or (c) corticosterone (10 nM). Cytosol was incubated with radioligand in the absence (cross-hatched bars) or presence (hatched bars) of the respective unlabeled ligand. At the same label concentration a BSA blank (open bars) was measured. To determine the most appropriate washing volume, varying numbers of washing steps with 3 ml of TEM buffer were performed.

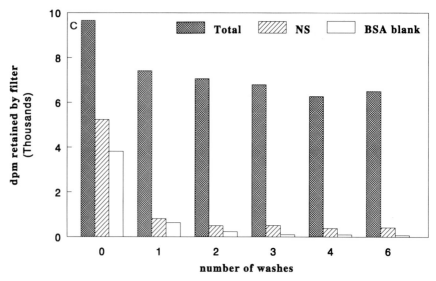

FIG. 4   (*Continued*)

site and the second term describes the nonspecific binding, which is supposed to be linear. The parameter estimation is based on a nonlinear regression fit. The nonspecific binding need not be assumed experimentally. Compared with Scatchard analysis, this approach generates better estimates of the binding parameters. Furthermore, because Scatchard analysis requires measurement of nonspecific binding for transformation of the data, an additional advantage of the routine employed here is a reduction in the use of tissue and chemicals.

InPlot 4.0 can estimate the parameter of only a single binding site. However, when MR is measured using [^{3}H]aldosterone in the presence of RU 28362, a disadvantage of this approach is the interaction of RU 28362 with the MRs. Although its affinity for MRs is low ($K_d \sim 5 \times 10^{-6} M$), with the concentrations of RU 28362 used to compete for binding of [^{3}H]aldosterone to GRs (up to $10^{-6} M$) a minor inaccuracy may be introduced in the estimation of the binding parameters for aldosterone to MRs.

Other routines can estimate all binding sites for one ligand present in a particular tissue. The most widely used routine, Ligand (19), is also based on a nonlinear regression fit. This routine enables estimation of binding sites for one ligand, provided that their affinities differ by at least one order of magnitude. In the case of corticosterone binding to MRs and GRs this differ-

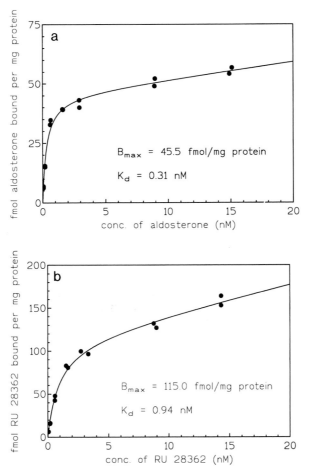

FIG. 5   Representative data from cytosol binding assays for MRs (a) and GRs (b), as determined by the PEI method. Assays were performed following the PEI method (Section III,C,2), using tritiated aldosterone or RU 28362 for estimation of the binding parameters of MRs and GRs, respectively.

ence may be smaller. A cross-displacement experiment, as described for opioid receptors (20), can facilitate the fit. Corresponding experiments are in progress. However, this model also does not account for phenomena such as cooperativity between and heterogeneity of receptor populations.

Two examplary graphs for MR and GR binding are shown in Fig. 5.

Conclusion

The use of PEI-coated filters for separating receptor-bound from free steroid may prove to be a reliable method. This method has several advantages over the use of Sephadex columns. Especially when handling larger amounts of samples, the preparation for the separation step is less laborious and the filtration procedure is much faster than the elution by Sephadex columns. Furthermore, in our experience the data points show less scatter with the PEI method. In addition, the costs of the materials used for the PEI method are considerably lower than those for the Sephadex method. Although a disadvantage of the filtration method, when compared with the use of Sephadex, is the volume of radioactive waste from the washing step, one should take into account that the rinsing of Sephadex may also lead to considerable amounts of nuclear waste.

The molar dissociation constants (K_d) for MRs and GRs we find with the present method are similar to those found with the Sephadex procedure. The estimations of B_{max} values, however, appear to be consistently lower with the PEI method, especially for GRs. The main reason for this discrepancy is the fact that in the Sephadex method cytosolic protein concentrations are measured in the eluent, which contains about 50% less protein than the initial incubate. In the PEI method, the receptor-bound ligand that is retained by the filter is expressed per milligram of protein, as measured in the incubate, and may thus explain the lower estimates for B_{max}. However, if the protein levels in the eluents from the Sephadex methods are corrected for proteins lost over the column, the estimates from both methods appear to be similar.

In contrast to a method like the Scatchard method, nonspecific incubations can be omitted from the assay procedure when applying the method of curve fitting to the untransformed data points, as we describe here. Thus, by reducing the amount of tissue needed for an assay, the efficiency of the assay is increased. In addition, nonlinear regression fitting has been described to be more accurate than methods requiring a linearization step.

References

1. E. R. de Kloet, *Front. Neuroendocrinol.* **12,** 95–164 (1991).
2. B. G. Yongue and E. J. Roy, *Brain Res.* **436,** 49–61 (1987).
3. B. S. McEwen and R. E. Zigmond, *Res. Methods Neurochem.* **1,** 140–161 (1972).
4. O. H. Lowry, N. J. Rosebrough, A. L. Farr, and R. J. Randall, *J. Biol. Chem.* **193,** 265–275 (1951).
5. H. D. Veldhuis, C. van Koppen, M. van Ittersum, and E. R. de Kloet, *Endocrinology (Baltimore)* **110,** 2044–2051 (1983).

6. E. R. de Kloet, G. Wallach, and B. S. McEwen, *Endocrinology (Baltimore)* **96,** 598–609 (1975).
7. B. van Steensel, A. D. van Haarst, E. R. de Kloet, and R. van Driel, *FEBS Lett.* **292,** 229–231 (1991).
8. T. J. Schmidt and G. Litwack, *Physiol. Rev.* **62,** 1131–1192 (1982).
9. P. R. Housley, M. K. Dahmer, and W. B. Pratt, *J. Biol. Chem.* **257,** 8615–8618 (1982).
10. J. P. Bennett and H. I. Yamamura, *in* "Neurotransmitter Receptor Binding" (H. I. Yamamura, S. J. Enna, and M. J. Kuhar, eds.), pp. 61–89. Raven, New York, 1985.
11. R. F. Bruns, K. Lawson-Wendling, and T. A. Pugsley, *Anal. Biochem.* **132,** 74–81 (1983).
12. G. Scatchard, *Ann. N. Y. Acad. Sci.* **51,** 660–672 (1949).
13. I. M. Klotz, *Science* **217,** 1247–1249 (1982).
14. I. M. Klotz, *Trends Pharmacol. Sci.* **4,** 253–255 (1983).
15. E. Bürgisser, *Trends Pharmacol. Sci.* **5,** 142–144 (1984).
16. R. J. Leatherbarrow, *Trends Biol. Sci.* **15,** 455–458 (1990).
17. D. A. Deranleau, *J. Am. Chem. Soc.* **91,** 4044–4049 (1969).
18. V. Pliska, *Ann. N.Y. Acad. Sci.* **248,** 480–493 (1975).
19. P. Munson, *in* "Immunochemical Techniques, Part E: Monoclonal Antibodies and General Immunoassay Methods" (J. Langone and H. Van Vunakis, eds.), Methods in Enzymology, Vol. 92, pp. 543–576. Academic Press, New York, 1983.
20. R. A. Lutz, R. A. Cruciani, P. Munson, and D. Rodbard, *Life Sci.* **36,** 2233–2238 (1985).

[7]   Steroid Hormone Binding to Membrane Receptors

Miles Orchinik and Thomas F. Murray

Introduction

The mechanisms that mediate some rapid effects of steroids on brain function appear to be distinct from the mechanisms that mediate the genomic responses to steroids. In the well-established genomic model of steroid action, the focus of this volume, steroids passively diffuse across plasma membranes and bind to intracellular receptors in target cells. The activated steroid–receptor complex binds to specific sequences of DNA to act as a ligand-dependent transcription factor regulating gene expression. While the details of this model undergo constant revision, there is growing recognition that certain steroid effects are incompatible with the genomic model of action. Genomically mediated responses generally occur with a latency of hours, even days. However, 50 years ago Hans Seyle discovered that certain steroids have rapid anesthetic and anticonvulsant effects, distinct from the typical hormonal actions of steroids (1, 2). Since that time, there have been numerous reports demonstrating neuronal responses that occur within seconds to minutes of steroid application, responses that occur in preparations lacking intracellular receptors, in the presence of RNA or protein synthesis inhibitors, or in response to steroids coupled to large molecules that block access to intracellular receptors (reviewed in Refs. 3–8). These rapid effects are presumed to be mediated through membrane-bound receptors rather than intracellular receptors.

An alternative mechanism for steroid action involves steroid binding to the γ-aminobutyric acid $(GABA)_A$ receptor–chloride channel complex in neuronal membranes (reviewed in Refs. 9 and 10). Other rapid steroid actions do not involve direct modulation of $GABA_A$ receptor function. One example is the rapid suppression of reproductive behavior by corticosterone or stress in *Taricha granulosa,* the roughskin newt (11, 12). Male *Taricha* sexual behavior is profoundly inhibited within 8 min of an intraperitoneal injection of corticosterone (13), and neurophysiological responses occur with similar latency (14). The pharmacology of these responses to corticosterone is not consistent with modulation of $GABA_A$ receptor function (15). Instead, there

Methods in Neurosciences, Volume 22

appears to be a functional receptor in neuronal membranes that is specific for corticosteroids (13, 16).

There have been numerous reports of steroid binding to plasma membranes of peripheral tissues (reviewed in Refs. 5, 6, 8, and 17), but fewer reports of steroid binding to brain membranes (13, 16, 18–25). A recurring problem appears to be that physiological concentrations of steroids in radioligand-binding assays result in prohibitively high levels of "nonspecific" binding, presumably due to the intercalation of amphipathic steroids into the lipid bilayer of neuronal membranes. In this chapter, we describe radioligand-binding procedures that we used to characterize a corticosteroid receptor in neuronal membranes, using *Taricha* brains.

Criteria for Physiologically Relevant Receptors

Radioligand binding can be a powerful tool for the discovery and characterization of new receptors, the understanding of receptor regulation, and the elucidation of transduction/effector mechanisms. However, it is also possible to generate spurious data that may be misinterpreted. Therefore, a number of standard pharmacological criteria must be satisfied in order to distinguish genuine, physiologically relevant membrane-bound receptors from artifactual or nonphysiological binding sites: (a) radioligand binding to receptors should be saturable, reflecting a finite number of binding sites, with appropriately high affinity (expressed as the K_d, the equilibrium dissociation constant); (b) the kinetically derived K_d should agree with the K_d derived from equilibrium saturation studies and the kinetics of binding should be consistent with physiological responses. The specific binding of the radioligand should be reversible. The kinetic and equilibrium studies provide an assessment of whether the observed radioligand–receptor interaction follows the law of mass action; (c) in control experiments, specific binding should be eliminated under conditions that denature or degrade proteins; (d) receptors should exhibit pharmacological specificity that parallels the specificity of compounds in eliciting functional responses; (e) the distribution of receptors should make sense, that is, receptors should be enriched in appropriate subcellular fractions derived from physiologically responsive tissues, but not unresponsive tissues; and (f) activation of the recognition site should result in a functional response.

Further studies might include receptor purification and reconstitution, or receptor gene cloning, sequencing, and expression. In addition, one might examine the transduction and effector mechanisms associated with the receptor.

Methods

Membrane Preparation

Crude Synaptosomal Membranes

We use a well-washed neuronal membrane preparation to characterize [³H]corticosterone binding to membranes (13, 16). Whole *Taricha* brains are homogenized in 25 vol (volume per original weight) of cold 0.32 *M* sucrose containing 5 m*M* N-2-hydroxyethylpiperazine-*N'*-2-ethanesulfonic acid (HEPES; pH 7.45), using a glass–Teflon homogenizer. The homogenate is centrifuged at 1000 g (15 min, 4°C), the pellet discarded, and the supernatant centrifuged at 30,000 g (40 min, 4°C). This P2 pellet is frozen, thawed, and resuspended in 150 vol (volume per original weight) of buffer [25 m*M* HEPES, 10 m*M* ethylenediaminetetraacetic acid (EDTA), bacitracin (60 μg/ml), pH 7.45] using a Dounce (Wheaton, Millville, NJ) glass–glass tissue homogenizer with a tight pestle. The suspension is maintained at 4°C for 2–3 hr to dissociate endogenous ligands and to remove cations. The prolonged wash step was necessary because adrenalectomizing amphibians is not feasible; a 15-min wash should be sufficient for preparing membranes from rats that have been adrenalectomized for 24 hr prior to sacrifice. The suspension is then centrifuged at 30,000 g (30 min, 4°C). The resulting pellet is washed in 150 vol (volume per original weight) of 25 m*M* HEPES (pH 7.45) and centrifuged again at 30,000 g (30 min, 4°C). The final pellet can be stored at -70°C for 2 months without significant loss of binding. However, the modulation of [³H]corticosterone binding by guanyl nucleotides and Mg^{2+} is adversely affected by prolonged freezing.

Partially Purified Synaptosomal Membranes

More purified membrane preparations may be employed to examine the subcellular distribution of receptors or to enhance the specific binding signal-to-noise ratios. We use discontinuous sucrose gradient centrifugation, as developed by Whittaker (26), to determine the subcellular distribution of [³H]corticosterone binding in brain tissue. Brains are homogenized in 0.32 *M* sucrose as described above, the homogenate is centrifuged at 1000 g (15 min, 4°C), and the resulting supernatant centrifuged at 17,000 g (40 min, 4°C). The P2 pellet (synaptosomes, myelin and membrane fragments, and mitochondria) is layered onto a discontinuous density gradient (equal volumes of 0.8 and 1.2 *M* sucrose) and centrifuged at 53,000 g (2 hr, 4°C). Mitochondria are recovered in the pellet, synaptosomes are recovered from the interface between the 0.8 and 1.2 *M* sucrose, and myelin is recovered between the 0.32 and 0.8 *M* sucrose. The synaptic membrane and myelin fractions are diluted in 25 mM HEPES and centrifuged at 30,000 g (30 min, 4°C) prior to resuspension for use in a binding assay. However, the yield of myelin from *Taricha* brains is too low for assay with the other fractions.

Another procedure for preparing enriched synaptosomal membranes involves combined flotation–sedimentation density gradient centrifugation (27). Briefly, brains are homogenized in 10% (w/v) sucrose, centrifuged at 800 g (12 min, 4°C) and the supernatant centrifuged at 9000 g (30 min, 4°C). This pellet is lysed in hypotonic buffer, and after 30 min a sufficient volume of 48% sucrose is added to yield a 34% sucrose solution. A 28.5% sucrose solution is layered onto the membrane fraction. The density gradient is centrifuged at 60,000 g (2 hr, 4°C). Synaptic membranes are recovered from the 34%–28.5% sucrose interface.

Radioligand-Binding Assays

We use tritiated corticosterone, the endogenous glucocorticoid in *Taricha,* in binding studies (72–101.6 Ci/mmol; New England Nuclear, Boston, MA). In addition to using tritiated steroids, other laboratories have reported specific binding of progesterone complexed to iodinated bovine serum albumin (BSA) to neuronal membranes (19, 23) and of aldosterone 3-(O-carboxymethyl)oximino-(2-[^{125}I]iodohistamine) binding to leukocyte membranes (28). Although the ^{125}I confers higher specific activity, we are able to obtain a sufficiently robust signal using the biologically active steroid with known stoichiometry. Owing to the limited abundance of brain tissue (*Taricha* brain weight, ~40 mg), we routinely use 300-μl binding assays. The final pellet obtained by the procedures outlined above is resuspended in assay buffer to a protein concentration of 300–600 μg/ml for use in binding studies. Protein concentration is determined by the Bradford method, using a Coomassie protein assay reagent (Pierce, Rockford, IL) and BSA standard. The binding reaction is initiated by adding 100 μl of [^{3}H]corticosterone to 100 μl of membrane preparation and 100 μl of drugs or competitors. The assay buffer contains 25 mM HEPES and 10 mM MgCl$_2$ (unless specified otherwise), pH 7.45. In early studies, we used 100 mM NaCl in the assay buffer, but subsequently found no consistent effect of Na$^+$, K$^+$, Ca^{2+}, or Cl$^-$ on [^{3}H]corticosterone-specific binding. In *Tarichia,* specific binding is optimal with a 2-hr incubation at 30°C, although many studies are performed at 15°C to approximate physiological conditions more closely.

Bound [^{3}H]corticosterone is separated from unbound radioligand by vacuum filtration, using a Brandel (Gaithersburg, MD) M-24 tissue harvester and Whatman (Clifton, NJ) GF-C glass fiber filters soaked in rinse buffer (25 mM Tris, pH 7.45) for 15 min prior to use. In this procedure, the binding reaction is terminated by the addition of 3 ml of ice-cold buffer and immediate filtration, followed by an additional 6-ml wash with cold buffer. The radioactivity bound to membranes is trapped by the filters and the free radioligand is removed by the washes. Radioactivity bound to filters is measured by standard liquid scintillation spectroscopy. Nonspecific binding or nonreceptor-bound ligand is determined by addition of 1–10 μM unlabeled corticosterone (CORT) to parallel tubes. Specific binding (total binding minus nonspecific binding) typically represents 75 to 80% of total binding at [^{3}H]corticosterone concentrations approximating the K_d. The binding of [^{3}H]corticosterone to filter blanks is negligible.

Binding to Soluble Intracellular Receptors

It is highly unlikely that binding of [^{3}H]corticosterone to intracellular corticoid receptors constitutes a significant proportion of the radioligand bound in this filtration assay, for several reasons. The nuclear and cytosolic fractions that contain intracellular receptors are discarded during membrane preparation; cell nuclei are pelleted in the first spin, and cytosolic components are removed with the subsequent pelleting, freeze-thawing, and prolonged wash with large volumes of buffer. Furthermore, one would not expect soluble receptor proteins to bind to untreated glass fiber filters (retention size of 1.2 μm) during vacuum filtration. Because the glass fiber filters possess a net negative charge, charge repulsion is likely to minimize the binding of soluble proteins to the filters. Moreover, the buffers used during tissue preparation and binding do not contain reducing agents or molybdate, which are used to stabilize intracellular corticoid receptors. On the contrary, we find that addition of 100 μM dithiothreitol (DTT), a disulfide bond-reducing agent, to the homogenization and assay buffers slightly decreases the equilibrium binding of [^{3}H]corticosterone and greatly decreases the potency of GTPγS to modulate [^{3}H]corticosterone binding (16). A 10- to 50-fold higher concentration of dithiothreitol is routinely included in assays of intracellular adrenal steroid receptors. Finally, the binding of [^{3}H]corticosterone to the nuclear and cytosolic fractions, in which intracellular receptors are enriched, is minimal when using these assay procedures (see below). This negative control suggests that the binding assay is not significantly contaminated by binding to soluble receptors.

Data Analysis

Radioligand-binding data are analyzed by computer-assisted, nonlinear regression analysis, using a variety of commonly available software programs for the personal computer: LUNDON I (Lundon Software, Cleveland, OH),

EBDA, KINETIC, and LIGAND (Elsevier-Biosoft, Cambridge, UK), and GraphPad InPlot (ISI Software, San Diego, CA). Nonlinear regression analysis provides more reliable parameter estimates than the estimates obtained by linear regression of transformed binding data, such as the Scatchard–Rosenthal and Hill plots. Each program generates binding parameter estimates (with approximate standard errors) by fitting the untransformed data to appropriate equations using iterative, weighted least-squares curve-fitting techniques. The programs assume that binding is in accordance with the principles of mass action.

Data from equilibrium saturation binding studies are fit to the equation for a rectangular hyperbola:

$$[B] = B_{max} [L]/(K_d + [L])$$

where [B] is the concentration of radioligand bound, B_{max} is the total receptor concentration, [L] is the amount of free radioligand, and K_d is the equilibrium dissociation constant. For binding to two or more sites, the data are fit to the following equation:

$$[B] = \sum_{i=1}^{i=n} \frac{B_{maxi} [L]}{K_{di} + [L]}$$

where K_{di} and B_{maxi} are the constants for radioligand binding to site i. The appropriateness of a one-site vs a multiple-site model may be tested statistically by an F test that determines whether the inclusion of additional parameters to the model significantly increases the goodness of fit.

Kinetic parameters are generated using the program KINETIC. The dissociation rate constant (k_{-1}) is derived by fitting the data to the first-order rate equation:

$$[B_t] = \sum_{i=1}^{i=n} [B_{0i}] e^{k_{-1i}t}$$

where $[B_t]$ is the amount of radioligand bound at time t after the start of dissociation, $[B_0]$ is the amount bound at time 0, and k_{-1i} is the dissociation rate constant for each site i out of a possible n sites. The association rate constant (k_{+1}) is estimated by initially fitting the data to a pseudo-first-order rate equation:

$$[B_t] = \sum_{i=1}^{i=n} [B_{Eqi}](1 - e^{k_{obsi}t})$$

where $[B_t]$ is the amount of radioligand bound at time t, k_{obs} is the observed association rate, and $[B_{Eqi}]$ is the amount of radioligand bound at equilibrium to site i out of a possible n sites. The pseudo-first-order equation assumes that ligand concentration is constant throughout the binding reaction. Therefore, to maintain pseudo-first-order reaction conditions, the ligand concentration used is 10-fold greater than that of receptor. The association rate constant, k_{+1}, is then calculated from the following relationship:

$$k_{+1i} = \frac{k_{obsi} - k_{-1i}}{[L]}$$

where k_{-1i} is the empirically determined dissociation rate constant for site i, and $[L]$ is the radioligand concentration used.

Competition experiments are analyzed using sigmoidal curve-fitting programs EBDA and LIGAND, or GraphPad InPlot. Displacement curves are fit to the following equation in EBDA:

$$[B] = B_{tot}/[1 + ([C]/IC_{50})^n] + NSB$$

where B_{tot} is the total amount of radioligand specifically bound in the absence of competitor, $[C]$ is the concentration of competitor, IC_{50} is the concentration of competitor that inhibits 50% of specific binding, n is the estimated Hill coefficient, and NSB is the estimate of nonspecific binding. If the Hill coefficient does not significantly deviate from unity, one can estimate the affinity of the unlabeled competitor for the receptor (K_i) from the IC_{50}, using the Cheng–Prusoff equation:

$$K_i = IC_{50}/[1 + ([L]/K_d)]$$

where $[L]$ is the concentration of radioligand and K_d is the empirically determined equilibrium dissociation constant for the radioligand.

Characterization of [³H]Corticosterone Binding to Amphibian Brain Membranes

Equilibrium Saturation Binding

Equilibrium saturation analysis demonstrated that the binding of [³H]corticosterone to brain membranes was specific, saturable, and of high affinity (Fig. 1). Using 0.03–5.0 nM [³H]corticosterone and 10 mM MgCl$_2$, a single

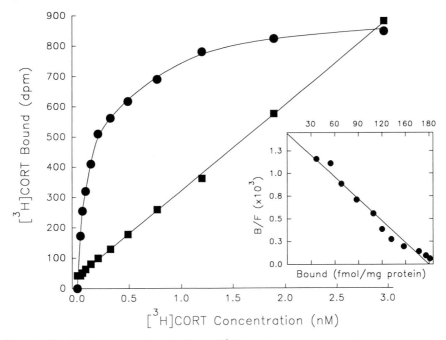

FIG. 1   Equilibrium saturation binding of [³H]corticosterone to crude synaptosomal membranes. Well-washed neuronal membranes were incubated with the concentrations of [³H]CORT shown for 2 hr at 30°C in buffer containing 10 mM MgCl₂. Specific binding (circles) equals total binding (not shown) minus nonspecific binding (squares). Data were best fit with a one-site model with K_d = 0.14 ± 0.01 nM and B_{max} = 183 ± 4 fmol/mg protein. *Inset:* Scatchard replot of specific binding data was linear; Hill coefficient = 1.08.

population of high-affinity binding sites was labeled at 30°C with a K_d of 0.14 ± 0.01 nM and a B_{max} of 183 ± 4 fmol/mg protein. Consistent with a one-site model, the Scatchard replot of the data was linear; the Hill coefficient was 1.08. In other experiments performed at 15°C (in the physiological range of *Taricha*), with no Mg^{2+} added, the K_d was estimated to be 0.51 ± 0.04 nM, the B_{max} 146 ± 4 fmol/mg protein, and the Hill coefficient 0.98. The subnanomolar K_d is similar to the affinity of corticosterone for the type I intracellular mineralocorticoid receptor, but represents much higher affinity binding than that described for rat brain membranes by Towle and Sze (18) and Chen *et al.* (25). In the experiments that followed, we used [³H]CORT concentrations close to the K_d, which typically resulted in 75–80% specific binding.

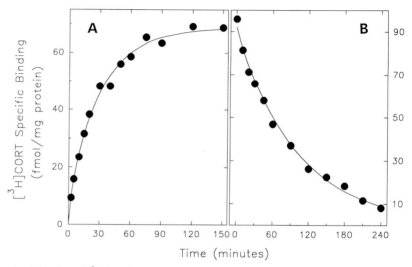

FIG. 2   Kinetics of [³H]corticosterone binding to neuronal membranes. (A) Association of [³H]CORT-specific binding to brain membranes. Membranes were incubated
with 0.5 nM [³H]CORT at 15°C for the intervals shown. Data were fit by a one-site
model with k_{obs} = 0.054 ± 0.005 min^{-1}. (B) Dissociation of [³H]CORT from neuronal
membranes. Membranes were incubated for 2 hr at 15°C with 0.5 nM [³H]CORT
prior to addition of 100 μM cold CORT. Data best described with a one-site model
yielding k_{-1} = 0.013 ± 0.001 min^{-1}. [Reproduced from Orchinik et al. (13), with
permission of Science (copyright 1991 by the AAS).]

Kinetic Studies

The association of 0.5 nM [³H]corticosterone (no MgCl$_2$ added to buffer)
with neuronal membrane recognition sites reached equilibrium within 90 min
at 15°C (Fig. 2A). Specific binding was completely reversible (Fig. 2B);
dissociation initiated by the addition of unlabeled corticosterone was described by a monophasic equation yielding a k_{-1} of 0.013 ± 0.001 min^{-1}.
[³H]Corticosterone association was also monophasic, with an estimated k_{obs}
of 0.054 ± 0.005 min^{-1}; the calculated k_{+1} was 0.079 min^{-1} nM^{-1}. The
kinetically derived K_d (k_{-1}/k_{+1}) was 0.16 nM, which is in reasonable
agreement with the K_d derived from the equilibrium study under the same
conditions (0.5 nM). This concordance supports the validity of the binding
methodology.

Control Experiments

In control experiments, specific binding was eliminated by heat (60°C) denaturation of membranes (Fig. 3A), and inhibited in a dose-dependent manner
by protease (trypsin) treatment (Fig. 3B). In addition, the specific binding

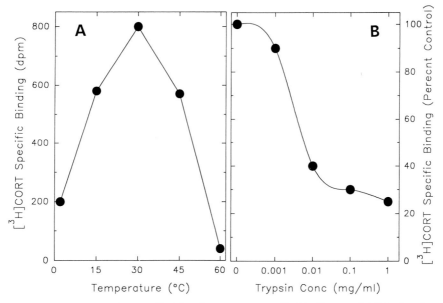

FIG. 3   Control experiments. [³H]Corticosterone-specific binding was inhibited under conditions that denatured or degraded membrane proteins. (A) Effect of heating. Membranes were treated at the temperatures shown for 20 min prior to a 2-hr incubation with 0.5 n*M* [³H]corticosterone at each temperature. (B) Effect of protease treatment. Membranes were treated for 30 min at 30°C with trypsin at the concentrations indicated prior to addition of trypsin inhibitor (1 mg/ml) and a 2-hr incubation at 30°C with 0.5 n*M* [³H]corticosterone. Trypsin inhibitor alone had no effect on specific binding. Incubation with trypsin and [³H]corticosterone for 2 hr without addition of trypsin inhibitor produced no further inhibition of specific binding.

of [³H]corticosterone was linearly related to tissue concentration (data not shown). When brains were saline perfused prior to decapitation and membrane preparation, [³H]corticosterone binding persisted, suggesting that plasma corticosteroid-binding globulins (CBG) do not play a major role in the binding of corticosterone to brain membranes.

Specificity

Equilibrium competition experiments indicated that the binding site displays distinct pharmacological specificity. The pharmacological profile of the membrane-binding site differed from the intracellular corticoid receptors in being highly specific for corticosterone and cortisol, with either low or negligible

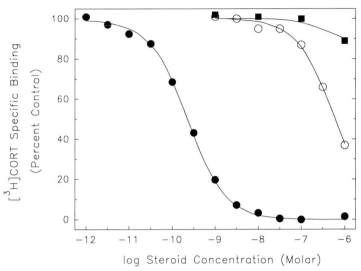

FIG. 4   Specificity of binding to recognition site in neuronal membranes. Membranes were incubated for 2 hr at 15°C with 0.5 nM [³H]corticosterone and the concentrations of competitor shown. Steroids were dissolved in solvent (ethanol or ethanol : dimethyl sulfoxide, 3 : 1, v/v) at 3–30 mM and then diluted in buffer. If solvent concentration exceeded 0.1%, the highest concentration of solvent used was added to control tubes. (●) Corticosterone; (○) aldosterone; (■) dexamethasone.

affinity for both aldosterone and dexamethasone (Fig. 4). A series of unlabeled steroids were tested as competitors for [³H]corticosterone-specific binding to membranes (Table I). In addition, we have also tested a number of neurotransmitter receptor ligands, including ligands for GABA, glutamate, acetylcholine, dopamine, serotonin, cannabinoid, and peripheral benzodiazepine receptors, but have not found a potent modulator of [³H]corticosterone binding.

Together, these data suggested that the binding of [³H]corticosterone to neuronal membranes reflects a ligand–receptor interaction with a unique pharmacological profile that follows the principle of mass action.

Distribution of Binding Sites

Subcellular Fractionation

We examined the distribution of binding sites, using partially purified subcellular fractions prepared by discontinuous sucrose gradient centrifugation (Table II). The succinate cytochrome-c reductase assay indicated that the

TABLE I   Potency of Steroids as Inhibitors of
          [³H]Corticosterone Binding[a]

Steroid	K_i (nM)	Maximal inhibition (%)[b]	Slope
Corticosterone	0.11 ± 0.006	100	0.96 ± 0.03
Cortisol	3.75 ± 0.56	100	0.92 ± 0.08
Aldosterone	293 ± 13	69	1.06 ± 0.06
5α-THDOC[c]	297 ± 14	70	0.96 ± 0.05
RU 28362	569 ± 40	36	1.14 ± 0.24
Progesterone	759 ± 113	37	0.99 ± 0.10
Testosterone	1138 ± 63	32	0.96 ± 0.03
ZK 91587	>5000	13	
Dexamethasone	>5000	11	
RU 38486	>5000	8	
3α-OH-DHP[c]	>5000	5	

[a] Experiments were performed as described in Fig. 4. Data are mean ± SEM values.
[b] Maximal inhibition reported for 1 μM competitor, RU 38486 courtesy of Roussel-UCLAF, Romainville, France.
[c] 5α-THDOC, 3α,21-hydroxy-5α-pregnan-20-one; 3α-OH-DHP, 3α-hydroxy-5α-pregnan-20-one. Potent modulator of GABA_A receptors.

TABLE II   Binding of [³H]Corticosterone and [³H]Quinuclidinyl Benzilate to
           Partially Purified Synaptosomes[a]

Fraction	Relative enrichment[b]		Succinate cytochrome-c reductase specific activity[e]
	[³H]CORT[c]	[³H]QNB[d]	
Homogenate	1.00	1.00	1.24
Supernatant	0.06	0.06	1.18
Nuclei	0.59	0.68	0.44
Mitochondria	7.04	3.06	10.58
Synaptosomes	11.55	11.19	2.80

[a] Fractions were prepared by discontinuous sucrose gradient centrifugation.
[b] Data are reported as the enrichment of [³H]CORT- and [³H]QNB-specific binding per milligram of protein relative to specific binding in the homogenate. Specific binding in the synaptosomal fraction was 87.6 fmol/mg protein for [³H]CORT, and 1066 fmol/mg protein for [³H]QNB.
[c] Tissue was incubated with 0.5 nM [³H]CORT for 2 hr at 15°C.
[d] Parallel tubes were incubated with 0.5 nM [³H]QNB in the absence or presence of 100 μM scopalamine, to determine nonspecific binding.
[e] Succinate cytochrome-c reductase assay was performed as in Feyereisen et al. (36) with minor modifications (13). Eight determinations were made for each fraction in the presence and absence of succinate. Specific activity is reported as units of change per minute per milligram of protein.

mitochondrial fraction separation was largely successful. As a marker for synaptic membranes, we measured the binding of quinuclidinyl benzilate ([³H]QNB) to muscarinic cholinergic receptors. As expected, [³H]QNB binding was most enriched in the synaptosomal fraction. The specific binding activity of [³H]corticosterone was also enriched more than 11-fold in the synaptosomal fraction relative to the brain homogenate. The finding that [³H]corticosterone binding was enriched to the same degree as [³H]QNB in the synaptosomal fraction provided compelling evidence that corticosterone-binding sites are found on synaptic membranes. Under these assay conditions, [³H]corticosterone binding to the cytosolic and nuclear fractions was negligible; this negative control suggested that binding in the synaptosomal fraction was not due to contamination by soluble intracellular receptors.

Receptor Autoradiography

In vitro receptor autoradiography (13) provided independent corroborating evidence for the presence of [³H]corticosterone-binding sites in brain regions enriched in synaptic terminals. By measuring [³H]corticosterone binding in the presence of unlabeled ligands specific for intracellular corticoid receptors, membrane-bound sites were preferentially labeled. As shown in Fig. 5, the specific binding of [³H]corticosterone was concentrated in the neuropil surrounding the preoptic area neurons, clearly outside regions where intracellular receptors would be enriched. It is also significant that binding sites are located in the preoptic area, a region of the brain that regulates male sexual behavior in vertebrates.

Functional Studies

Behavioral Studies

To determine if the corticosterone receptor mediated the rapid response to corticosterone injection or stress, five steroids with a wide range of affinities for the membrane-bound receptor were used in behavioral experiments. Males were injected intraperitoneally with one of six doses of steroid or vehicle. Five minutes later, male sexual behavior (amplectic clasping) was monitored for 20 min to detect rapid behavioral changes in response to steroid administration. The potencies of the five steroids to inhibit male sex behavior rapidly were linearly related to their potencies to inhibit [³H]corticosterone binding to neuronal membranes (13). The strong correlation between the half-maximal effective doses to inhibit behavior and to inhibit binding suggests that the receptor is behaviorally relevant.

Electrophysiological Studies

Another approach to determining receptor functionality was to implant extra-cellular electrodes into the brain of intact male *Taricha* and record the electro-physiological responses of single neurons to corticosterone. Within 3–7 min following intraperitoneal injection of behaviorally active doses of corticoste-rone, the neurophysiological properties of several types of hindbrain neurons were altered (14). Spontaneously active medullary neurons showed a rapid decrease in firing rate following injection. Quiescent reticulospinal neurons, identified by backfiring, exhibited reduced excitability in response to anti-dromic stimulation following corticosterone injection. In addition, neurons that responded to cloacal pressure, a stimulus that elicits amplectic clasping, showed a rapid decrease in sensory responsiveness following corticosterone injection. Dexamethasone injection had little or no direct neurophysiological effect within this time frame. The rapid onset and pharmacological specificity of electrophysiological responses are consistent with the membrane-bound receptor.

Together, the behavioral data and the neurophysiological data support the conclusion that the corticosterone receptor in neuronal membranes is functionally relevant.

Transduction Mechanism

The modulation of [^{3}H]corticosterone binding is largely consistent with a receptor coupled to guanyl nucleotide-binding proteins (G proteins; see Ref. 15). The specific binding of [^{3}H]corticosterone was enhanced up to 50% in a concentration-dependent manner by $MgCl_2$, with an ED_{50} value of 0.5 mM. Mg^{2+} may enhance agonist binding to G protein-coupled receptors by promoting the formation and stabilization of a high-affinity ternary complex of hormone–receptor–G protein. In other studies, we found that [^{3}H]aldoste-rone binding to *Taricha* brain membranes has an absolute requirement for Mg^{2+} (29). The nonhydrolyzable GTP analog, GTPγS, was a potent inhibitor of [^{3}H]corticosterone binding, inhibiting close to 90% of corticosterone-spe-cific binding (Fig. 6). The negative modulation of [^{3}H]corticosterone binding by guanyl nucleotides is consistent with a G protein-coupled receptor because GTP promotes dissolution of the high-affinity ternary complex and dissocia-tion of the receptor from the G protein, thereby generating the uncoupled receptor with low affinity for agonist ligands (30). In kinetic experiments, dissociation initiated by the addition of GTPγS plus unlabeled CORT resulted in a biphasic dissociation with a rapid phase of CORT dissociation that was not seen with the addition of CORT alone (Fig. 7). The initial rapid phase of dissociation was described by a $k_{-1} = 0.515 \pm 0.134$ min^{-1}, in which 22%

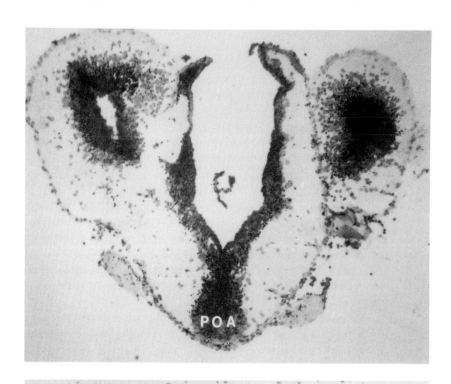

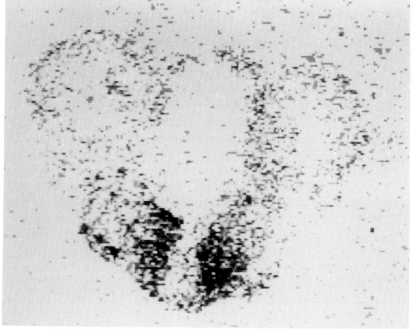

of [^{3}H]CORT dissociated. The subsequent slow dissociation was described by a rate constant ($k_{-1} = 0.015 \pm 0.0004$ min^{-1}) similar to the k_{-1} value obtained by the addition of CORT alone. These data suggest that GTPγS caused a rapid dissociation of the ternary complex with resultant loss of a fraction of high-affinity agonist-binding sites. The slower terminal dissociation phase presumably reflects dissociation of [^{3}H]corticosterone from the G protein-coupled receptor. Thus, the modulation of [^{3}H]corticosterone binding by guanyl nucleotides likely reflects a negative heterotropic interaction.

Concluding Remarks

Using a variety of radioligand-binding procedures, we have characterized a high-affinity binding site for corticosteroids in neuronal membranes. These studies satisfied a number of criteria used to distinguish physiologically relevant receptors from spurious nonreceptor binding. The receptor is enriched in synaptic membrane fractions and in the neuropil of *Taricha* brains, and has a unique pharmacological profile. Behavioral and neurophysiological studies suggest that the receptor may mediate some rapid behavioral responses to stress. The negative modulation of [^{3}H]corticosterone binding by guanyl nucleotides suggests that signal transduction through this receptor may be mediated by G proteins.

In-depth characterization of corticosteroid receptors in mammalian neuronal membranes has been hindered by limitations inherent in labeling low-affinity binding sites in amphipathic membranes, using amphipathic steroids. These methodological constraints do not negate the possibility of functionally relevant, low-affinity corticosteroid receptors in mammalian brains. An electron microscopic study has revealed glucocorticoid receptor immunoreactivity associated with plasma membranes and dendritic processes in the rat hippocampus (31). Membrane-resident transduction mechanisms would be well suited for mediating rapid neuronal responses to sudden, stress-induced fluctuations in adrenal steroid levels. Further, 120 nM K_d binding sites, as

---

FIG. 5   Distribution of putative, membrane-bound [^{3}H]corticosterone-binding sites in the preoptic area (POA). Histological section of *Taricha* brain (left) shows darkly stained perikarya of POA neurons. Autoradiogram (right) generated from the identical section shows localization of [^{3}H]corticosterone-specific binding sites in the neuropil surrounding perikarya of POA neurons. Sections were incubated with 2 nM [^{3}H]corticosterone in assay buffer containing 200 nM dexamethasone and ZK91587 to occupy nuclear steroid receptors. [Reproduced from Orchinik *et al.* (13), with permission of *Science* (copyright 1991 by the AAS).]

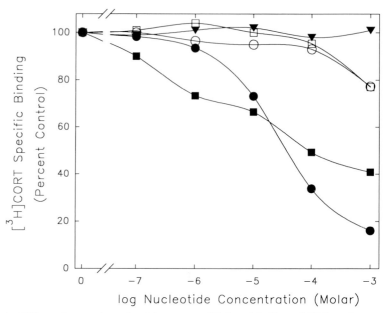

FIG. 6   Effect of guanyl nucleotides on equilibrium binding of [³H]corticosterone. *Taricha* brain membranes were incubated with 0.75 nM [³H]CORT and the concentrations of nucleotide shown. Nonhydrolyzable guanyl nucleotides inhibited specific binding with IC$_{50}$ estimates of 43.2 ± 11 μM for GTPγS and 104 ± 65 μM for Gpp(NH)p. (▼) ATP; (○) GDP; (□) GMP; (■) Gpp(NH)p; (●) GTPγS. [Reproduced from Orchinik *et al.* (15), with permission of the National Academy of Science.]

described by Towle and Sze (18) and Chen *et al.* (25), would be ideal for responding to stress levels of corticosterone in the rat. Stress levels of corticosterone in the laboratory rat are 20–50 times higher than in *Taricha*. Assuming that only 10% of circulating corticosterone is free to cross the blood–brain barrier, neurons would be exposed to 150 nM corticosterone during a typical stress response in the rat (31). This level of free corticosterone would occupy 55% of neuronal membrane receptors having a K_d of 120 nM, thereby allowing the brain to respond effectively to widely varying corticosterone concentrations.

One should be aware of the potential confound posed by the interaction of plasma steroid-binding proteins with neuronal membranes. We have found that [³H]corticosterone binds with high affinity (6.5 nM) to extensively washed prairie vole brain membranes (33). The high-affinity binding component is eliminated when brains are saline perfused prior to decapitation, and high-affinity

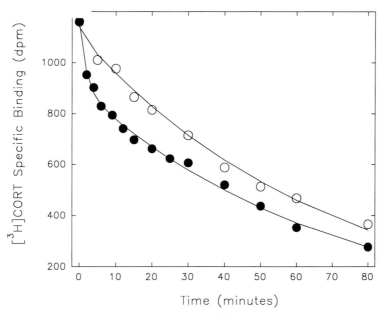

FIG. 7   Effect of GTPγS on dissociation of [³H]corticosterone. *Taricha* neuronal membranes were equilibrated with 0.75 n*M* [³H]CORT in buffer with 10 m*M* MgCl₂ for 2 hr at 30°C, at which time dissociation was initiated by the addition of 25 μl of unlabeled CORT (○) or CORT plus 1 m*M* GTPγS (●). For CORT alone, dissociation was monophasic, with $k_{-1} = 0.014 \pm 0.0006$ min⁻¹, similar to results shown in Fig. 2B. The addition of GTPγS plus CORT resulted in biphasic dissociation rates, with an initial dissociation with $k_{-1} = 0.515 \pm 0.134$ min⁻¹, and a terminal dissociation with $k_{-1} = 0.015 \pm 0.0004$ min⁻¹. [Reproduced from Orchinik *et al.* (15), with permission of National Academy of Science.]

binding is restored when plasma is added to the incubate. Because these membranes were subjected to numerous, prolonged washes with large volumes of buffer, it appears that plasma CBGs bind to nonperfused brain membranes during tissue homogenization. Once bound, the CBGs must dissociate slowly, allowing [³H]corticosterone to bind with high affinity to the membrane-bound CBG. Consistent with this idea, we found that [³H]corticosterone binds with similar affinity and specificity to vole brain membranes as it does to a preparation containing diluted vole plasma but no brain membranes. Also supporting this conclusion, iodinated steroid-binding globulins have been shown to bind to membranes in peripheral tissue, stimulate adenylate cyclase activity, and dissociate very slowly from membranes (33, 34). However, the significance of steroid binding to membrane-bound plasma CBGs in the brain is unknown and may be purely artifactual. These data underscore the importance of applying the rigorous pharmacological methodology described in this chapter to the characterization of novel steroid receptor systems.

Acknowledgments

The authors acknowledge the invaluable participation of Frank L. Moore in all of the studies presented, as well as the significant contributions of Paul F. Franklin and James D. Rose. Research was supported by NSF Grants BNS8901500 (F. L. Moore and M.O.) and BNS8909173 (Frank L. Moore), NIH Grant NS13748 (James D. Rose), and NIMH Grant MH41256 (Bruce S. McEwen).

References

1. H. Seyle, *Proc. Soc. Exp. Biol. Med.* **46,** 116 (1941).
2. H. Seyle, *J. Lab. Clin. Med.* **27,** 1051 (1942).
3. B. S. McEwen, *Trends Pharmacol. Sci.* **12,** 141 (1991).
4. E. Costa and S. M. Paul, eds., "Neurosteroids and Brain Function," Fidia Research Foundation Symposia Series, Vol. 8. Thieme, New York, 1991.
5. D. Duval, S. Durant, and F. Homo-Delarche, *Biochim. Biophys. Acta* **737,** 409 (1983).
6. D. J. Weiss and E. Gurpide, *J. Steroid Biochem.* **31,** 671 (1988).
7. M. Schumacher, *Trends Neurosci.* **13,** 359 (1990).
8. M. Orchinik and B. S. McEwen, *Neurotransmissions* **9,** 1 (1993).
9. L. D. McCauley and K. W. Gee, this volume [13].
10. S. M. Paul and R. H. Purdy, *FASEB J.* **6,** 2311 (1992).
11. F. L. Moore and L. J. Miller, *Horm. Behav.* **18,** 400 (1984).
12. F. L. Moore and M. Orchinik, *Semin. Neurosci.* **3,** 489 (1991).
13. M. Orchinik, T. F. Murray, and F. L. Moore, *Science* **252,** 1848 (1991).
14. J. D. Rose, F. L. Moore, and M. Orchinik, *Neuroendocrinology* **57,** 815 (1993).
15. M. Orchinik, T. F. Murray, and F. L. Moore, *Brain Res.* **646,** 258 (1994).
16. M. Orchinik, T. F. Murray, P. H. Franklin, and F. L. Moore, *Proc. Natl. Acad. Sci. U.S.A.* **89,** 3830 (1992).
17. M. Wehling, M. Christ, and R. Gerzer, *Trends Pharmacol. Sci.* **14,** 1 (1993).
18. A. C. Towle and P. Y. Sze, *J. Steroid Biochem.* **18,** 135 (1983).
19. F.-C. Ke and V. D. Ramirez, *J. Neurochem.* **54,** 467 (1990).
20. M. D. Majewska, S. Demirgören, and E. D. London, *Eur. J. Pharmacol. Mol. Pharm. Sect.* **189,** 307 (1990).
21. S. Demirgören, M. D. Majewska, C. E. Spivak, and E. D. London, *Neuroscience* **45,** 127 (1991).
22. P. M. Rosenblum, P. W. Sorensen, N. E. Stacey, and R. E. Peter, *Chem. Senses* **16,** 143 (1991).

23. S. A. Tischkau and V. D. Ramirez, *Proc. Natl. Acad. Sci. U.S.A.* **90,** 1285 (1993).
24. P. Y. Sze and A. C. Towle, *Int. J. Dev. Neurosci.* **11,** 339 (1993).
25. Y.-Z. Chen, H. Fu, and Z. Guo, this series, Vol. 11, p. 16, 1993.
26. V. P. Whittaker, *in* "Handbook of Neurochemistry" (A. Lajtha, ed.), Vol. 2, p. 327. Plenum, New York, 1969.
27. D. H. Jones and A. I. Matus, *Biochim. Biophys. Acta* **356,** 276 (1974).
28. M. Wehling, M. Christ, and K. Theisen, *Am. J. Physiol.* **263,** E974 (1992).
29. F. L. Moore, C. S. Bradford, and M. Orchinik, *Soc. Neurosci. Abstr.* **18,** 895 (1992).
30. L. Birnbaumer, J. Abramowitz, and A. M. Brown, *Biochim. Biophys. Acta* **1031,** 163 (1990).
31. Z. Liposits, and M. C. Bohn, *J. Neurosci. Res.* **35,** 14 (1993).
32. M. F. Dallman, S. F. Akana, C. S. Cascio, D. N. Darlington, L. Jacobson, and N. Levin, *Recent Progr. Horm. Res.* **43,** 113 (1987).
33. M. Orchinik, D. M. Witt, and B. S. McEwen, *Soc. Neurosci. Abstr.* **18,** 821 (1993); in preparation (1994).
34. W. Rosner, *Endocr. Rev.* **11,** 80 (1990).
35. O. A. Strel'chyonok and G. V. Avvakumov, *J. Steroid Biochem. Mol. Biol.* **40,** 795 (1991).
36. R. Feyereisen, G. D. Baldridge, and D. E. Farnsworth, *Comp. Biochem. Physiol. B* **82B,** 559 (1985).

[8] *In Vitro* Autoradiography for Steroid Receptors

Neil J. MacLusky, He Yuan, Debbie Bowlby, Richard B. Hochberg, and Theodore J. Brown

Introduction

For more than two decades, autoradiography has remained one of the principal methods for the study of the distribution of steroid receptors (1). In the brain, autoradiographic studies of steroid binding have almost exclusively used the approach of infusing labeled ligands into the circulation of living animals, allowing a short period of time to elapse to let the majority of free and nonspecifically bound steroid to be cleared from the tissues, then rapidly freezing the brain for sectioning in a cryostat. Although there are many theoretical and practical disadvantages to this approach, it was widely used because until recently no viable alternatives were available. The problems can be summarized as follows: large amounts of isotope are required to occupy the receptors *in vivo;* metabolism of the labeled steroids may occur, so that the observed cellular uptake of radioactivity may not in fact represent the injected compound; and, most importantly, because of competition from endogenous steroid hormones, it is impossible to interpret *in vivo* autoradiographic studies in quantitative terms unless the principal steroid-producing organs, the gonads and adrenals, are removed prior to the experiment.

Theoretically, at least some of these problems could be avoided if the techniques of *in vitro* autoradiography pioneered by Kuhar *et al.* (2) for neurotransmitter and neuropeptide receptor systems could be adapted for use with steroids. *In vitro* autoradiography entails incubation of tissue sections with labeled ligand, obviating the need to inject isotope into the animal. The sections are then washed to remove free ligand prior to exposure to photographic emulsion. Unfortunately, application of this methodology to steroids presents particularly difficult technical problems. In contrast to the membrane receptors for which *in vitro* autoradiography was originally developed, steroid receptors are relatively unstable proteins, loosely associated (in their unoccupied state) with the cell nucleus from which they are readily solubilized after freeze-thaw-induced disruption of the cell. The majority of unoccupied steroid receptors in tissue sections do not survive *in vitro* incubation (reviewed in Ref. 3). Furthermore, the lipophilic nature of steroid hormones makes it difficult to wash out free and nonspecifically

Methods in Neurosciences, Volume 22

bound ligand, so that the small amounts of specific receptor-bound steroid present may be undetectable against background nonspecific labeling. Because of these problems, section-based histochemical steroid receptor assays have never been fully accepted, even in nonneural tissues (4, 5).

In the central nervous system (CNS), numerous attempts have been made to adapt *in vitro* labeling techniques to localization of steroid receptors. As early as 1973, McEwen and Wallach (6) demonstrated *in vitro* nuclear labeling of 300-μm-thick rat hippocampal slices with [^{3}H]corticosterone. In the 1980s, Sarrieau *et al.* (7, 8) and Reul and de Kloet (9) reported successful autoradiographic measurement of rat brain glucocorticoid receptors using cryostat tissue sections and ^{3}H-labeled steroids. Apart from some pioneering studies by Vann *et al.* (10), who demonstrated the feasibility of applying *in vitro* autoradiography to the detection of estrogen receptors in the brain, very little additional developmental work on *in vitro* autoradiography for steroid receptor systems in the CNS has been reported in the last 7 years. Three factors have contributed to this slow progress. First, the advent during the 1980s of specific antibodies against the receptor proteins dramatically altered the way in which investigators could approach localization of steroid receptors in target tissues. Autoradiography, with its requirement for use of radiolabeled steroids that could be detected (at least using the tritiated ligands then available) only after extended exposure periods of months or years (1), was no longer the only method available for localization of steroid receptors at the cellular level. Second, autoradiography depends on the ability to get the radiolabeled steroid to the ligand-binding domain of the receptor protein. As mentioned previously with respect to *in vivo* labeling, if this domain is already occupied by endogenous steroids, isotopic labeling of the receptors may be difficult or impossible. This problem cannot be obviated simply by carrying out the labeling reaction *in vitro:* conditions must be utilized that will ensure complete dissociation of the occupied receptor complexes, so that the receptor can be quantitatively relabeled with isotope. This may take some time, because of the relatively slow dissociation rate constants for occupied steroid receptor complexes. Finally, the technical problems of *in vitro* autoradiography for steroid hormones are particularly severe in the brain. Steroid receptor concentrations are relatively low, and nonspecific binding to lipophilic macromolecules is higher than in many nonneural tissues.

Nevertheless, there are sound theoretical and practical reasons for continued development of autoradiographic procedures. Immunocytochemistry is much more difficult to quantitate precisely than an autoradiographic image. Although antibody-based methods should theoretically circumvent the problems associated with receptor occupation by endogenous steroids, in fact problems of differential detection of occupied and unoccupied receptors remain even with immunocytochemistry. Changes in the conformational state of the receptor after hormone-induced receptor activation may make the antigenic sites on the molecule more or less accessible, resulting in apparent changes in immunostaining intensity that have nothing to do with actual tissue receptor concentrations (11–13). Ligand binding-based methods, by contrast, can be made absolutely quantitative and, so long as the available receptor steroid-binding sites can be stoichiometrically labeled, they are relatively unaffected by the molecular configuration of the receptor. Study of the kinetic parameters of the steroid-binding reaction may yield insights into receptor function. Moreover, ligand-binding methods offer the opportunity to address directly the issue of the extent to which the receptors are occupied under different physiological conditions. Because occupied receptor complexes must dissociate before they can be labeled, selective measurement of occupied and unoccupied receptor populations is theoretically possible using different incubation conditions designed either to prevent or encourage receptor complex dissociation. This approach has been widely utilized in biochemical "exchange" assays for measurement of occupied steroid receptors (14–16), but such biochemical methods lack sufficient resolution for studies in the brain, even when combined with microdissection tissue sampling techniques (17). Although conventional autoradiographic techniques have the necessary resolution, they cannot be used to define regional patterns of receptor occupation. *In vivo* autoradiographic attempts to study receptor occupation indirectly by measuring the suppression of ^{3}H-labeled ligand uptake in intact or steroid-treated, as compared to gonadectomized, animals (see, e.g., Refs. 18–23) must be interpreted cautiously, because they are dependent on the unproved and almost certainly unwarranted assumption that total receptor concentrations in the brain do not change with physiological state.

These considerations led us to reevaluate the potential of *in vitro* autoradiography as a technique for quantitation of occupied and unoccupied steroid receptors in the central nervous system. Two critical breakthroughs facilitated the development of the modified techniques now in use in our laboratories. The first was the development of new, high-affinity iodinated steroid receptor ligands of high specific activity, allowing autoradiographic exposure times to be reduced to hours or days, instead of months (24–27). The second was the recognition that labeling of nuclear-bound steroid–receptor complexes might be possible in tissue sections, so long as sufficiently efficient washing procedures could be developed to facilitate removal of nonspecifically bound steroid. This chapter briefly reviews application of the available *in vitro* steroid autoradiographic techniques to the brain, concluding with a discussion of possible future directions in this now rapidly developing field.

Estrogen Receptors

Previous work from our laboratories has extensively explored combinations of conventional biochemical assay procedures with microdissection in order to achieve sufficient anatomical specificity in studies of steroid receptor distribution in the brain (17, 28–33). Although these methods provide a means of unequivocally measuring steroid receptor concentrations, and in some cases the extent of nuclear receptor occupation, in different brain regions, they suffer from serious drawbacks. In addition to being arduous and requiring considerable technical skill if they are to be reproducible, microdissection procedures are severely limited by the requirement to obtain at least a reasonable amount of tissue if the biochemical receptor assay is to be successful. In the case of estrogen receptors, it is difficult to measure receptor occupation resulting from physiological levels of circulating gonadal steroids in microdissected brain regions, and more detailed information about the binding characteristics (equilibrium binding affinity, binding capacity, or concentration) of the receptors can be obtained only by pooling tissues from relatively large numbers of animals (31). In some regions of the brain, precise dissection is essentially impossible because the structures involved are too diffuse or irregularly shaped to allow them to be delineated reliably by freehand dissection. Moreover, in some cases microdissection is simply too crude to allow correlations to be made between receptor occupation and behavioral or neuroendocrine responses, because the steroid target structures involved are too small to be identified and dissected in unfixed tissue (17).

During the course of studies aimed at improving the specificity and sensitivity of microdissection-based receptor assays, we observed that quantitative measurement of nuclear-bound estrogen receptors could be achieved in relatively crude subcellular fractions, containing other cellular organelles in addition to cell nuclei, so long as the homogenization buffer volume was relatively large in comparison to the tissue weight (17). One interpretation of these findings was that, at extreme buffer dilutions, protease activity was reduced sufficiently to allow estrogen receptors to survive during the exchange incubation step. This led to the idea that exchange labeling of occupied estrogen receptors might in fact be possible even without subcellular fractionation, using whole tissue sections. Because it was already known that unoccupied receptors are easily eluted from tissue sections into hypotonic buffer (3) it seemed reasonable to suppose that incubation of cryostat sections through the brain with radiolabeled estrogen under exchange conditions would lead to selective localization of receptor complexes that had been occupied at the time of death. Consistent with this hypothesis, attempts to label estrogen receptors in cryostat sections from the brains of ovariecto-

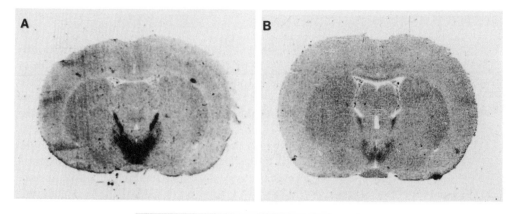

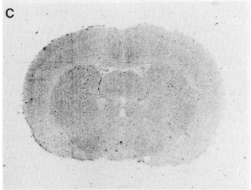

FIG. 1   Localization of [^{125}I]MIE$_2$ uptake in tissue sections from the brains of OVX rats. Cryostat sections (20 μm) through the preoptic area were incubated with [^{125}I]MIE$_2$ and then washed to remove low-affinity bound radioactivity as described in Table I. Sections were exposed against Hyperfilm-^{3}H (Amersham) for 24 hr to generate autoradiograms. In sections from animals injected with unlabeled E$_2$ (36 μg/kg) 1 hr before sacrifice (A), dense accumulations of silver grains are apparent over the periventricular and medial preoptic area and the bed nucleus of the stria terminalis. Labeling in OVX animals that did not receive estrogen prior to sacrifice is much weaker (B). (C) Autoradiogram from a section adjacent to that shown in (A), incubated with [^{125}I]MIE$_2$ in the presence of excess unlabeled diethylstilbestrol (DES). Note the absence of regional concentration of radioactivity in (C). [Reproduced from Walters *et al.* (36), with permission from the *Journal of Histochemistry & Cytochemistry*, **41**, 1279 (1993).]

mized (OVX) rats using an *in vitro* labeling incubation resulted in only low levels of binding, presumably as a result of receptor elution and/or degradation (Fig. 1B). However, after *in vivo* estrogen injection strong *in vitro* regional labeling was observed (Fig. 1A), distributed in a pattern resembling

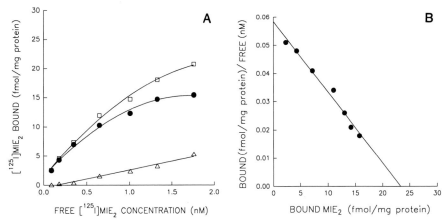

FIG. 2   Binding isotherms for labeling of the periventricular preoptic area on cryostat sections through the brain of a rat injected with unlabeled E_2 (36 μg/kg, iv) 1 hr before sacrifice. Sections were incubated for 2 hr at 37°C with varying concentrations of $[^{125}I]MIE_2$, washed, and levels of specific $[^{125}I]MIE_2$ retention measured by quantitative autoradiography. (A) Estimates of total (□), nonspecific (△), and specific (●) binding. (B) Specific binding data from (A) expressed in the form of a Scatchard (64) plot. [Reproduced from Walters *et al.* (36), with permission from the *Journal of Histochemistry & Cytochemistry,* **41,** 1279 (1993).]

that previously reported for estrogen uptake in the rat brain (31, 35). The binding sites observed under these conditions exhibited estrogen specificity (Fig. 1C) (36) and high ligand affinity (Fig. 2).

To develop a successful tissue section-based estrogen receptor exchange assay, however, it was not sufficient simply to devise incubation conditions capable of labeling the receptors. The most critical aspect of the method was the washing procedure required to reduce tissue nonspecific steroid binding sufficiently to allow visualization of the receptor-bound radioactivity. This procedure was developed by combining some of the components of the buffer washes used by Roy and McEwen (15) to prepare purified cell nuclei prior to estrogen receptor exchange assay, with several additional steps designed to stabilize the tissue sections and minimize residual free steroid. The complete procedure is summarized in Table I. After incubation of cryostat sections with the labeled ligand for 2 hr at 37°C, a length of time sufficient to effect complete exchange of bound ligand (36), the slides on which the sections are mounted are chilled to 0–4°C and the residual incubation buffer is decanted. The sections are then washed sequentially in buffer (5 min), buffered 4% (wt/v) paraformaldehyde (5 min), buffer containing 0.1% (v/v) Triton X-100 (three times, 5 min each), and buffer without Triton X-100 for the final wash (twice, 5 min each). After a brief rinse in ice-cold distilled water,

TABLE I   *In Vitro* Autoradiographic Procedure for Exchange Labeling of Brain
Estrogen Receptors[a]

1. Cryostat sections (10–20 μm thick) cut, mounted on subbed glass slides, and stored at
   −80°C until use

Incubation procedure

2. Slide-mounted sections transferred to desiccator, to equilibrate to 0–4°C
3. Sections covered with ice-cold incubation buffer (100 μl/section) containing labeled
   steroid. Adjacent, control sections, used in assessing nonspecific binding, contain 1 μM
   diethylstilbestrol (DES) in addition to the labeled steroid
4. Sections incubated for 2 hr in a humidified incubator preheated to 37°C, then returned
   to the cold room. All subsequent procedures are performed at 0–4°C

Washing procedure

5. Slides loaded into a slide rack and rinsed in cold, circulating PM buffer for 5 min, then
   postfixed using 4% paraformaldehyde in 0.1 *M* phosphate buffer for 5 min
6. Sections washed three times (5 min each) in cold, circulating PMTx buffer, followed by
   two 5-min washes in PM buffer
7. Sections dipped in double-distilled water and fan dried overnight. Exposure times
   against LKB Ultrofilm or Amersham Hyperfilm-³H 18–36 hr using iodinated ligands;
   3–4 months using tritiated ligands

Buffers

Incubation	3 mM MgCl$_2$, 1 mM KH$_2$PO$_4$, 0.32 M sucrose, pH 6.8; 1 mM dithiothreitol, 0.5 mM bacitracin added on day of use
PM	3 mM MgCl$_2$, 1 mM KH$_2$PO$_4$, pH 6.8
PMTx	PM buffer with addition of 0.1% (v/v) Triton X-100, pH 6.8

[a] Adapted from Walters *et al.* (36).

the sections are air dried in the cold and placed against autoradiography film
for exposure. Each of these steps contributes in an important way to the
overall success of the method. The initial wash immediately before the paraformaldehyde fixation step removes the majority of unbound steroid from
the sections. Paraformaldehyde fixation improves the mechanical stability
of the sections throughout the subsequent washing procedures, and also
reduces nonspecific binding. Omission of the paraformaldehyde fixation step
has no significant effect on specific labeling in the tissue sections, but leaves
a higher background and the sections are more prone to mechanical damage.
The Triton X-100 buffer washes remove the great majority of residual nonspecifically labeled steroid. The final double rinse in distilled water is necessary
to eliminate residual traces of buffer salts that otherwise produce chemogra-
phy in contact with the uncoated film used for the autoradiographic exposure.
    This technique can be applied equally well to tritiated as well as iodinated
estrogen ligands. Thus, excellent labeling patterns are observed after incuba-

tion of tissue sections with [³H]estradiol or [³H]moxestrol (Fig. 3), comparable to those obtained using the ¹²⁵I-labeled synthetic estrogen, 11β-methoxy-16α-iodoestradiol (MIE₂). However, much longer exposure times are required to obtain usable images with tritiated ligands, on the order of 3–4 months, as compared to less than 24 hr using [¹²⁵I]MIE₂. The only ligand so far tested that appears to be unusable is the iodinated estrogen 16α-[¹²⁵I]iodoestradiol (37), which is commercially available but gives unacceptably high nonspecific labeling backgrounds (36).

Quantitation of images can be achieved in the same way as for other *in vitro* autoradiographic methods, using either commercially available radiolabeled plastic film standards [e.g., Amersham (Arlington Heights, IL) microscales] or brain "paste" standards (2). We have preferred to use standards that more closely approximate the experimental material. Sections through the same brain regions as the experimental sections are incubated for a short time with a range of concentrations of radiolabeled isotope, then briefly buffer-washed to remove excess free steroid. This results in a series of sections in which nonspecific labeling predominates. Because nonspecific labeling is essentially constant throughout the brain and is directly related to free steroid concentration, this results in evenly labeled standard sections that can be exposed alongside the experimental tissue. After exposure, the standard sections are scraped from the slides, and the radioactivity in them is determined and used to construct standard curves for each experiment relating autoradiographic film density to tissue isotope concentrations (31).

Although the data from OVX animals suggested that relatively little unoccupied estrogen receptor survives the *in vitro* autoradiographic procedure, low levels of binding were still apparent in the brains from some OVX animals (36) (Fig. 1). There are two possible explanations for this residual binding: either there may be low levels of estrogen in OVX rats, sufficient to occupy some of the available estrogen receptor sites; or, alternatively, a small fraction of the unoccupied sites present in the tissue might have survived in the sections. The latter possibility is favored by the observations that adrenalectomy did not further reduce the levels of binding observed in OVX animals (T. J. Brown, unpublished observations); and that labeling at least as strong as that observed after incubation for 2 hr is observed in sections from OVX rats incubated for only 10 min at 37°C (Fig. 4). Because this time period is too short to achieve significant displacement of estradiol (E₂) from occupied receptor complexes (Fig. 4), this suggests that the low level of binding sites observed in OVX animals probably represents residual unoccupied receptors.

The high degree of selectivity for occupied estrogen receptors of the *in vitro* autoradiographic procedure summarized in Table I provides the opportunity for regional mapping of occupied estrogen receptors in the brain under

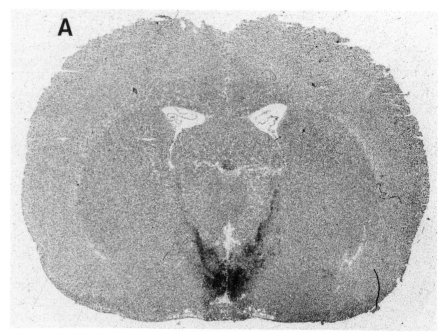

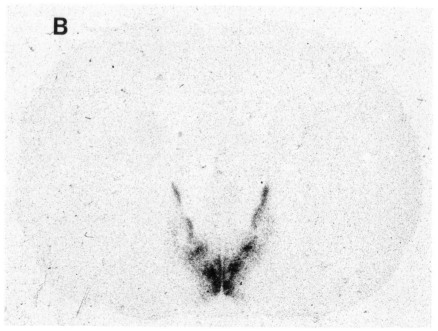

different physiological conditions. An example of such an application of the technique is presented in Fig. 5, which shows the regional pattern of estrogen receptor occupation in female rats at different stages of the estrous cycle, in comparison to ovariectomized rats. Higher levels of binding are clearly observed at proestrus, coincident with the elevated circulating estrogen levels observed at this stage of the cycle (38) and consistent with previous biochemical studies indicating that estrogen receptor occupation in the brain increases markedly at proestrus (39).

A second major advantage of this method is that, for the first time, it becomes possible to ask questions about the regional control of estrogen receptor concentrations in the brain without having to remove the gonads. The only requirement is that the tissue must be exposed to saturating concentrations of estrogen a short time before death, to convert the receptors into their occupied, tightly bound form. The issue of whether estrogen receptor concentrations in the brain are subject to physiological regulation has never been satisfactorily resolved. Previous biochemical studies have suggested that estrogen receptor concentrations in the brain are relatively constant, remaining virtually unchanged after gonadectomy (40, 41), in contrast to the situation in peripheral estrogen target tissues such as the uterus, in which estrogen receptor concentrations decline markedly after removal of the ovaries (41, 42). A direct test of the hypothesis that central estrogen receptor concentrations are unaffected by ovarian steroid hormone concentrations, however, was not possible, because of the difficulties inherent in simultaneously measuring occupied and unoccupied receptor populations. *In situ* hybridization studies showing changes in the levels of estrogen receptor mRNA in the brain after exposure to estrogen have raised the possibility that the assumption of stability in brain estrogen receptor concentrations in different physiological states may, in fact, be erroneous. Some studies have demonstrated a negative effect of estrogen treatment on estrogen receptor mRNA expression in the hypothalamus (43, 44). Shughrue *et al.* (45) suggested that the effects of rising endogenous estrogen levels during the estrous cycle on brain estrogen receptor mRNA concentrations may be region specific. Thus, in the preoptic area estrogen receptor mRNA levels decline during proestrus and increase following ovariectomy, whereas in the arcuate and ventromedial hypothalamic nuclei mRNA concentrations increase between estrus and pro-

---

FIG. 3   Sections through the preoptic area of a rat injected 1 hr prior to sacrifice with unlabeled E_2 (36 μg/kg, iv), then incubated *in vitro* for 2 hr at 37°C with either [^{3}H]E_2 (A) or [^{3}H]moxestrol (B). Sections were exposed against Hyperfilm-^{3}H for 3 months. [Reproduced from Walters *et al.* (36), with permission from the *Journal of Histochemistry & Cytochemistry*, **41**, 1279 (1993).]

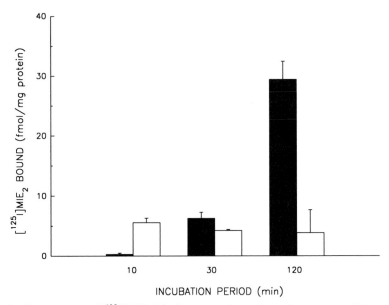

FIG. 4   Time course of $[^{125}I]MIE_2$ binding to the hypothalamic ventromedial nucleus in sections from OVX (open bars) or OVX–E_2 injected (36 $\mu g/kg$, iv, 1 hr before death; solid bars) rats. Sections were incubated with $[^{125}I]MIE_2$ at 37°C for 10, 30, or 120 min, then washed as described in Table I. After autoradiographic exposure (24 hr) against Hyperfilm-^{3}H, binding was assessed by computer-assisted densitometry (H. Yuan, T. J. Brown, and N. J. MacLusky, unpublished observations).

estrus. Our data suggest that changes in tissue estrogen receptor concentrations do occur in the brain during the estrous cycle, but that these changes may not precisely follow the changes reported at the mRNA level. Figure 6 depicts the results of the *in situ* hybridization studies presented by Shughrue *et al.* (45), in comparison to the levels of estrogen receptor at the metestrus and proestrus stages of the rat estrous cycle in different regions of the rat brain determined by *in vitro* autoradiography. In all regions of the brain examined, total estrogen binding is slightly lower during proestrus than at metestrus. In percentage terms, the decline in binding in the arcuate and ventromedial nuclei appears to be similar to that in the preoptic area (~15–20%), despite the fact that at the estrogen receptor mRNA level the changes for these three cell groups as determined by Shughrue *et al.* (45) are obviously different. The reason for this apparent disparity remains to be established, but it may reflect contributions from translational or posttranslational factors in the control of estrogen receptor synthesis. In this context, it is noteworthy that previous studies in developing animals have also suggested that there may not be a simple relationship between cellular estrogen receptor mRNA levels as assessed by *in situ* hybridization and estrogen receptor binding in different regions of the brain (46). Whatever the explanation, taken together these data reinforce the necessity for studies of the regional control of estrogen receptor synthesis in the brain to evaluate changes at both the mRNA and protein levels.

Selective Measurement of Unoccupied Estrogen Receptors

The data in Fig. 4 raise the possibility of using a short incubation period to label unoccupied estrogen receptors selectively. However, such an approach would not be feasible using untreated cryostat sections as a result of losses of unoccupied receptors during incubation. Theoretically, because occupied receptor complexes result from binding of the steroid to the unoccupied sites, in OVX rats levels of estrogen binding in the brain should be at least as high as those observed 1 hr after *in vivo* estrogen administration. In fact, as the data in Fig. 4 indicate, using the standard *in vitro* exchange method outlined in Table I, binding in sections from uninjected OVX rats is less than 20% of that observed in estradiol-injected animals. We therefore investigated the possibility of treating cryostat sections chemically to trap the unoccupied receptors so that they are immobilized immediately on rehydration of the tissue sections, to allow measurement of all the estrogen receptors present, whether occupied or unoccupied. Attempts to use nonaqueous tissue fixatives, such as acetone or ethanol, to achieve this goal were unsuccessful. The alternative route was to use agents to complex with and precipitate the receptors. Previous biochemical assays for steroid receptors have utilized many different techniques to immobilize steroid receptors to allow their physical separation from free steroid. These include precipitation methods as well as adsorption onto a variety of different insoluble matrices. One such method utilized incubation with the polybasic peptide, protamine, which rapidly precipitates steroid receptors by complexing with acidic domains on their surfaces (47, 48). The simple expedient of adding protamine sulfate to the incubation buffer used for *in vitro* autoradiography results in quantitative recovery of unoccupied estrogen receptors in brain sections from OVX rats. Sections from OVX animals incubated with [^{125}I]MIE$_2$ for 15 min at 37°C give specific labeling patterns virtually identical to those obtained from estradiol-injected rats after a 2-hr incubation at the same temperature (Fig. 7). The only problem occasionally noted with protamine incorporation is a slight increase in nonspecific myelin labeling.

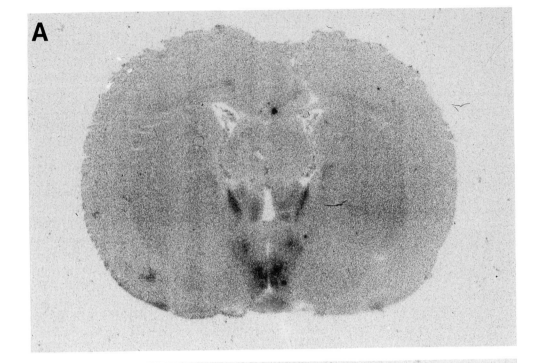

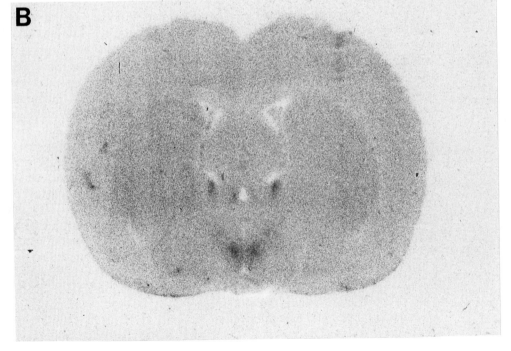

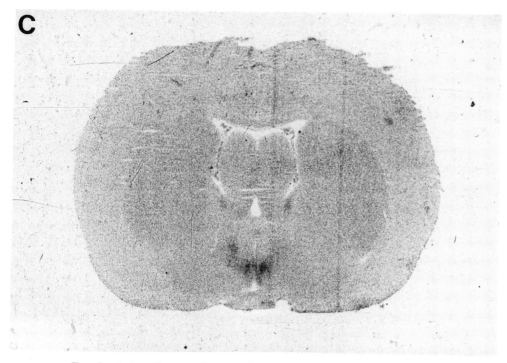

Fɪɢ. 5 Autoradiograms from sections through the preoptic area of untreated normal adult cycling female rats at either proestrus (A) or metestrus (B) compared to the same region of the brain from an OVX rat (C). Sections were labeled *in vitro* with $[^{125}I]MIE_2$, as described in Table I, then exposed against Hyperfilm-^3H for 24 hr. [Reproduced from Walters *et al.* (36), with permission from the *Journal of Histochemistry & Cytochemistry*, **41**, 1279 (1993).]

Corticosteroid, Androgen, and Progestin Receptors

Until recently, the only central nervous steroid receptors for which practical *in vitro* autoradiographic methods had been developed were the corticosteroid receptors. Sarrieau *et al.* demonstrated with unfixed rat (8, 49, 50) and human (51) brain cryostat sections that measurable regional labeling could be observed after incubation of the sections with $[^3H]$glucocorticoids. In addition, by making use of the affinity of the tritiated antiprogestin, RU 486, for both glucocorticoid and progestin receptors, the same group of workers reported labeling of progestin-specific binding sites in human brain after selective competitive inhibition of the binding of ^3H-labeled RU 486 to glucocorticoid receptors (51). Similarly, Reul and de Kloet (9) found that selective measurement of type I (mineralocorticoid) or type II (glucocorticoid) cortico-

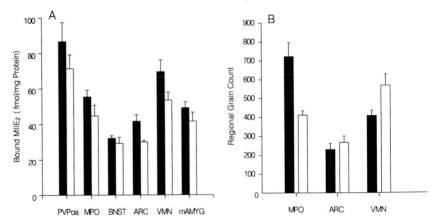

FIG. 6 Comparison of mRNA levels (B) for the estrogen receptor in the hypothalamus and preoptic area, measured by *in situ* hybridization, compared to levels of estrogen binding (A) measured by *in vitro* autoradiography in intact female rats at the metestrus (filled bars) and proestrus (open bars) phases of the estrous cycle. The mRNA data are drawn from the work of Shughrue *et al.* (45). The cycling female rats for the *in vitro* autoradiographic receptor measurements were injected with unlabeled E_2 (36 μg/kg, iv) 1 hr prior to death to occupy the available estrogen receptor sites maximally (D. Bowlby, T. J. Brown, and N. J. MacLusky, unpublished observations). PVPoa, Periventricular preoptic area; MPO, medial preoptic area; BNST, bed nucleus of the stria terminalis; ARC, arcuate nucleus; VMN, hypothalamic ventromedial nucleus; mAMYG, medial nucleus of the amygdala.

steroid receptors in the brain could be achieved by *in vitro* autoradiography in combination with appropriate specific labeled and unlabeled ligands. The procedures used by these workers are summarized in Table II.

Although these methods have been extensively validated, they remain problematic in one respect. Utilizing tritiated ligands, long autoradiographic exposures are required—on the order of 6–12 weeks for binding of [^{3}H]corticosterone to rat brain sections. The need for such long exposures is puzzling, in view of the fact that biochemical studies have demonstrated that concentrations of corticosterone receptors in the brain are at least 10 times higher than those of estrogen receptors. Because our studies (36) (Fig. 3) have shown that estrogen receptors can be visualized using tritiated ligands after a 3-month exposure, we postulated that the procedures in Table II might not recover all of the receptors present. In particular, it seemed possible that, by analogy with the situation for unoccupied estrogen receptors, losses of unoccupied corticosteroid receptor might occur during incubation through receptor solubilization. Because protamine has also been shown to precipi-

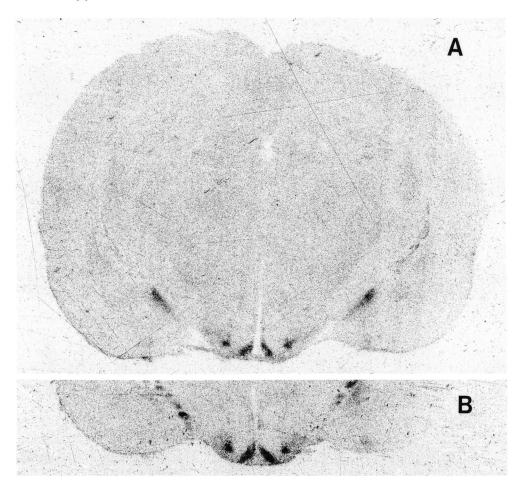

FIG. 7 Autoradiographic localization of estrogen-binding sites in the mediobasal hypothalamus and medial amygdala of OVX rats. (A) Section from an OVX rat injected 1 hr prior to sacrifice with estradiol (36 μg/kg, iv): labeling was performed by incubation with 2 nM [^{125}I]MIE$_2$ for 2 hr at 37°C. (B) Section from an uninjected OVX rat. Incubation was performed with 2 nM [^{125}I]MIE$_2$ for 15 min at 37°C, in the presence of protamine sulfate (1 mg/ml). Compare this result with the sections from OVX rats incubated without protamine sulfate depicted in Figs. 1 and 5. Both sets of sections were washed after incubation, using the procedure shown in Table I.

TABLE II   *In Vitro* Autoradiographic Procedure for Labeling of Brain
Corticosteroid Receptors[a]

---

1. Cryostat sections (10–20 μm thick) cut, mounted on subbed glass slides, and stored at $-80°C$ until use

Incubation procedure

2. Slide-mounted sections transferred to desiccator, to equilibrate to 0–4°C
3. Sections covered with ice-cold incubation buffer (100 μl/section) containing appropriate labeled steroid. Adjacent, control sections, used in assessing nonspecific binding, contain excess unlabeled corticosteroid in addition to the labeled steroid
4. Sections incubated for 20–30 min at room temperature, then returned to the cold room. All subsequent procedures are performed at 0–4°C

Washing procedure

5. Slides loaded into a slide rack and rinsed three times (5 min each) in cold, circulating Tris-HCl buffer
6. Sections dipped in double-distilled water and fan dried overnight. Exposure time against LKB Ultrofilm or Amersham Hyperfilm-^{3}H, from 2–6 weeks (9) to up to 5 months (7, 8) using tritiated ligands

---

Buffers

Tris-HCl	50 mM Tris-HCl, pH 7.4
Incubation	Tris-HCl buffer; plus 5% (v/v) glycerol, 2 mM EGTA, 10 mM dithiothreitol, 6 mM sodium molybdate, and 5 mM ATP

---

[a] Data from Sarrieau *et al.* (7, 8) and Reul and de Kloet (9).

tate corticosteroid receptors quantitatively from rat brain cytosol fractions (48), it seemed possible that improved labeling of these receptors might also be achieved in the presence of protamine. We therefore evaluated the effect of adding protamine sulfate (1 mg/ml) to the incubation buffer used by Sarrieau *et al.* (8, 49), with the additional minor modification of omitting ATP from the incubation buffer, to obviate the coprecipitation of ATP and protamine that otherwise occurs.

Protamine dramatically improves the labeling efficiency of rat brain sections incubated with [^{3}H]dexamethasone. Figure 8 shows an autoradiogram from a 20-μm section through the brain of an ovariectomized–adrenalectomized rat, labeled *in vitro* with 5 nM [^{3}H]dexamethasone (43.9 Ci/mmol) in the presence of protamine sulfate (1 mg/ml). After incubation, the sections were washed as described above, then dried and exposed against Amersham Hyperfilm-^{3}H for 21 days. Control sections for nonspecific binding contained 5 μM unlabeled dexamethasone in addition to the labeled steroid. Autoradiograms from these control sections are not shown: under the conditions used, nonspecific labeling was not detectable above the film background. Specific binding after incubation with [^{3}H]dexamethasone alone (Fig. 8A) is well

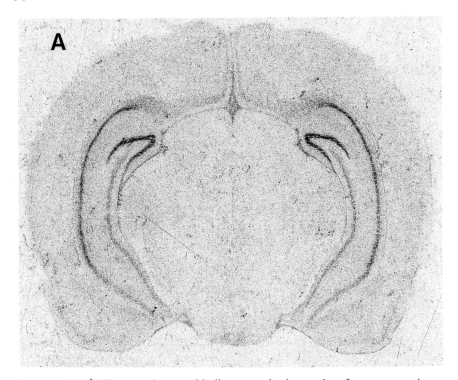

FIG. 8 (A) [³H]Dexamethasone binding to a brain section from an ovariecto-mized–adrenalectomized rat, with incorporation of protamine sulfate into the incubation buffer and first wash buffer of the procedure summarized in Table II. The section was exposed for 21 days against Amersham Hyperfilm-³H. (B) Binding measured by liquid scintillation counting of serial sections through an adrenalectomized rat brain incubated either under the conditions outlined in Table II ("−protamine, +ATP"), under the same conditions but with omission of ATP ("−protamine, −ATP"), or omitting ATP but adding protamine sulfate (1 mg/ml) to the incubation buffer ("+protamine, −ATP"). Results represent the means ±SEM of six to eight observations and are expressed as a percentage of the binding observed in the presence of protamine. (*) $p < 0.05$ versus +protamine, −ATP group; Mann–Whitney U test.

defined, with regional localization of radioactivity distributed in a manner resembling that previously reported for glucocorticoid receptors in the rat brain (52). By contrast, labeling in sections incubated without protamine is considerably reduced, whether or not ATP is present in the incubation buffer (Fig. 8B). Further work will be necessary to determine whether protamine addition has a similar beneficial effect on the binding of other corticosteroid receptor ligands in the central nervous system.

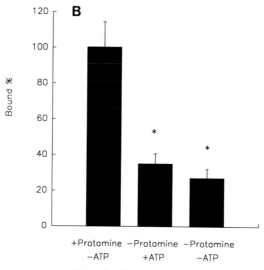

FIG. 8   *(Continued)*

By comparison, *in vitro* autoradiography for androgen receptors is far less well developed. Constraints include a lack of radiolabeled ligands that are both selective for the androgen receptor and resistant to metabolism, as well as the inherent lability of the androgen receptor itself. Tritiated methyltrieno-lone (R 1881), currently the most widely used androgen receptor radioligand, is resistant to metabolism; however, it also binds with high affinity to the progestin receptor. With biochemical assays, this problem is easily avoided by conducting incubations in the presence of triamcinolone acetonide (TA), a synthetic glucocorticoid that binds the glucocorticoid and progestin recep-tor but has low affinity for the androgen receptor (53). Thus ^{3}H-labeled R 1881 binding to the progestin receptor is effectively masked without affect-ing binding to the androgen receptor. With conventional *in vivo* autoradio-graphic procedures, however, complete and selective masking of the proges-tin receptor cannot be completely assured.

An *in vitro* autoradiographic method using ^{3}H-labeled R 1881 for the detec-tion of androgen binding in tissue sections was introduced by Peters and Barrack (54) for prostate tissue sections. Although this method allows high-resolution localization of the receptors, and has the advantage of being usable with paraformaldehyde-fixed tissue, it does not distinguish between occupied and unoccupied receptor. By combining the basic principles of our method for occupied estrogen receptors (36) with the method of Peters and Barrack (54), we developed a procedure that allows assay of occupied androgen

TABLE III *In Vitro* Autoradiographic Procedure for Labeling of Androgen Receptors[a]

---

1. Cryostat sections (10–20 μm thick) cut, mounted on subbed glass slides, and stored at $-80°C$ until use

Incubation procedure

2. Slide-mounted sections transferred to desiccator, to equilibrate to 0–4°C
3. Sections covered with ice-cold incubation buffer (100 μl/section) containing labeled steroid (^{3}H-labeled R 1881). Adjacent, control sections, used in assessing nonspecific binding, contain excess unlabeled androgen (DHT or R 1881) in addition to the labeled steroid
4. Sections incubated for 72 hr at 4°C. All subsequent procedures are performed at 0–4°C

Washing procedure

5. Slides loaded into a slide rack and rinsed in cold, circulating PM buffer for 5 min
6. Sections washed twice (5 min each) in cold, circulating PMTx buffer, followed by two 5-min washes in PM buffer
7. Sections dipped in double-distilled water and fan dried overnight. Exposure time against LKB Ultrofilm or Amersham Hyperfilm-^{3}H, 12 weeks using tritiated ligands

---

Buffers

Incubation	1.0 mM KH$_2$PO$_4$, 3.0 mM MgCl$_2$, 10% (v/v) glycerol, pH 7.0; 1.0 mM phenylmethylsulfonyl fluoride, 20 mM Na$_2$MoO$_4$, 0.5 mM bacitracin, 1.0 mM dithiothreitol, 10 mM triamcinolone acetonide, and 20 mM indomethacin added on day of use
PM	3 mM MgCl$_2$, 1 mM KH$_2$PO$_4$, pH 6.8
PMTx	PM buffer with addition of 0.1% Triton X-100, pH 6.8

---

[a] Data from Brown *et al.* (34).

receptors in nonneural target tissues, using slide-mounted tissue sections (34). The assay procedure (outlined in Table III) consists of incubating unfixed cryostat sections with buffer containing ^{3}H-labeled R 1881 at 0–4°C. Sections incubated in the presence of a 200-fold molar excess of unlabeled dihydrotestosterone (DHT) are used to assess nonspecific binding. After incubation, the sections are washed through several changes of cold buffer (two changes including Triton X-100), dipped in distilled water, dried, and exposed to film.

This procedure is selective for the occupied form of the androgen receptor. Essentially no binding is measured in prostate sections from a castrated male rat whereas testosterone injection results in high levels of binding. The binding is of high affinity, to a single class binding site, and is displaceable only by androgen receptor ligands (34). Figure 9 illustrates autoradiographic images obtained using this method for ^{3}H-labeled R 1881 binding in coronal sections through the male rat brain. Labeling of androgen binding sites can

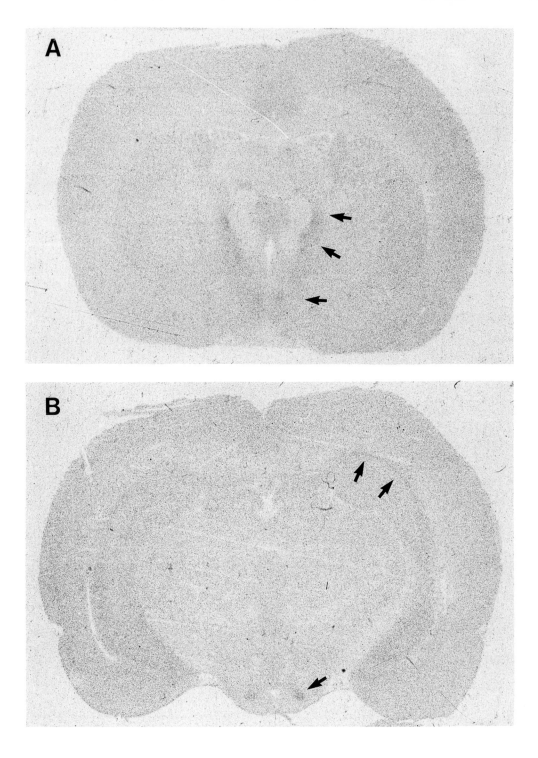

be observed in the hippocampus, mediobasal hypothalamus, septum, and bed nucleus of the stria terminalis, areas where *in vivo* autoradiographic and immunocytochemical studies have indicated the presence of androgen receptor (55, 56). This labeling is not observed in castrated animals and is displaced by the addition of excess unlabeled DHT (data not shown). However, the distinction between specific regional ^{3}H-labeled R 1881 binding and background nonspecific tissue retention of radioactivity is relatively poor (cf Figs. 7–9). Additional work will be necessary to improve the ratio of specific to nonspecific binding, as well as to evaluate other possible modifications to the technique that may enhance detection of androgen receptor binding in brain tissue.

Conclusions and Prospectus

Autoradiography remains valuable for study of the regional distribution of steroid receptors in the brain because, unlike any other procedure currently available, it provides a quantitative measure of the capacity of the receptor systems to interact with their appropriate ligands, without sacrificing regional tissue anatomy. In this sense, it combines many of the advantages of conventional biochemical and histochemical techniques. Adaptation of autoradiographic techniques to allow *in vitro* labeling of the receptors circumvents many of the problems associated with *in vivo* autoradiography. Isotope usage is drastically reduced, biochemical manipulations to allow measurement of receptor kinetic binding parameters are possible, and the problems of steroid metabolism are reduced.

The *in vitro* autoradiographic methods available at the present time, however, remain far from perfect and a considerable amount of additional work will be required before optimal procedures for all steroid receptors are available. In the case of the estrogen receptor, whereas quantitation of unoccupied receptors can be achieved using short incubation periods in the presence of protamine, occupied receptors present a more difficult problem. Although it appears to be possible to label occupied estrogen receptors in brain sections

---

FIG. 9   Localization of ^{3}H-labeled R 1881 binding in coronal sections taken from an intact male rat treated with DHT (500 μg, iv) 1 hr before sacrifice. Arrows indicate DHT displaceable binding in the medial preoptic area and bed nucleus of the stria terminalis (A), and in the ventral premammillary nucleus and CA1 region of the hippocampus (B). The procedure used is outlined in Table III. Sections were exposed for 12 weeks against Amersham Hyperfilm-^{3}H.

under "exchange" incubation conditions, even in the absence of protamine some unoccupied receptor survives in the tissue sections; thus, measurements made using the procedure summarized in Table I necessarily reflect contributions from both occupied and unoccupied receptor sites. Under conditions of high receptor occupancy (e.g., during proestrus, or after high-dose estrogen injection) the contribution from unoccupied sites is relatively small (Fig. 4); but, under physiological conditions of low endogenous estrogen secretion, measurement of occupied estrogen receptors using *in vitro* autoradiography will be confounded by residual unoccupied receptor binding. This currently remains a significant deficiency of the procedure outlined in Table I. It seems likely, however, that further modifications of this methodology will be possible in the future (e.g., preincubation or prewashing of the sections) to resolve completely the contributions from occupied and unoccupied receptor sites.

Methods for the autoradiographic detection of the other major classes of steroid receptor are less well developed. Problems still remain in the *in vitro* autoradiographic identification of androgen receptors in the brain, as a result of poor recovery of specific (receptor) binding and relatively high nonspecific labeling. Unoccupied corticosteroid receptors can be detected autoradiographically, but measurement of occupied corticosteroid receptors will require additional developmental work. As is the case with androgen receptors, the lability of the corticosteroid receptors presents difficulties that will probably necessitate carefully designed exchange labeling conditions. Autoradiographic measurement of progestin receptors in the brain may well be possible, although this has been attempted so far only in the human brain, using the antiprogestin/antiglucocorticoid RU 486 (7). Methods for autoradiographic detection of progestin receptors in peripheral target tissues have been reported (57), and because the dissociation of progesterone–receptor complexes is relatively rapid (58) there would seem to be no major theoretical problems standing in the way of an *in vitro* autoradiographic approach to localization of progestin receptors in the brain.

For all steroid receptors other than the estrogen receptor, the development of new methodology is inextricably linked to the availability of new receptor ligands. Use of tritium-labeled ligands is severely constrained by the low specific activity and low-emission energy of the isotope, effectively precluding autoradiography except under analytical circumstances in which lengthy exposure times can be tolerated. The introduction of storage-phosphor technology (59, 60) promises considerable reductions in exposure times, but this technology is expensive and the reduction in exposure times is still not sufficient to allow completion of an assay in less than a few days, using tritiated ligands. Ligands labeled with ^{125}I, on the other hand, offer drastically reduced exposure times, even using conventional autoradiographic film. The problem is that, except for estrogen receptors (25, 37), ^{125}I-labeled steroids suitable for *in vitro* autoradiography have not yet been developed, although progress is being made in this area (24, 61, 62).

One other area of potential future development should be mentioned here, even though it may represent a way in which *in vitro* autoradiography will eventually be superseded. The major disadvantage of *in vitro* autoradiography is clearly the fact that radioactivity is involved and, notwithstanding the development of new iodinated ligands, there must be an autoradiographic exposure period that makes it difficult to obtain results rapidly. Theoretically, these problems could be overcome by use of nonisotopically "tagged" ligands, which might allow much faster detection of the receptors at the end of the ligand-binding incubation, using either fluorescence or nonisotopic histochemical (e.g., avidin–biotin–peroxidase) procedures. Over the years, numerous attempts have been made to develop nonisotopic ligand-binding procedures for detection of steroid receptors in tissue sections, with varied success and a great deal of controversy (5). The problems involved are enormous. Use of inherently fluorescent ligands, such as coumestrol, 4-hydroxytamoxifen, or 12-oxoestradiol, is constrained by the fact that image intensification techniques are required to achieve sufficient sensitivity because the fluorescence emission intensity of these ligands is not sufficient to allow detection using fluorescence microscopy alone (5), although the more recently developed tetrahydrochrysenes may be more useful in this regard (63). Modifications to the steroid molecule, through attachment of either a strongly fluorescent functional group or another easily recognized moiety, such as biotin, are likely to disrupt binding to the receptors unless some way can be found to distance the steroid from the added functional group(s). These problems have in the past precluded the development of any widely acceptable histochemical steroid-binding assay (3, 5). Nevertheless, with the *in vitro* labeling procedures presented in this and preceding (8, 9, 49) papers, it may be opportune to reassess the potential for development of such nonisotopic ligands, because if the problems of chemical synthesis could be overcome, there would seem to be no practical reason why such ligands could not be used under the same incubation conditions to achieve rapid and selective histochemical detection of steroid receptor-binding sites.

In summary, *in vitro* autoradiographic methods continue to offer a useful approach for the detection and regional quantitation of steroid receptors in the brain. Complementing immunocytochemical and *in situ* hybridization procedures, *in vitro* autoradiography provides a simple, practical method for measurement of steroid receptor concentrations in tissue sections, allowing regional correlation of rates of steroid receptor mRNA synthesis and expression of the functional receptor protein. In the specific case of the estrogen receptor, the ability to distinguish between occupied and unoccupied receptors by taking advantage of the slow apparent "on" rate for labeling of the occupied receptor population and the relative lability of unoccupied receptors in hypotonic buffers presents unique opportunities to use autoradiography to provide quantitative mapping of regional receptor occupation under physiological conditions. At least theoretically, similar approaches should be possible in future for the receptors for other steroid hormones, providing methodology with sufficiently high anatomical resolution to allow precise regional definition of patterns of steroid–receptor interaction within the brain.

Acknowledgments

Work in the authors' laboratories is supported by operating grants from the Medical Research Council of Canada (MT-11235 to T. J. B., and PG-11115 to N. J. M.) and by a grant from the USPHS (CA37799, to R. B. H.). T. J. B. is a Scholar of the Medical Research Council of Canada.

References

1. W. E. Stumpf and G. E. Duncan, this series, Vol. 3, p. 35, 1990.
2. M. J. Kuhar, E. B. De Souza, and J. R. Unnerstall, *Annu. Rev. Neurosci.* **9,** 27 (1986).
3. J. C. E. Underwood, *in* "Steroid Hormone Receptors: Their Intracellular Localization" (C. R. Clark, ed.), p. 172. VCH, New York, 1987.
4. J. C. Underwood, E. Sher, M. Reed, J. A. Eisman, and T. J. Martin, *J. Clin. Pathol.* **35,** 401 (1982).
5. G. C. Chamness, W. D. Mercer, and W. L. McGuire, *J. Histochem. Cytochem.* **28,** 792 (1980).
6. B. S. McEwen and G. Wallach, *Brain Res.* **57,** 373 (1973).
7. A. Sarrieau, M. Dussaillant, F. Agid, D. Philibert, Y. Agid, and W. Rostene, *J. Steroid Biochem.* **25,** 717 (1986).
8. A. Sarrieau, M. Vial, D. Philibert, and W. Rostene, *Eur. J. Pharmacol.* **98,** 151 (1984).
9. J. M. Reul and E. R. de Kloet, *J. Steroid Biochem.* **24,** 269 (1986).
10. V. R. Vann, T. O. Fox, D. Blank, and W. F. White, *Soc. Neurosci. Abstr.* **12,** 168.3 (1986).
11. J. D. Blaustein, *Endocrinology (Baltimore)* **132,** 1218 (1993).
12. L. L. DonCarlos, G. L. Greene, and J. I. Morrell, *Neuroendocrinology* **50,** 613 (1989).
13. T. J. Brown, N. J. MacLusky, C. Leranth, M. Shanabrough, and F. Naftolin, *Mol. Cell. Neurosci.* **1,** 58 (1990).
14. J. N. Anderson, E. J. Peck, Jr., and J. H. Clark, *Endocrinology (Baltimore)* **93,** 711 (1973).
15. E. J. Roy and B. S. McEwen, *Steroids* **30,** 757 (1977).

16. M. Y. McGinnis, P. G. Davis, M. J. Meaney, M. Singer, and B. S. McEwen, *Brain Res.* **275**, 75 (1983).

17. N. J. MacLusky, T. J. Brown, E. Jones, C. Leranth, and R. B. Hochberg, this series, Vol. 3, p. 3, 1990.

18. R. P. Michael, R. W. Bonsall, and H. D. Rees, *Endocrinology* (*Baltimore*) **118**, 1935 (1986).

19. R. W. Bonsall, D. Zumpe, and R. P. Michael, *Neuroendocrinology* **51**, 474 (1990).

20. H. D. Rees, R. W. Bonsall, and R. P. Michael, *Brain Res.* **452**, 28 (1988).

21. R. P. Michael, R. W. Bonsall, and H. D. Rees, *Neuroendocrinology* **46**, 511 (1987).

22. R. P. Michael, H. D. Rees, and R. W. Bonsall, *Brain Res.* **502**, 11 (1989).

23. H. D. Rees, R. W. Bonsall, and R. P. Michael, *Exp. Brain Res.* **63**, 67 (1986).

24. R. B. Hochberg, R. M. Hoyte, and W. Rosner, *Endocrinology* (*Baltimore*) **117**, 2550 (1985).

25. J. E. Zielinski, H. Yabuki, S. L. Pahuja, J. M. Larner, and R. B. Hochberg, *Endocrinology* (*Baltimore*) **119**, 130 (1986).

26. P. J. Shughrue, W. E. Stumpf, N. J. MacLusky, J. E. Zielinski, and R. B. Hochberg, *Endocrinology* (*Baltimore*) **126**, 1112 (1990).

27. P. J. Shughrue, W. E. Stumpf, W. Elger, P. E. Schulze, and M. Sar, *Brain Res. Dev. Brain Res.* **59**, 143 (1991).

28. J. D. Kranzler, E. E. Jones, N. J. MacLusky, H. Sakamoto, and F. Naftolin, *J. Neurochem.* **43**, 895 (1984).

29. B. Parsons, T. C. Rainbow, N. J. MacLusky, and B. S. McEwen, *J. Neurosci.* **2**, 1446 (1982).

30. T. C. Rainbow, B. Parsons, N. J. MacLusky, and B. S. McEwen, *J. Neurosci.* **2**, 1439 (1982).

31. T. J. Brown, R. B. Hochberg, J. E. Zielinski, and N. J. MacLusky, *Endocrinology* (*Baltimore*) **123**, 1761 (1988).

32. T. J. Brown, N. J. MacLusky, M. Shanabrough, and F. Naftolin, *Endocrinology* (*Baltimore*) **126**, 2965 (1990).

33. T. J. Brown, F. Naftolin, and N. J. MacLusky, *Brain Res.* **578**, 129 (1992).

34. T. J. Brown, M. Sharma, and N. J. MacLusky, *Steroids,* in press, (1994).

35. D. W. Pfaff and M. J. Keiner, *J. Comp. Neurol.* **151**, 121 (1973).

36. M. J. Walters, T. J. Brown, R. B. Hochberg, and N. J. MacLusky, *J. Histochem. Cytochem.* **41**, 1279 (1993).

37. R. H. Hochberg, *Science* **205**, 1138 (1979).

38. F. Naftolin, K. Brown-Grant, and C. Corker, *J. Endocrinol.* **53**, 17 (1972).

39. M. Y. McGinnis, L. C. Krey, N. J. MacLusky, and B. S. McEwen, *Neuroendocrinology* **33**, 158 (1981).

40. C. R. Clark, N. J. MacLusky, B. Parsons, and F. Naftolin, *Horm. Behav.* **15**, 289 (1981).

41. J. Barley, M. Ginsburg, N. J. MacLusky, I. D. Morris, and P. J. Thomas, *Brain Res.* **129**, 309 (1977).

42. J. H. Clark and E. J. Peck, "Sex Steroids: Receptors and Function." Springer-Verlag, New York, 1979.

43. R. B. Simerly and B. J. Young, *Mol. Endocrinol.* **5**, 424 (1991).

44. A. H. Lauber, C. V. Mobbs, M. Muramatsu, and D. W. Pfaff, *Endocrinology* (*Baltimore*) **129**, 3180 (1991).

45. P. J. Shughrue, C. D. Bushnell, and D. M. Dorsa, *Endocrinology (Baltimore)* **131,** 381 (1992).

46. C. D. Toran-Allerand, R. C. Miranda, R. B. Hochberg, and N. J. MacLusky, *Brain Res.* **576,** 25 (1992).

47. G. C. Chamness, K. Huff, and W. L. McGuire, *Steroids* **25,** 627 (1975).

48. B. S. McEwen, C. Magnus, and G. Wallach, *Endocrinology (Baltimore)* **90,** 217 (1972).

49. A. Sarrieau, M. Vial, B. McEwen, Y. Broer, M. Dussaillant, D. Philibert, M. Moguilewsky, and W. Rostene, *J. Steroid Biochem.* **24,** 721 (1986).

50. A. Sarrieau, M. Vial, D. Philibert, M. Moguilewsky, M. Dussaillant, B. McEwen, and W. Rostene, *J. Steroid Biochem.* **20,** 1233 (1984).

51. A. Sarrieau, M. Dussaillant, R. M. Sapolsky, D. H. Aitken, A. Olivier, S. Lal, W. H. Rostene, R. Quirion, and M. J. Meaney, *Brain Res.* **442,** 157 (1988).

52. J. M. Reul and E. R. de Kloet, *Endocrinology (Baltimore)* **117,** 2505 (1985).

53. D. T. Zava, B. Landrum, K. B. Horwitz, and W. L. McGuire, *Endocrinology (Baltimore)* **104,** 1007 (1979).

54. C. A. Peters and E. R. Barrack, *J. Histochem. Cytochem.* **35,** 755 (1987).

55. W. E. Stumpf and M. Sar, *Am. Zool.* **18,** 435 (1978).

56. M. Sar, D. B. Lubahn, F. S. French, and E. M. Wilson, *Endocrinology (Baltimore)* **127,** 3180 (1990).

57. A. F. De Goeij, H. M. Scheres, M. J. Rousch, G. G. Hondius, and F. T. Bosman, *J. Steroid Biochem.* **29,** 465 (1988).

58. M. T. Vu Hai, F. Logeat, and E. Milgrom, *J. Endocrinol.* **76,** 43 (1978).

59. M. J. Kuhar, D. G. Lloyd, N. Appel, and H. L. Loats, *J. Chem. Neuroanat.* **4,** 319 (1991).

60. N. M. Appel, S. A. Mathews, G. M. Storti, and M. J. Kuhar, *Soc. Neurosci. Abstr.* **17,** 156.14 (1991). Abstr.

61. R. M. Hoyte, N. J. MacLusky, and R. B. Hochberg, *J. Steroid Biochem.* **36,** 125 (1990).

62. R. M. Hoyte, T. J. Brown, N. J. MacLusky, and R. B. Hochberg, *Steroids* **58,** 13 (1993).

63. K.-J. Hwang, K. E. Carlson, G. M. Anstead, and J. A. Katzenellenbogen, *Biochemistry* **31,** 11536 (1992).

64. G. Scatchard, *Ann. N.Y. Acad. Sci.* **51,** 660 (1949).

[9] Immunocytochemical Studies on Glucocorticoid Receptor

Antonio Cintra, Gunnar Akner, Rafael Coveñas,
Mercedes de León, Ann-Charlotte Wikström,
Luigi F. Agnati, Jan-Åke Gustafsson, and Kjell Fuxe

Monoclonal Antibodies Demonstrate Glucocorticoid Receptor in Rat Central Nervous System

Glucocorticoid Receptors: Neuronal Localization

The localization of the glucocorticoid receptor (GR)-immunoreactive neurons in the rat central nervous system (CNS) has been studied using an anti-GR monoclonal mouse antibody, MAb 7 (28), directed against an epitope in the N-terminal domain of the rat GR (16–18, 23). Double-immunoperoxidase or immunofluorescence techniques using different chromogens or fluorophores (Table I) (13) allowed the identification of the neuronal populations containing GRs. A strong nuclear GR immunoreactivity has been demonstrated *inter alia* in the stress-sensitive monoaminergic neurons of the brain stem (24) and in several hypothalamic neurons containing hypophysiotropic hormones, but not in the luteinizing hormone-releasing hormone (LHRH) system of the male rat (12, 13). Also, GR was not demonstrated in cholinergic neurons (37). Another monoclonal antibody, BUGR-2, which identifies an epitope adjacent to the DNA-binding domain of the rat GR, has demonstrated a widespread distribution of neuronal and glial GRs in the central nervous system of the rat (1, 7). Using this antibody glucocorticoid receptor immunoreactivity has been shown in catecholaminergic neurons (31). Glucocorticoid receptor immunoreactivity has even been demonstrated in a small population of LHRH-immunoreactive neurons of the preoptic region (3).

There is a great similarity between the pattern of distribution of GR immunoreactivity shown by MAb 7 and BUGR-2 monoclonal antibodies. Table II summarizes the regions where there are some discrepancies. MAb 7 shows strongly the GR-immunoreactive neurons, for example, within layers II/III of the cerebral cortex (Fig. 1), the hippocampal formation (Fig. 2), and the arcuate, periventricular, and paraventricular hypothalamic nuclei. Also some thalamic nuclei, for example, the parafascicular nucleus, revealed a strong GR immunoreactivity. In addition, BUGR-2 revealed strong GR immunoreactivity in the CA3 pyramidal cells as well as within the mitral cells of the

TABLE I   Staining of Glucocorticoid Receptor Using
Avidin–Biotin–Peroxidase Technique

Step	Comments[a]
Fixation	4% (v/v) Paraformaldehyde, 0.2% (w/v) picric acid, 0.1 M phosphate buffer
Cryoprotection	30% (w/v) sucrose, 0.1 M phosphate-buffered saline (PBS)
Sectioning	20-μm-thick cryostat sections, collected free floating
Mouse monoclonal antibody against rat liver GR	Diluted in 0.1 M PBS containing 0.3% (v/v) Triton X-100 (Sigma) 1% (v/v) goat whole serum (Cappel). Incubated at room temperature, overnight
Biotinylated horse anti-mouse antibody	Diluted in 0.3 M PBS. Incubated at room temperature, 2 hr
Avidin–biotin complex	Diluted in 0.1 M PBS. Incubated at room temperature, 1 hr
Color-generating substrate	0.03–0.05% (w/v) 3,3'-diaminobenzidine tetrahydrochloride (Sigma), 0.015% (v/v) H_2O_2, 0.03% (w/v) ammonium nickel sulfate (BDH Chemicals, Ltd.), 50 mM Tris-HCl (pH 7.4)
Mounting	Rinsed in Tris-HCl; mounted on gelatin-coated slides; dehydrated in alcohols/xylene; coverslipped with Enthelan (Merck)

[a] Supplier locations: Sigma (St. Louis, MO); Cappel (West Chester, PA); BDH Chemicals, Ltd. (Poole, England); Merck (Rahway, NJ).

olfactory bulb, the substantia nigra, and the Purkinje cells of the cerebellar cortex (1, 26), where MAb 7 showed slight or no GR immunoreactivity. Important contributions to the central GR localization and regulation were also made by the use of another monoclonal antibody (35).

Semiquantitative Evaluations of Glucocorticoid Receptor Immunoreactivity

Subjective analyses reporting various degrees of intensity/density of GR immunoreactivity can be improved by computer-assisted image analysis to provide semiquantitative microdensitometric evaluations. We have used the IBAS system (Zeiss Kontron image analysis; Kontron Elektronik, Germany). Glucocorticoid receptor immunoreactivity in different areas of the rat CNS can be expressed in mean gray values (MGVs). Figure 2 contains measurements of MGVs in different areas of the hippocampal formation. Another parameter provided by computer-assisted image analysis is the density of GR-immunoreactive neurons or glial cells present in different CNS areas (Fig. 1).

TABLE II   Glucocorticoid Receptor Immunoreactivities Using MAb 7 and
BUGR-2 Monoclonal Antibodies against Rat Liver
Glucocorticoid Receptor

	Antibody	
Brain region	MAb 7[a]	BUGR-2[b]
Telencephalon		
Internal granular layer of olfactory bulb	+	+++
Mitral cell layer of olfactory bulb	0	++
Islands of Calleja	+	++
Nucleus of horizontal diagonal band	0/+	+/++
Nucleus of vertical diagonal band	0/+	++
CA3 hippocampal area	+	+++
Subfornical organ	+++	nd
Diencephalon		
Median preoptic nucleus	+++	+
Anterior hypothalamic area	+	++
Suprachiasmatic nucleus	0/+	+
Paraventricular hypothalamic nucleus, parvocellular part	+++	++
Periventricular nucleus	+++	+
Arcuate nucleus	+++	+
Tuber cinereum	+++	+/++
Dorsomedial thalamic nucleus	+++	++
Laterodorsal thalamic nucleus	+	+++
Parafascicular thalamic nucleus	+++	nd
Zona incerta	0/+	+/++
Mesencephalon		
Substantia nigra, pars compacta	+	++/+++
Substantia nigra, pars lateralis	+	++
Central gray, dorsal	++	+++
Central gray, medial	++	0/+
Central gray, lateral	+/++	+++
Oculomotor nucleus	0/+	+++
Darschewitsch nucleus	0/+	+++
Pons		
Nucleus ambiquus	+	++/+++
Dorsal nucleus of the lateral lemniscus	+/++	nd
Parabraquial nucleus, medial	+	+++
C1 adrenaline cells	++	+++
C2 adrenaline cells	++	+++
C3 adrenaline cells	++	+++
Cerebellum		
Purkinje cell layer	0	++
Granular cell layer	++	+++

(continued)

TABLE II   (*continued*)

Brain region	MAb 7[a]	BUGR-2[b]
	Antibody	
Spinal cord		
Layer I	0/+	nd
Layer II	+ +	+ + +
Layer 8, 9	+	+/+ +
Layer 10	+	nd
White matter	+/+ +	nd
Intermediolateral cell column	+	+ +

[a] From Okret *et al.* (28); Fuxe *et al.* (17); Fuxe *et al.* (18).
[b] From Antakly and Eisen (7); Ahima and Harlan (1).

Glucocorticoid Receptors: Glial Localization

Fuxe *et al.* (17) and Ahima and Harlan (1) referred to the nuclear localization of GR in glial cells, assuming that it is present within small nuclear profiles. Within the white matter of the rat CNS, especially within the white matter of the lower brain stem and the spinal cord, low-intensity GR immunoreactivity was observed in glial profiles organized in rows. It seems likely that these glial profiles of the white matter represent both oligodendrocytes (mainly telencephalon) as well as fibrillary astrocytes (mainly lower brain stem and spinal cord). On the basis of the scattered distribution without relationship to nerve cell bodies, the GR-immunoreactive glial profiles within the gray matter might represent protoplasmatic astrocytes. Nevertheless, in some cranial nerve cell nuclei of the brain stem, such as the motor nucleus of the facial and trigeminal nerve, a perineuronal location could be found, suggesting that some of these glial profiles also represent perineuronal oligodendrocytes. In line with these results glucocorticoids have been demonstrated to influence the metabolism of glia cells and the degree of myelinization (21). On the basis of the low to moderate intensity of GR immunoreactivity and relatively high density of GR-immunoreactive glial profiles within restricted regions of the gray matter, it seems likely that several glial cell populations found especially in the brain stem and spinal cord represent a major target for the action of the glucocorticoids.

An *in vitro* study using BUGR-2 (33) revealed the presence of GR immunoreactivity in primary cultures of mixed glial cells and within glial cell lines. This receptor was found in astrocytes, oligodendrocytes, and Schwannoma cells (33).

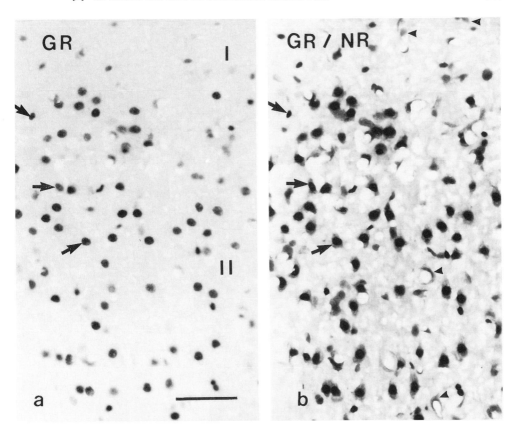

FIG. 1 Glucocorticoid receptor immunoreactivity is shown in a coronal section of the rat parietal cortex stained by the GR (MAb 7) antibody in combination with the immunoperoxidase technique with diaminobenzidine as the chromogen. (a) GR immunoreactivity in the molecular (I) and the granular cell (II) layers; (b) same field after counterstaining with neutral red (NR). The majority of the granular cells have nuclear GR (arrows). Arrowheads point to cells devoid of GR immunoreactivity in the molecular layer or to endothelial cells in the granular cell layer. Bar: 100 μm. Using computer-assisted image analysis (IBAS system), the GR nuclear profiles within layers II/III were found to be distributed in two classes: profiles with a mean diameter of 13 $\pm$ 4.5 μm with a density of 1.3/10^4 μm^2 (glial profiles) and profiles with a mean diameter of 32 $\pm$ 13 μm with a density of 11.7/10^4 μm^2 (neuronal profiles). Sample area: 70,750 μm^2.

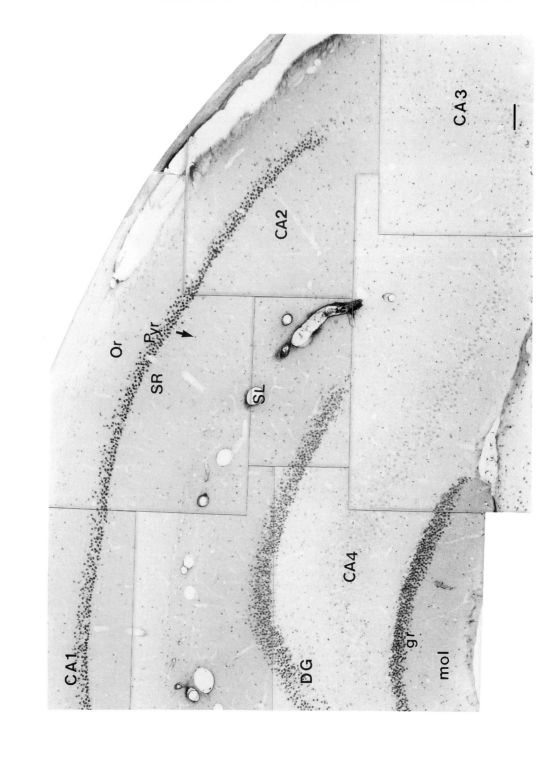

Subcellular Distribution of Glucocorticoid Receptors

The intracellular distribution of GR is still a matter of controversy. An accurate knowledge of the subcellular localization of GR would be of major importance in the elucidation of signal transduction by glucocorticoid hormones. A large number of biochemical cell fractionation reports have shown unanimously that treatment with glucocorticoid hormones induced a shift in GR distribution from the high-speed supernatant (designated "cytosol") to the pellet (designated "nuclear fraction"). The concept of "cytosol" was (erroneously) regarded as representing the cytoplasm and thus, the "two-step model" of GR action was founded: glucocorticoid hormones induce a translocation of the hormone–GR complex from the cytoplasm to the nucleus followed by regulation of target gene transcription (20). However, the result of immunolocalization studies of individual cells, using mono- or polyclonal antibodies, has been inconsistent and has formed the basis for several notions regarding GR localization with or without the presence of receptor agonist: (a) GR is a nuclear protein both in the absence and presence of glucocorticoid hormones (9, 29); (b) GR is present both in the cytoplasm and nucleus in the absence or presence of ligand. Glucocorticoid treatment does not change the relative intracellular GR distribution (19, 25); and (c) GR is present both in the cytoplasm and nucleus in the absence or presence of ligand. Glucocorticoid treatment induces a nuclear translocation of a part of or the whole cytoplasmic GR pool (11, 30) Autoradiographic studies using radiolabeled GR agonists have also given various results, but suffer from the potential problem of ligand binding affecting the intracellular receptor localization.

Some of the discrepancies may be explained by technical circumstances such as different species, cell types, anti-GR antibodies, fixation/permeabilization procedures, and detection techniques. Other problems may have in-

Fig. 2 Glucocorticoid receptor immunoreactivity is shown in a coronal section of the rat brain at bregma level −3.3 mm. The MAb 7 antibody was employed and the labeling was performed by the immunoperoxidase technique, using diaminobenzidine as chromogen. The strongest GR immunoreactivity is found in the pyramidal cell layer (Pyr) of the CA1–CA2 area. The following microdensitometric values for the intensity of the immunostaining (expressed as mean gray values) were obtained by computer-assisted image analysis: stratum oriens (Or), 79; pyramidal cell layer of the CA1 area, 108; stratum radiatum (SR), 74; stratum lacunosum (SL) moleculare, 61; pyramidal cell layer of the CA2 area, 111; pyramidal cell layer of the CA3 area, 44; dentate gyrus (DG) molecular layer (mol), 64; and dentate gyrus granular cell layer (gr), 90. Bar: 200 μm.

volved (a) direct comparison between cytology and histology, because these methods imply different means of cell or tissue sample processing prior to immunostaining; (b) comparison between the distribution of endogenous GR in target cells and exogenous, transfected, often heterologous, GR in nontarget cells; and (c) selection bias when choosing fields of view for microphotography.

Nuclear versus Cytoplasmic Location of Glucocorticoid Receptor

In histological sections of the rat brain and spinal cord, GR was mainly restricted to the nuclei of neuronal and glial cells. This preferential localization of the GR into nuclear profiles can have several explanations; one is that cytoplasmic but not nuclear GR immunoreactivity is abolished by the detergent Triton X-100 used in the immunocytochemical experiments (36). Indeed, a weak cytoplasmic GR immunoreactivity is observed only in some large nerve cells such as the pyramidal nerve cells of the cerebral cortex. The monoclonal antibodies may also preferentially recognize the liganded form of the GR. Conformational changes of the GR after binding the ligand may expose antigenic determinants on the GR. The disappearance of nuclear GR immunoreactivity in the absence of glucocorticoids, for example, following adrenalectomy (ADX) (1, 17), would be in line with this explanation. The effect of long-term ADX was studied in detail (2, 32). Parallel to the disappearance of nuclear GR immunoreactivity after ADX and reappearance after corticosterone administration some groups of neurons also displayed a cytoplasmic GR immunoreactivity. Most GRs recognized by antibodies in histological sections are activated GRs. Thus, GR immunoreactivity in a brain area will be dependent not only on the number of receptors found in a cell population but also on the affinity of these receptors to corticosterone and on the relative amount of hormone that can reach the receptors in that brain area. The degree of labeling of the various GR-immunoreactive neuronal and glial cell populations as well as the GR distribution pattern will therefore vary with the physiological state of the animal. Indeed, a circadian variation of neuronal GR immunoreactivity has been reported (32).

Cytological Studies of Glucocorticoid Receptor in Nonneuronal Cell Cultures

We have studied the intracellular immunocytological distribution of GR in a number of different mammalian cell types, both primary cultures and established cell lines, originating from the entodermic and mesodermic em-

TABLE III   Immunocytological Staining of Glucocorticoid Receptor Using Indirect
Immunofluorescence Technique

Step	Comments[a]
Fixation/permeabilization	Techniques compared
	Cross linking
	F/T: 4% (v/v) formaldehyde in 0.14 M (PBS), 4°C, 10–15 min followed by 0.05% (v/v) Triton X-100, room temperature, 30 min
	Precipitating
	M: methanol (100%), −20°C, 10 min
Mouse monoclonal antibody against rat liver GR	Diluted in 0.14 M PBS, yielding a final protein concentration of 10–20 μg/ml. Incubated at room temperature, 1–2 hr
Secondary antibody	Goat anti-mouse IgG, FITC labeled, diluted in 0.14 M PBS, yielding a final protein concentration of 25 μg/ml. Incubated at room temperature, 1 hr
Mounting	Washed in 0.14 M PBS. Mounted upside down on glass slides in 50% (v/v) glycerol–0.14 M PBS

[a] PBS, phosphate-buffered saline; IgG, immunoglobulin G; FITC, fluorescein isothiocyanate.

bryonal germinal layers, focusing on human primary culture of gingival fibroblasts. In these experiments we have used several different monoclonal mouse anti-rat liver GR antibodies that recognize different epitopes in the N-terminal *trans*-activation domain of the GR (28), a number of different fixation/permeabilization techniques, and several detection systems. The antibodies cross-react well with human GR (10). Of major interest was the comparison between various fixation techniques employing different chemical principles, such as cross-linking (aldehyde/detergent) and precipitating (organic solvents) methods in various combinations and concentrations.

Emphasis was placed on comparison between a standard fixation technique previously used in our laboratory (Table III), that is, *cross-linking* followed by 0.05% (v/v) Triton X-100 (36), and a *precipitation technique* (4, 5). As an indirect immunodetection system we have used fluorescence, peroxidase, and gold techniques, and have obtained similar results. We have concentrated on indirect immunofluorescence because of the sharp contrast in the images, the potential for double-staining of individual cells using two different fluorochromes with nonoverlapping excitation and emission spectra, and the use of confocal microscopy (see below).

Besides conventional transmission light microscopy, we have also used confocal laser scanning microscopy (CLSM) in collaboration with the Department of Physics IV (Royal Institute of Technology, Stockholm, Sweden).

The stained cell monolayer is excited one point at a time by a focused laser beam and the emitted light is detected by a photomultiplier tube (PM tube). This CLSM technique has several advantages over conventional transmission light microscopy: (a) CLSM provides thin optical sections of the cells by reducing the out-of-focus contributions of the emitted light. The "thickness" of each optical section in these studies ranges from 1 to 2 μm and depends on the numerical aperture (NA) of the objective, the refractive index of the immersion medium, and the wavelength of the light. Optical sectioning reduces the risk of projection artifacts; (b) CLSM provides better resolution than conventional microscopy; (c) CLSM presents data in a digitized form that allows various kinds of direct image analysis such as subtraction of one image from another and quantitation of relative fluorescence intensities; and (d) CLSM allows serial sectioning of individual cells followed by two- or three-dimensional reconstruction. (For a more detailed description of the CLSM technique see Refs. 27 and 38.)

The data recorded by the microscope and the PM tube are stored as digital images, for which the value of each image element, lying between 0 and 255, is proportional to the fluorescent light emitted from corresponding points in the specimen. To calculate the relative intensity distribution in the nucleus and cytoplasm in the same transverse optical section through the central part of a particular cell, the image is shown on a TV monitor and the whole cell, the cell nucleus, and a part of the background are circled. The intensity of GR immunoreactivity is measured by the PM tube and expressed in relative units. The cytoplasmic intensity is defined by subtracting the signal due to the nucleus from that of the whole cell. The mean photometric immunointensity of GR in the nuclear, cytoplasmic, and whole cell compartments, respectively, is quantified in 40–60 cells for each variable, followed by statistical analysis.

Cytoplasmic versus Nuclear Localization

Without the addition of any exogenous glucocorticoid, GR was found to be localized in both the nucleus (n) and the cytoplasm (c). On average, the relative $c:n$ ratio $\approx 9:1$ with a significant heterogeneity of GR intensity in both compartments. This heterogeneity was more easily recognized visually after methanol (M) rather than formaldehyde/Triton (F/T) fixation, but was photometrically detected in both instances. F/T yielded a two to three times higher intracellular GR concentration, indicating that M caused a much higher extraction of GR from all cellular compartments compared to F/T.

The subcellular distribution of GR was observed to be similar using both F/T and M fixations (Fig. 3). During interphase, cytoplasmic GR was observed in three distinct distributions: (a) along cytoplasmic microtubules,

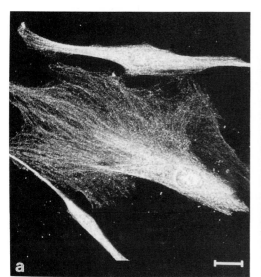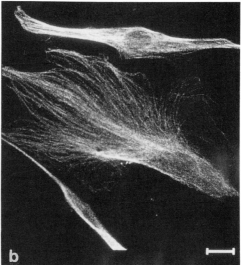

FIG. 3   An interphase human fibroblast primary culture was fixed with methanol and stained for glucocorticoid receptor (a) and tubulin (b) by means of the double-immunofluorescence technique. Confocal laser scanning microscopy provided a 1-μm-thick optical, transversal section through the central part of the cell. Glucocorticoid receptor immunoreactivity exhibits a cytoplasmic fibrillar staining pattern that colocalizes well with cytoplasmic microtubules. In the nucleus there is both a diffuse and granular GR signal, but little, if any, tubulin immunoreactivity. Bar: 20 μm.

(b) in individual centrioles, and (c) in the plasma membrane. The nuclear GR was found to be diffuse in all, and in addition granular in ~30% of the nuclei in a cell monolayer, leaving the nucleoli unstained. During mitosis, GR was concentrated in the centriolar regions and in the mitotic spindle apparatus during all stages of mitosis, together with a diffuse staining pattern in the surrounding cytoplasm. (For a more detailed description of the intracellular GR distribution see Refs. 4–6.)

Treatment with dexamethasone (10^{-6} M for 1 hr) caused a visually observable increase in nuclear GR staining in ~30% of the nuclei; however, this was found only after M, not F/T, fixation (see below). Photometry of GR on thin optical CLSM sections revealed a significant hormone-induced increase in the average GR intensity in both the nucleus and the cytoplasm, again detectable only after M but not F/T fixation. The increment in GR intensity was ~60% in the nucleus and ~35% in the cytoplasm. Also, when expressing the average GR intensity per picture element (pixel), we found ligand-induced increments in both compartments (~60% and ~13% in the nucleus and cytoplasm, respectively), thereby eliminating the impact of differences in compartment size. The qualitative subcellular distribution of GR in the cytoplasm and nucleus remained unchanged after hormone treatment (see above).

Gasc *et al.* (19) studied the distribution of GR in rat liver by immunohistology, using two different monoclonal anti-GR antibodies, I GR 49/4 (35) and MAb 7 (28), and various fixation techniques. Their observation of an increase in nuclear GR after dexamethasone treatment was interpreted as indicating less loss of hormone–receptor complexes during the fixation procedure as compared to hormone-free GR, rather than demonstrating nuclear translocation of GR. These results are in agreement with ours.

On the basis of these results, we have formulated an "extraction hypothesis" regarding GR localization: during basal conditions, GR is present both in the cytoplasmic and nuclear compartments, exhibiting a strong heterogeneity between the individual cells. Treatment with glucocorticoid hormones does not induce any detectable compartment shift, but changes the affinity of GR for its intracellular docking sites, both in the nucleus and in the cytoplasm. This affinity change can be detected (visually or photometrically) only after precipitating fixatives such as methanol, which extract less GR from hormone-treated than nontreated cells, but not after cross-linking fixatives such as aldehyde, which locks both nonliganded and liganded GR to its docking sites and does not allow any detergent-induced extraction. Because there is less methanol-induced extraction of GR from the nucleus than the cytoplasm, this may give rise to a false visual or photometric impression of a hormone-induced apparent nuclear translocation of GR.

In summary, our results illustrate the fundamental role of the cell or tissue sample processing prior to the immunostaining procedure. The diffuse and partly granular nucleus GR population probably participates in nuclear gene transcriptional regulation. The extranuclear GR pool does not seem to shift toward the nucleus following hormone administration. This cytoplasmic GR may constitute a storage form of GR, but may also reflect the potential of GR to exert effect(s) on sites in the cytoplasm without involving nuclear genomic transcription. One may speculate that the observed distribution of GR along cytoplasmic and mitotic microtubules as well as in cytoplasmic and mitotic centrioles may reflect direct sites of GR action.

Evidence has been presented in support of an association between another member of the steroid hormone receptor superfamily, that is, the vitamin D_3 receptor, and microtubules (8).

Glucocorticoid Receptors and Pain Mechanisms

Glucocorticoid receptor-immunoreactive neurons are particularly abundant in brain and spinal cord areas related to the nociceptive pathway and are suggested to be under glucocorticoid regulation. High densities of moderately to strongly GR-immunoreactive neurons exist within the substantia gelatinosa of the dorsal horn, the central gray, the midline thalamic nuclei, as well as within the posterior nuclear group and the ventral posterior nuclei of the thalamus. Pain pathways are known to project into these areas. High densities of GR-immunoreactive neurons are found all over the cerebral cortex, with the exception of layer IV and the outer part of layer V. Thus, pain, which may be globally perceived at the cortical level, may be under glucocorticoid control. Furthermore, the antinociceptive systems projecting into the substantia gelatinosa from the lower brain stem, such as the 5-hydroxytryptamine and noradrenaline descending systems, exhibit strong GR immunostaining, showing that systems mediating or reducing pain may be modulated by glucocorticoids in an integrated way. This view is supported by the phenomenon of stress-induced analgesia, which may involve both the nociceptive and the antinociceptive systems.

We designed some experiments to understand how glucocorticoids are implicated in pain mechanisms (see ref. 39 and 40).

Glucocorticoid Receptor and Neuropeptide Contents of Primary Afferent Neurons

In the spinal and sensitive cranial nerves, several neuropeptides are implicated in pain mechanisms: substance P (SP), calcitonin gene-related peptide (CGRP), somatostatin (SOM), galanin (GAL), and neuropeptide Y (NPY). We performed a study on the presence of GR immunoreactivity in the ganglionic neuronal populations containing one or more of these peptides. Most of these cells are small neurons giving rise to amyelinic fibers. The double-immunofluorescence technique was employed (14). One-third of both the CGRP- and SP-containing neurons of the spinal and trigeminal ganglia demonstrated nuclear GR (Fig. 4), whereas 50% of the GAL immunoreactive neurons of spinal ganglion, but only occasionally in the trigeminal ganglion, showed GR. Neither SOM- nor NPY-immunoreactive neurons demonstrated GR. The results suggest that glucocorticoids may regulate the synthesis of peptides via GR in some chemically identified neuronal populations of the spinal and trigeminal ganglia.

Regulation of Neuropeptide Contents of Spinal Ganglion by Glucocorticoids

The hypothesis of glucocorticoid regulation of peptide-containing nerve cell populations within the spinal ganglia was tested by performing adrenalectomy

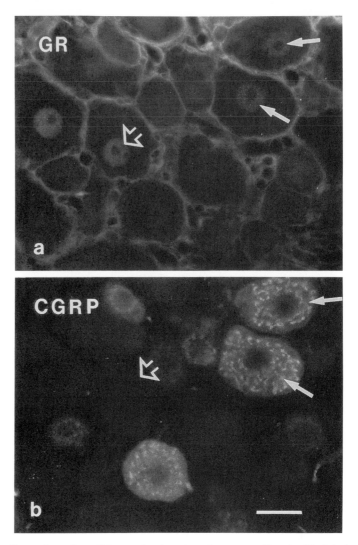

FIG. 4 Coronal section of the rat lumbar dorsal root ganglion double labeled for (a) GR and (b) calcitonin gene-related peptide (CGRP). After an incubation with both GR (MAb 7) and CGRP (Milab, Malmö, Sweden) antibodies at 4°C overnight, the secondary antibodies were linked to fluorescein-conjugated donkey anti-mouse and Texas Red-conjugated donkey anti-rabbit immunoglobulins, respectively (Jackson, West Grove, PA) at 37°C for 2 hr. Arrows point to large ganglion cells containing both nuclear GR and cytoplasmic CGRP. An open arrow shows a neuron with GR but lacking CGRP. Bar: 50 μm.

on different groups of animals and treating them with corticosterone. The immunoreactivity for these neuropeptides was assessed in a semiquantitative way by means of unbiased counts of immunoreactive ganglion cells, allowing the comparison between the experimental groups. This stereological method developed by the Danish group lead by Gundersen (22) involves systematic sampling using an optical fractionator (34). Serial sections of three lumbar ganglia per animal were analyzed using a computer-assisted stereological toolbox. Briefly, an Olympus BH2 microscope (Olympus, Silkeborg, Denmark) was interfaced with a computer (Amiga 2000) and a color video camera (CCD-Iris, Sony, Tokyo, Japan), both linked to a color video monitor (BT-D2000; Panasonic, Tokyo, Japan). The GRID software package (Olympus) was used to generate sampling frames. A microcator (MT12; Heidenhain, Germany) was linked to the microscope to monitor movements in a vertical (Z) direction. For counting the cells, a $100\times$ oil-immersion objective was used. Seven days after ADX the number of cells containing SP and SOM significantly increased (99 and 90%, respectively; $p < 0.01$). Adrenalectomy plus corticosterone treatment (10 mg/kg body weight, intraperitoneal injection, daily) increased the number of CGRP neurons above control and ADX levels. The number of GAL neurons was not altered in the experimental groups. The results suggest a negative regulation of SP and SOM synthesis in spinal ganglion neurons by glucocorticoids. The analgesic and antiinflammatory actions of glucocorticoids may be related to effects on SP transcription rates.

Glucocorticoid Receptor and c-Fos Immunoreactive Neurons of Dorsal Horn

Two types of nociceptive stimulations of the sciatic nerve were used: a 10-min pulse train (5 Hz, 0.5 msec, 10 mA), or 5% (v/v) mustard oil in paraffin oil applied to the hindlimb feet (15). After 2 hr we analyzed the presence of GR immunoreactivity in the activated cells of the spinal cord dorsal horn. Neuronal activity was estimated by the induction of nuclear immunoreactivity to the protein encoded by the protooncogene c-fos (Fig. 5). Colocalization of GR and c-Fos was found in a restricted neuronal population of the dorsal horn, which may be one site of action for glucocorticoids at the spinal cord level. Thus, a large number of pain pathways may not be directly affected by adrenocortical hormones, at the dorsal horn level.

Conclusions

Immunocytochemical analysis has demonstrated the existence of various amounts of glucocorticoid receptors in a large number of nerve and glial cell

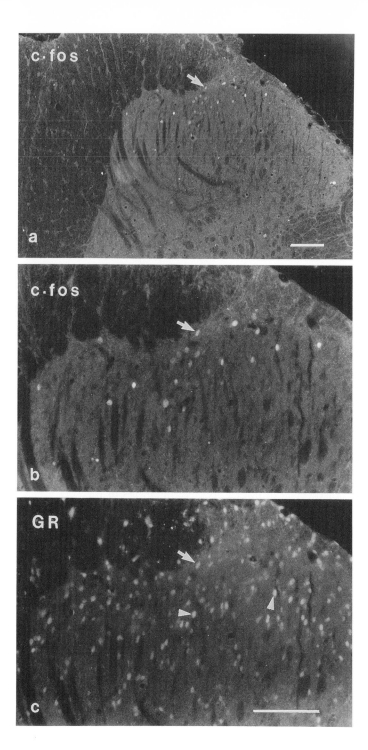

Fig. 5   Immunoreactivities of c-Fos and GR are shown in a coronal section of the rat spinal cord by means of a double-immunofluorescence procedure. The immunostaining was performed in two steps: first, binding of the GR antibody (MAb 7), at 4°C overnight, and labeling with fluorescein-conjugated donkey anti-mouse antibody (Jackson): second, binding of the c-Fos antibody (Affiniti, Derbyshire, UK), at 4°C overnight, and labeling with Texas Red-conjugated donkey anti-goat immunoglobulins (Jackson). Electric stimulation of the sciatic nerve 2 hr before sacrifice elicited c-Fos. (a) shown in large magnification in (b). Few (less than 10%) of the c-Fos-induced neurons of the superficial layers (I–II) of the dorsal horn exhibit GR (arrows). Arrowheads in (c) point to GR in neurons without c-Fos. Bars: 100 μm.

populations in the CNS. The results are compatible with the existence of both a cytoplasmic and nuclear pool of GR but a translocation of cytoplasmic GR to the nucleus still remains to be clearly demonstrated. Glucocorticoid receptors may directly modulate pain pathways at various levels (primary sensory neurons, periaqueductal gray matter, thalamus, and cerebral cortex) but not importantly at the level of the dorsal horn. The results emphasize the role played by the glucocorticoids in nuclear genomic transcription.

Acknowledgments

This work has been supported by Grants 04X-715 and 13X-2819 from the Swedish Medical Research Council, from the Swedish Medical Society, and from the Lars Hierta Minne Foundation.

References

1. R. S. Ahima and R. E. Harlan, *Neuroscience* **39**, 579 (1990).
2. R. S. Ahima and R. E. Harlan, *Endocrinology (Baltimore)* **129**, 226 (1991).
3. R. S. Ahima and R. E. Harlan, *Neuroendocrinology* **56**, 845 (1992).
4. G. Akner, K. G. Sundqvist, M. Denis, A. Wikström, and J.-Å. Gustafsson, *Eur. J. Cell Biol.* **53**, 390 (1990).
5. G. Akner, K. Mossberg, A.-C. Wikström, K. Sundqvist, and J.-Å. Gustafsson, *J. Steroid Biochem. Mol. Biol.* **39**, 419 (1991).
6. G. Akner, A.-C. Wikström, K. Mossberg, K. Sundqvist, and J.-Å. Gustafsson, *J. Histochem. Cytochem.* **42**, 645 (1994).
7. T. Antakly and H. J. Eisen, *Endocrinology* **115**, 1984 (1984).
8. J. Barsony and W. McKoy, *J. Biol. Chem.* **267**, 24457 (1992).
9. M. Brink, B. M. Humbel, E. R. de Kloet, and R. Van Driel, *Endocrinology* **130**, 3575 (1992).

10. M. Brönnegård, L. Poellinger, S. Okret, A.-C. Wikström, O. Bakke, and J.-Å. Gustafsson, *Biochemistry* **26,** 1697 (1987).

11. J. A. Cidlowski, D. Bellingham, F. E. Powell-Oliver, D. Lubahn, and M. Sar, *Mol. Endocrinol.* **4,** 1427 (1990).

12. A. Cintra, K. Fuxe, A. Härfstrand, L. Agnati, A.-C. Wikström, S. Okret, W. Vale, and J.-Å. Gustafsson, *Neurosci. Lett.* **77,** 25 (1987).

13. A. Cintra, K. Fuxe, V. Solfrini, L. Agnati, B. Tinner, A.-C. Wikström, W. Staines, S. Okret, and J.-Å. Gustafsson, *J. Steroid Biochem. Mol. Biol.* **40,** 93 (1991).

14. A. Cintra, R. Coveñas, M. de León, B. Bjelke, J.-Å. Gustafsson, L. F. Agnati, and K. Fuxe, *Neuroprotocols* **1,** 77 (1992).

15. A. Cintra, C. Molander, and K. Fuxe, *Brain Res.* **632,** 334 (1993).

16. K. Fuxe, A. Härfstrand, L. F. Agnati, Z.-Y. Yu, A. Cintra, A.-C. Wikström, S. Okret, E. Cantoni, and J.-Å. Gustafsson, *Neurosci. Lett.* **60,** 1 (1985).

17. K. Fuxe, A.-C. Wikström, S. Okret, L. F. Agnati, A. Härfstrand, Z.-Y. Yu, L. Granholm, M. Zoli, W. Vale, and J.-Å. Gustafsson, *Endocrinology* (*Baltimore*) **177,** 1803 (1985).

18. K. Fuxe, A. Cintra, A. Härfstrand, L. F. Agnati, M. Kalia, M. Zoli, A.-C. Wikström, S. Okret, M. Aronsson, and J.-Å. Gustafsson, *Ann. N.Y. Acad. Sci.* **512,** 362 (1987).

19. J. M. Gasc, F. Delahaye, and E. E. Baulieu, *Exp. Cell Res.* **181,** 492 (1989).

20. W. Grody, W. T. Schrader, and B. W. O'Malley, *Endocr. Rev.* **3,** 141 (1982).

21. M. Gumbinas, M. Oda, and O. Huttenlocher, *Biol. Neonate* **22,** 355 (1973).

22. H. C. G. Gundersen and E. B. Jensen, *J. Microsc.* (*Oxford*) **147,** 229 (1987).

23. J. Honkaniemi, M. Pelto-Huikko, L. Rechardt, J. Isola, A. Lammi, K. Fuxe, J.-Å. Gustafsson, A.-C. Wikström, and T. Hökfelt, *Neuroendocrinology* **55,** 451 (1992).

24. A. Härfstrand, K. Fuxe, A. Cintra, L. Agnati, I. Zini, A.-C. Wikström, S. Okret, Z.-Y. Yu, M. Goldstein, H. Steinbusch, A. Verhofstad, and J.-Å. Gustafsson, *Proc. Natl. Acad. Sci. U.S.A.* **83,** 9779 (1986).

25. R. E. LaFond, S. Kennedy, R. Harrison, and C. Villee, *Exp. Cell Res.* **175,** 52 (1988).

26. A. Lawson, R. Ahima, Z. Krozowski, and R. Harlan, *Neuroendocrinology* **55,** 695 (1991).

27. K. Mossberg and M. Ericsson, *J. Microsc.* (*Oxford*) **158,** 215 (1990).

28. S. Okret, A.-C. Wikström, Ö. Wrange, B. Andersson, and J.-Å. Gustafsson, *Proc. Natl. Acad. Sci. U.S.A.* **81,** 1609 (1984).

29. A. Pekki, J. Koistinaho, T. Ylikomi, P. Vilja, H. Westphal, and P. Touhimaa, *J. Steroid Biochem. Mol. Biol.* **41,** 753 (1992).

30. D. Picard and K. Yamamoto, *EMBO J.* **6,** 3333 (1987).

31. P. Sawchenko and M. C. Bohn, *J. Comp. Neurol.* **285,** 107 (1989).

32. J. van Eekelen, J. Kiss, H. Westphal, and E. R. de Kloet, *Brain Res.* **436,** 120 (1987).

33. U. Vielkind, A. Walencewicz, J. Levine, and M. Bohn, *J. Neurosci. Res.* **27,** 360 (1990).

34. M. West, L. Slomianka, and H. Gundersen, *Anat. Rec.* **231,** 482 (1991).

35. H. Westphal, G. Moldenhauer, and M. Beato, *EMBO J.* **1,** 1467 (1982).
36. A.-C. Wikström, O. Bakke, S. Okret, M. Brönnegård, and J.-Å. Gustafsson, *Endocrinology (Baltimore)* **120,** 1232 (1987).
37. M. Zoli, A. Cintra, I. Zini, L. Hersh, J.-Å. Gustafsson, K. Fuxe, and L. Agnati, *J. Chem. Neuroanat.* **3,** 355 (1990).
38. N. Åslund, K. Carlsson, A. Liljeborg, and L. Majlöf, *Proc. Scand. Conf. Image Anal., 3rd* pp. 338–343. Studentlitteratur, Lund, 1983.
39. R. Coveñas, M. de León, G. Chadi, A. Cintra, J.-Å. Gustafsson, J. A. Narvaez, and K. Fuxe, *Brain Res.* **640,** 352 (1994).
40. M. de León, R. Coveñas, G. Chadi, J. A. Narvaez, K. Fuxe, and A. Cintra, *Brain Res.* **636,** 338 (1994).

[10] Confocal Scanning Laser Microscopy of Steroid Receptors in Brain

Bas van Steensel, Erica P. van Binnendijk,
and Roel van Driel

Introduction

Immunocytochemistry has contributed significantly to our understanding of the working mechanisms of steroid receptors. At the tissue level *in situ* hybridization and immunohistochemistry have provided qualitative and quantitative data on expression of steroid receptors in different brain areas (1–5). At the subcellular level immunolabeling studies have had great impact on our views on the molecular working mechanism and sites of action of steroid receptors (6–10).

During the past few years confocal scanning laser microscopy (CSLM) has been developed as a new and powerful tool for the analysis of immunofluorescently labeled preparations. So far, only a few CSLM studies on steroid receptors have been reported (11–14).

In this chapter a brief outline is presented of the features of CSLM and its applications in steroid receptor research. Because other chapters in this volume (15, 16) deal more extensively with general aspects of immunocytochemistry of steroid receptors, we focus on practical considerations concerning immunolabeling of steroid receptors for CSLM purposes.

Confocal Scanning Laser Microscopy

A major drawback of conventional fluorescence microscopy is that images are blurred by out-of-focus information. Light from regions of the specimen above and below the focal plane is collected by the objective lens and thus contributes as out-of-focus blur to the image of the specimen. This seriously reduces the contrast and sharpness of the image. For this reason conventional immunofluorescence microscopy studies are generally restricted to well-flattened cultured cells and relatively thin (<10 μm) tissue sections.

In CSLM images, out-of-focus blur is essentially absent. Owing to the effective elimination of all information from above and below the focal plane, optical sections can be made of cells and tissues. This has the advantage

Methods in Neurosciences, Volume 22

that cells can be visualized even if they are inside thick tissue slices. Under optimal conditions the resolution of CSLM is $0.2 \times 0.2 \times 0.8 \mu$m ($X$, Y, Z), thus allowing detailed analyses of subcellular structures. The principles of CSLM are described elsewhere (17, 18).

An important feature of CSLM is the possibility to construct three-dimensional images by making serial optical sections. Because images are directly stored as digital information, they can readily be visualized by computer-based three-dimensional display techniques, such as the generation of stereoscopic pairs of images, rotating images, or other methods (19).

Although confocal imaging is possible in reflection mode, CSLM is generally used to detect fluorescent probes. Most commercial confocal microscopes are equipped with a two-channel detection system, allowing simultaneous recording of two different fluorochromes in one specimen. Therefore, CSLM is extremely useful for double-labeling experiments, because the two images obtained are in exact spatial register. This facilitates any subsequent image processing involving both images, such as dual-channel display and quantitative colocalization analysis.

Digital CSLM images can be subjected to image processing and image analysis. Image processing refers to software routines that improve the crude image. Common examples of image processing are background subtraction, contrast enhancement, and removal of noise. Software for image processing is generally available and is often purchased with the confocal microscope hardware. However, these routines should be used with great caution, because "improvement" of an image is often subjective. Image analysis procedures allow quantitative assessment of specific aspects of an image. Examples of image analysis are (a) a quantitative estimate of the overlap of two images of a double-labeled specimen and (b) quantitative analysis of the distribution of a fluorescent probe over different cell compartments. Often, software must be developed specifically for each image analysis problem. In all cases, one should realize that the amount of fluorescence that is measured is not necessarily linearly related to the number of fluorescent molecules, let alone (in the case of immunofluorescence) to the number of antigen molecules.

Immunolabeling for Confocal Scanning Laser Microscopy: Special Requirements

The general methodology of immunocytochemistry of steroid receptors is described elsewhere in this volume (15, 16). This chapter focuses on special requirements for CSLM analysis of immunolabeled specimen.

Fluorochromes

Because CSLM is designed primarily for detection of fluorescently labeled probes, primary or secondary antibodies, conjugated to a fluorochrome, are used. Common fluorochromes for immunofluorescence microscopy are fluorescein isothiocyanate (FITC), tetramethylrhodamine isothiocyanate (TRITC), Texas Red (sulforhodamine 101 acid), and 7-amino-4-methylcoumarin-3-acetic acid (AMCA). A wide range of antibodies conjugated to these and other fluorochromes is commercially available.

Three physical properties of a fluorochrome should be taken into consideration for CSLM: (a) absorption spectrum, (b) emission spectrum, and (c) photostability. The absorption (excitation) spectrum of the fluorochrome must match the wavelength of the excitation light, that is, the laser light. Most commercial confocal microscopes are equipped with either an argon laser (emission, 488 and 514 nm) or an argon–krypton laser (emission, 488 and 568 nm). Figure 1a shows the absorption spectra of the most commonly used fluorochromes. At 488 nm FITC is excited efficiently, but TRITC and Texas Red are not. At 514 nm TRITC is excited reasonably well, whereas FITC and Texas Red are excited only very weakly. The 568-nm laser line is near-optimal for TRITC and Texas Red excitation. Therefore, when the confocal microscope is equipped with an argon laser, FITC or TRITC is the primary choice. When using an argon–krypton laser, Texas Red may be used instead of TRITC.' AMCA is excited at 350 nm; therefore this fluorochrome cannot be used.

The emission spectrum of the fluorochrome (Fig. 1b) should match the dichroic mirrors and filters that are used in the confocal microscope. Careful selection of these mirrors and filters is of crucial importance. As can be seen in Fig. 1b, a considerable overlap occurs between the emission spectra of FITC and TRITC and to a lesser extent between the spectra of FITC and Texas Red. In dual-labeling experiments this may result in a substantial amount of cross-talk between the two signals. Cross-talk is due to incomplete separation of the two fluorescent signals in a dual-labeling setup. When an argon–krypton laser is available, Texas Red is preferred to TRITC, because its emission spectrum allows better separation from the FITC signal. Also, excitation laser light that is scattered by the specimen should be suppressed. For this and for separation of the fluorochrome signals in double labeling experiments, highly selective filters and dichroic mirrors are required. However, improvement of signal separation at the same time inevitably results in decrease of signal strength. A typical filter setup is indicated in Fig. 1b.

Fluorochromes are prone to photobleaching. Because the laser light used in CSLM is intense, this fading can be a significant practical problem. Photobleaching is much faster for fluorescein conjugates than for rhodamine conju-

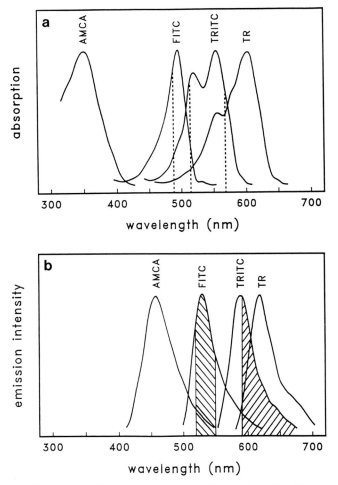

FIG. 1   Normalized absorption (a) and emission (b) spectra of antibody-conjugated AMCA, FITC, TRITC, and Texas Red (TR). Vertical dotted lines in (a) represent the commonly used laser light wavelengths of 488, 514, and 568 nm. Shaded areas in (b) represent a typical filter setup for dual-channel detection of FITC and TRITC (respectively, a 520 to 550-nm bandpass filter and a 590-nm longpass filter) when using a 488 nm/514 nm dual-wavelength argon laser.

gates (20). Thus, for single-labeling experiments, TRITC and Texas Red are more favorable than FITC. In any case, bleaching retardants such as *n*-propyl gallate (21) or *p*-phenylenediamine (22) should be added to the mounting medium.

Double Labeling

The procedure for double labeling is basically the same as for single labeling. However, in dual-labeling experiments one should be aware of artifacts due to immunological cross-reactivity and to optical cross-talk.

Double labeling is easiest when the two primary antibodies to be used are from different species or are of different immunoglobulin subclasses. Direct fluorescent labeling of the primary antibodies can be considered in case they cannot be discriminated by secondary antibodies. However, this may considerably reduce signal strength, because normally one primary antibody molecule binds several fluorescent secondary antibody molecules, resulting in a much greater number of fluorescent probes per antigen molecule. To minimize cross-reactivity secondary antibodies should be affinity purified and preadsorbed against serum proteins from other species. For instance, when the primary antibodies are from rabbit and mouse the secondary anti-rabbit antibody should be preadsorbed against mouse serum proteins and the anti-mouse antibody should be preadsorbed against rabbit serum proteins. Affinity-purified and preadsorbed secondary antibodies are commercially available.

In double-labeling experiments the amount of optical cross-talk should be determined. Careful selection of filters and dichroic beam splitters reduces this problem considerably, but never completely. Residual cross-talk can be removed by image processing (23, 24). Such procedures require a carefully standardized imaging protocol and appropriate image-processing software.

Antibody labeling can also be combined with nonimmunological staining methods. Many fluorescent dyes that specifically bind nucleic acids are available. For instance, mithramycin is excited at 488 nm and can be used for specific labeling of DNA. Propidium iodide is excited at 514 nm and binds to both DNA and RNA. Fluorochrome-conjugated phalloidin can be used for staining actin filaments. These dyes are useful for monitoring cell shape and location of the cell nucleus.

Confocal Scanning Laser Microscopy of Steroid Receptors in Cultured Cells

The success of an immunolabeling protocol depends primarily on the antibodies that are used. The antibodies should be specific and must be able to recognize their epitope in fixed biological material. The optimal antibody concentration must be determined experimentally for each batch of antibody.

The following protocol is a good starting point for immunofluorescent labeling of steroid receptors in cultured cells, for example, primary neuronal cultures, neuroblastoma cells, and other adherent cell types. With this method we have successfully labeled the glucocorticoid receptor, the mineralocorticoid receptor, the androgen receptor, and the progesterone receptor. However, other fixation and labeling methods have been described for steroid receptors (11, 25). In general, each step in a labeling protocol must be optimized.

Steroid receptors are present in low concentrations in the cell. To increase the intensity of the fluorescent signal, signal enhancement protocols can be applied. A well-known example is the biotin–streptavidin system. Here, the secondary antibody is conjugated to biotin, which is then detected with fluorochrome-conjugated streptavidin. This results in a larger number of fluorochrome molecules per antigen molecule. Similarly, the secondary antibody can be conjugated to digoxygenin, which is then detected by a fluorochrome-conjugated anti-digoxygenin antibody. These and other enhancement methods often result in increased sensitivity, but have the drawback that with each extra step nonspecific labeling may also increase.

Phosphate-buffered saline (PBS; 10×): 82 g of NaCl, 2.0 g of KCl, 11.5 g of $Na_2HPO_4 \cdot 2H_2O$, and 2.0 g of KH_2PO_4 in 1 liter of H_2O

PBS: Dilute 10× PBS 10 times; the pH should be 7.4

Formaldehyde fixative: Add 2.0 g of paraformaldehyde to 20 ml of H_2O, heat to 60°C in a fume hood, and add 80 μl of 1 N NaOH. When the solid has completely dissolved, add 10 ml of 10× PBS and 70 ml of H_2O. Adjust the pH to 7.4 at room temperature. This solution should be prepared fresh on the day of the experiment

Permeabilization buffer: 0.5% (w/v) Nonidet P-40 in PBS

PBG: 0.5% (w/v) bovine serum albumin (BSA) and 0.1% (w/v) gelatin (from cold-water fish skin; Sigma, St. Louis, MO) in PBS

Mounting medium: Dissolve 20 mg of p-phenylenediamine (Sigma) in 2 ml of 10× PBS. Add 18 ml of glycerol (87%; Merck, Rahway, NJ) and mix well. Adjust the pH to 8.0. Store in 0.5-ml aliquots under nitrogen at −20°C. Discard when the solution becomes brownish

Procedure

Grow cells on sterilized microscope coverslips. To improve adherence of the cells, the coverslips may be precoated with gelatin, Alcian Blue, or polylysine. If required, treat the cells with steroid hormone. The presence or absence of steroid hormone may affect the subcellular localization of steroid receptors. For the glucocorticoid receptor the observed, apparent hormone-dependent localization has been reported to depend on the method of fixation (10).

All steps are carried out at room temperature. Remove the culture medium and gently wash the cells twice in PBS. Fix the cells for 10 min in formaldehyde fixative. Wash twice in PBS. Treat the fixed cells for 10 min with permeabilization buffer, then wash twice in PBS. Incubate for 5 min with 100 mM glycine in PBS to inactivate remaining free aldehyde groups from the fixative.

Incubate twice (10 min each time) in PBG. Incubate for 1 hr at room temperature or overnight at 4°C with the primary antibody diluted in PBG. Wash six times (5 min each time) in PBG. Incubate for 1 hr at room temperature with a secondary, fluorochrome-conjugated antibody. Wash four times (5 min each time) in PBG. Wash twice (5 min each) in PBS. Drain excess fluid and embed in a small drop of mounting medium. Seal the coverslip onto a microscope slide with colorless nail polish.

Example: Labeling of Glucocorticoid Receptor in HeLa Cells

Figure 2a shows a CSLM optical section of the nucleus of a HeLa cell that was labeled with a monoclonal antibody against the glucocorticoid receptor (26). Figure 2b shows the same cell nucleus stained with propidium iodide, labeling primarily the nucleoli. The glucocorticoid receptor is present in clusters which are scattered throughout the nucleoplasm. No labeling of the glucocorticoid receptor is observed in the nucleoli (14).

Confocal Scanning Laser Microscopy of Steroid Receptors in Brain Tissue

Immunolabeling of brain tissue slices is basically identical to that of cultured cells. However, because of the relatively large tissue dimensions compared to a single layer of cultured cells, special attention should be paid to the fixation and permeabilization methodology. On the one hand, the tissue should be fixed properly to maintain structural integrity of the cells. On the other hand, fixation should not be so strong that after permeabilization the tissue cannot be penetrated by the antibodies.

Fixation and Sectioning of Brain

For rapid and thorough fixation it is essential to fix the brain by intracardiac perfusion with fixative solution. Although other fixatives have been described (27), we concluded after comparing different fixatives that 4% formaldehyde results in excellent preservation of cellular structures, combined with good permeability to antibodies during the labeling procedure. In very thick sections penetration of antibodies can be insufficient. We found that antibody penetration is adequate when using 30-μm sections. For making these brain sections, Vibratome tissue sectioning is strongly preferred to cryosectioning, because the latter is likely to cause freezing artifacts. This is particularly relevant for subcellular localization studies.

For CSLM the thickness of the brain sections is not critical. Owing to the confocal effect excellent images can be made up to several tens of micrometers deep into the tissue. However, if the focal plane is positioned too deep in the tissue, light scattering and diffraction by the tissue between the objective and the focal plane have an adverse effect on the resolution of the images.

PBS (10×): 82 g of NaCl, 2.0 g of KCl, 11.5 g of $Na_2HPO_4 \cdot 2H_2O$, and 2.0 g of KH_2PO_4 in 1 liter of H_2O

PBS: Dilute 10× PBS 10 times; the pH should be pH 7.4 at room temperature

Perfusion solution: 0.9% (w/v) NaCl in H_2O; prepare 100 ml for one rat

Formaldehyde fixative: Prepare 300 ml for one rat. Add 6.0 g of paraformaldehyde to 60 ml of H_2O, heat to 60°C in a fume hood, and add 240 μl of 1 N NaOH. When the solid has completely dissolved, add 30 ml of 10× PBS and 210 ml of H_2O. Adjust the pH to 7.4 at room temperature. This solution should be prepared fresh on the day of the experiment

Storage buffer: 78% (v/v) glycerol in PBS

Procedure

The solutions should be at room temperature. Anesthetize a rat with Nembutal. Fix the brain by intracardial perfusion with perfusion solution for about 2 min, followed by fixative for 15 min. Remove the brain and store it overnight in 20 ml of fixative. Then replace the fixative with 20 ml of PBS, and let it stand for 24 hr. Cut 30-μm slices with a Vibratome. Collect the brain sections in PBS. The sections can be stored at least several weeks at −20°C in storage buffer.

FIG. 2   CSLM optical section of a dual-labeled HeLa cell nucleus. Cells were fixed in 2% formaldehyde, permeabilized in 0.5% Nonidet P-40, and labeled with a mouse monoclonal antibody against the GR (26), followed by biotin-conjugated sheep anti-mouse antibody and FITC-conjugated streptavidin (a). The same cells were stained with propidium iodide, which mainly labels the nucleoli (b).

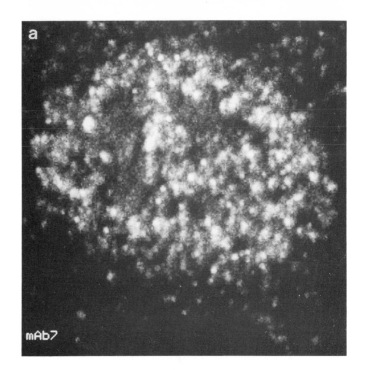

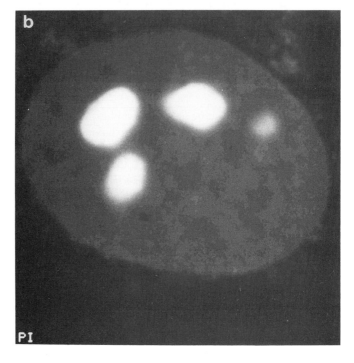

Labeling of Brain Sections

In many labeling protocols an antibody incubation time of 48 hr is recommended. However, we found that overnight incubation with the primary antibody is sufficient. Longer incubation times may result in degradation or denaturation of antibodies and deterioration of the tissue.

> Blocking buffer: 5% (v/v) nonimmune donkey serum or goat serum (depending on whether donkey or goat secondary antibodies are used, respectively) and 0.1% (w/v) Nonidet P-40 in PBS
> PBGN buffer: 0.5% (w/v) BSA, 0.1% (w/v) gelatin (from cold-water fish skin; Sigma) and 0.1% (w/v) Nonidet P-40 in PBS
> Mounting medium: Dissolve 20 mg of p-phenylenediamine (Sigma) in 2 ml of $10\times$ PBS. Add 18 ml of glycerol (87%; Merck) and mix well. Adjust the pH to 8.0. Store in 0.5-ml aliquots under nitrogen at $-20°C$. Discard when the solution becomes brownish

Procedure

All steps are carried out at room temperature, unless indicated otherwise. When brain sections have been stored in glycerol–PBS, they should be washed first in excess PBS. Incubate the floating sections 1 hr in blocking buffer, followed by overnight incubation at 4°C with the primary antibody diluted in PBGN. Wash the sections four times (10 min each) in PBGN buffer and subsequently incubate for 1 hr with a secondary antibody diluted in PBGN. Wash four times (10 min each). If a biotin–streptavidin enhancement protocol is selected, incubate for 1 hr with fluorochrome-conjugated streptavidin diluted in PBGN. Wash 10 min in PBGN. Wash twice (10 min each) in PBS. Rinse briefly in H_2O, and immediately transfer the brain sections to microscope slides. Drain excess water and let the surface dry briefly in air until the brain sections just stick to the slide. Avoid complete drying of the tissue. Embed in mounting medium. Seal with colorless nail polish. Best results are obtained when preparations are examined immediately. Labeled specimens can generally be stored for approximately 1 week at $-20°C$.

Example: Localization of Glucocorticoid Receptor in Rat Hippocampus

Figure 3 shows a CSLM optical section of the rat hippocampus CA1 region, labeled with a mouse monoclonal antibody (26) against the glucocorticoid receptor, followed by a TRITC-conjugated donkey anti-mouse antibody. A clear labeling of the neuronal cell nuclei is observed.

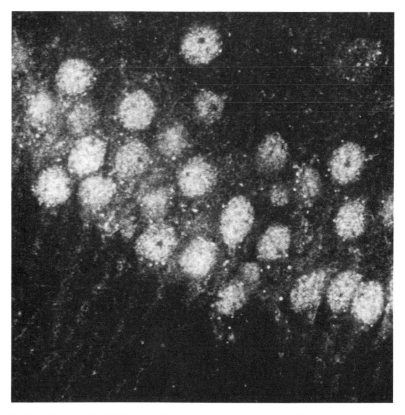

FIG. 3 Labeling of the GR in rat hippocampus. A rat was subcutaneously injected with 1 mg of corticosterone. After 1 hr the brain was fixed by perfusion with 4% formaldehyde and 30-μm Vibratome sections were made. Hippocampus sections were labeled with a mouse monoclonal antibody against the GR (26), followed by a TRITC-conjugated donkey anti-mouse antibody. A CSLM optical section is shown of the hippocampal CA1 area, displaying clear labeling of the neuronal cell nuclei.

Concluding Remarks

Confocal scanning electron microscopy in combination with immunofluorescent labeling is a powerful technique with which to study the localization of steroid receptors in cells and tissues. Optical sections of the labeled specimen are made in which out-of-focus blur is virtually absent. In this way three-dimensional images can be obtained. The technique is particularly suited to the analysis of receptor distributions in intact tissue. Digital data acquisition and storage allows direct computer-based image processing and quantitative image analysis. Because most confocal microscopes are equipped with a dual detection system, CSLM is particularly useful for double-labeling studies. In this way the distribution of steroid receptors can be compared to that of other cell components. In conclusion, the CSLM technique can contribute significantly to our understanding of the mechanisms underlying the modulation of brain function by steroid receptors.

References

1. J. A. M. Van Eekelen, W. Jiang, E. R. De Kloet, and M. C. Bohn, *J. Neurosci. Res.* **21**, 88 (1988).
2. K. Fuxe, A. Härfstrand, L. F. Agnati, Z.-Y. Yu, A. Cintra, A.-C. Wikström, S. Okret, E. Cantoni, and J.-Å. Gustafsson, *Neurosci. Lett.* **60**, 1 (1985).
3. K. Fuxe, A. C. Wikström, S. Okret, L. F. Agnati, A. Härfstrand, Z.-Y. Yu, L. Granholm, M. Zoli, W. Vale, and J.-Å. Gustafsson, *Endocrinology* (*Baltimore*) **117**, 1803 (1985).
4. A. N. Clancy, R. W. Bonsall, and R. P. Michael, *Life Sci.* **50**, 409 (1992).
5. R. Ahima, Z. Krozowski, and R. Harlan, *J. Comp. Neurol.* **313**, 522 (1991).
6. J. M. Gasc and E.-E. Baulieu, *Biol. Cell.* **56**, 1 (1986).
7. M. Beato, *Biochim. Biophys. Acta* **910**, 95 (1987).
8. J.-Å. Gustafsson, J. Carlstedt-Duke, L. Poellinger, S. Okret, A.-C. Wikström, M. Brönnegard, M. Gillner, Y. Dong, K. Fuxe, A. Cintra, A. Härfstrand, and L. Agnati, *Endocr. Rev.* **8**, 185 (1987).
9. M. Beato, *FASEB J.* **5**, 2044 (1991).
10. M. Brink, B. M. Humbel, E. R. De Kloet, and R. Van Driel, *Endocrinology* (*Baltimore*) **130**, 3575 (1992).
11. G. Akner, K. Mossberg, A. C. Wikström, K. G. Sundqvist, and J.-Å. Gustafsson, *J. Steroid Biochem. Mol. Biol.* **39**, 419 (1991).
12. V. R. Martins, W. B. Pratt, L. Terracio, M. A. Hirst, G. M. Ringold, and P. R. Housley, *Mol. Endocrinol.* **5**, 217 (1991).
13. M. Perrot-Applanat, P. Lescop, and E. Milgrom, *J. Cell Biol.* **119**, 337 (1992).
14. B. van Steensel, M. Brink, I. Van der Meulen, E. P. Van Binnendijk, D. G. Wansink, L. De Jong, E. R. De Kloet, and R. Van Driel, submitted.
15. A. Cintra, G. Akner, A.-C. Wikström, L. F. Agnati, J.-Å. Gustafsson, and K. Fuxe, this volume [9].
16. K. Fuxe, A. Cintra, R. Diaz, G. Chadi, J.-Å. Gustafsson, and L. F. Agnati, this volume [23].
17. G. J. Brakenhoff, H. T. M. Van der Voort, E. A. Van Spronsen, and N. Nanninga, *J. Microsc.* (*Oxford*) **153**, 151 (1989).
18. D. M. Shotton, *J. Cell Sci.* **94**, 175 (1989).
19. H. T. M. Van der Voort, G. J. Brakenhoff, and M. W. Baarslag, *J. Microsc.* (*Oxford*) **153**, 123 (1989).

20. R. Y. Tsien and A. Waggoner, *in* "Handbook of Biological Confocal Microscopy" (J. B. Pawley, ed.), p. 169. Plenum, New York, 1989.

21. H. Giloh and J. W. Sedat, *Science* **217,** 1252 (1982).

22. G. D. Johnson and D. C. Nogueira Araujo, *J. Immunol. Methods* **32,** 349 (1981).

23. E. M. M. Manders, F. J. Verbeek, and J. A. Aten, *J. Microsc. (Oxford)* **169,** 375 (1993).

24. K. Carlsson and K. Mossberg, *J. Microsc. (Oxford)* **167,** 23 (1992).

25. A. Guiochon-Mantel, H. Loosfelt, P. Lescop, S. Sar, M. Atger, M. Perrot-Applanat, and E. Milgrom, *Cell* **57,** 1147 (1989).

26. S. Okret, A. C. Wikström, O. E. Wrange, B. Andersson, and J.-Å. Gustafsson, *Proc. Natl. Acad. Sci. U.S.A.* **81,** 1609 (1984).

27. J. A. M. Van Eekelen, J. Z. Kiss, H. M. Westphal, and E. R. De Kloet, *Brain Res.* **436,** 120 (1987).

[11] Ultrastructural Aspects of Steroid Receptor Localization: Immunocytochemical Perspective

Zsolt Liposits and Imre Kalló

Introduction

Steroid molecules influence the operation of neurons and supporting cells constituting the nervous system. Their effects are accomplished at different cellular levels targeting the membrane, different cytoplasmic organelles, and nuclei (1, 2). Some of the effects might be executed by the steroid molecules themselves; others require the binding of steroids to specific receptors that subsequently triggers their activation and in certain cases translocation to the site(s) of action. Although the membrane and genomic effects of some steroids have been previously characterized by electrophysiology and biochemistry, the morphological mapping of sites of steroid action has become available only recently, utilizing immunocytochemical techniques and highly specific antibodies against steroid receptors. Because cellular components of the nervous system display various pheno- and chemotypes, and neuronal functions are dependent on the cellular organization and network connections of brain regions, the *en masse* biochemical sampling of the neural tissue conceals the type and connectivity of steroid receptor-possessing cells. Receptor immunocytochemistry makes possible the *in situ* characterization of individual, steroid-sensing cells and the morphological mapping of neuronal circuits directly regulated by steroids. Ultrastructural immunocytochemistry reveals the intracellular distribution, organelle-specific association, and trafficking mechanisms of steroid receptors. This chapter presents electron microscopic methods suitable for detecting glucocorticoid receptor (GR)- and estrogen receptor (ER)-immunoreactive elements of the nervous system.

Preexperimental Considerations

In comparison with regular peptide and transmitter immunocytochemistry, the morphological detection of steroid receptors might be more problematical at the beginning owing to a lower rate of expression and potential vulnerability to fixatives. Because circulating steroids determine the synthesis rate of their corresponding receptors, the humoral status of the experimental animals has great influence on the outcome of the labeling; therefore, an animal model possessing theoretically the highest levels of steroid receptors should be chosen for standardization of the method. Brain regions and cell populations display disproportionate amounts of the receptors that might account for negative staining results. Evaluation of receptor mRNA expression maps of *in situ* hybridization studies (3, 4) prior to the experiment might help to estimate the receptor content in a given area. In immunocytochemistry, not all of the well-characterized, specific antibodies can recognize the fixative-altered steroid receptors of the tissue; therefore, the use of antibodies with previously documented applicability to immunocytochemical staining is recommended. Although the techniques presented here were applied to neural tissue, they also detect steroid receptors in other tissues.

Ultrastructural Detection of Glucocorticoid Receptor Immunoreactivity in Neural Tissue

Animal Model

The hippocampal formation of the rat brain is recommended for test purposes because of its high GR content (5, 6). A satisfactory signal can be achieved in the intact animal (Fig. 1). After chronic adrenalectomy (ADX), the GR immunoreactivity is attenuated and difficult to trace. Corticosterone treatment (10 mg/kg intraperitoneally, 4 hr before sacrifice) of adrenalectomized rats results in a restoration of the immunoreactivity, and the intensity of the nuclear staining seems even to exceed that of the intact animal.

Fixation and Section Preparation

1. Use Nembutal for anesthesia at a dose of 40 mg/kg ip.
2. Perfuse animals transcardially, first with 50 ml of phosphate-buffered saline (PBS, pH 7.4), followed by 500 ml of 4% (w/v) paraformaldehyde (PFA) dissolved in PBS containing 20 mM molybdic acid, 1.0 nM dithiothreithol (DTT), and 1 mM ethylenediaminetetraacetic acid (EDTA) (7). Postfix the brains overnight. Alternate fixatives are picric acid–paraformaldehyde solution (8) or PFA at double pH (9). After the first successful stainings, the effect of adidtion of glutaraldehyde to the fixative in gradual increments might be tested; however, the tolerance depends greatly on the nature of the GR antibody.
3. Prepare 25- to 30-μm thick Vibratome (Technical Products Inc., St. Louis, MO) sections.

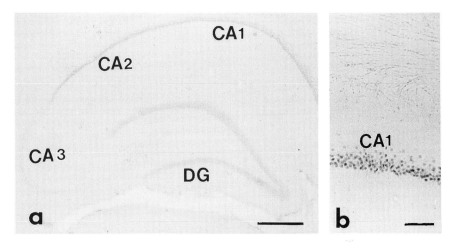

FIG. 1   Glucocorticoid receptor-immunoreactive cells in the hippocampal formation. (a) Intense staining in the CA1 and CA2 sectors and the dentate gyrus (DG) and faint staining in the CA3 area; (b) GR-immunopositive nuclei in pyramidal neurons of the CA1 area. Bar: (a) 150 μm; (b) 50 μm.

Immunocytochemical Labeling

1. Treat sections with 0.2% (v/v) Triton X-100 in PBS for 15 min to facilitate antibody penetration.

2. Block the specimens in 2% (v/v) normal goat serum for 20 min.

3. Apply specific GR antibody for 36 hr. The BuGr 2 monoclonal antibody was raised in mouse against purified rat liver GR (10) and used at a 1:750 dilution. (For other specific primary antibodies see Refs. 11–15.)

4. Incubate the sections in biotinylated anti-mouse IgG and streptavidin–peroxidase for 1 hr each, at 1:500 dilutions (Jackson ImmunoResearch Laboratories, Inc., West Grove, PA).

5. Visualize the immunoreactive loci with 3,3'-diaminobenzidine (DAB) chromogen as described earlier (16).

Signal Amplification

Steroid receptors occur in nuclei of cells exhibiting highly electron-dense chromatin material, making the recognition of the DAB label difficult. A chromogen significantly exceeding the electron density of the chromatin can be achieved by silver–gold postintensification of the DAB end product labeling GR-immunoreactive sites as follows.

1. Treat sections with 10% (v/v) thioglycolic acid for 2 hr to suppress endogenous argyrophilia of the tissue.
2. Intensify sections in a special physical developer (17) and allow them to react until the brown DAB turns black.
3. Tone the silver with gold.

For the step-by-step protocol of silver enhancement see Refs. 16 and 18.

Processing for Electron Microscopy

1. Postfix sections in 2% (w/v) OsO_4 for 20 min, dehydrate in a graded series of ethanol, and flat embed into Epon.
2. Cut ultrathin sections and contrast them with Reynolds' lead citrate and 2% (w/v) aqueous uranyl acetate.

Results

The preembedding labeling of glucocorticoid receptor allows a combined light and electron microscopic evaluation of the specimen (19). Loci displaying GR immunoreactivity contain silver–gold grains that are prominent in light and electron microscopic (EM) preparations. In intact animals, the cell nuclei possess most of the immunolabel, apparent in Vibratome (Fig. 1) and semithin (Fig. 2a) sections. At the ultrastructural level, the GR immunoreactivity appears to be associated with the chromatin (Fig. 3a and b). In glucocorticoid-deprived rats, the cytoplasm of the cells demonstrates relatively faint GR activity (Fig. 2b and c), and the nuclei are free of immunoreaction. Accordingly, ribosomes display most of the immunolabel in adrenalectomized animals (Fig. 3c and d), and the receptor immunoreactivity spreads into the dendritic processes. Corticosterone treatment of chronically adrenalectomized rats results in a shift in immunoreactivity from the cytoplasm to the nuclei (5, 18, 20), which become intensely labeled (Fig. 2d). In the latter experimental parardigm, GR immunoreactivity is traceable in conjunction with the plasma membrane and transport vesicles (Fig. 4) (21).

The ultrastructural verification of GR in association with nuclear chromatin (5, 19) further supports the view of genome targeting by the activated receptor, whereas demonstration of GR trafficking between the plasma membrane

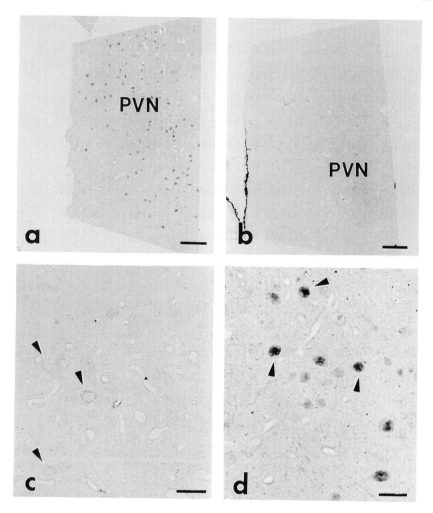

FIG. 2   Ligand-dependent shifting of GR immunoreactivity in parvicellular neurons of the hypothalamic paraventricular nucleus (PVN): semithin sections. (a) Nuclear expression in intact rat. (b) Diminution of GR staining after adrenalectomy. (c) Weak cytoplasmic signal (arrowheads) in adrenalectomized (ADX) rat. (d) Restored nuclear staining pattern (arrowheads) in corticosterone-treated ADX rat. Bar: (a and b) 50 μm; (c and d) 80 μm.

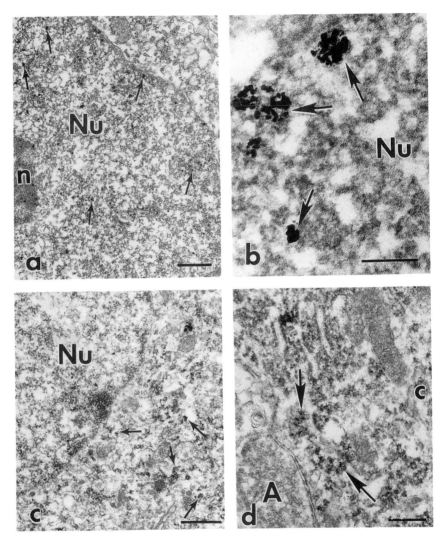

FIG. 3   Ultrastructural detection of GR immunoreactivity in intact (a and b) and adrenalectomized rats (c and d). Nu, Nucleus. (a) Scattered silver–gold grains (arrows) distributed in the nucleus. n, Nucleolus. (b) GR immunoreactivity (arrows) associated with chromatin. (c) Slight staining in the cytoplasm (arrows) and lack of labeling in the nucleus. (d) GR-immunoreactive ribosomes (arrows) in the cytoplasm, c. A, Axon terminal. Bar: (a) 0.5 μm; (b) 0.1 μm; (c) 0.5 μm; (d) 0.25 μm.

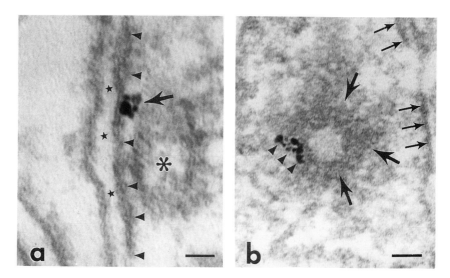

FIG. 4   Coupling of GR immunoreactivity to plasma membrane (a) and coated vesicles (b) in CA1 pyramidal neurons. (a) A cluster of silver grains (arrow) labels the inner surface (arrowheads) of the cell membrane. Asterisk, hollow vesicle; stars, intercellular space. (b) A coated vesicle (large arrows) demonstrates GR immunoreactivity beneath the cell membrane (small arrows). Bar: (a) 50 nm; (b) 100 nm.

and the cell nucleus (21) favors a role for the receptor in nongenomic membrane events evoked by glucocorticosteroids.

Further Applications

By means of ultrastructural double labeling, the hormone (19) or transmitter content of glucocorticosteroid-sensing neurons can be identified and the origin and chemotype of their afferents determined.

Ultrastructure of Estrogen Receptor-Immunoreactive Neural Elements

Animal Model

The intensity of estrogen receptor (ER) staining varies among species and different regions of the central nervous system. In the rodent brain, the medial preoptic area and the medial basal hypothalamus demonstrate a substantial ER immunoreactivity. The receptor staining is exaggerated after ovariectomy, providing a useful model for standardizing the receptor-labeling technique. Estradiol administration blunts the immunodetectability of ER (22). In intact female animals, the change in ER immunoreactivity during the estrous cycle should also be considered.

Choice of Fixatives

The wide spectrum of fixatives for ultrastructural localization of ER includes buffered picric acid–paraformaldehyde (PFA) containing 0.02% (v/v) glutaraldehyde (GA), (23) 4% (w/v) PFA containing 0.2% (v/v) GA (24), 4% (w/v) PFA at dual pH (25), and a 1% (w/v) PFA–1% (v/v) GA solution (26).

Fixation and Sectioning

1. Perfuse deeply anesthetized animals through the ascending aorta with 30 ml of PBS followed by 500 ml of the selected fixative. Depending on the nature of the fixative, the tissue might be postfixed further by immersion for 2–48 hr.
2. Prepare 25-μm Vibratome sections.

Immunolabeling of Estrogen Receptor

1. Treat sections with 1% (w/v) sodium borohydride in PBS for 30 min, provided the fixative contains more than 0.02% (v/v) glutaraldehyde. Agitate the floating sections frequently.
2. Rinse thoroughly in PBS and permeabilize the sections with detergent [0.2% (v/v) Triton X-100 in PBS] for 15 min.
3. Block potential nonspecific antibody-binding sites by incubating the specimens in 2% (v/v) normal goat serum.
4. Incubate the sections in the primary antibody for 36–48 hr. The published EM data are based on the use of the H222 rat monoclonal antibody raised against purified human ER (27), generously distributed by Abbott Laboratories (North Chicago, IL). The working dilution is in the range of 200 to 1000. An antibody against amino acids 270–284 of the rat ER (28) has become available under the aegis of the National Hormone and Pituitary Program (Baltimore, MD).
5. Complete the immunolabeling by incubating the sections in biotinylated goat anti-rat IgG followed by streptavidin–peroxidase at 1 : 500 dilutions for 1 hr in each (Jackson ImmunoResearch Laboratories).

6. Visualize the ER-containing profiles with the stable 3,3'-diaminobenzidine (DAB) chromogen. Another choice of chromogen might be 3,3',5,5'-tetramethylbenzidine (TMB) (29), which, however, requires further stabilization steps.

Silver–Gold Enhancement of 3,3'-Diaminobenzidine Label

Amplify the DAB end product of the enzymatic reaction by silver intensification (30) as described for the enhancement of GR immunostaining.

Electron Microscopic Procedure

Osmicate the sections and apply any of the routine flat-embedding techniques. Use contrasted ultrathin sections.

Results of Estrogen Receptor Labeling

The examples presented below were chosen from the medial preoptic area of the female rat brain, which demonstrates a moderate ER immunoreactivity in the intact animal (Fig. 5a). Ovariectomy leads to an increased ER immunoreactivity, manifested in the appearance of more labeled neurons (Fig. 5b and c). In both cases, the receptor immunoreactivity is confined to the nuclei of the cells. Acute estradiol treatment of ovariectomized rats diminishes the ER staining (Fig. 5d). Ultrastructural analysis (23) shows that the ER is bound to chromatin in the cell nuclei without representation in the nucleoli. The diaminobenzidine chromogen ensures the recognition of the labeled nuclei of ER-positive cells that show up in an immunonegative neuropil and the neighborhood of nonlabeled cells (Fig. 6). The silver–gold-intensified chromogen enhances the recognition of ER-bearing elements (Fig. 7a) and provides a higher resolution of the organelle immunopositive for ER (Fig. 7b).

The predominant nuclear residence of the unoccupied and saturated forms of ER (31) observed in the rat (23) seems to contrast with that of the guinea pig. In that species, the receptor was detected not only in the nuclei but also in the perikaryal cytoplasm, dendritic processes (32), and, occasionally, in axons of neurons of the ventromedial hypothalamus (24). It should be determined whether methodological differences account for the discrepancies or whether variations in subcellular receptor residence exist among different species. Other, nonneuronal cell types have also been reported as immunoreactive for ER in the guinea pig brain (33). In the human endometrium, the majority of the receptors exist in the nuclear pool bound to euchromatin (34).

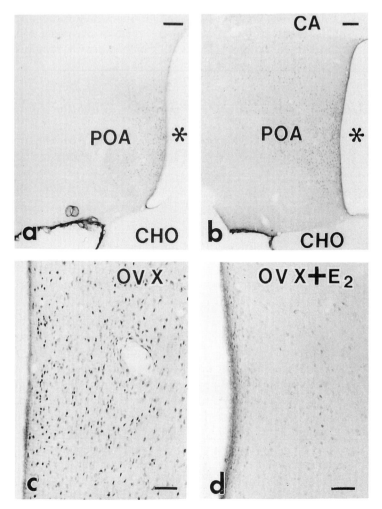

FIG. 5   Estrogen receptor-immunopositive cells in the medial preoptic area (POA) of the female rat brain. CHO, Optic chiasma; CA, anterior commissure; asterisk, third ventricle. (a) Intact animal; (b) ovariectomized (OVX) rat; (c) nuclear staining pattern in OVX rat; (d) abolishment of ER immunoreactivity by estradiol (E_2) administration to OVX rat. Bar: (a and b) 200 μm; (c and d) 100 μm.

Indications

These ultrastructural findings support the view that the majority of estrogen receptors reside in the cell nucleus even in the absence of the ligand (35), and that the activated receptors display an anatomical position required for

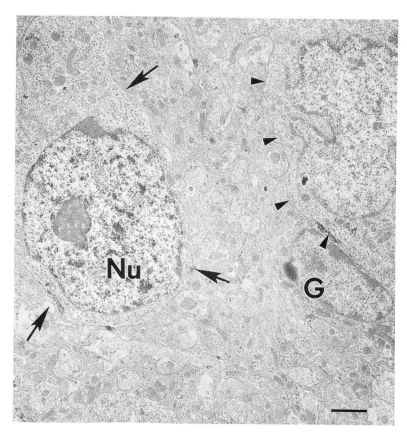

FIG. 6   Localization of ER at the ultrastructural level. Immunoreactive cell (arrows) contains a diaminobenzidine-labeled nucleus (Nu). Arrowheads point to an immuno-negative neuron. G, ER-free glial cell. Bar: 2 μm.

interaction with the genome. The cytoplasmic occurrence of the receptor, apparently low in the rat, seems to reflect sites of synthesis and degradation (23). The cytoplasmic ER pool detected in the guinea pig (24) might indicate loci for nongenomic actions.

Further Applications

By the use of chromogens of contrasting physicochemical properties, the neuromessenger content of afferents to ER-synthesizing neurons (Fig. 8) can be revealed in the framework of ultrastructural dual-antigen localization techniques (25, 29).

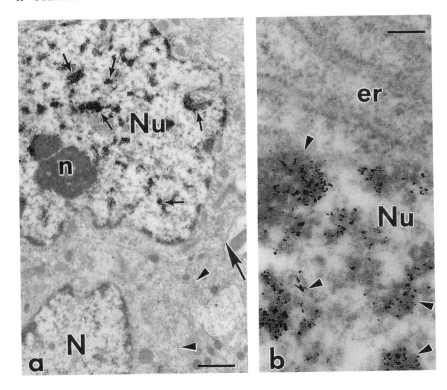

FIG. 7   Detection of ER immunoreactivity with silver–gold-intensified DAB chromogen. Nu, nucleus. (a) The chromatin of the ER-immunopositive neuron (large arrow) is labeled by metallic grains (small arrows). The nucleolus (n) and cytoplasm are immunonegative. N, Nonlabeled neuron (arrowheads). (b) High-power demonstration of the chromatin-coupled ER (arrowheads) and nonreactive rough endoplasmic reticulum (er). Bar: (a) 1 μm; (b) 200 nm.

Concluding Remarks

Ultrastructural immunocytochemistry offers insight into the subcellular distribution of steroid receptors; reveals the cellular organelles that participate in the synthesis, processing, and transport of the receptors; and provides information about the putative intracellular sites of receptor action. The electron microscopic receptor localization technique, which has the highest morphological resolution, is an equal counterpart of biochemical and molecular biological methods in addressing the mechanisms of receptor-mediated steroidal effects within the nervous system.

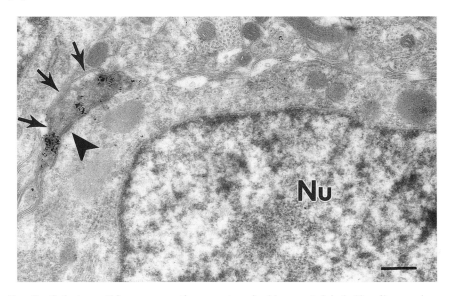

FIG. 8   Substance P-immunoreactive axon terminal (arrows), labeled by silver grains, forms a synapse (arrowhead) with an ER-immunopositive neuron whose nucleus (Nu) is DAB stained. Bar: 0.5 μm.

Acknowledgments

We thank Dr. R. W. Harrison for the kind donation of the BuGr 2 antibody and the Abbott Laboratories for the gift of the H222 monoclonal antibody. This work was supported by grants from the National Science Foundation of Hungary (OTKA 17 and 5512). This review is dedicated to Professor Béla Flerkó on the occasion of his 70th birthday.

References

1. B. S. McEwen, H. Coirini, and M. Schumacher, *in* "Steroids and Neuronal Activity" (D. Chadwick and K. Widdows, eds.), p. 3. Wiley, New York, 1990.
2. R. M. Evans and J. L. Arriza, *Neuron* **2,** 1105 (1989).
3. M. Aronsson, K. Fuxe, Y. Dong, L. F. Agnati, S. Okret, and J.-Å. Gustafsson, *Proc. Natl. Acad. Sci. U.S.A.* **85,** 9331 (1988).
4. R. B. Simerly, C. Chang, M. Muramatsu, and L. W. Swanson, *J. Comp. Neurol.* **294,** 76 (1990).
5. K. Fuxe, A. Cintra, A. Härfstrand, L. F. Agnati, M. Kalia, M. Zoli, A.-C. Wikström, S. Okret, M. Aronsson, and J.-Å Gustafsson, *Ann. N.Y. Acad. Sci.* **512,** 362 (1987).

6. J. M. H. M. Reul and E. R. De Kloet, *Endocrinology (Baltimore)* **117,** 2505 (1985).
7. R. Uht, J. F. McKelvey, R. W. Harrison, and M. C. Bohn, *J. Neurosci. Res.* **19,** 405 (1988).
8. L. Zamboni and C. D. Martino, *J. Cell Biol.* **35,** 148 (1967).
9. A. Berod, B. Hartman, and J. F. Pujol, *J. Histochem. Cytochem.* **29,** 844 (1981).
10. B. Gametchu and R. W. Harrison, *Endocrinology (Baltimore)* **114,** 274 (1984).
11. P. A. Bernard and T. H. Joh, *Arch. Biochem. Biophys.* **229,** 466 (1984).
12. J.-Å. Gustafsson, S. Okret, A.-C. Wikström, B. Anderson, M. Radojcic, Ö. Wrange, W. Sachs, A. J. Doupe, P. H. Patterson, B. Cordell, and K. Fuxe, *in* "Steroid Hormone Receptors: Structure and Function" (H. Eriksson and J.-Å. Gustafsson, eds.), p. 355. Elsevier, Amsterdam, 1983.
13. S. Okret, A.-C. Wikström, Ö. Wrange, B. Anderson, and J.-Å Gustafsson, *Proc. Natl. Acad. Sci. U.S.A.* **81,** 1609 (1984).
14. W. C. McGimsey, J. A. Cidlowski, J. A. Stumpf, and M. Sar, *Endocrinology (Baltimore)* **129,** 3064 (1991).
15. H. M. Westphal, G. Moldenhauer, and M. Beato, *EMBO J.* **1,** 1467 (1982).
16. Z. Liposits, D. Sherman, C. Phelix, and W. K. Paull, *Histochemistry* **85,** 95 (1986).
17. F. Gallyas, *Acta Morphol. Acad. Sci. Hung.* **19,** 57 (1971).
18. Z. Liposits, *Progr. Histochem. Cytochem.* **21,** 1 (1990).
19. Z. Liposits, R. M. Uht, R. W. Harrison, F. P. Gibbs, W. K. Paull, and M. C. Bohn, *Histochemistry* **87,** 407 (1987).
20. R. S. Ahima, C. N. B. Tagoe, and R. E. Harlan, *Neuroendocrinology* **55,** 683 (1992).
21. Z. Liposits and M. C. Bohn, *J. Neurosci. Res.* **35,** 14 (1993).
22. A. Cintra, K. Fuxe, A. Härfstrand, L. F. Agnati, L. S. Miller, J. L. Greene, and J.-Å. Gustafsson, *Neurochem. Int.* **8,** 587 (1986).
23. Z. Liposits, I. Kalló, C. W. Coen, W. K. Paull, and B. Flerkó, *Histochemistry* **93,** 233 (1990).
24. J. D. Blaustein, M. N. Lehman, J. C. Turcotte, and G. Greene, *Endocrinology (Baltimore)* **131,** 281 (1992).
25. I. Kalló, Z. Liposits, B. Flerkó, and C. W. Coen, *Neuroscience* **50,** 299 (1992).
26. I. Kalló and Z. Liposits, unpublished observations (1992).
27. G. L. Greene, C. Nolan, J. P. Engler, and E. V. Jensen, *Proc. Natl. Acad. Sci. U.S.A.* **77,** 5115 (1980).
28. J. D. Furlow, H. Ahrens, G. C. Mueller, and J. Gorski. *Endocrinology (Baltimore)* **127,** 1028 (1990).
29. M. C. Langub and R. E. J. Watson, *Brain Res.* **573,** 61 (1992).
30. Z. Liposits, G. Setalo, and B. Flerko, *Neuroscience* **13,** 513 (1984).
31. W. J. King and G. L. Greene, *Nature (London)* **30,** 745 (1984).
32. A.-J. Silverman, L. L. DonCarlos, and J. I. Morrell, *J. Neuroendocrinol.* **3,** 623 (1991).
33. M. C. Langub and R. E. J. Watson, *Endocrinology (Baltimore)* **130,** 364 (1992).
34. M. F. Press, N. A. Nousek-Goebl, and G. L. Greene, *J. Histochem. Cytochem.* **33,** 915 (1985).
35. J. Gorski, D. Furlow, F. E. Murdoch, M. Fritsch, K. Kaneko, C. Ying, and J. R. Malayer, *Biol. Reprod.* **48,** 8 (1993).

[12] Hybridization Studies of Adrenocorticosteroid Receptors in the Central Nervous System

James P. Herman, Seung P. Kwak, and Stanley J. Watson

Introduction

Steroid hormone receptors are prime modulators of protein biosynthesis in all vertebrate organisms. These receptors translate endocrine signals into actions at the individual cell by selective binding to steroid hormone molecules. Following ligand binding, receptor–ligand complexes undergo conformational changes that culminate in direct and/or indirect interactions with gene transcription and/or mRNA half-life of a wide variety of genes, resulting in potentially profound changes in the assortment of proteins available to receptive cells (1). Given their widespread influence, it is critical to understand mechanisms governing the cellular placement and regulation of these receptor molecules.

The present chapter summarizes the methodological approaches taken by our laboratories for study of the mRNAs and genes encoding adrenocorticosteroid hormone receptors. Adrenocorticosteroid receptor molecules bind glucocorticoid (e.g., corticosterone) and mineralocorticoid (e.g., aldosterone) hormones. These receptors have been traditionally viewed as modulators of the neuroendocrine stress response and renal sodium : potassium balance, respectively; however, the advent of molecular biological approaches to the study of these receptors has revealed the potential for considerable functional complexity. The glucocorticoid receptor (GR) (also referred to as the "type II corticosteroid receptor") is the prime mediator of traditional glucocorticoid effects at the liver and pituitary, yet is present in a diverse population of central nervous system (CNS) neurons, indicating the potential for important glucocorticoid actions on CNS gene transcription and negative feedback regulation of the hypothalamo–pituitary–adrenal (HPA) stress axis (2–4). The mineralocorticoid receptor (MR) (also referred to as the "type I corticosteroid receptor") is also localized in a large subset of CNS neurons (5–7) and indeed binds the endogenous glucocorticoid corticosterone with a higher affinity than the GR (8) and may, at the level of brain, conduct the physiological effects of low circulating corticosterone doses (9). In addition, the two receptors may have interactive influences on cellular function, serving to influence the efficacy of other protein factors on gene transcription (10).

One of the prime requirements for understanding GR and MR protein actions in the CNS is a knowledge of factors promoting their synthesis. Over the past several years, many laboratories, including our own, have been involved in analyzing the regulation of mRNA molecules encoding these proteins *in vivo,* using a variety of hybridization approaches. What has emerged from these studies is a complex picture of adrenocorticosteroid receptor biosynthesis, involving regulation at the level of mRNA expression and RNA processing. We summarize below the methodology used to uncover this evidence and to begin to pursue the implications of this interesting body of data on corticosteroid receptor synthesis.

Northern Blot Hybridization

Northern blot hybridization techniques have not been employed extensively for analysis of GR and MR mRNA regulation in brain for two primary reasons: first, GR and MR mRNAs are of low to intermediate abundance in brain, and can require large amounts of tissue for analysis, and second, both GR and MR mRNAs are large (up to 7 kb), and thus do not transfer efficiently from agarose gels to Nytran or nitrocellulose membrane. However, it should be noted that initial characterization of GR mRNA was accomplished using this technique, and indeed revealed some interesting data especially amenable to analysis by this method, namely, that multiple mRNA forms exist for the GR that vary according to selection of polyadenylation site (11).

In Situ Hybridization

Overview

In situ hybridization essentially applies the logic of hybridization in solution to individual tissue sections (see Ref. 12 for review). This method involves application of radiolabeled probe to fixed tissue sections in which the mRNA molecules of interest are known or believed to be localized. Sections and probe are incubated under conditions appropriate for RNA:RNA duplex formation, and the sections subsequently digested with RNase to eliminate single-stranded RNA molecules (including unbound probe). What is left behind are thus "protected" double-stranded RNAs localized to individual cells in the tissue sections. The anatomical resolution of this method, and its amenability to relative quantitation, render it especially well suited to the study of GR and MR mRNA regulation in the heterogeneous brain.

Probes

As in any hybridization experiment, choice of probe is essential for accurate and valid assessment of corticosteroid receptor mRNA expression in brain. There are three basic types of probes, each of which has its own set of advantages and disadvantages: (a) oligonucleotide probes are comparatively short DNA constructs (generally less than 50 nucleotides) that are designed to be complementary to a carefully chosen stretch of the target RNA. In that the target can be carefully specified, and that they are easy to use, oligonucleotides can be useful for a number of applications. However, the labeling method employed for oligonucleotides (tailing with deoxynucleotidyltransferase) does not produce probes with sufficient specific activity to visualize low-abundance mRNAs, into which category the corticosteroid mRNAs fall. (b) cDNA probes can be synthesized by any of several methods, including nick translation, random priming, or polymerase chain reaction (PCR). These probes can be labeled to varying degrees of specific activity (with PCR being the highest), and are significantly more sensitive than oligonucleotides. Most work to date suggests that nick-translated or random-primed probes are less sensitive than cRNA probes; the relative sensitivity of PCR-generated probes has yet to be assessed. (c) cRNA probes are synthesized *in vitro* using viral RNA polymerases. By virtue of their ability to incorporate labeled nucleotides into growing RNA strands, these probes can be labeled to as high a specific activity as desired or feasible, affording maximal sensitivity, and have indeed been employed most often in the study of corticosteroid receptors. Note that cRNA probes are not without disadvantages, in that the "stickiness" of RNA molecules can generate high levels of background signal.

Choice of target sequences within the RNA under analysis is critical. Clearly, selection of a sequence having high nucleotide homology with other mRNAs can produce probes that cross-hybridize, yielding erroneous signal localization. Nowhere is this consideration more relevant than in the study of MR and GR mRNAs, which share regions of high homology (i.e., DNA-binding domain) not only with each other but with a whole family of DNA-binding proteins (13). Thus, probes should be designed to base pair with sequences unique to the molecule of interest (verified against a nucleic acid database, such as GenBank). In addition, riboprobe sequences derived from one species and used on tissue of a heterologous species may increase the incidence of "cross-hybridization," because the hybridization conditions must be less stringent to allow for nucleotide mismatch. When attempting to hybridize cross-species, it is therefore essential to choose carefully sequences either possessing or likely to possess highly homologous regions.

For *in situ* hybridization studies of MR and GR mRNA regulation we have used cRNA probes exclusively. Therefore, our remarks are focused on this particular method; however, it should be noted that other approaches may be employed with good prospects for success.

Protocol 1: Labeling of cRNA Probes

1. Following selection of an appropriate target sequence, plasmid constructs are linearized with restriction endonucleases yielding blunt ends or 5′ overhangs (3′ overhangs are not recommended, owing to formation of hairpins and the potential for double-back transcription).

2. Immediately prior to transcription, a "transcription cocktail" is mixed. Depending on the total counts needed, we run reactions in either 12.5- or 5-μl volumes. Reaction conditions for 5-μl reactions are in parentheses. The transcription cocktail contains the following, added in order: 5 μl of 5× reaction buffer; 2 μl of 100 mM dithiothreitol (added to maintain reduced sulfhydryl groups); 1.5 μl of a mixture of 2.5 mM ATP, CTP, and GTP; 1 μl of 100 μM UTP [assures that the final UTP concentration is maintained above the K_m of the polymerase molecules (e.g., 13 μM for SP6 polymerase)]; 1 μl (20–40 U) of RNase inhibitor [e.g., placental RNase inhibitor (Boehringer Mannheim, Indianapolis, IN)]; 10.5 (9.5) μl diethyl pyrocarbonate (DEPC)-treated water. Premixed cocktail is good for two 12.5-μl (or five 5-μl) reactions, but should be used immediately.

3. ^{35}S UTP [125 μCi (50 μCi)] is evaporated to dryness and resuspended in 10.5 μl (4 μl) of transcription cocktail; to this cocktail are added, in order, 1 μl (0.5 μl) of linearized template DNA, corresponding to 1 μg (0.5 μg), and 1 μl (0.5 μl) of appropriate RNA polymerase (10–20 U).

4. Transcription reactions are incubated at 37°C for 60–90 min. Following incubation, labeling reactions are brought to a 100-μl total volume and precipitated with a 0.5-vol of 7.5 M ammonium acetate and 3 vol of ice-cold 100% ethanol.

5. Probes are resuspended in 100 μl (50 μl) of DEPC-treated water and a 1-μl aliquot counted to determine incorporation of label. The 12.5-μl reactions should yield 100–400 million total counts per minute (cpm); 5-μl reactions should yield 30–150 million cpm.

Tissue Preparation

Appropriate pretreatment of tissue sections is required for all *in situ* hybridization applications. Pretreatment consists of several steps, including fixation, permeabilization, and acetylation. Fixation is required to ensure retention of RNA in the tissue section throughout hybridization. There are a number of fixation protocols employed for *in situ* hybridization histochemistry (ISHH), ranging from perfusion fixation to postfixation of fresh-frozen tissue with acetone : alcohol. We routinely postfix fresh-frozen tissue in 4% (w/v) paraformaldehyde for 30 min immediately following removal of slide-mounted sections from the freezer; in our hands, this technique is optimal for visualization of GR and MR mRNAs in brain. Both RNAs can be detected in perfusion-fixed sections, albeit with lesser sensitivity; aggressive fixation with stronger cross-linking agents (glutaraldehyde and acrolein) tends to dampen ISHH signal significantly (probably due to poor probe penetration) and is not recommended.

Permeabilization is required to allow access of RNA probes to RNAs immobilized in tissue sections. This step can be carried out using either washes in 0.2 *N* HCl or incubation with proteases (e.g., proteinase K). We have found the latter to be most effective in promoting visualization of corticosteroid receptor mRNAs, using concentrations of 0.1–0.2 μg/ml. Somewhat higher concentrations are necessary when using perfusion-fixed material (1 μg/ml).

As mentioned above, one of the primary problems with cRNA probes is nonspecific electrostatic binding of probe to positively charged moieties in the tissue section. To reduce this problem, an acetylation step is frequently employed to reduce the amount of free positive charge. The acetylation reaction involves brief incubation of tissue sections with acetic anhydride.

Following pretreatment, tissue is dehydrated through graded ethanols and stored at $-20°C$ until hybridized. We have maintained pretreated tissue at $-20°C$ for up to 6 months with no loss of signal intensity.

Protocol 2: Tissue Pretreatment (Postfixation)

1. Sections are removed from the freezer and immediately immersed in a cold phosphate-buffered 4% paraformaldehyde solution for 30 min. Tissue is then rinsed twice with nanopure water, followed by two more prolonged (10-min) washes in 2× SSC (1× SSC is 150 m*M* sodium chloride, 15 m*M* sodium citrate, pH 7.0).

2. Tissue is then incubated in 0.1–0.2 μg of proteinase K (PK) per milliliter in 10 m*M* Tris–5 m*M* ethylenediaminetetraacetic acid (EDTA), pH 8.0, for 15 min at 37°C in prewarmed glassware dedicated to this purpose. (*Note:* use 1.0 μg of PK per milliliter for perfusion fixed tissue.)

3. After PK treatment, tissue is rinsed once in nanopure water (1 min) and once in 0.1 *M* triethanolamine (TEA), pH 8.0 (1 min). Following rapid equilibration in TEA, acetic anhydride is added under rapid agitation to a final concentration of 0.25% and stirred for 10 min.

4. The acetic anhydride solution is then decanted, and the sections incubated in 2× SSC for 5 min and dehydrated through graded ethanols.

Hybridization

Hybridization conditions are set to promote formation of stable, "full-length" probe : RNA hybrids. Owing to the stability of RNA : RNA molecules, optimal riboprobe hybridization temperatures can be in excess of 80°C in aqueous solution. Probes are thus diluted in a buffer containing 50–75% (v/v) formamide, which interferes with hydrogen bonding and effectively decreases the optimal temperature for the probe : RNA duplex (approximately 0.35°C/% formamide) (14). Specificity of hybridization can be compromised by nonspecific binding of probe to highly abundant tissue constituents, including tRNAs, DNA, proteins, and polysaccharides. Such nonspecific binding is typically blocked by inclusion of representative molecules (e.g., yeast tRNA, salmon sperm DNA, polyvinylpyrrolidone, Ficoll) in the hybridization mix. Finally, the speed of hybrid formation is greatly increased by the inclusion, in the reaction mix, of dextran sulfate, a large polyanionic molecule that acts to concentrate probe at the mRNA by an excluded volume effect.

Posthybridization

Following hybridization of probes for an appropriate time period (at least 8 hr), tissue sections are rinsed and incubated in an RNase A solution. As RNase A digests only single-stranded RNA, cRNA–mRNA duplexes are protected and remain *in situ*, whereas the remaining RNAs, including probe, are digested. This protocol in essence eliminates labeling associated with probe trapped or nonspecifically bound to the tissue section, whereas the cRNA : mRNA hybrid is unmodified. Tissue is then exposed to a high-temperature wash, generally under highly stringent conditions (0.1× SSC, 15–20°C below calculated melting temperature), to eliminate hybrids formed between probe and RNAs having partial homology to the RNA of interest.

Protocol 3: Hybridization and Posttreatment

1. Probe is denatured at 65°C for 10 min and diluted in an appropriate hybridizaton buffer to a concentration of 500,000 to 1 million disintegrations per minute (dpm)/30 μl. Our standard hybridization buffer consists of 50% (v/v) formamide, 3× SSC, 10% (w/v) dextran sulfate, 1× Denhardt's solution yeast tRNA (200 μg/ml), and salmon sperm DNA (100 μg/ml) in 50 mM Tris-HCl, pH 7.5.

2. Probe is applied to coverslips and the coverslips apposed to tissue sections. Slides are placed in bioassay dishes (Nunc, Roskilbe, Denmark)

atop absorbent paper (G5000; Schleicher & Schuell, Keene, NH) moistened with 50% formamide. Slides are kept off the surface of the filter paper by plastic dowels. Sections are hybridized overnight (16 hr) at 55–60°C.

3. Soak off coverslips in 2× SSC. Wash uncoverslipped slides in fresh 2× SSC (10–20 min).

4. Immerse slides in prewarmed 100 μg/ml RNase A in 2.5 M NaCl–50 mM Tris, pH 8.0; incubate at 37°C for 30 min.

5. Rinse slides three times in room-temperature 0.1× SSC; place in 0.1× SSC warmed to 60–65°C for 1 hr.

6. Remove slides to room-temperature 0.1× SSC and dehydrate through graded ethanols.

Detection and Analysis

There are several alternatives for detection and quantitation of steroid receptor mRNA expression by *in situ* hybridization. Selection of the appropriate technique depends in large part on the experimental design; for example, when analyzing GR and/or MR expression in large regions of brain where receptor mRNA is particularly abundant (such as the hippocampus), it is often most useful to generate high-quality X-ray autoradiographs for densitometric analysis. Alternatively, cells from nuclei of interest can be visualized on photographic emulsion and analyzed by grain-counting or areal density measures. Finally, for visualization or colocalization applications nonradioactive labeling and detection methods can be employed.

Each analytical method listed above has its own set of assets and limitations. Areal density measures are applicable to cases in which cells are densely packed, as is the case for both GR and MR mRNA in the hippocampal formation and of GR mRNA in the medial parvocellular paraventricular nucleus. However, when labeled cells are widely scattered, areal measurements are diluted by nonhybridized regions, potentially masking effects at the cellular level. Measures at the cellular level (i.e., grain counting) can effectively define changes at the single cell, and are applicable to regions showing scattered GR and MR neurons (e.g., amygdaloid nuclei). However, this method is time and labor intensive, and in areas of high packing density it can be difficult to clearly define individual neurons. Nonradioactive methods suffer from sensitivity limitations, and indeed in our experience have proved inconsistent for visualization of GR and MR mRNA, showing positive signal only in regions of high abundance (e.g., hippocampus). In addition, nonradioactive *in situ* hybridization using immunochemical detection steps is not truly quantitative, limiting applicability for study of corticosteroid receptor mRNA regulation.

For our applications, we have found low-power areal densitometric analysis most useful for study of corticosteroid receptor regulation. Following appropriate exposure times [generally, 5–7 days on Kodak (Rochester, NY) XAR5 film, 30–45 days on NTB2 emulsion], sections through regions of interest, for example, the hippocampus, are visualized (light box or microscope under low power) and digitized using NIH Image software (Wayne Rasband; Bethesda, MD). Individual subregions are resolved on the basis of morphology, the mean gray level of the sample determined, and background gray level subtracted. Mean gray level is calculated for each region in each individual of the group under study, and groups then analyzed for experimentally relevant changes at the regional level.

Protocol 4: Densitometric Analysis of in Situ Hybridization Autoradiographs

Equipment

Densitometric analysis of *in situ* hybridization data requires a computer equipped with a video capture board and sufficient RAM to store and process images (recommend at least 8 Mbytes) interfaced with a high-resolution video camera. Images are generated from autoradiographs using an illuminator with a macro lens (mounted on a copy stand) or a microscope (with dark-field and bright-field illumination). An optical drive for image storage is recommended, but not required.

Procedure

1. Expose photographic medium (either film or emulsion) to achieve desired signal intensity (i.e., avoid saturation), processing sections and standards (see ref. 12) concomitantly.

2. Set appropriate illumination parameters: first, select "darkest" nonsaturated sample image in set to be analyzed. Adjust video gain and illumination to achieve maximal gray level (we recommend mean gray level of 150–200 (black being 255), to avoid the possibility of saturation). Select a homogeneous nonhybridized region of film or slide (i.e., away from tissue section), and adjust video gain and illumination to a low but nonzero gray level (10–20 units). Continue adjusting the video interface to satisfy both high-level and low-level criteria. On a "background" image, use the illumination correction option to adjust for any uneven illumination resulting from the light source.

3. Construct a standard curve. Digitize brain paste or commercial standards and measure the gray level over a homogeneous area, using the sampling option of the imaging software. Using calibration options available in

the software, enter radioactivity levels corresponding to the appropriate standard to generate the standard curve. Alternatively, gray-level values can be exported to other software capable of performing curve-fitting functions. Establish the linear range of the standard curve; accept only data values falling within this range.

4. Sample images using one of the following protocols:

Simple Densitometry   Capture a sample image; sample either by drawing an image boundary or by using a selection tool to obtain the RAW MEAN GRAY LEVEL. Capture the background image over the unhybridized area, and subtract from sample determination to arrive at the CORRECTED MEAN GRAY LEVEL. If the sample is taken from equivalent areas of tissue, one may choose to multiply the CORRECTED MEAN GRAY LEVEL by the area encompassed by signal to obtain the INTEGRATED GRAY LEVEL.

Redirected Sampling

Counterstained or double stained image (emulsion applications): Images are captured under illumination conditions ideal to visualize the reference image (i.e., a counterstained region delineating a region of interest). The region of interest is defined with a selection tool. Sample images are then captured under conditions appropriate for visualization (factoring out the counterstain) without moving slide. A redirect option is then used to apply selection to the sample image; proceed as above.

Aligned images (film or emulsion): Capture the reference image as described above and store. Using the overlay option (e.g., Paste Control in Image), align the sample image as near as possible to the reference image. Redirect selection tool as above.

Ratio imaging (film or emulsion): Align the internal standard and sample image as described above. Using overlay control, instruct the computer to divide the sample gray levels by standard gray levels derived from the same pixel. The net result will be an image expressing the ratio of the two images.

When necessary, redirected imaging protocols may be used to sample specific cell populations. An example is shown in Fig. 1. Three consecutive sections demonstrating Nissl substance, GR mRNA, and MR mRNA were aligned manually at low power. The sampling tool was then used to select subfield CA1 on the basis of the pattern of Nissl staining, and the sampled region "redirected," or applied, to the two autoradiographic images. This method allows for accurate assessment of localization and/or regulation of GR and MR mRNA in tightly defined cell regions.

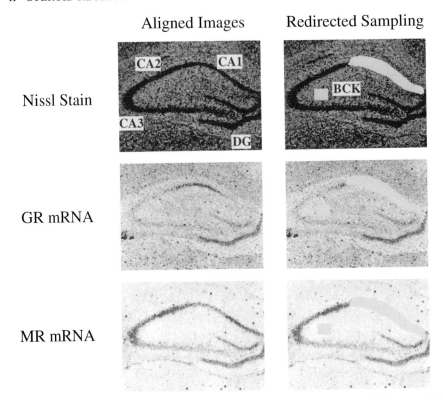

FIG. 1 Example of sampling protocol for densitometric analysis of GR and MR mRNA expression in the hippocampal formation. *Top:* Digitized Nissl stain. *Middle* and *lower:* Digitized representations of autoradiographs images obtained following hybridizations with GR and MR probes, respectively. Simple densitometry is conducted by visual comparison of the cytoarchitectonic subdivisions identifiable by Nissl stain and GR and MR hybridization signal, followed by manual sampling. Redirected densitometry is accomplished by alignment of the serial sections, superimposing a live image over one previously saved. As pixels will be mapped precisely, several consecutive images can be saved and aligned. In this case, three serial sections are manually aligned. Subfield CA1 is delineated on the Nissl-stained section, and the selected area superimposed on each of the two *in situ* hybridization autoradiographs, using a redirect option. In this manner, specific anatomical regions can be defined on the basis of cytoarchitectonic criteria and used to assess regional steroid receptor mRNA expression accurately. DG, Dentate gyrus; BCK, background.

Applications and Caveats

In situ hybridization analysis has yielded a wealth of information on numerous aspects of corticosteroid mRNA regulation. In addition to straightforward densitometric analyses conducted as above (reviewed in Ref. 12), this technique is also useful for addressing issues of mRNA abundance and cytoplasmic and nuclear RNA processing. The latter, "nontraditional" uses of ISHH have been applied extensively to the study of corticosteroid receptors, by virtue of both accident and intent. The results of these analyses, summarized below, reveal both the limitations and values of ISHH as a tool for the study of steroid receptor mRNAs.

Relative Abundance of Corticosteroid Receptor mRNAs

Most reports agree that GR and MR mRNAs exhibit highest abundance in the hippocampal formation. Distribution within the hippocampus differed for the two receptor types, with GR distributed as CA > dentate gyrus (DG) ≫ CA3, and MR as CA2–CA3a > CA1 = DG > CA3b (15–18). Both GR and MR mRNAs were shown to be widely distributed in numerous other brain regions (15–18) as well as glial cell cultures (19), albeit at significantly lower abundance than in hippocampus. By matching probe-specific activity, length, and G-C content, we were able to demonstrate that MR mRNA is present in higher abundance than GR mRNA, with MR/GR abundance ratios varying from 1.5 to 1.7 in CA1 and from 2.5 to 5.0 in CA3 (20). Relative abundance levels were verified by RNase protection analysis of regional hippocampal microdissections. It should be noted that MR/GR ratios vary with CNS region; indeed, with some notable exceptions (e.g., lateral septum, and cortical amygdala), extrahippocampal regions appear to exhibit higher expression of GR than MR.

Key inconsistencies in the literature regarding relative abundance of GR in different hippocampal subfields have led to substantial rethinking of the mechanism of corticosteroid receptor mRNA regulation. Whereas several reports, including our own, indicate low levels of GR mRNA in CA3 and a high MR/GR ratio in hippocampus (20–22), other groups have reported homogeneous GR distribution across the hippocampus (with CA3 levels rivaling those of CA1 or DG) and a low MR/GR ratio (23). Comparison across laboratories revealed that reports showing high CA3 expression employed probes directed against 5' coding sequences, whereas low CA3 expression is observed with probes against the 3' coding region and/or 3' untranslated region. Analysis using both probes verified this observation, and regulatory studies revealed that the 3' probe was significantly downregulated in aged rats, whereas the 5' probe showed no significant age regulation (24, G. Cizza, personal communication). The 5' probe revealed substantially greater hybridization intensity than the 3′ probe, further suggesting posttranscriptional processing of GR mRNA within hippocampal neurons. The low relative abundance of CA3 3′ sequences indicates that this processing event is cell specific as well; in CA3, a greater proportion of GR mRNA may be processed (or degraded) into a form that does not result in translation of full-length protein (inferred from low expression of 3′ coding sequences).

Heterogeneous Regulation of Glucocorticoid and Mineralocorticoid Receptor mRNA Expression

One of the prime advantages of semiquantitative ISHH analysis is the ability to visualize changes in mRNA expression in heterogeneous cell populations. This approach has been used by our laboratories to assess regulation of corticosteroid receptors by their appropriate ligands, that is, glucocorticoids themselves. The most striking result of such analyses is the consistent observation that GRs and MRs are regulated by glucocorticoids in a site-specific fashion. For example, glucocorticoid depletion results in twofold increases in GR mRNA in subfields CA1 and CA3 in the hippocampus, a 50% increase in dentate gyrus, and no change in cerebral cortex, thalamus, or hypothalamic paraventricular nucleus. Mineralocorticoid receptor mRNA was increased by only 30–40% in subfields CA1 and CA2, and not at all in CA3, dentate gyrus, or cortex (6, 20). These data clearly indicate that, although susceptible to steroid regulation, CNS GR and MR mRNA regulation is probably governed by actions at the neural circuit level as well, and that the balance of regulatory influences is summated differently by different classes of neurons.

Data from our laboratories have indicated that steroid regulation of corticosteroid receptor mRNA expression interfaces considerably with endogenous diurnal rhythms. Notably, marked decreases in both GR and MR mRNA expression can be observed in CA1, CA3, and DG at a point in the diurnal cycle where steroid secretion is actively rising. These drops are sufficiently precipitous that simple zero-order decay curves predict mRNA half-lives on the order of 6.5 (MR) to 8 (GR) hr (6). Although these simple estimates do not take into account variables influencing cellular mRNA stability (e.g., degradation), it is clear that regulation of corticosteroid mRNA levels can occur across narrow time windows, and that biosynthesis of these receptors in hippocampus is susceptible to short-term changes in glucocorticoid levels.

Gene Transcription, Splicing, and mRNA Regulation

Issues related to tissue specific regulation, mRNA half-life and processing, and neural regulation have lead us to begin a study of regulation at the level of gene transcription and splicing at the cellular level. Our basic methodological approach has employed the design of probes complementary to introns lying between exons encoding the MR proteins and various alternatively spliced mRNA forms found mainly in 5′ untranslated exons.

Intronic *in situ* hybridization, by virtue of using complementary gene rather than mRNA probes, allows visualization of MR gene transcripts [e.g., heteronuclear RNA (hnRNA)] in the nucleus prior to RNA splicing. Validation of this technique has been worked out in other systems (25). Initial studies using MR intron probes reveals the presence of intron-containing transcripts in nuclei of appropriate populations of hippocampal neurons (Fig. 2). Unlike other introns present in peptide-encoding genes (e.g., CRH), MR

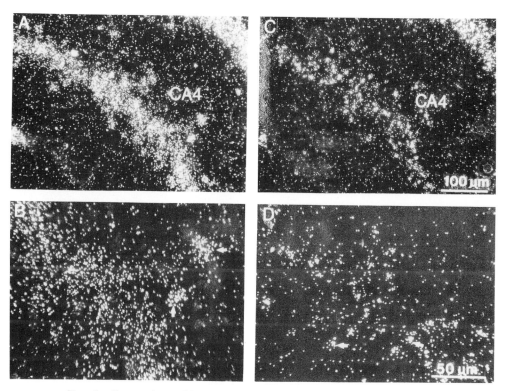

FIG. 2 Comparison of images generated from adjacent hippocampal sections, using cRNA probes complementary to exonic (A and B) or intronic (C and D) sequences of rat MR RNA. MR mRNA (exon) is scattered throughout the cytoplasm of neurons in hippocampal subfield CA4, resulting in a diffuse signal roughly defining neuronal cell bodies [arrows in (B)]. MR gene transcripts, defined using a probe complementary to a coding region intron (located between exons 2 and 3), result in a more sharply defined signal corresponding to confinement of signal to the cell nucleus [arrows in (D)].

intron is present in substantial quantities under normal, resting conditions, suggesting either a high ongoing rate of transcription or the presence of a persistent pool of MR hnRNA in the cell nucleus.

Studies employing probes against alternatively spliced 5′ variants of the MR reveal heterogeneous regulation of expression of α, β, and γ forms (see Fig. 3 for a schematic of splicing variants). In response to steroid depletion, the α variant shows a twofold increase in expression in all subfields except the dentate gyrus, whereas the β form shows little or no increase. Parallel sections processed with a probe against the 3′ coding region and 3′ UT show intermediate changes, reflecting a "weighted average" of the two variants (26, 27). It has yet to be determined whether the preferential change in the α variant is related to RNA processing or translatability.

The issue to be considered most from all the above ISHH applications is the overwhelming importance of the nucleotide sequence used to generate probes. Clearly, the differential splicing of the MR, the apparent processing of the GR, and differential selection of polyadenylation sites suggest that probes complementary to unique sequences at the 3′ coding region of the mRNA represent perhaps the most prudent approach to "straightforward" analysis of changes in mRNA level. However, the RNA processing occurring in corticosteroid receptor precursors indicates that mRNA regulation may be anything but straightforward, and future research is left to determine the role played by processing in synthesis of steroid receptors in the CNS.

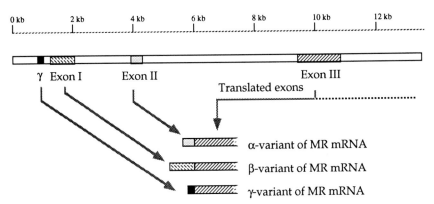

FIG. 3   Schematic illustrating splicing of 5′ untranslated sequences of the MR gene. The MR is differentially spliced to yield either α, β, or γ forms in a tissue- and stimulus-specific manner. To assay regulation of the respective splicing variants, probes specific for each of the respective variants are employed. Note that exon III is translated, and probes directed against this species will recognize all variants.

Solution Hybridization: Nuclease Protection Assay

Overview

The nuclease protection assay has been the method of choice for analysis of GR and MR mRNA content in tissue extracts, primarily because of its high sensitivity. This method relies on solution-phase hybridization between the mRNA molecule of interest and a radiolabeled complementary RNA molecule. Complementary RNAs are synthesized by *in vitro* transcription. The 1:1 stoichiometry of the probe mRNA reaction, and the (theoretical) availability of all RNA to hybridize with probe, renders this technique the most sensitive hybridization approach yet applied to quantitation of cortico-steroid receptor mRNAs in tissue extracts.

Probes

We employ single-stranded cRNA probes for our solution hybridization experiments. cRNAs are generated by *in vitro* transcription in a fashion similar to that described above for *in situ* hybridization, and are directed against sequences within the coding region of mRNAs under study. In addition to the standard labeling protocol, following transcription reactions probes are sized on agarose gels, excised, and eluted prior to assay. In that the solution hybridization paradigm is less tolerant of probe degradation or incomplete transcription than other hybridization methods, it is generally important to verify probe integrity prior to protection assay to ensure the presence of full-length transcripts. Furthermore, this paradigm necessitates the complete removal of template plasmid DNA after cRNA synthesis, as vector DNA:RNA interactions can generate spurious hybridization.

Protocol 5: Probe Preparation: Solution Hybridization Assay

1. *In vitro* transcription is performed as described above using linearized DNA templates, substituting high specific activity [^{32}P]UTP for [^{35}S]UTP. Transcription reactions consist of 4 μl of 5× transcription buffer, 2 μl of 100 mM DTT, 4.0 μl of a mixture of 2.5 mM ATP, CTP, and GTP, 1 μl of 0.5 mM UTP, 1 μl of RNase inhibitor, 1 μl of plasmid DNA (100 to 300-ng total), and 7.0 μl of dried-down, high specific activity [^{32}P]UTP (3000 Ci/mmol) (total, 250 μCi). Reactions are started with the addition of 1.0 μl of SP6, T3, or T7 RNA polymerase and incubated at 37°C for 60–90 min.

2. Load the reaction mixture on a 1% low-melt agarose gel and electropho-rese for 5–10 min, or enough to separate incorporated nucleotides from free

nucleotides and plasmid DNA. Cut out the RNA band (identified by ethidium bromide stain) and add 200 μl of diethyl pyrocarbanate (DEPC)-treated H_2O. Heat the tube at 65°C for 1 min to melt agarose and extract twice with phenol, once with chloroform. Precipitate in 2.5 vol of ethanol with 1/20 vol of 3 M sodium acetate (pH 5.2).

3. Resuspend the probe in 50 μl of DEPC-treated H_2O; run 1 μl on a 4% urea-polyacrylamide gel to verify probe length and absence of unincorporated free nucleotide. Count an additional 1-μl aliquot in a scintillation counter; use the probe within 2 days for assay.

Hybridization

Hybridization conditions for protection assays are conducted under conditions appropriate for specific formation of cRNA–mRNA hybrids. Again, owing to the stability of hybrids all reactions are conducted in a buffer containing formamide (generally 50%). RNase-free yeast RNA is included in the reaction mixture to saturate nonspecific RNA interactions; note that because protein and DNA have been extracted from RNA samples, inclusion of salmon sperm DNA and Denhardt's solution is unnecessary. In addition, because reactions are conducted in solution, dextran sulfate is unnecessary as well. Following hybridization, reactions are treated with either S1 nuclease or RNase A (the latter either alone or in combination with RNase T1) to digest single-stranded RNAs, leaving only protected hybrids intact. Note that the RNase digestion step is the most sensitive in this protocol, in that overdigestion of samples can have deleterious results. The concentration of RNases must be optimized for each cRNA : RNA hybrid.

We generally use a semiquantitative approach to analysis of solution hybridization data. For these purposes, we routinely ethanol precipitate hybridization mixtures without phenol–chloroform extraction, obviating unequal recovery of aqueous phase during the extraction procedure. Resuspended pellets are then electrophoresed through nondenaturing gels and exposed to X-ray film, and radioactive bands quantitated by simple densitometric analysis (see above).

Protocol 6: Solution Hybridization (RNase Protection) Assay

1. RNA pellets (1–10 μg total RNA) are resuspended in 10 μl of DEPC-treated H_2O and kept on ice. To a separate tube add 5 μl of 5× hybridization buffer [0.2 M PIPES (pH 6.4), 2.0 M NaCl, 5.0 mM EDTA] and 15 μl of deionized formamide.

2. Riboprobe is added to the hybridization buffer–formamide mixture to reach a final concentration of 5×10^4 cpm/μl.

3. The probe mixture is then pipetted into tubes containing 5 μl of RNA sample, vortexed, and denatured at 85°C for 3 min.

4. For hybridization, probe/RNA tubes are incubated at 60°C for >4 hr.

5. Following hybridization, tubes are allowed to cool to room temperature. A total of 100 μl of RNase A digestion buffer [10 mM Tris-HCl (pH 7.5), 5.0 mM EDTA, 0.2 M NaCl, 0.1 M LiCl], containing 1–40 μg of RNase A, is added to each tube and the digestion incubated at room temperature for 30 min.

6. Ten microliters of 10% SDS and 10 μl of proteinase K (10-mg/ml stock) is added to each tube to digest the remaining protein/RNase. Tubes are mixed and incubated at 37°C for 30 m.

7. For analytical purposes (i.e., accurate sizing of fragments), samples are extracted twice with phenol–chloroform–isoamyl alcohol (25 : 24 : 1) and precipitated with 2.5 vol of ethanol. If bands are to be quantitated densitometrically, samples are directly precipitated with 2.5 vol of ethanol. These eliminate the extraction step, which can lead to differential recovery across samples.

8. Analytical samples are run on 4% polyacrylamide–urea gels. Gels are fixed in 10% methanol : 10% acetic acid for 10 min, dried, and exposed to film.

9. For semiquantitative analysis, samples are run on 4% nondenaturing polyacrylamide gels, transferred onto Whatman (Clifton, NJ) 3MM paper, dried, and exposed to X-ray film. Resulting bands and background are digitized and quantitated off of film as described above for *in situ* hybridization.

Applications and Caveats

The RNase protection assay has several important applications for the study of corticosteroid receptors. First, the quantitative nature of this protocol can provide accurate estimates of GR and MR mRNA content in extracted RNA samples. These numbers can be useful in comparisons across experiments and laboratories, and perhaps more importantly, can provide a comparison of relative levels of receptor subtypes and/or variants derived from homogenates.

Quantitative analysis using RNase protection protocols is well established, and has been used by our laboratory and others to illustrate negative regulation of hippocampal GR and MR mRNA by adrenocorticosteroids (26–29). Although the sensitivity of this technique is unquestioned, its major drawback is lack of anatomical resolution. Unfortunately, for the study of GR and MR mRNA this flaw is particularly magnified, in that the anatomical region possessing the greatest levels of both species, the hippocampus, has a heterogeneous and laminar organization not readily susceptible to accurate dissection. The problem of anatomical resolution is further complicated by the fact that *in situ* hybridization analyses have repeatedly shown that GR and MR mRNA synthesis is clearly regulated in a cell-specific fashion (3). For example, we have shown that MR mRNA levels are upregulated by steroid depletion in hippocampal subfields CA1 and CA2, but not in CA3–4 or the dentate gyrus. Protection analysis performed on whole hippocampal extracts samples both upregulated and unregulated cell populations, resulting in a net "dilution" of the magnitude of change that can thereby mask significant changes (cf. 26). This may explain some of the discrepancies observed between *in situ* hybridization and protection studies of glucocorticoid regulation of corticosteroid receptor mRNA. In addition, homogenates sample glial elements as well as neurons; glia are known to contain both GR and MR protein (19) and are more than likely regulated in a different fashion than neurons.

We have found the RNase protection assay to be useful for the study of the organization of the MR gene. Our initial MR cDNA, isolated from rat hippocampus, corresponded to the clone published by Arriza *et al.* (15) throughout the coding region of the molecule, but differed substantially in the 5′ untranslated region. RNase protection analyses using probes overlapping homologous and nonhomologous regions of the two clones (α and β) revealed that this difference was due to the presence of MR splicing variants, resulting in expression of distinct MR mRNAs having identical coding regions but different 5′ sequences. Semiquantitative analysis of the protection data also

---

FIG. 4   Schematic depicting assets and liabilities of *in situ* and solution hybridization methodologies in terms of anatomical resolution, fragment sizing, and quantitation. Both *in situ* and solution hybridization can yield semiquantitative information on relative amounts of RNA in tissues under study [e.g., see control vs treatment 1 (T1) and 2 (T2) under Solution Hybridization]. *In situ* hybridization affords excellent anatomical resolution, but cannot report fragment sizing or errors associated with cross-hybridization (partial hybridization can look like a "weak" signal). Owing to numerous physical barriers and the presence of tissue background, the signal generated probably underestimates tissue content by a considerable amount. Solution hybridization can yield information on fragment sizing [important in addressing questions relevant to RNA processing (e.g., splicing variants)], and can indicate potential problems related to cross-hybridization (cross-hybridization now shows up as an additional, "nonregulated" band on polyacrylamide gel electrophoresis). Because all extracted RNA is theoretically available for analysis, solution hybridization is probably able to render more accurate estimates of tissue content than *in situ* hybridization. However, note that this method is limited by the efficiency of the extraction process, and also underestimates actual content. Solution hybridization loses all anatomical information, and may not be appropriate for accurate analysis of heterogeneously organized tissues.

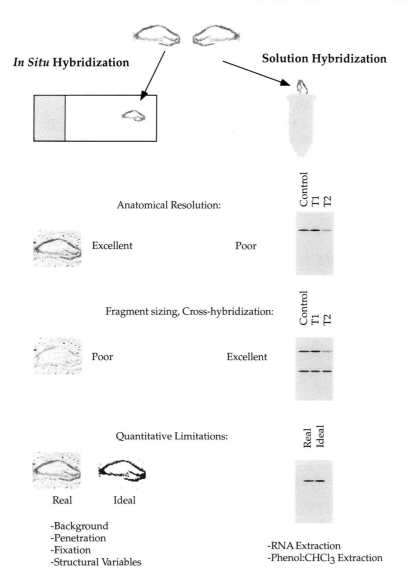

revealed that when compared together with a probe recognizing all MR mRNA forms, the α and β variants accounted for only 60% (30% each) of the total MR mRNA in hippocampus, suggesting the existence of additional variants. This was subsequently confirmed using RACE polymerase chain reaction, whereby one additional subtype has been characterized to date (γ

variant), accounting for an additional 3% of total MR mRNA (26, 27). The significance of the splicing variants has yet to be determined.

Solution hybridization is a valuable tool for numerous applications above and beyond quantitation of mRNA changes *in vivo*. As described above, information concerning the size of protected fragments can reveal information concerning mRNA heterogeneity and splicing. Protection analysis is perhaps the most sensitive method of quantifying mRNA production in neuronal or glial cultures. Finally, the pattern of labeled bands can provide insight into potential problems regarding mRNA integrity or potential cross-hybridization with other RNA species.

Concluding Remarks

In situ and solution hybridization techniques have unique advantages and disadvantages for the study of steroid receptors in brain, some of which are shown schematically in Fig. 4. *In situ* hybridization is anatomically oriented, and can yield a wealth of meaningful information about location and local regulation of steroid receptor molecules. Careful selection of probe can further resolve regulation of differentially spliced and/or processed RNA and even events at the transcriptional level. However, this technique cannot reveal splicing or processing (because no sizing is afforded) and owing to methodological limitations cannot be considered truly "quantitative." Solution hybridization reveals fragment size information and provides more accurate estimates of receptor concentrations, but cannot resolve receptor mRNA regulation at anything less than a gross regional level. Clearly, selection of one technique compromises certain classes of information, and it is thus critical to know the technical limitations and minimize potential sources of error (i.e., choose probes carefully!).

In all, the body of data derived from *in situ* and solution hybridization analyses of corticosteroid receptor mRNA expression, aimed originally at simply quantifying mRNA levels in response to experimental manipulations, has led to a wealth of provocative data speaking to issues of heterogeneous expression, RNA processing and half-life, gene transcription, and gene splicing. It is clear from these data that steroid receptor regulation is complex, and that hybridization approaches can supply a rich source of information for further study of corticosteroid receptor systems.

Acknowledgments

We would like to thank Dr. Paresh Patel for his substantial contribution to our knowledge of MR regulation and gene structure, and to Dr. Keith Yamamoto for the gift of the rat GR clone used in many of our experiments. This research effort has been supported by MH49698 (J. P. H.) and MH42251 (S. J. W.).

References

1. K. R. Yamamoto, *Annu. Rev. Genet.* **19**, 205 (1985).
2. B. S. McEwen, E. R. deKloet, and W. Rostene, *Physiol. Rev.* **66**, 1121 (1986).
3. J. P. Herman, *Cell. Mol. Neurobiol.* **13**, 349 (1993).
4. K. Fuxe, A. Cintra, A. Härfstrand, L. F. Agnati, M. Kalia, M. Zoli, A.-C. Wikström, S. Okret, M. Aronsson, and J.-Å. Gustafsson, *Ann. N.Y. Acad. Sci.* **512**, 362 (1987).
5. J. M. Reul and E. R. deKloet, *J. Steroid Biochem.* **24**, 269 (1986).
6. J. P. Herman, S. J. Watson, H. M. Chao, H. M. Coirini, and B. S. McEwen, *Mol. Cell. Neurosci.* **4**, 181 (1993).
7. R. S. Ahima, Z. Krozowski, and R. E. Harlan, *J. Comp. Neurol.* **313**, 522 (1991).
8. J. M. Reul and E. R. deKloet, *Endocrinology (Baltimore)* **117**, 2505 (1985).
9. M. F. Dallman, N. Levin, C. S. Cascio, S. F. Akana, L. Jacobson, and R. W. Kuhn, *Endocrinology (Baltimore)* **124**, 2844 (1989).
10. D. Pearce and K. R. Yamamoto, *Science* **259**, 1161 (1993).
11. R. Miesfeld, S. Rusconi, P. Godowski, B. A. Maler, S. Okret, A.-C. Wikström, J.-Å. Gustafsson, and K. R. Yamamoto, *Cell* **46**, 389 (1986).
12. M. K.-H. Schäfer, J. P. Herman, and S. J. Watson, *in* "Imaging Drug Action in the Brain" (E. D. London, eds.), p. 337. CRC Press, Boca Raton, Florida, 1993.
13. R. M. Evans, *Science* **240**, 889 (1988).
14. L. M. Angerer, M. H. Stoler, and R. C. Angerer, *in* "*In situ* Hybridization: Applications to Neurobiology" (K. Valentino, J. Barchas, and J. Eberwine, eds.), p. 42. Oxford Univ. Press, New York, 1987.
15. J. L. Arriza, C. Weinberger, G. Cerelli, T. M. Glaser, B. L. Handelin, D. E. Housman, and R. M. Evans, *Science* **237**, 268 (1987).
16. M. Aronsson, K. Fuxe, Y. Dong, L. F. Agnati, S. Okret, and J.-Å. Gustafsson, *Proc. Natl. Acad. Sci. U.S.A.* **85**, 9331 (1988).
17. R. J. Sousa, N. H. Tannery, and E. M. Lafer, *Mol. Endocrinol.* **3**, 481 (1989).
18. G. Yang, M. F. Matocha, and S. I. Rapoport, *Mol. Endocrinol.* **2**, 682 (1988).
19. M. C. Bohn, E. Howard, U. Vielkind, and Z. Krozowski, *J. Steroid Biochem. Mol. Biol.* **40**, 105 (1991).
20. J. P. Herman, P. D. Patel, H. Akil, and S. J. Watson, *Mol. Endocrinol.* **3**, 1886 (1989).
21. J. A. M. VanEekelen, W. Jiang, E. R. deKloet, and M. C. Bohn, *J. Neurosci. Res.* **21**, 88 (1988).
22. J. R. Seckl, K. Dickson, and G. Fink, *J. Neuroendocrinol.* **2**, 911 (1990).
23. H. R. Whitfield, Jr., L. S. Brady, M. A. Smith, E. Mamalakis, R. J. Fox, and M. Herkenham, *Cell. Mol. Neurobiol.* **10**, 145 (1990).
24. G. Cizza, A. E. Calogero, L. S. Brady, G. Bagdy, E. Bergammi, M. Blackman, G. P. Chrovsos, and P. W. Gold, *Endocrinology* **134**, 1611 (1994).
25. J. P. Herman, T. G. Sherman, M. K.-H. Schäfer, and S. J. Watson, *Mol. Endocrinol.* **5**, 1447 (1991).
26. S. P. Kwak, P. D. Patel, R. C. Thompson, H. Akil, and S. J. Watson, *Endocrinology (Baltimore)* **133**, 2344 (1993).

27. P. D. Patel, S. P. Kwak, J. P. Herman, E. A. Young, H. Akil, and S. J. Watson, *in* "Stress and Reproduction" (K. E. Sheppard, J. H. Boublik, and J. W. Funder, eds.), p. 1. Raven, New York, 1992.
28. J. M. Reul, P. T. Pearce, J. W. Funder, and Z. S. Krozowski, *Mol. Endocrinol.* **3,** 1674 (1989).
29. H. M. Chao, P. H. Choo, and B. S. McEwen, *Neuroendocrinology* **50,** 365 (1989).

[13] Detection and Characterization of Epalon Receptors: Novel Recognition Sites for Neuroactive Steroids That Modulate the GABA_A Receptor Complex

Linda D. McCauley and Kelvin W. Gee

GABA$_A$ Receptor Complex Active Neurosteroids

Background

The genomic effects of steroid hormone binding to intracellular receptors is well documented (for review see Ref. 1). Ligand-bound steroid hormone receptors interact with hormone response elements on DNA, producing long-term effects on cell structure and function. This is in sharp contrast to the rapid and reversible steroid actions on neuronal membranes that have long been observed but only recently studied in detail. Indeed, certain neurosteroids (i.e., defined as steroids found in the brain) were found to be potent modulators of brain excitability over 50 years ago when Selye first described the anesthetic activity of steroids in rodents (2). On the basis of these early *in vivo* observations by Selye and others, a class of steroid anesthetics was introduced clinically in the 1970s (3, 4). Even so, relatively little progress was made to elucidate the mechanism by which these rapid behavioral effects of steroids are mediated until much later. It was not until 1980, when Schofield reported that the synthetic anesthetic steroid 5α-pregnan-3α-ol-11,20-dione (alphaxalone) enhanced γ-aminobutyric acid (GABA)-mediated inhibitory events in rat olfactory neurons using electrophysiological techniques, that the GABA$_A$ receptor complex (GRC) was implicated as a site of action for neuroactive steroids in the brain (5). This hypothesis was further supported by the demonstration by Harrison and Simmonds that alphaxalone potentiated GABA-stimulated currents in the rat cuneate nucleus and allosterically modulated the binding of [^{3}H]muscimol to the GABA$_A$ receptor in rat brain membranes (6).

Initially, *in vitro* biochemical experiments with GRC-active neurosteroids showed barbiturate-like pharmacology in that they prolonged channel open time as opposed to increasing the frequency of channel opening like the benzodiazepines (BZs). This single observation led investigators to suggest that the two classes of compounds, GRC-active neurosteroids and barbiturates, shared a common site of action (7). Subsequent biochemical experiments demonstrated that GRC-active neurosteroids were acting at a site distinct from that of the barbiturates. For example, it was observed that 5α-pregnan-3α-ol-20-one (epiallopregnanolone, 3α,5α-P) did not competitively antagonize the sodium pentobarbital enhancement of [³H]flunitrazepam ([³H]FLU) binding to the BZ receptor (8). Furthermore, sodium pentobarbital was found to potentiate the effect of saturating concentrations of 3α,5α-P on the dissociation of steady state *tert*-butylbicyclophosphoro[³⁵S]thionate ([³⁵S]TBPS) binding to the chloride ionophore. Together these observations were indicative of independent binding sites. Electrophysiological data describing barbiturate potentiation of neuroactive steroid-stimulated chloride channel conductance further supported the notion of distinct sites (9). Interestingly, stringent structure–activity requirements were observed for this new class of neuroactive steroids. Most notably, a 5α- or 5β-reduced steroid A ring and a 3α-hydroxyl group conferred the greatest potency whereas 3β-hydroxyl analogs were primarily devoid of activity (8, 10, 11). The most potent neuroactive steroid, measured by allosteric modulation of sites on the GRC, to date is 3α,5α-P, a naturally occurring metabolite of progesterone.

The definitive evidence supporting a specific neuroactive steroid/receptor–protein interaction stems from biochemical and electrophysiological studies in recombinantly expressed GABA_A receptors. Several laboratories using different techniques have found that these neuroactive steroids modulate GABA action in transiently expressed recombinant GRCs (12–14). In the last several years, researchers have continued to define the structure–activity relationships of both endogenous and novel steroids that are active at the GRC. This specificity of action has opened up a new avenue of research into the role of these neuroactive steroids as endogenous modulators of the GRC. Indeed, the remarkable specificity exhibited by these GRC-active steroids combined with their potent pharmacological effects suggest that these neuroactive steroids belong to a novel and distinct class of steroids, hence the need for a new name to define this class. Thus, the term *Epalon*, an acronym derived from *epi*allopregnano*lon*e, to refer to those steroids both natural and synthetic that specifically modulate GRC activity has been proposed (15).

Physiological and Pharmacological Significance

In light of the abundant evidence demonstrating that Epalons are among the most potent modulators of GRC-mediated inhibition and that they have an implicit ability to influence brain excitability, the importance of the central nervous system (CNS) as a potential steroidogenic tissue becomes apparent.

The rate-limiting step in steroidogenesis is the cytochrome P450 side-chain cleavage conversion of cholesterol to pregnenolone. This steroid is the precursor of endogenous Epalons that are metabolites of progesterone and other hormonally active steroids. This steroidogenic pathway has been reported to occur in CNS-derived tissues (16, 17). Moreover, CNS-derived tissue has been shown to biosynthesize both progesterone and $3\alpha,5\alpha$-P from pregnenolone, and all measures indicate that this is occurring within the confines of the CNS (16, 17). Studies have shown that the requisite enzymes for the complete conversion of cholesterol to progesterone and its reduced metabolites are present in the brain. The concept that steroid biosynthesis may occur in tissue other than the classic steroidogenic tissues (e.g., adrenal gland and gonads) has prompted speculation that endogenous Epalons may indeed be endogenous modulators of the GRC. Collectively, these studies provide compelling evidence that the brain may support *de novo* synthesis of Epalons, which may in turn act via the GRC to alter brain excitability. The clinical success of the GRC-active BZs leads one to speculate that with further characterization of Epalon receptors, a new therapeutically useful class of compounds can be developed for the treatment of disorders that respond to positive modulation of the GRC. The following is meant to be a brief practical guide to *in vitro* binding assays useful for the detection and characterization of the putative Epalon receptor(s).

Methods of Detection of GABA$_A$ Receptor Complex-Active Epalons

Cooperativity between Known Recognition Sites on the GABA$_A$ Receptor Complex

Cooperativity is a term used to describe the interaction between two physically separate ligand-binding sites. A classic example of *homotropic* cooperativity is observed in the binding of oxygen to hemoglobin. It has been shown that the binding of oxygen to each of the four heme moieties in hemoglobin does not follow simple mass action kinetics. On the contrary, the oxygen saturation curve was much steeper than predicted; the binding of the first oxygen facilitated the binding of the second and so on. This type of cooperativity is referred to as homotropic because it involves two identical binding sites that are physically distinct. However, this model does not distinguish between two individual sites on a monomeric protein or one site on each of two subunits of a multimeric protein.

Homotropic cooperativity describes the interaction between two sites for the same class of ligands; *heterotropic* cooperativity refers to a similar interaction between sites for different ligands. The model developed by Monod,

Wyman, and Changeux proposes that the binding of the first ligand induces a conformational change in the binding protein that confers cooperativity between the two sites (18, 19). This hypothesis assumes the existence of an oligomeric protein. In general, the potential exists for all ligand-gated neurotransmitter receptors composed of multiple subunits to exhibit cooperativity. This type of cooperativity has certain practical advantages in drug therapy (20). By acting at a site distinct from the neurotransmitter, a heterotropic ligand would have activity only when the endogenous transmitter is present, thus maintaining the normal pattern of neurotransmission. In addition, this effect would be moderated by the degree of cooperativity between the neurotransmitter receptor and the heterotropic site. The nature of this relationship may be either positive or negative, enhancing the functional activity or diminishing it, respectively.

Current data suggest that the GRC is a pentameric protein on which binding sites for several independent classes of compounds reside (for molecular biology review see Ref. 21). On the GRC, it appears that all of the known recognition sites exhibit cooperativity with each other. The term *allosteric modulation* is often used interchangeably with heterotropic cooperativity. Simply stated, it refers to the modulation of a ligand binding by the binding of a second, different ligand. Using this type of interaction, it is possible to characterize the recognition properties of one ligand site by modulating the binding of a specific radioligand for the second site. The initial detection and characterization of the Epalon site was performed in this manner.

Allosteric Modulation of GABA$_A$ Receptor Binding

There are some practical problems to overcome before obtaining reliable data on GABA$_A$ receptor binding. Endogenous GABA is present in sufficient quantities to interfere with binding to the GABA$_A$ receptor in the brain. High nanomolar to low micromolar concentrations of GABA have been measured in standard tissue preparations (22), which can interfere with the accurate detection of GABA$_A$ receptor binding. Consequently, extensive washing of the tissue to reduce endogenous GABA levels is necessary. In addition, [^{3}H]muscimol, a GABA agonist, is preferred over [^{3}H]GABA for several reasons: (a) it has less GABA$_B$ receptor activity (23); (b) it is not metabolized by GABA/2-oxoglutarate aminotransferase (24); and (c) it is not a substrate for the sodium-dependent GABA uptake system (25). For either [^{3}H]GABA or [^{3}H]muscimol binding, a standard P$_2$ synaptosomal tissue preparation is used (25). Briefly, freshly harvested tissue from the brain region of interest is homogenized as 10% (w/v) in 0.32 M sucrose (pH 7.4), using a Teflon homogenizer. The homogenate is centrifuged at 1000 g for 10 min at 0–4°C and the supernatant collected. The supernatant is centrifuged again at 0–4°C at 9000 g for 20 min. The resultant pellet contains the synaptosomes. One of two methods may be chosen to reduce the endogenous GABA. To induce the release of sequestered GABA, the membranes can be subjected to osmotic shock (e.g., >100 mM K$^+$) and then frozen, thawed, and washed. Alternatively, extensive washing in excess volumes of cold buffer (more than four times) may follow the freeze-thaw process. Harsh treatment of the membranes may result in the loss or alteration of allosteric coupling between GABA$_A$ receptors and the allosteric site of interest. Care must be taken to preserve the sometimes delicate coupling between binding sites, which can be eliminated by overenthusiastic washing of the tissue and result in the incorrect conclusion that coupling is not present.

One can use various incubation conditions for [^{3}H]muscimol binding. When incubating at room temperature or 37°C, equilibrium is typically achieved in approximately 10 min. At 4°C, [^{3}H]muscimol does not reach equilibrium until closer to 30 min. Under normal assay conditions, 5–10 nM [^{3}H]muscimol is necessary owing to the generally low specific activity of commercially available radioligand, whereas 1 mM GABA is used to define nonspecific binding. The choice of assay buffer is not particularly critical [i.e., it can be phosphate-, Tris-, or N-2-hydroxyethylpiperazine-N'-2-ethanesulfonic acid (HEPES) based] because investigators have reported reasonable specific binding using various buffers although the pH was always physiological (6, 11, 26, 27). Allosteric modulation of the most commonly measured GRC sites have all been demonstrated in 50 mM sodium/potassium phosphate buffer with 200 mM NaCl (PBS, pH 7.4). Consequently, all procedures from this point forward use PBS unless specifically stated otherwise. The assay protocol described below encompasses those details associated with Epalon modulation of GABA$_A$ receptor binding. For each receptor, subtle differences in protocol exist to adapt the assay to optimal conditions for that particular receptor assay. Both centrifugation and filtration methods for separating bound radioactivity from free radioactivity have been used successfully.

1. Prepare tissue as previously described and resuspend as a 10% (original wet weight/volume) homogenate.
2. Total incubation volume is 1 ml.
3. Add 845 μl of assay buffer to culture tubes, except those defining nonspecific binding, which receive 745 μl.
4. Add GABA (dissolved in buffer) in 100-μl aliquots to nonspecific tubes.
5. Add 5-μl aliquots of dimethyl sulfoxide (DMSO) or the solvent chosen to dissolve the test Epalon to nonspecific and total binding tubes.
6. Prepare a dose (sufficient to produce a response) of the Epalon of interest in DMSO (or solvent used) and add to assay tubes (5 μl/tube).
7. Prepare radioligand in assay buffer and add in a 50-μl volume per assay tube.

8. Add 100-μl aliquots of tissue homogenate to each tube and mix well.
9. Incubate for 30 min at 4°C (or adjust to condition chosen).
10. Collect tissue-bound radioactivity by filtration through #32 glass fiber filters (Schleicher & Schuell Inc., Keene, NH) and wash three times with 4 ml of ice-cold phosphate buffer.
11. Quantitate bound radioactivity by liquid scintillation spectrophotometry.

Typically, compounds that allosterically enhance the binding of GABA$_A$ receptor radioligands (i.e., GABA-agonist like) will favor an open chloride channel conformation, whereas those having opposite effects on binding will either block the channel or prefer the closed channel conformation. Nearly all of the Epalons reported to date enhance GABA$_A$ receptor binding. As such, they are presumed to be Epalon site agonists. However, Epalon modulation of [^{3}H]muscimol binding alone is insufficient evidence to suggest that a specific site for Epalons exists on the GRC. Early studies attempting to define the neuronal action of adrenal steroids reported an apparent enhancement of [^{3}H]muscimol binding to the GABA$_A$ receptor in various regions of the rat brain (28). This effect, detected with both natural and synthetic glucocorticoids, was evaluated only with [^{3}H]muscimol binding yet this type of allosteric modulation has not been observed with binding to other sites on the GRC. This case exemplifies the need to evaluate allosteric interactions between other known allosteric sites on the GRC before an accurate prediction of the presence of a unique site on the GRC can be made.

Linear transformations of nonlinear data collected from radioligand-binding experiments can magnify any experimental errors and cause difficulties in resolving multiple binding sites or allosteric interactions (29, 30). This problem can be solved by the use of computer-based nonlinear curve-fitting techniques (several of which are commercially available) to analyze radioligand-binding site interactions in a statistically rigorous and reliable manner. Epalon modulation of GRC recognition site binding is plotted as the percentage specific radioligand bound versus the logarithm of Epalon concentration. Iterative nonlinear curve fitting (or nonlinear regression) results in a sigmoidal curve extending either above or below 100%. Because the lowest Epalon concentration(s) typically chosen show virtually no allosteric effect, they are not different from the total binding measured and define the initial plateau as 100% of specifically bound [^{3}H]muscimol. As the Epalon concentration increases, the percentage bound also increases, shaping the curve above 100%, or defining an enhancement curve. This is usually observed in Epalon modulation of agonist binding to the GABA$_A$ and BZ receptors. As a result of the curve-fitting procedure, the computer programs calculate an EC$_{50}$, or the concentration at which half-maximal enhancement is observed. A low EC$_{50}$ implies high potency and vice versa. The same regression analysis can be applied to inhibition data. The parameter in inhibition experiments corresponding to an EC_{50} is an IC_{50}, or the concentration of ligand necessary to produce half-maximal inhibition. Similarly, a low IC_{50} indicates high potency and vice versa. Epalon agonists such as $3\alpha,5\alpha$-P will inhibit [^{35}S]TBPS binding to sites on the GRC (see Modulation of [^{35}S]TBPS Binding, below).

Modulation of Benzodiazepine Receptor Binding

In contrast to measuring $GABA_A$ receptor radioligand binding, the central BZ receptor provides a relatively easy and reliable method for observing Epalon effects via allosteric modulation of the BZ site, using a BZ agonist (i.e., GABA-agonist like) radioligand. Several radiolabeled BZs with high affinity are commercially available. Furthermore, a BZ receptor subtype-selective radioligand may be selected for evaluating subtype selective interactions. Although BZ receptor binding is in fact influenced (i.e., enhanced in the case of BZ agonists) by the GABA concentration, one need not employ extensive washing procedures to rid the tissue preparation of endogenous GABA. Most often a standard P_2 preparation will yield excellent specific BZ binding. One can choose to wash the tissue in order to examine the effect of Epalons on BZ binding in the relative absence of GABA, or add GABA back into the assay to observe the effects of a known GABA concentration. Alternatively, no washing allows one to evaluate Epalon modulation of BZ binding in the presence of endogenous GABA. One of the more commonly used BZ agonist ligands is [^{3}H]FLU. The specific activity of commerically available [^{3}H]FLU is usually 80–90 Ci/mmol (Du Pont-New England Nuclear, Boston, MA). With this range of activity and a standard 10% (w/v) homogenate one observes 90% or greater specific [^{3}H]FLU binding. Nonspecific binding is defined as the binding in the presence of 1 μM clonazepam, and 0.5 nM [^{3}H]FLU is sufficient for most experiments. Using the basic assay protocol outlined in the previous section, the total incubation volume in PBS buffer is again 1 ml.

1. Prepare tissue as outlined for $GABA_A$ receptor binding.
2. Add 745 μl of buffer if adding GABA, or 845 μl of buffer in the absence of GABA, to each tube.
3. If using exogenous GABA, add 100-μl aliquots of GABA dissolved in assay buffer to each tube for a final concentration of 1 μM. This concentration will influence Epalon modulation but will not maximally enhance agonist BZ binding alone.
4. Add 5-μl aliquots of clonazepam or DMSO to each nonspecific or total binding tube, respectively.
5. Add Epalons in 5-μl aliquots.

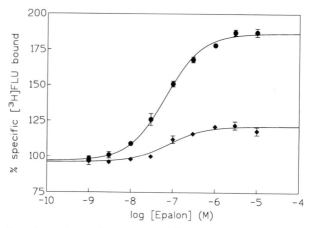

FIG. 1   The dose-dependent enhancement of [³H]FLU (0.5 nM) binding by Epalons in the presence of 1 μM GABA shows differential efficacy. (●) 3α,5α-P shows a much greater effect when compared to (◆) 5α-pregnan-3α,20α-diol (5α-pregnanediol) in washed rat cortical P₂ homogenates. Each curve represents the mean ± SEM of at least three or four independent observations, and is expessed as the percentage of [³H]FLU bound in the absence of an Epalon.

6. Mix well and incubate for 60 min at room temperature.
7. Harvest bound radioactivity by filtration through #32 glass fiber filters (Schleicher & Schuell) and quantitate by liquid scintillation spectrophotometry.

Epalon agonists will produce a dose-dependent enhancement of agonist BZ receptor binding with varying degrees of efficacy. Because in enhancement studies the effect is limited only by the ability of the receptor to respond (which may be greater than 100%), the varying degrees of efficacy of neurosteroids may be more readily discernible as shown in Fig. 1.

Recently, Prince and Simmonds have proposed 5β-pregnan-3β-ol-20-one (3β,5β-P, an endogenously occurring stereoisomer of 3α,5α-P) as a putative antagonist on the basis of their observations that 3β,5β-P competitively inhibited 5β-pregnan-3α-ol-20-one (3α,5β-P) potentiation of [³H]FLU binding to rat brain membranes (31). Furthermore, 3β,5β-P alone showed minimal enhancement of [³H]FLU binding. Subsequently, they have extended their findings to show that 5α-pregnan-3β-ol-20-one (3β,5α-P) competitively antagonized 3α,5α-P potentiation of [³H]FLU binding to rat brain membranes (32). Again, it may be premature to suggest that 3β,5β-P is an endogenous antagonist at the Epalon site on the basis of allosteric modulation of BZ binding in the absence of any assays of GRC function. Nevertheless, Epalon modulation of BZ binding provides important information regarding the coupling of the BZ and Epalon sites and the nature of their interaction. Allosteric Epalon modulation of BZ binding is limited to those Epalon sites that are coupled to BZ sites. This may represent a subset of Epalon sites (i.e., only those coupled to BZ sites). Evidence obtained from autoradiographic studies suggest that in certain brain regions the BZ and Epalon sites may not be coupled (33). This further supports the contention that evaluation of Epalon–GRC interactions by a single allosteric modulation assay may not be sufficient to characterize the Epalon site accurately.

Modulation of [^{35}S]TBPS Binding

[^{35}S]TBPS is thought to bind to a site on or near the chloride channel (34). Owing to the close proximity of this site to the channel, [^{35}S]TBPS binding appears to be influenced by the conductance state of the channel (i.e., open or closed). Indeed, the allosteric modulation of [^{35}S]TBPS binding to a site on or near the chloride ionophore has been used as an indicator of the nature of interaction between a ligand and the GABA$_A$ receptor-gated chloride channel (34–36). Under specific conditions the inhibition of [^{35}S]TBPS binding is characteristic of an Epalon agonist (or GABA agonist-like effect) whose physiological correlate may be the enhancement of chloride conductance. An example of the stringent structure–activity requirements for the modulation of [^{35}S]TBPS binding is shown in Fig. 2. As with BZ receptor binding, [^{35}S]TBPS binding can be performed in washed or unwashed tissue homogenates with or without exogenous GABA. The protocols for the two binding assays are nearly identical, therefore only the differences are outlined below.

1. [^{35}S]TBPS binding cannot be measured in the absence of chloride ions (34).
2. The radioactive decay of ^{35}S is relatively rapid ($t_{1/2}$ of 87 days) and the specific activity ranges from 60 to 150 Ci/mmol (New England Nuclear).
3. Nonspecific binding is defined in the presence of 2 μM cold TBPS.
4. When desired, GABA is usually included at 5 μM (the IC$_{50}$ of GABA against [^{35}S]TBPS binding under the conditions used).
5. Incubate for 90 min at room temperature (23°C).
6. Harvest and quantity as with BZ binding.

The [^{35}S]TBPS binding assay can be used to differentiate between true allosteric agonists that modulate GABA-induced effects on channel function and those with direct actions on the chloride channel. As shown in Fig. 3, in the absence of exogenously applied GABA, 5α-pregnan-3α,20α-diol (5α-pregnanediol) shows minimal if any inhibition of [^{35}S]TBPS. However, under

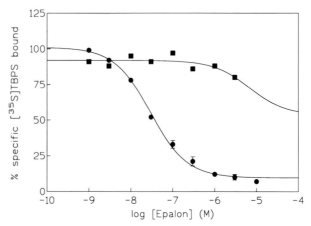

FIG. 2 An illustration of the structure–activity requirements of the Epalon site in the presence of 5 μM GABA. 3β-Hydroxy, (■) 3β,5α-P, compounds show little if any efficacy in modulating [^{35}S]TBPS (2 nM) binding even in the presence of GABA whereas the 3α-hydroxy, (●) 3α,5α-P, epalons are potent modulators. The 3α,5α-P curve represents the mean ± SEM of at least three or four independent observations, and is expressed as the percentage of [^{35}S]TBPS bound in the absence of an Epalon. The 3β,5α-P curve is the average of two experiments.

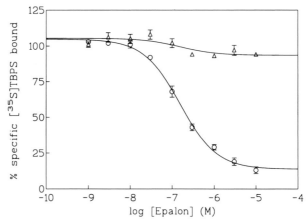

FIG. 3 An illustration of the absolute GABA dependence of a true allosteric modulator of the GRC-coupled chloride channel in washed rat cortical P$_2$ homogenates. In the absence of GABA (△) 5α-pregnanediol has minimal efficacy in modulating [^{35}S]TBPS binding in contrast to 3α,5α-P (○). Each curve represents the mean ± SEM of at least three or four independent observations, and is expressed as the percentage of [^{35}S]TBPS bound in the absence of an Epalon.

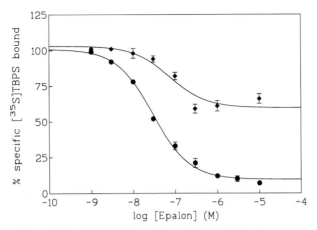

FIG. 4   The effect of 5 μM GABA on the efficacy of Epalons as modulators of [^{35}S]TBPS binding. ($\blacklozenge$) 5α-Pregnanediol shows limited efficacy yet high potency in modulating [^{35}S]TBPS binding. The efficacy of 3α,5α-P ($\bullet$) is unchanged in the presence of GABA; however, the effect is observed with greater potency. Each curve represents the mean $\pm$ SEM of at least three or four independent observations, and is expressed as the percentage of [^{35}S]TBPS bound in the absence of an Epalon.

similar conditions, 3α,5α-P inhibits [^{35}S]TBPS binding in the absence of exogenously applied GABA (i.e., a direct effect of 3α,5α-P) in washed cortical P$_2$ homogenates. The electrophysiological/functional correlate of this observation is the finding that Epalons such as 3α,5α-P will directly activate chloride channels in the absence of GABA (37, 38) whereas 5α-pregnanediol will not (J. Lambert, personal communication). Accordingly, when GABA is added to the assay, 5α-pregnanediol inhibits [^{35}S]TBPS binding with high potency but limited efficacy. Figure 4 illustrates the differences in efficacy between 5α-pregnanediol and 3α,5α-P in washed rat cortical P$_2$ homogenates. Thus, the modulation of [^{35}S]TBPS binding is useful in separating direct effects (i.e., GABA-independent) on the chloride channel from those mediated exclusively through the potentiation of GABA. Conversely, the allosteric enhancement of [^{35}S]TBPS binding is typically associated with chloride channels that are nonconducting (i.e., a GABA antagonist-like effect).

In the absence of a direct Epalon-binding assay, the combination of all three allosteric modulatory assays (i.e., [^{3}H]muscimol, [^{3}H]FLU, and [^{35}S]TBPS) and their corresponding electrophysiological assays provides the most accurate means to characterize the Epalon site. By evaluating and correlating Epalon effects on these three independent recognition sites on the GRC one can better define the structure–activity of the site that is coupled to what is pharmacologically defined as the GRC. Differential coupling of the Epalon site to the other recognition sites on the GRC has already suggested apparent Epalon site heterogeneity on the basis of allosteric modulatory assays (39). However, the amount of information gathered from these assays is indirect and subject to validation by a direct Epalon site-binding assay. A direct binding assay is essential for the resolution of Epalon receptor subtypes suggested by indirect assays and to facilitate our detailed understanding of the recognition properties of the Epalon site(s) and their physiological significance. Nevertheless, the use of the allosteric modulatory assays described is more than adequate to provide a first approximation of the characteristics of the Epalon receptor.

Development of Direct Epalon Site-Binding Assay

Inherent Problems Associated with Direct Epalon Site-Binding Assay

To date, efforts at obtaining a reasonable direct Epalon site-binding assay have been hampered by the high lipophilicity and "stickiness" of the radiolabeled Epalons used thus far. Specifically, the high level of nonspecific binding observed even in enriched synaptosomal preparations using $[^3H]3\alpha,5\alpha$-P (New England Nuclear) has been prohibitive. In our experience, the use of 1 nM $[^3H]3\alpha,5\alpha$-P results in nonspecific binding of approximately 75% of the total binding regardless of incubation temperature. Nevertheless, the data obtained were consistent with the presence of an Epalon site that is functionally coupled to the $GABA_A$ and benzodiazepine receptors. The limited amount of specific binding under these conditions makes definitive characterization of the binding site difficult. Furthermore, the counts per minute (cpm) specifically bound at this $[^3H]3\alpha,5\alpha$-P concentration is 5000–10,000. In this range experimental variation often masked potential specific effects when the specific binding is $\leq 25\%$.

It is noteworthy that if the level of nonspecific binding of the nonradioactive $3\alpha,5\alpha$-P in allosteric modulation assays is similar to that revealed by direct $[^3H]3\alpha,5\alpha$-P binding, it would imply that these Epalons are significantly more potent than suggested by their IC_{50}/EC_{50} values derived from allosteric modulatory assays. One can address the issue of Epalons sticking to the binding assay materials (e.g., culture tubes and pipette tips) by silanization of all surfaces the radioligand contacts to minimize adherence to these surfaces. In addition, the use of a steroid displaying no activity at the Epalon site can be added to all assay tubes to saturate the "nonspecific steroid" binding. For example, progesterone or other pregnanes that show minimal if any activity at the Epalon site may be chosen to saturate these nonspecific sites.

In light of the high potency with which Epalons have been shown to modulate GABA, BZ, and TBPS binding allosterically under standard assay conditions, one would expect that there are a sufficient number of Epalon receptors present and ligands of adequate affinity to label Epalon sites directly. This suggests that the limiting problems are associated with the radioligand chosen and not necessarily the relative abundance of receptor. There may be several avenues one can follow to address this issue. One potential solution is to employ a water-soluble salt of a high-affinity Epalon, which may aid in two ways. First, a water-soluble radioligand may have a reduced tendency to bind nonspecifically to neuronal membranes. Second, during the filtration procedure, radioligand not specifically bound but water soluble may be readily removed from the filter paper by the buffer washes. We have achieved some success in improving the signal-to-noise ratio by using a water-soluble Epalon derivative with a hemisuccinate group in the 21 position of $[^3H]5\alpha$-pregnan-3α,21-diol-20-one (5α-$[^3H]$THDOC or epiallotetrahydrodeoxycorticosterone). By incubating for 2 hr at 0–4°C, 10.5 ± 2.9 fmol/mg protein of specifically bound radioligand was observed using rat cortical homogenates washed three times and 1 nM 5α-$[^3H]$THDOC hemisuccinate. Nonspecific binding was defined as the amount of radioligand bound in the presence of 3 μM 5αTHDOC. This binding was enhanced by 5 μM GABA, resulting in 23.8 ± 1.9 fmol bound per milligram of protein. Consistent with the notion of cooperativity among GRC recognition sites, the combination of 5 μM GABA and a 1 μM concentration of the BZ agonist clonazepam resulted in the enhancement of specific binding beyond that of GABA alone, to 51.8 ± 8.5 fmol radioligand bound per milligram of protein.

Concluding Remarks

Current direct binding assays for the characterization of the Epalon site on the GRC are unsatisfactory. Nevertheless, the indirect evidence overwhelmingly supports the notion that Epalon action at the GRC is a receptor-mediated event. Techniques that remove components from membrane homogenates (e.g., solubilization of the receptor) not critically involved in Epalon binding to its receptor remain a hopeful prospect for reducing nonspecific binding. The use of soluble receptors in experiments involving BZs and TBPS have been fruitful and have yielded data comparable to that of brain homogenates (K. W. Gee, unpublished observations). As such, solubilization of the GRC and accompanying Epalon sites may rid the tissue preparation of membrane components that may contribute to high nonspecific binding. A variation on this theme is the use of recombinantly expressed GABA$_A$ receptors. It may be possible to express GRCs and Epalon sites with such high density that nonspecific binding is a relatively small percentage of total binding. Such strategies are currently being implemented. With increased efforts to design more potent and hydrophilic Epalons, coupled with the efficient recombinant expression of GRCs, the development of a useful and reliable direct binding assay for Epalon receptors is imminent.

Acknowledgments

Portions of the work described in this chapter were supported by grants from the USPHS, NS 25986 and NS 24645, and by a grant from CoCensys, Inc., Irvine, CA.

References

1. Baniahmad, A., and Tsai, M. J. (1993). *J. Cell Biochem.* **51,** 151–156.
2. Selye, H. (1941). *Proc. Soc. Exp. Biol. Med.* **46,** 116–121.
3. Gyermek, L., and Soyka, L. F. (1975). *Anesthesiology* **42,** 331–344.
4. Holzbauer, M. (1976). *Med. Biol.* **54,** 227–242.
5. Schofield, C. N. (1980). *Pfluegers Arch.* **383,** 249–255.
6. Harrison, N. L., and Simmonds, M. A. (1984). *Brain Res.* **323,** 287–292.
7. Majewska, M. D., Harrison, N. L., Schwartz, R. D., Barker, J. L., and Paul, S. M. (1986). *Science* **232,** 1004–1007.
8. Gee, K. W. Bolger, M. B., Brinton, R. E., Coirini, H., and McEwen, B. S. (1988). *J. Pharmacol. Exp. Ther.* **246,** 803–812.
9. Peters, J. A., Kirkness, E. F., Callachan, H., Lambert, J. J., and Turner, A. F. (1988). *Br. J. Pharmacol.* **94,** 1257–1269.
10. Gee, K. W., Brinton, R. E., Chang, W. C., and McEwen, B. S. (1987). *Eur. J. Pharmacol.* **136,** 419–423.
11. Harrison, NL, Majewska, M. D., Harrington, J. W., and Barker, J. L. (1987). *J. Pharmacol. Exp. Ther.* **241,** 346–353.
12. Puia, G., Santi, M. R., Vicini, S., Pritchett, D. B., Purdy, R. H., Paul, S. M., Seeburg, P. H., and Costa, E. (1990). *Neuron* **4,** 759–765.
13. Lan, N. C., Bolger, M. B., and Gee, K. W. (1991). *Neurochem. Res.* **16,** 347–356.
14. Lan, N. C., Chen, J.-S., Belelli, D., Pritchett, D. R., Seeburg, P.H., and Gee, K. W. (1990). *Eur. J. Pharmacol. Mol. Pharmacol.* **188,** 403–406.
15. McNeil, R. G., Gee, K. W., Bolger, M. B., Lan, N. C., Wieland, S., Belelli, D., Purdy, R. H., and Pual, S. M. (1992). *Drug News Perspect.* **5,** 145–152.
16. Hu, Z. Y., Bourreau, E., Jung-Testas, I., Robel, P., and Baulieu, E. (1987). *Proc. Natl. Acad. Sci. U.S.A.* **84,** 8215–8219.
17. Jung-Testas, I., Hu, Z. Y., Baulieu, E., and Robel, P. (1989). *J. Steroid Biochem.* **34,** 511–519.
18. Monod, J., Wyman, J., and Changeux, J. P. (1965). *J. Mol. Biol.* **12,** 88–118.

19. Changeux, J.-P., and Podleski, T. P. (1968). *Proc. Natl. Acad. Sci. U.S.A.* **59,** 944–950.
20. Ehlert, F. J. (1986). *Trends Pharmacol. Sci.* **7,** 28–32.
21. Olsen, R. W., and Tobin, A. J. (1990). *FASEB J.* **4,** 1469–1480.
22. DeLorey, T. M., and Brown, G. B. (1992). *J. Neurochem.* **58,** 2162–2169.
23. Hill, D. R., and Bowery, N. G. (1981). *Nature (London)* **290,** 149–152.
24. Beart, P. M., and Johnson, G. A. R. (1973). *Brain Res.* **49,** 459–462.
25. White, W. F., and Snodgrass, S. R. (1983). *J. Neurochem.* **40,** 1701–1708.
26. Turner, D. M., Ransom, R. W., Yang, J. S.-J., and Olsen, R. W. (1989). *J. Pharmacol. Exp. Ther.* **248,** 960–966.
27. Jossofie, A. (1993). *Biol. Chem. Hoppe-Seyler* **374,** 61–68.
28. Majewska, M.D., Bisserbe, J.-C., and Eskay, R. L. (1985). *Brain Res.* **339,** 178–182.
29. Limbird, L. (1986). "Cell Surface Receptors: A Short Course on Theory and Methods," pp. 33–48. Nijhoff, The Hague.
30. Munson, P. J. (1984). *In* "Brain Receptor Methodologies," Part A (P. J. Marangos, I. C. Campbell, and R. M. Cohen, eds.), pp. 33–47. Academic Press, New York.
31. Prince, R. J., and Simmonds, M. A. (1992). *Neurosci. Lett.* **135,** 273–275.
32. Prince, R. J., and Simmonds, M. A. (1993). *Neuropharmacology* **32,** 59–63.
33. Belelli, D., Wieland, S., Lan, N. C., Brinton, R. E., and Gee, K. W. (1991). *Soc. Neurosci. Abstr.* Part 2, p. 1343.
34. Squires, R. F., Casida, J. E., Richardson, M., and Saederup, E. (1983). *Mol. Pharmacol.* **23,** 326–336.
35. Gee, K. W., Lawrence, L. J., and Yamamura, H. I. (1986). *Mol. Pharmacol.* **30,** 218–225.
36. Maksay, G., and Simonyi, M. (1986). *Mol. Pharmacol.* **30,** 321–328.
37. Lambert, J. J., Peters, J. A., and Cottrell, G. A. (1987). *Trends Pharmacol. Sci.* **8,** 224–227.
38. Hill-Venning, C., Lambert, J. J., Peters, J. A., Hales, T. G., Gill, C., Callachan, H., and Sturgess, N. C. (1992). *In* "GABAergic Synaptic Transmission" (G. Biggio, A. Concas, and E. Costa, eds.), pp. 93–102. Raven, New York.
39. Gee, K. W., and Lan, N. C. (1991). *Mol. Pharmacol.* **40,** 995–999.

[14] Mutation Analysis of Steroid Hormone Receptors

Michael Karl, Heinrich M. Schulte, and George P. Chrousos

Introduction

Steroid hormone receptors are members of the ligand-activated nuclear receptor superfamily. These receptors bind to specific consensus DNA sequences called *hormone response elements* and exert control of gene expression either in a stimulatory or inhibitory fashion (1).

The cloning of the human glucocorticoid receptor cDNA by Hollenberg and colleagues in 1985 allowed for the first time the deduction of the complete primary structure of a prototype steroid hormone receptor (2). This was followed by the elucidation of cDNA clones encoding the remaining members of the steroid hormone receptor family, including the estrogen, progesterone, mineralocorticoid, and androgen receptors (3–8). Mutational analysis revealed the functional organization of these receptors and demonstrated that all of them consist of three major domains: an amino-terminal region harboring immunogenic properties, a DNA-binding domain in the center, and a ligand-binding domain located in the carboxy-terminal portion (9–12) (Fig. 1).

Qualitative and/or quantitative alterations of steroid hormone receptors are capable of influencing their transcriptional activity, leading to clinical syndromes of steroid hormone resistance. The elucidation of steroid hormone receptor alterations as the cause of generalized end-organ steroid resistance syndromes provides important evidence for the roles of these ligand-activated transcription factors in gene regulation, physiology, and development (13).

Steroid hormone receptors whose roles are not important for survival appear to have a higher incidence of mutations. Such mutations may be inherited, in contrast to those whose defect would result in failure to develop or death. For example, there have been no reports of generalized insensitivity to estrogens. This would probably be incompatible with life, because implantation and early embryogenesis would be affected. Similarly, insensitivity to progesterone has been reported only once, in a woman suffering from infertility (14). In contrast, end-organ resistance to androgens is frequent. Testicular feminization resulting from androgen receptor defects ranges from 46,XY individuals with a female phenotype, to subjects with ambiguous genitalia, to men with mild hypospadias and infertility (15). The incidence

Methods in Neurosciences, Volume 22

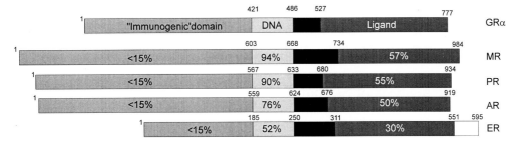

FIG. 1   The three domains of human steroid hormone receptors are represented. The homologies of four steroid hormone receptors as compared to the glucocorticoid receptor are expressed as percent identity in primary sequence (GRα, glucocorticoid receptor α; MR, mineralocorticoid receptor; AR, androgen receptor; PR, progesterone receptor; ER, estrogen receptor).

of mutations in familial glucocorticoid resistance is intermediate (16–18). This state has been described in eight families and five independent subjects, always as a partially compensated and generalized form of end-organ insensitivity. Complete steroid hormone resistance syndromes would be expected only with hormones whose absence would not interfere with survival.

The characterization of alterations in steroid hormone receptors from patients with generalized steroid-resistant syndromes provided important insights into the target cell function of these nuclear transcription factors. Extensive studies demonstrated the influence of single amino acid substitutions caused by single nucleotide mutations on steroid hormone receptor action. The examination of naturally occurring mutations has, thus, been valuable in explaining the mechanism(s) by which these nuclear receptors regulate gene transcription, and in understanding the pathophysiology of disease states (19–22).

For example, analysis of the large number of point mutations identified in the androgen receptor gene allowed the identification of two regions in the ligand-binding domain, with a disproportionately high incidence of gene alterations. More than 90% of the substitution mutations cluster between amino acids 726 and 772 and between 826 and 864, two stretches accounting for 30% of the total region of the hormone-binding domain (23).

Advances in molecular biology, especially the use of the polymerase chain reaction (PCR) (24), have provided a first step in the analysis of naturally occurring steroid hormone receptor mutations. Our laboratory investigations focus on the detection of mutations in steroid hormone receptors causing generalized or tissue-specific steroid hormone resistance syndromes. Specifically, alterations in the glucocorticoid receptor, mineralocorticoid receptor, and androgen receptor genes leading to generalized glucocorticoid, aldo-sterone, or androgen resistance, respectively, have been studied using the following techniques.

Mutational Analysis of Steroid Hormone Receptors

Sources of Genomic DNA and RNA

Genomic DNA and total RNA are extracted from Epstein–Barr virus (EBV)-transformed lymphoblasts or skin fibroblasts (25, 26). The former are derived from peripheral leukocytes of patients or controls. Establishment of permanent cell lines that express the desired steroid hormone receptor permits studies of its protein, mRNA, and gene. Therefore, EBV-transformed lymphoblasts are useful for the investigation of glucocorticoid and mineralo-corticoid receptors (27), whereas cultures of genital skin fibroblasts are most suitable for studies of the androgen receptor (28).

Polymerase Chain Reaction Amplification of Steroid Hormone Receptor Genes

Androgen and glucocorticoid receptor genes consist of 8–10 exons, respectively, which are interrupted by intron sequences of invariant length ranging from a few hundred to several thousand base pairs (29–31). Therefore, entire steroid hormone receptor genes span regions 60–100 kb in size. Investigations of mutations leading to the classic steroid hormone resistance syndromes focus mainly on the analysis of changes within an exon such as point muta-tions, alterations in adjacent splice site regions, or deletions concerning parts of or complete exons. As a consequence, the strategy for the investigation of steroid hormone receptor genes, as described below, assumes that the gene structure as well as the sequences of the 5' and 3' intron regions flanking each exon are known (21, 30).

For our investigations of the glucocorticoid receptor gene, we employ primer pairs for the amplification of entire exons by selecting oligonucleotides complementary to sequences in the 5' or 3' exon-flanking regions, respectively (21). Polymerase chain reaction amplification is carried out in the presence of 0.5–1.0 μg of genomic DNA, 30 pmol of each primer, a 200 mM concentration of each dNTP, 50 mM KCl, 10 mM Tris-HCl (pH 8.3), 3 mM MgCl$_2$, 0.01% (w/v) gelatin, and 2.5 U of AmpliTaq (Perkin-Elmer Cetus, Emeryville, CA) in a total volume of 100 μl. The reaction mixture is overlaid with 100 μl of mineral oil. Thermal cycling parameters consist of 1 min at

94°C for denaturation, 1.5 min at 55 or 60°C for primer annealing, and 1.5 min at 72°C for oligonucleotide extension. The initial denaturation period is lengthened to 5 min and the temperature is increased to 95°C. The extension period of the final cycle lasts 5 min.

Genomic DNA for PCR amplification of glucocorticoid receptor exons 1 and 6 requires additional treatment. This is necessary to assure specific amplification. First, we pretreat genomic DNA samples with restriction enzymes possessing no recognition sites within the sequence to be amplified. DNA samples used for amplification of exon 1 are digested with the restriction enzyme *Sac*I, and for exon 6 with *Bam*HI and *Xho*I. Without further purification, aliquots of the digested DNA are then used as templates, for PCR amplification in the presence of 5% (v/v) formamide and 5 U of AmpliTaq. The other thermal cycling parameters remain the same as described above.

After PCR amplification we routinely purify the PCR samples by chloroform extraction. The mineral oil layer is removed from the sample after centrifugation for 1 minute at maximum speed (Eppendorf centrifuge, 5415C, 14,000 rpm/min) and 100 μl of chloroform is added. After vortexing and centrifugation, the aqueous phase, containing the PCR product, is taken off and transferred to a fresh tube. It should be emphasized that, after the addition of chloroform, the upper layer is formed by the aqueous phase. A 5- to 10-μl aliquot of each PCR product is then analyzed by Tris–acetate Na_2 EDTA $2H_2O$ (TAE)– or Tris–burate–ethylenediaminetetraacetic acid (EDTA) Tris base boric acid EDTA (TBE)–agarose gel electrophoresis to check for purity and length of the desired fragment. Major changes in the size of an amplified product can be detected by this procedure.

Synthesis and Amplification of cDNA by Combination of Reverse Transcription and Polymerase Chain Reaction

For the sequence analysis of mRNA transcripts, a combination of reverse transcription (RT) of RNA followed by PCR (RT-PCR) is the method of choice to yield sufficient amounts for further investigation (32). This technique is extremely helpful for the investigation of steroid hormone receptors for which only the cDNA sequence is known. However, only RNA of cells expressing the desired steroid hormone receptor is suitable for this technique. Leukocytes from peripheral blood are relatively easy to obtain for RT-PCR of the glucocorticoid and mineralocorticoid receptor mRNA. In contrast, for the investigation of androgen receptor transcripts, cultured fibroblasts from genital skin biopsies are the specimens of choice. Total RNA or mRNA extracted from such sources is then incubated with specific antisense primers, random hexamer sequences, or an oligomer of polydeoxythymidylate [oligo(dT)] in a reaction with reverse transcriptase. This enzyme, originally isolated from retroviruses, copies RNA molecules into DNA.

For reverse transcription of glucocorticoid receptor transcripts, we heat aliquots of total RNA (2.5–10 μg) at 70°C for 10 min. The reaction mix is then chilled on ice and reverse transcribed at 42°C for 1 hr. The reaction is performed in a total volume of 20 μl containing 5 mM MgCl$_2$, RNase inhibitor (1 U/ml), a 1 mM concentration of each dNTP, 1× PCR buffer, 50 U of reverse transcriptase (Perkin-Elmer Cetus Corp.), and 50 pmol of antisense oligonucleotide primers. The specific 3′ primers are located at different cDNA positions. This leads to the polymerization of single-stranded antisense DNA molecules complementary to the RNA strand (cDNA). After transcription, the RNA–DNA hybrid is denatured by heat at 99°C for 5 min and cooled to 4°C, leaving the newly synthesized single-stranded cDNA product as template. In the presence of sense primers that serve as start sites, a sense DNA strand is synthesized during the first cycle of a successive polymerase chain reaction. The locations of these sense primers are at different cDNA positions, respectively. The double-stranded cDNA product is amplified in subsequent PCR cycles. The PCR is performed as described above, in the presence of 2 mM MgCl$_2$, 1× PCR buffer, 2.5 U of AmpliTaq, and 50 pmol of sense primer in a final volume of 100 μl. This results in the amplification of the 2.331-kb coding region of the glucocorticoid receptor cDNA in three overlapping segments.

Generation of Single-Stranded DNA for Dideoxy Sequencing by Asymmetric Polymerase Chain Reaction

In our hands, the combination of asymmetric PCR and Sanger dideoxy sequencing has proved to be an excellent procedure for the detection of mutations in steroid hormone receptors. Asymmetric PCR yields single DNA strands required by the Sanger method. Therefore, the combination of both techniques circumvents time- and labor-consuming cloning steps, which are necessary for the introduction of double-stranded PCR-amplified fragments into vectors for subsequent sequence analysis. Direct sequencing of asymmetric PCR products also offers the advantage of simultaneous investigation of both steroid hormone receptor alleles. During PCR amplification both alleles are amplified and, therefore, two single-stranded DNA populations, each representing one allele, serve as templates in one sequencing reaction. Otherwise, several plasmid clones must be sequenced to ensure the analysis of both alleles.

Asymmetric PCR represents a variation of the symmetric PCR technique, described above (33). It is based on an asymmetric primer ratio during a

PCR, leading to the preferential amplification of a single DNA strand. The chloroform-purified product of a symmetric PCR serves as template for reamplification. A 1-μl aliquot of the first symmetric PCR still containing low concentrations of incorporated primers from the initial reaction is amplified in the presence of 50 pmol of a sense or antisense primer. Therefore, only one DNA strand, either sense or antisense, is amplified selectively. The asymmetric PCR is carried out for 20 thermal cycles with unchanged denaturing, annealing, and extension conditions derived from the first symmetric PCR. Even though the single-stranded PCR product increases arithmetically with each cycle, high enough yields of single strands for sequencing are achieved. At the end of the asymmetric PCR, the single-stranded products are purified from any remaining primer. Some laboratories use salt–ethanol precipitation. In our hands the spin-dialysis technique works best. The chloroform-purified sample is filtered through a membrane by centrifugation at low speed (UFC3 THK 00; Millipore, Bedford, MA). Depending on the size of the membrane pores, the amplified fragment products are retained above the filter while small oligonucleotides flow through. The retained products are resuspended in water and the procedure is repeated three times. Finally, the purified product is reconstituted in 40 μl of water, transferred into a new tube, and dried under vacuum.

DNA Sequencing

DNA sequencing is the central technique in the investigation of genetic defects leading to abnormal steroid hormone receptor function. Two methods are regularly used to determine the order of nucleotides: the chemical degradation technique of Maxam and Gilbert (34) and the enzymatic dideoxynucleotide-mediated chain-termination method by Sanger *et al.* (35). The latter has become the method of choice for the detection of mutations.

The Sanger sequencing method is based on the ability of a DNA polymerase to use 2'-3'-dideoxynucleoside triphosphates (ddNTPs) as substrates. After the incorporation of such a ddNTP, the elongation of a DNA strand is terminated at the 3' end of a nucleotide chain, because of the lack of a 3'-hydroxyl group. Single-stranded DNA molecules serve as templates, and a complementary primer serves as the start site for the chain elongation. In four separate reactions only one of four possible ddNTPs (ddGTP, ddATP, ddTTP, or ddCTP) is added to a mixture of unmodified nucleotide triphosphates. Therefore, in each of four parallel reactions, the extension of a fraction of elongating primer chains is specifically terminated at a site determined for the incorporation of each specific nucleotide. The resulting elongated primer chains are characterized by a fixed 5' end determined by the start site of the original primer and variable 3′ end extensions as a result of the ddNTP incorporation.

Labeling of the newly synthesized single-stranded oligonucleotides is carried out in several ways, with either radioactive substrates or more recently with nonisotopic reagents. Incorporation of labeled nucleoside triphosphates into the nascent DNA strand is a commonly used technique; another possibility is the application of labeled primers for the sequencing reaction.

The sequencing products are finally resolved in a high-resolution denaturing polyacrylamide gel. This procedure allows oligonucleotides differing by 1 base in length to be separated. Four populations of single-stranded sequencing products exist, each of them terminated at all possible residues of guanine, adenosine, thymidine, or cytosine. Electrophoresis of these products on adjacent lanes will, therefore, reveal a complete sequence information.

We regularly perform direct sequencing of PCR-amplified DNA, using asymmetric PCR products as templates and a modified T7 DNA polymerase (Sequenase; U.S. Biochemical, Cleveland, OH). Following purification and vacuum drying of the single-stranded product, the amplified DNA is reconstituted in 7 μl of doubly distilled H_2O. According to manufacturer instructions, 2 μl of a 5× reaction buffer containing 200 mM Tris-HCl (pH 7.5), 100 mM $MgCl_2$, and 250 mM NaCl is added, followed by 1 μl of the desired sequencing primer (10 pmol). The mixture is vortexed and centrifuged and the capped tube is heated for 5 min at 95°C. The tube is immediately centrifuged to collect the condensate, and allowed to cool at room temperature for 30 min. Ideally, the sequencing primer should anneal to the template during this period. Following the cooling interval, 1 μl of dithiothreitol (DTT, 0.1 M), 2 μl of 1:5 diluted labeling mix (dGTP, dCTP, and dTTP, each in a 5× concentration of 7.5 μM), 1 μl of [α-^{35}S]dATP [1000 Ci/mmol; Amersham (Arlington Heights, IL) or New England Nuclear (Boston, MA)] and 2 μl of diluted Sequenase version 2.0 (U.S. Biochemical) is added, as recommended by the manufacturer. For sequencing close to the primer, we routinely include 1 μl of Mn buffer (0.15 M sodium isocitrate, 0.1 M $MnCl_2$). The solution is mixed, centrifuged, and incubated for 3 min at room temperature. For the final termination reaction, 4-μl aliquots of the previous reaction mixture are transferred into four separate reaction tubes. Each contains 2.5 μl of a nucleotide termination mix of either 8 μM ddATP, ddTTP, ddCTP, or ddGTP in a solution of 80 μM dGTP, dATP, dTTP, and dCTP. The termination mixture must be prewarmed at 37°C for a minimum of 3 min. These four new reactions containing 4 μl of the original mixture are incubated at 37°C for 5 min. Subsequently, 4 μl of a stop solution is added [95% (v/v) formamide, 20 mM EDTA, 0.05% (v/v) bromophenol blue, and 0.05% (v/v) xylene cyanol FF] and, after mixing, the samples are kept on ice for immediate use or can be stored at −20°C for up to 1 week. The sequencing reaction is then electrophoresed on a 6% (w/v) acrylamide/8 M urea denaturing gel at 55°C for several hours. Before loading the samples onto the gel, we heat up the reaction mixture to 95°C for exactly 3 min and snap-cool on ice. We use gels of 0.4-mm thickness with a Bio-Rad (Richmond, CA) gel electrophoresis system. Following electrophoresis, the gel is transferred to Whatman (Clifton, NJ) 3MM paper without prior soaking in acetic acid–methanol, and dried at 80°C for about 1 hr. For overnight exposure, Kodak (Rochester, NY) XAR-5 films or Amersham films are routinely employed.

Allele-Specific Oligonucleotide Hybridization and Restriction Enzyme Analysis

We test family members as well as control subjects for specific mutations detected in a patient by sequence analysis, using the techniques of allele-specific oligonucleotide hybridization and restriction enzyme analysis. These methods enable investigators to confirm or exclude specific base changes in related family members and allow rapid assessment of the frequency of such changes in a control group.

Allele-Specific Oligonucleotide Hybridization

The allele-specific hybridization technique is simply based on the annealing of a labeled synthetic oligonucleotide to blotted DNA (36).

For the investigation of a point mutation in the glucocorticoid receptor, we amplify genomic DNA of relatives and control subjects by PCR to yield sufficient amounts of the specific DNA fragment to be tested. The amplified DNA is then electrophoresed on either an TAE– or TBE–agarose gel and subsequently transferred to a membrane by Southern techniques (37). Alternatively, the amplified PCR product is heat denatured and directly blotted on a nitrocellulose filter, using a slot or dot-blot system (Minifold; Schleicher & Schuell, Keene, NH). Two nearly identical hybridization probes are used for the analysis of a base change. The sequences of these two oligonucleotides differ in just one base, one probe having the wild-type sequence, the other bearing the identified mutation at the corresponding position. Mismatches in the middle of an oligonucleotide are more deleterious for annealing to the investigated DNA under stringent conditions than are mismatches at either end. Therefore, usually 19-mer probes are synthesized containing the mutant or wild-type base in their central 10-position.

In two separate reactions, the membrane-bound DNA is either incubated with the labeled wild-type or mutant probe for annealing. For hybridization, we use ^{32}P-labeled oligonucleotides (10^6 cpm/ml) in a solution containing $5 \times$ SSPE ($20 \times$ SSPE is 3.0 M NaCl; 0.2 M NaH$_2$PO$_4 \cdot$ H$_2$O, 0.02 M disodium EDTA, pH 7.4), $5 \times$ Denhardt's solution, and 0.1% (w/v) sodium dodecyl sulfate (SDS). Subsequently, the filters with mutant or wild-type oligonucleotide–DNA hybrids are washed twice at room temperature in $2 \times$ SSPE and 0.1% SDS. This is followed by a 10-min wash at 55°C in $5 \times$ SSPE and 0.1% SDS, allowing only perfect matches to remain annealed. DNA of subjects heterozygous for a point mutation show hybridization signals with both oligonucleotides. Hybrid-positives received exclusively with the mutant or wild-type probe are homozygotic for either the mutant or wild-type sequence, respectively.

Restriction Enzyme Analysis

Restriction enzymes are enzymes that bind to specific recognition sequences to cleave double-stranded DNA (38). Mutations creating or abolishing such recognition sites can, therefore, be investigated by employment of restriction enzymes. Once a change in a recognition site has been identified in the DNA sequence of a patient, specific PCR-amplified fragments of the DNA obtained from relatives or controls, potentially harboring the base change, are incubated with the specific restriction enzyme. The cut or uncut segments can be visualized by gel electrophoresis to determine whether an individual is a carrier of such a mutation or bears the wild-type sequence. Unfortunately, only a small number of mutations can be identified by this technique.

Functional Testing of Receptor Mutations

Following the identification of mutations in steroid hormone receptor genes leading to amino acid substitutions, these alterations must be tested for an effect on steroid hormone receptor action. Tests of immunogenicity, ligand-binding affinity, or DNA-binding capacity of a mutant receptor are useful measurements of isolated steroid hormone receptor properties. However, these studies do not demonstrate the integrated influence of a mutation on receptor function, or the ability of mutant receptors to *trans*-activate or modulate the transcription of specific hormone-responsive genes.

Cloning of Mutant Steroid Hormone Receptor Expression Vectors

In vitro studies require expression of either wild-type or mutant steroid hormone receptors, preferably in a mammalian cell system. For example, Giguere and colleagues have cloned an expression vector for the glucocorticoid receptor (9). This plasmid contains the complete cDNA of the glucocorticoid receptor α form (pRShGRα) (2, 9). For the analysis of glucocorticoid receptor mutations, we have employed this vector, generously supplied by R. M. Evans of the La Jolla Institute for Biological Studies, as a template. Different techniques depending on the location of the mutation within the glucocorticoid receptor cDNA and the availability of mutant cDNA fragments can be used to clone a mutant cDNA into an expression vector. We apply either PCR-based site-directed mutagenesis or subcloned cDNA cassettes originating from the mRNA of a patient to introduce the desired mutation into the wild-type glucocorticoid receptor sequence. We describe here the construction of an expression vector for the mutant glucocorticoid receptor Ser-363, as an example. This mutation, nucleotide substitution G-1220, has been identified in patients suffering from glucocorticoid resistance, as well as in unaffected subjects (21). It resides in the amino terminus of the glucocorticoid receptor, downstream of the τ_1 region and upstream of the DNA-binding domain. For subcloning, we replace a *Hin*dIII/*Bst*XI fragment, including glucocorticoid receptor nucleotides 1044–1240 of plasmid pRShGRα, with the corresponding cassette derived from the cDNA of the proband. Therefore, we digest the expression vector for the wild-type glucocorticoid receptor, pRShGRα, with the restriction endonucleases *Sal*I and *Bst*XI (Boehringer Mannheim Biochemicals, Indianapolis, IN), which yields two fragments 6011 and 715 bp in size. In addition, we cut the expression vector pRShGRα with the restriction enzymes *Sal*I and *Hin*dIII in a separate reaction. We obtain three fragments and isolate the 519-bp-long *Sal*I/*Hin*dIII segment. A 196-bp fragment bearing nucleotide substitution G-1220 is received by digestion of PCR-amplified cDNA of the patient. The reverse-transcribed product derived from the patient RNA is cut with the restriction enzymes *Hin*dIII and *Bst*XI and includes nucleotides 1044–1240 of the human glucocorticoid receptor cDNA. For subsequent ligation, we incubate the 6011-bp *Sal*I/*Bst*XI fragment of the first digestion, the 519-bp *Sal*I/*Hin*dIII segment of the second digestion, and the 196-bp *Hin*dIII/*Bst*XI-digested cDNA product in a single reaction, using T4 DNA ligase (Boehringer Mannheim Biochemicals). To confirm the G-1220 substitution without any other base changes, we sequence the cDNA insert including the ligation sites of the recombinant plasmid by double-stranded dideoxynucleotide-mediated sequencing. We designate the G-1220 recombinant plasmid as pRShGR-

Ser[363], which is then employed in a cotransfection assay to test the transcriptional activity of the mutant receptor expressed from this vector.

Cotransfection Studies

Giguere and colleagues were the first to develop an ingenious method for testing the impact of steroid hormone receptor mutations on gene activation in mammalian cells (9). In their monkey kidney cell system (CV-1, COS cells), wild-type and mutant forms of the glucocorticoid receptor, expressed from a transfected vector, induce transcription of a reporter gene by binding to a steroid hormone responsive promoter. CV-1, COS-1, or COS-7 cells are widely used as host cell lines because they do not carry any detectable amounts of endogenous steroid hormone receptors. The parental CV-1 cells are fibroblasts derived from simian green monkey kidneys. COS-1 and COS-7 cell lines are established from the former by transformation with an origin-defective mutant simian virus 40 (SV 40), which encodes the wild-type T antigen (39). Expression vectors bearing an SV40 origin will therefore replicate to high copy numbers, providing sufficient minichromosomes for the expression of the encoded protein (40).

The transient expression system used for testing glucocorticoid receptor constructs requires two vectors to be transfected into the cell line. The first construct, which serves as an expression vector, encodes the cDNA of the wild-type or mutant glucocorticoid receptor and bears an SV40 origin. After transfection, this plasmid replicates up to several thousand new vector copies representing minichromosomes, from which the cDNA of the wild-type or mutant glucocorticoid receptor will be transcribed and translated into functional protein. The second vector bears the mouse mammary tumor virus (MTV) long terminal repeat coupled to a reporter gene such as the gene for chloramphenicol transferase (CAT) or luciferase (41). The MTV promoter encodes a cluster of four binding sites or hormone response elements for glucocorticoid, mineralocorticoid, and androgen receptors (42).

Wild-type or mutant receptors, expressed from the first plasmid, are activated by adding dexamethasone, aldosterone, or androgens to the cell medium. The activated receptors then bind to the hormone responsive elements in the MTV promoter of the second plasmid, activating the transcription of the CAT or luciferase gene. The CAT or luciferase transcripts are translated into CAT or luciferase protein within the cells. Therefore, measurement of CAT or light generation serves as an indirect indicator for the transcription of activity of the glucocorticoid receptor.

In our studies to examine the influence of glucocorticoid receptor mutations on gene transcription, we use COS-7 cells obtained from the American Type

Culture Collection (ATCC; Rockville, MD). These cells are cultured in Dulbecco's modified Eagle's medium (DMEM) supplemented with 10% (v/v) charcoal-treated fetal bovine serum (HyClone, Logan, UT). The cells are plated on 60-mm tissue culture dishes (7.5×10^5 cells/plate) 18 hr prior to transfection and grown to 60% confluence. For transfections, we incubate the cells with 3 ml of Optimem I containing 30 μl of lipofectin reagent (GIBCO-BRL, Gaithersburg, MD) (43), 1 μg of the expression vector encoding either the wild-type hGRα, the Ser-363 mutant glucocorticoid receptor, or the truncated human glucocorticoid receptor β, and 5 μg of the reporter plasmid pMTVCAT. After 5 hr, we stop the transfections by replacing the medium with DMEM supplemented with 20% fetal bovine serum (FBS). After 18 hr, we feed the cells with DMEM (10% FBS) or the same medium supplemented with dexamethasone at concentrations ranging from 10^{-11} to 10^{-6} M. Forty-eight hours after transfection, cells are harvested by scraping and cell extracts are prepared by three freeze-thaw cycles. The Coomassie protein assay (Pierce Chemical Co., Rockford, IL) is used to assure equivalence of lysate protein. We assay CAT activity by monitoring the transfer of acetyl groups from acetyl-CoA (Pharmacia–LKB Biotechnology, Uppsala, Sweden) to [^{14}C]chloramphenicol (Du Pont–New England Nuclear Research Products, Boston, MA), using thin-layer chromatography. The results are quantified with a Betascope 603 blot analyzer (Betagen, Waltham, MA). We determine the CAT activities as the percentage of chloramphenicol acetylated per hour per 10 μg of protein lysate. To ensure reliable results, at least two transfections are carried out in triplicate on separate days.

Polymerase Chain Reaction-Based Screening Methods

The extremely powerful techniques of PCR-based DNA amplification and subsequent sequence analysis of the amplified products provide the elementary tools for the investigation of gene mutations. However, their application to investigate an entire gene for steroid hormone receptor mutations is limited to individual cases by the labor and cost intensity, excluding more widespread research applications and their use in clinical routine. Therefore, the development of procedures allowing simple, rapid detection of single point mutations in DNA fragments will have a tremendous impact on molecular genetics. Currently, three techniques are the methods of choice for screening for potential mutations: denaturing gradient gel electrophoresis (DGGE) (44), temperature gradient gel electrophoresis (TGGE) (45), and the analysis of single-strand conformation polymorphisms (SSCPs) (46). These methods are based on physicochemical changes evoked by single base mutations.

Denaturing gradient gel electrophoresis and temperature gradient gel electrophoresis rely on differences in the melting properties caused by base substitutions in the DNA fragment to be investigated (44, 45). Double-stranded DNA fragments consist of the so-called melting domains separated by fairly sharp boundaries. These domains that range from 25 to several hundred nucleotides in length deanneal at distinct temperatures in a cooperative way. In addition, the melting temperature of two adjacent domains may differ by several degrees. The melting temperature of a domain is determined by its nucleotide sequence and the stacking interactions between adjacent bases on the same DNA strand. Therefore, even a single point mutation is able to change the melting behavior of a DNA fragment significantly, regardless of its location within the melting domain. Polyacrylamide gel electrophoresis of double-stranded DNA fragments through a gradient of linearly increasing temperature (TGGE) or increasing concentrations of denaturing substances, such as urea and formamide (DGGE), are associated with a DNA branching process. First, the domain with the lowest melting temperature starts to deanneal, leading to a Y-shaped configuration of the branched DNA fragment, which slows down the electrophoretic migration in the gel. Single nucleotide substitutions change the melting characteristics of DNA fragments and, therefore, alter their branching behavior. During electrophoresis the slowing down of wild-type and mutant sequences occurs at different positions in the gel, leading to a distinct separation from each other. However, base changes in all but the last or highest melting domain are generally detectable by DGGE and/or TGGE, because the loss of the branched Y configuration as an essential requirement prevents the detection of mutations after complete strand separation.

The polymerase chain reaction is used for the specific amplification of fragments to be investigated by these techniques. The length of fragments should be limited to 500 bp. For PCR amplification, oligonucleotides are employed with the addition of a 40-GC base overhang leading to the introduction of a GC-rich 5' or 3' clamp. This synthetic section of the PCR-amplified fragment serves as the domain exhibiting the highest melting temperature for the use in DGGE or TGGE (47, 48). Both of these techniques require special equipment for electrophoresis. DGGE is performed in a 6% (w/v) polyacrylamide gel with a linearly increasing gradient of denaturants, such as formamide and urea, in a vertical gel apparatus at a constant temperature of 60°C. TGGE, in contrast, requires a temperature gradient generated by a heating plate for horizontal gel electrophoresis. Reaction conditions for each PCR-amplified fragment of a steroid hormone receptor must be determined separately. Listing each condition specifically used is beyond the scope of this chapter.

SSCP analysis requires the formation of single-stranded DNA, obtained by PCR performed in the usual manner with labeled deoxynucleotides or primers, followed by the addition of denaturants and heat (46). The samples are run at room temperature or at 4°C on nondenaturing 6% polyacrylamide gels without or with glycerol (10–20%, v/v). Samples showing aberrant migration in DGGE, TGGE, or SSCP analysis must be further analyzed by sequencing for the exact determination of the mutation.

Concluding Remarks

The application of techniques described in this chapter has enabled different laboratories, including ours, to identify a multitude of mutations in the androgen and glucocorticoid receptors. Analysis of these mutations has allowed investigators to determine amino acids playing key roles in receptor function. The development of useful methods is fast advancing, and is bound to simplify further the approach to steroid receptor gene defects. The investigation of specific organs, tissues, and cells for the presence of somatic steroid hormone receptor mutations may therefore yield important clinical information (49, 50).

References

1. R. M. Evans, *Science* **240**, 889–895 (1988).
2. S. M. Hollenberg, C. Weinberger, E. S. Ong, G. Cerelli, A. Oro, R. Lebo, E. B. Thompson, M. G. Rosenfeld, and R. M. Evans, *Nature* (*London*) **318**, 635–641 (1985).
3. S. Green, P. W. Walter, V. Kumar, A. Krust, J.-M. Bornert, P. Argos, and P. Chambon, *Nature* (*London*) **320**, 134–139 (1986).
4. G. Greene, P. Gilna, M. Waterfield, A. Baker, Y. Hort, and J. Shine, *Science* **231**, 1150–1154 (1986).
5. M. Misrahi, M. Atger, L. d'Auriol, H. Loosfelt, C. Meriel, F. Fridlansky, A. Guiochon-Mantel, F. Galibert, and E. Milgrom, *Biochem. Biophys. Res. Commun.* **143**, 740–748 (1987).
6. J. L. Arriza, C. Weinberger, G. Cerelli, T. M. Glaser, B. L. Handelin, D. E. Housman, and R. M. Evans, *Science* **237**, 268–274 (1987).
7. C. Chang, J. Kokontis, and S. Liao, *Science* **240**, 324–327 (1988).
8. D. B. Lubahn, D. R. Joseph, P. M. Sullivan, H. F. Willard, F. S. French, and E. M. Wilson, *Science* **240**, 327–330 (1988).
9. V. Giguere, S. M. Hollenberg, M. G. Rosenfeld, and R. M. Evans, *Cell* **46**, 645–652 (1986).
10. S. M. Hollenberg, V. Giguere, P. Segui, and R. M. Evans, *Cell* **49**, 39–46 (1987).
11. S. M. Hollenberg and R. M. Evans, *Cell* **55**, 899–906 (1988).

12. G. Jenster, H. A. G. M. van der Korput, C. van Vroonhoven, T. H. van der Kwast, J. Trapman, and A. O. Brinkmann, *Mol. Endocrinol.* **5**, 1398–1404 (1991).
13. G. P. Chrousos, D. L. Loriaux, and M. B. Lipsett, eds., *Exp. Med. Biol.* **196** (1986).
14. D. W. Keller, W. G. Wiest, F. B. Askin, L. W. Johnson, and R. C. Strickler, *J. Clin. Endocrinol. Metab.* **48**, 127–132 (1979).
15. J. E. Griffin and J. D. Wilson, *in* "The Metabolic Basis of Inherited Disease" (C. R. Scriver, A. L. Beaudet, W. S. Sly, and D. Valle, eds.), 6th Ed., pp. 1919–1944. McGraw-Hill, New York, 1989.
16. G. P. Chrousos, A. Vingerhoeds, D. Brandon, C. Eil, M. Pugeat, M. DeVroede, D. L. Loriaux, and M. B. Lipsett, *J. Clin. Invest.* **69**, 1261–1269 (1982).
17. G. P. Chrousos, A. C. M. Vingerhoeds, D. L. Loriaux, and M. B. Lipsett, *J. Clin. Endocrinol. Metab.* **56**, 1243–1245 (1983).
18. S. W. J. Lamberts, J. W. Koper, P. Biemond, F. H. den Holder, and F. H. de Jong, *J. Clin. Endocrinol. Metab.* **74**, 313–321 (1992).
19. M. J. McPhaul, M. Marcelli, S. Zoppi, J. E. Griffin, and J. D. Wilson, *J. Clin. Endocrinol. Metab.* **76**, 17–23 (1993).
20. D. M. Hurley, D. Accili, C. A. Stratakis, M. Karl, N. Vamvakopoulos, E. Rorer, K. Constantine, S. I. Taylor, and G. P. Chrousos, *J. Clin. Invest.* **87**, 680–686 (1991).
21. M. Karl, S. W. J. Lamberts, S. Detera-Wadleigh, I. J. Encio, C. A. Stratakis, D. M. Hurley, D. Accili, and G. P. Chrousos, *J. Clin. Endocrinol. Metab.* **76**, 683–689 (1993).
22. D. M. Malchoff, A. Brufsky, G. E. Reardon, P. McDermott, E. C. Javier, C. H. Bergh, D. Rowe, and C. D. Malchoff, *J. Clin. Invest.* **91**, 1918–1925 (1993).
23. M. J. McPhaul, M. Marcelli, S. Zoppi, C. M. Wilson, J. E. Griffin, and J. D. Wilson, *J. Clin. Invest.* **90**, 2097–2101 (1992).
24. R. K. Saiki, D. H. Gelfand, S. Stoffel, S. J. Scharf, R. Higuchi, G. T. Horn, K. B. Mullis, and H. A. Erlich, *Science* **239**, 487–491 (1988).
25. J. Sambrook, E. F. Fritsch, and T. Maniatis, *in* "Molecular Cloning: A Laboratory Manual," 2nd Ed., pp. 9.14–9.23. Cold Spring Harbor Lab. Press, Cold Spring Harbor, New York, 1989.
26. P. Chomczynski, and N. Sacchi, *Anal. Biochem.* **162**, 156–159 (1987).
27. M. Tomita, G. P. Chrousos, D. D. Brandon, S. Ben-Or, C. M. Foster, L. De-Vougn, S. Taylor, D. L. Loriaux, and M. B. Lipsett, *Horm. Metab. Res.* **17**, 674–678 (1985).
28. T. R. Brown and C. J. Migeon, *Mol. Cell. Biochem.* **36**, 3–22 (1981).
29. D. B. Lubahn, T. R. Brown, J. A. Simental, H. N. Higgs, C. J. Migeon, E. M. Wilson, and F. S. French, *Proc. Natl. Acad. Sci. U.S.A.* **86**, 9534–9538 (1989).
30. M. Marcelli, W. D. Tilley, C. M. Wilson, J. E. Griffin, J. D. Wilson, and M. J. McPhaul, *Mol. Endocrinol.* **4**, 1105–1116 (1990).
31. I. J. Encio and S. D. Detera-Wadleigh, *J. Biol. Chem.* **266**, 7182–7188 (1991).
32. E. Kawasaki and A. M. Wang, *in* "PCR Technology" (H. A. Erlich, ed.), pp. 89–97. Stockton, New York, 1986.
33. U. B. Gyllensten and H. A. Erlich, *Proc. Natl. Acad. Sci. U.S.A.* **85**, 7652–7656 (1988).

34. A. M. Maxam and W. Gilbert, *Proc. Natl. Acad. Sci. U.S.A.* **74,** 560–564 (1977).

35. F. Sanger, S. Nicklen, and A. R. Coulson, *Proc. Natl. Acad. Sci. U.S.A.* **74,** 5463–5467 (1977).

36. D. Accili, C. Frapier, L. Mosthaf, C. McKeon, S. C. Elbein, M. A. Permutt, E. Ramos, E. Lander, A. Ullrich, and S. I. Taylor, *EMBO J.* **8,** 2509–2517 (1989).

37. E. M. Southern, *J. Mol. Biol.* **98,** 503–515 (1975).

38. R. J. Roberts, *Nucleic Acids Res.* **16,** Suppl., r271–313 (1988).

39. Y. Gluzman, *Cell* **23,** 175–182 (1981).

40. C. M. Gorman, G. T. Merlino, M. C. Willingham, I. Pastan, and B. H. Howard, *Proc. Natl. Acad. Sci. U.S.A.* **79,** 6777–6781 (1982).

41. C. M. Gorman, L. F. Moffat, and B. H. Howard, *Mol. Cell. Biol.* **2,** 1044–1051 (1982).

42. M. Pfahl, *Cell* **31,** 475–482 (1982).

43. P. L. Felgner, T. R. Gadek, M. Holm, R. Roman, H. Chan, M. Wenz, J. Northrop, G. Ringold, and M. Danielsen, *Proc. Natl. Acad. Sci. U.S.A.* **84,** 7413–7417 (1987).

44. R. M. Myers, N. Lumelsky, L. S. Lerman, and T. Maniatis, *Nature (London)* **313,** 495–498 (1985).

45. V. Rosenbaum and D. Riesner, *Biophys. Chem.* **26,** 235–246 (1987).

46. M. Orita, H. Iwahana, H. Kanazawa, K. Hayashi, and T. Sekiya, *Proc. Natl. Acad. Sci. U.S.A.* **86,** 2766–2770 (1989).

47. R. M. Myers, S. G. Fischer, T. Maniatis, and L. S. Lerman, *Nucleic Acids Res.* **13,** 3111–3130 (1985).

48. R. M. Myers, S. G. Fischer, L. S. Lerman, and T. Maniatis, *Nucleic Acids Res.* **13,** 3131–3146 (1985).

49. C. Jonat, H. J. Rahmsdorf, K.-K. Park, A. C. B. Cato, S. Gebel, H. Ponta, and P. Herrlich, *Cell* **62,** 1189–1204 (1990).

50. M. C. Poznansky, A. C. H. Gordon, I. W. B. Grant, and A. H. Wyllie, *Clin. Exp. Immunol.* **61,** 135–142 (1985).

Section III

Molecular Effects of Steroids

[15] Protein–DNA-Binding Assay for Analysis of Steroid-Sensitive Neurons in Mammalian Brain

Yuan-shan Zhu and Donald W. Pfaff

Introduction

To study the mechanisms of gene expression requires the understanding of protein–nucleic acid interactions. Multiple assays are now available to assess these interactions, including X-ray crystallography, nuclear magnetic resonance (NMR), the filter-binding assay, footprinting, and the gel retardation assay (GRA). Although X-ray crystallography and NMR can provide critical information about the structural details of these interactions, they cannot explore the dynamic properties and the nature of such interactions, and they are complicated and costly. The filter-binding assay is rapid and simple, but it is not suitable for crude cell extracts and is less sensitive (Chodosh, 1988). The GRA, also called the electrophoretic mobility shift assay (EMSA) or gel mobility shift assay, is a powerful approach to study the affinity, stoichiometry, and cooperativity of protein–nucleic acid interactions, which is comparable to a receptor-binding assay. The basis of the GRA is the reduction in the electrophoretic mobility of a nucleic acid molecule through a nondenaturing gel when it is bound to protein(s). The mobility of a protein–nucleic acid complex is dependent not only on the sizes of the protein(s) and nucleic acid, but also on the physical conformations of the complex and the charge on the protein (Carey, 1988, 1991; Chodosh, 1988; Fried, 1989; Stone et al., 1991). In combination with chemical modification and/or mutation in the nucleic acid molecules, this basic technique has provided precise information on protein–nucleic acid interactions, and it is well suited for the study of single- or double-stranded DNA, linear or circular DNA, as well as RNA. The advantages of the GRA have been reviewed (Carey, 1991; Fried, 1989) and include detecting specific binding protein(s) in crude cell extracts, identifying a specific nucleic acid target site of a given binding protein, removing only the active protein in a complex, and resolving complexes by the differences in their stoichiometries and/or in the physical arrangement of their components. In addition, the GRA has greater sensitivity than other methods, as it may detect femtomole quantities of specific protein(s) (Chodosh, 1988).

One of the most attractive features of GRA is its simplicity. It is surprising that so much information can be obtained by simply incubating protein(s) with small amounts of radioactive DNA probe(s) and separating them in a gel. However, establishing the validity of a particular assay requires considerable time in which to run the requisite controls. In the present paper, we discuss in some detail technical points of the GRA used for the discovery of an estrogen response element (ERE) in the preproenkephalin promoter in mammalian steroid-sensitive neurons.

Estrogen regulates gene expression through the binding of estrogen receptor (ER) interacting with EREs of specific genes (Gronemeyer, 1991; Lucas and Granner, 1992). It has been shown that estrogen has profound effects on female reproductive behaviors (Pfaff, 1980; Pfaff *et al.*, 1994), which depend on RNA synthesis, and at least six behaviorally relevant neurochemical systems have been identified (Pfaff, 1989). For the progesterone receptor (PR) gene, estradiol induces the mRNA robustly (Romano *et al.*, 1989) and for the first time causally links a transcription factor to a specific behavior. It is intriguing that the estrogen induction occurs in female rats but not males (Lauber *et al.*, 1991). Using gel shift methods Lauber *et al.* (1993) have compared characteristics of ER binding to a consensus (vitellogenin) ERE with that of a transcriptionally active ERE on the rabbit PR gene.

This chapter discusses the preproenkephalin gene (PPE), turned on by estrogen (Romano *et al.*, 1988) rapidly (Romano *et al.*, 1989), in females but not males (Romano *et al.*, 1990). Although we are considering a putative ERE in the well-recognized PPE promoter, there may also be an alterative transcription start site in the first intron (Brooks *et al.*, 1933a), which is also used as part of an estrogen effect (Brooks *et al.*, 1993b). Indeed, interesting DNase hypersensitivity sites have been discovered in the promoter, as well as in the first intron, whose DNA is unmethylated (Funabashi *et al.*, 1992, 1993). Preproenkephalin mRNA levels in VMN are tightly correlated with female rat reproductive behaviors (Lauber *et al.*, 1990). Acting through δ-opioid receptors enkephalin can promote such behaviors (Pfaus and Pfaff, 1992).

Methodology for Brain Tissue

Materials and Buffers

Duncan homogenizer (Thomas Scientific, Swedesboro, NJ)
Proteinase inhibitors: Phenylmethylsulfonyl fluoride (PMSF), pepstatin A, leupeptin, and *p*-aminobenzamidine are obtained from Sigma (St. Louis, MO)

Klenow fragment of *Escherichia coli* DNA polymerase I (2 units/μl), G-25 DNA purification column, and poly(dI–dC) are obtained from Boehringer Mannheim Corp. (Indianapolis, IN)

Protein A–agarose (PAA) is obtained from Pierce Chemical Co. (Rockford, IL)

17β-Estradiol, β-estradiol 3-benzoate (EB), and progesterone are obtained from Sigma

Oligonucleotides are synthesized by Oligos, Etc., Inc. (Wilsonville, OR) Vit-ERE, a 41-bp synthetic oligomer, contains the vitellogenin A$_2$ ERE. PRE, a 27-bp oligonucleotide, contains the consensus progesterone response element (PRE) sequences *

Vit-ERE: 5′ AATTCGTCCAAAGTCA**GGTCACAGTGACC**TGATCAAAGTTG 3′
           GCAGGTTTCAGT **CCAGTGTCACTGG** ACTAGTTTCAACCTAG
PRE: 5′ TCGAC**TGTACAGGATGTTCT** AGCTACT 3′
      AGCTG**ACATGTCCTACAAGA**TCGATGA

[α-^{32}P]dATP is obtained from Du Pont-New England Nuclear (Boston, MA)

Buffers

Nuclear extraction buffer A: 10 mM N-2-hydroxyethylpiperazine-N'-2-ethanesulfonic acid (HEPES; pH 7.9), 1.5 mM MgCl$_2$, 10 mM KCl, 0.5 mM dithiothreitol (DTT), 0.5 mM PMSF, pepstatin A (1 μg/ml), leupeptin (10 μg/ml), and 0.1 mM p-aminobenzamidine. Prepare fresh each time

Nuclear extraction buffer C: 20 mM HEPES (pH 7.9), 0.42 M NaCl, 1.5 mM MgCl$_2$, 0.2 mM ethylenediaminetetraacetic acid (EDTA), 0.5 mM DTT, 0.5 mM PMSF, pepstatin A (1 μg/ml), leupeptin (10 μg/ml), 0.1 mM p-aminobenzamidine, and 25% (v/v) glycerol. Prepare fresh each time

Klenow buffer (10×): 500 mM Tris-HCl (pH 7.2), 100 mM MgSO$_4$, 1 mM DTT

Stock solution (2 mM) of dTTP, dCTP, dGTP

GRA binding buffer (10×): 100 mM Tris-HCl (pH 7.6), 10 mM EDTA, and 40% (v/v) glycerol

Tris–EDTA (TE) buffer (1×): 10 mM Tris-HCl (pH 7.6) and 1 mM EDTA

Tris–borate–EDTA buffer (5×): 0.45 M Tris–borate, 0.01 M EDTA (pH 8.3)

Loading dye: 0.5% (v/v) bromphenol blue

Animals and Drug Treatment

Female Sprague-Dawley rats (175–200 g; Charles River, Wilmington, MA) are used. The ovariectomy (OVX) is performed by the supplier. The animals are maintained in a 12 hr/12 hr light/dark cycle. Hormone treatments are carried out 2 weeks after the OVX.

Dissection of Tissues and Preparation of Nuclear Extracts

After decapitation of the rats, the brains are removed, submerged in ice-cold saline for 1 min, and then placed in a Jacobowitz brain slicer (Zivic-Miller) that has been chilled in ice-cold phosphate-buffered saline (PBS). The hypothalamus is dissected as follows: A 4-mm slice with a rostral border at the level of the optic chiasm is removed, placed on an iced Petri dish, and dissected with a razor blade. Sagittal cuts are made beside the fornix to the base of the brain, and a horizontal cut is made just above the fornix. The uterus and lung are dissected. Nuclear extracts are prepared from the tissues of ovariectomized rats that are either untreated or have been treated with various hormones. The preparation of nuclear extracts is performed as previously described (Dignam *et al.*, 1983; Korner *et al.*, 1989) with modifications.

1. Keep dissected tissues in Eppendorf tubes with 5 vol of buffer A on ice for at least 10 min and then centrifuge in an Eppendorf microcentrifuge for 3 min at 4°C.
2. After decanting the supernatant, homogenize the samples in a Duncan homogenizer with 2 vol of buffer A for 10–15 strokes [the uterus is homogenized by the use of a Polytron (Kinematica, Switzerland) for 15–20 sec].
3. Transfer samples from the homogenizer to Ependorf tubes and centrifuge in an Eppendorf microcentrifuge for 20 min at 4°C.
4. Remove the supernatants (cytosol fraction), which can be used to prepare the cytosol fraction (Dignam *et al.*, 1983).
5. Add 1–1.5 vol of buffer C to the nuclear pellet tubes and break down the pellet by homogenizing, or by pipetting and vortexing.
6. Keep the samples on ice for 30 min, and then centrifuge in an Eppendorf microcentrifuge for 30 min at 4°C.
7. Aliquot the supernatants, freeze them in dry ice, and store at −70°C until analysis.

Synthesis of Labeled DNA Probe

The labeling of oligonucleotides or nucleic acid fragments can be carried out either by filling-in or end-labelling reactions.

1. Anneal synthetic complementary oligonucleotides containing the protein-binding site and overhanging ends by heating to 85°C for 2 min, followed by incubating at 65°C for 15 min, 37°C for 15 min, and at room temperature for 15 min, and then on ice for 15 min (Sambrook *et al.*, 1989). The 5' overhanging ends should have bases complementary to the ^{32}P-labeled dNTP(s) used during the fill-in labeling, if Klenow fragment of DNA polymerase I is used.

2. Mix the annealed oligonucleotides with 2 μl of 10× Klenow buffer, 2 μl of a stock solution of dCTP, dGTP, and dTTP (2 mM), 3 μl of [α-^{32}P]dATP (3000 Ci/mmol), and 2 units of Klenow enzyme.

3. Incubate the mixture at room temperature for 30 min and separate the oligonucleotides from the free nucleotides by centrifugation through a Sephadex G-25 spin column.

4. Precipitate the ^{32}P-labeled oligonucleotides by adding 0.1 vol of 3 M sodium acetate and 3 vol of ethanol and keep at −70°C for 1 hr or at −20°C overnight.

5. Centrifuge the precipitated sample in a microcentrifuge for 20 min, wash the pellet with 75% (v/v) ethanol, and resuspend it in TE buffer with a concentration of 1 ng/μl.

6. Count a 1-μl aliquot from the beginning reaction, after the separation on the Sephadex G-25 column, and the final resuspended solution to estimate the incorporation, recovery, and specific activity of the probe.

DNA-Binding Reaction

The binding reaction is performed in a total volume of 30 μl containing 10 mM Tris-HCl (pH 7.6), 1 mM EDTA, 4% (v/v) glycerol, 5 mM DTT, 2 μg of poly(dI-dC), 5 μl of nuclear extracts, and 0.2 ng of ^{32}P-labeled oligonucleotides probe.

1. In a series of autoclaved Eppendorf tubes, add 3 μl of 10× binding buffer, 2 μg of poly(dI-dC), 1.5 μl of 0.1 M DTT, 1.5 μl of 1 M KCl (if KCl is used), and an amount of distilled H$_2$O to bring up the total volume to 30 μl. It is more convenient to make a binding mixture and then aliquot to each tube.

2. Add 5 μl of nuclear extracts; if less protein is needed, compensate with nuclear extract buffer C.

3. Mix the samples by flipping, centrifuge briefly (1 second in an Eppendorf Microcentrifuge at room temperature) to bring down the samples to the bottoms of the tubes, and preincubate at room temperature for 15 min.

4. Add the radioactive probe, mix the samples by flipping (do not vortex the samples, which may perturb the complex formation), centrifuge briefly, and continue to incubate for another 45 min.

5. Add 2 μl of loading dye, mix by pipetting, and load the samples on the prepared gel.

For competition assay or supershift assay, unlabeled oligonucleotides or antibodies, respectively, are added in the preincubation step.

Gel Electrophoresis and Detection

1. Assemble gel plates and make the gel solution (100 ml): 10 ml of 5× TBE, 10 ml of 40% (v/v) acrylamide : bisacrylamide (37.5 : 1) solution, 10 ml of 50% (v/v) glycerol (glycerol is viscous; prepare a 50% glycerol solution by mixing 1 vol of glycerol with 1 vol of water), 250 μl of 30% (w/v) ammonium persulfate, 100 μl of N,N,N',N'-tetramethylethylenediamine (TEMED), and distilled H_2O to bring up the volume to 100 ml.

2. Pour the gel (avoiding bubbles) and let it polymerize at room temperature for at least 30 min.

3. Wash the wells of the gel with reservoir buffer and prerun the gel at a constant 150 V at room temperature until the current is stable (usually more than 1 hr).

4. Immediately before loading the samples, the wells are cleaned out again with reservoir buffer. After loading the samples, run the gel at 200 V for about 5 min to let the samples rapidly enter the gel, and then electrophorese at a constant 150 V at room temperature for 2.5–3 hr in 0.5× TBE buffer.

5. Take out the gel by use of Whatman 3 MM paper (Maidstone, England), cover the gel with plastic wrap, and dry it.

6. Expose the gel to Kodak (Rochester, NY) X-Omat film with an intensifying screen at −70°C.

Immunoprecipitation

The immunoprecipitation is carried out as previously described with minor modification (Sambrook *et al.*, 1989).

1. Add nuclear extracts and a polyclonal anti-ER antibody (a gift from H. Okamura, Department of Anatomy and Embryology, Tokyo Metropolitan Institute for Neuroscience, Tokyo, Japan; see Okamura *et al.*, 1992) or nonspecific control antibody (Abbott Laboratories, Abbott Park, IL) in Eppendorf tubes, mix the samples by tapping on the bottoms of the tubes, and briefly centrifuge to bring down solutions to the bottom of the tubes.

2. Incubate the mixtures on ice for 1 hr, and then add 1 vol of protein A–agarose (1 mg/ml in 1× GRA binding buffer) to each. Mix the samples by flipping.

3. Continue to incubate the samples on ice for 1 hr; tap the bottoms of tubes to mix the samples every 10–15 min.

4. Centrifuge the samples in an Eppendorf microcentrifuge for 2 min, and pipette out the supernatants for gel shift binding assay as described above.

Methodological Observations

Selection of Conditions for Gel Retardation Assay

Several reviews and protocols have provided guidelines for starting a new GRA (Chodosh, 1988; Carey, 1991; Stone *et al.*, 1991). From our experience in studying protein–DNA interactions in crude nuclear extracts of steroid-sensitive neurons in mammalian brain, the following theoretical and technical issues should be considered.

Types of Gels

The types of gels and the conditions of gel electrophoresis can affect the results of a GRA. In our experience, a 4% polyacrylamide gel (37.5:1, acrylamide:bisacrylamide) with 5% glycerol works well for oligonucleotides for DNA fragments from 21 to 500 bp in length. If a longer DNA fragment is used, it can be separated in the same percentage gel, but the ratio of acrylamide:bisacrylamide should be increased to 50:1 or even 80:1, although these gels may be difficult to handle. In addition, agarose gels or agarose mixed with polyacrylamide gel have also been used in GRA (Revzin, 1989; Stone *et al.*, 1991) for long DNA fragment(s) and/or high molecular weight protein(s). Samples in the present chapter were electrophoresed in a 4% polyacrylamide gel (37.7:1, acrylamide:bisacrylamide) with 5% glycerol (see Figs. 1–4).

Electrophoresis Buffer

Both TBE and TAE buffers are commonly used in GRAs. A 0.5× TBE buffer was chosen for our assay on the basis of multiple tests of buffer systems. Low ionic strength TAE buffer was also tested, but the results were not as good as with the 0.5× TBE buffer. The choice of a buffer is empirical, but a 0.5× TBE buffer has been used in several GRAs and is recommended for a pilot study (Carey, 1991; Stone *et al.*, 1991). The high salt concentration of a buffer may increase heating during electrophoresis and may make the protein–DNA complex carry less current and move more slowly. On the other hand, a low salt concentration may lead to a broadening of the bands and may result in a shift of pH during electrophoresis, severely interfering with the stability of the protein–DNA complexes. Because the mobility of a complex in a gel is affected by protein charges (Fried, 1989), variation in the pH of the electrophoretic buffer may sometimes be enough to resolve different complexes, resulting in a high-resolution GRA, as previously reported (Carey, 1988). In addition, some cofactors, such as Mg^{2+} and cAMP, can be added to the binding mix, the gel, and the buffer, if such cofactors are required for complex formation (Stone *et al.*, 1991).

Conditions of Binding Reaction

The ideal condition for the binding assay is to reproduce the *in vivo* condition for the binding of a protein to DNA or RNA. Because this is impossible, the next best choice is to use the *in vitro* transcription conditions. The most appropriate conditions for a specific binding assay are usually discovered by trial and error. The following factors should be considered in a binding reaction.

Salt Concentration

The salt concentration in the assay influences most, if not all, binding interactions. Ionic strength in binding assays has profound effects on the affinity, specificity, and formation of ER–ERE complexes (Murdoch *et al.*, 1991). As shown in Fig. 1, the binding activity of ER–ERE in uterus nuclear extracts treated with EB plus progesterone was dependent on the concentration of KCl added to the binding assay. The binding activity was decreased with increases in KCl concentration (see lanes 1–3 of Fig. 1). We have chosen 50 mM KCl in the assay to ensure proper binding and specificity. The total ionic strength [50 mM KCl plus 70 mM NaCl (from nuclear extracts) is 120 mM] in our assay is in agreement with a previously reported optimum salt concentration for ER–ERE interactions (Murdoch *et al.*, 1991), which is about 100–150 mM, a concentration similar to the physiological condition.

Other factors also affect a particular binding assay as previously discussed by Carey (1991) and Stone *et al.* (1991), including divalent cations (Mg^{2+}, Zn^{2+}), nonionic detergent, carrier DNA, pH, and glycerol. Some cofactors such as cAMP or ribonucleoside triphosphates can be added if they are required in the reaction. Most of the DNA-binding assays reported are carried

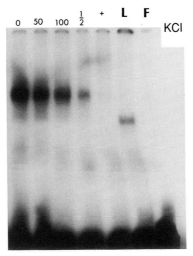

FIG. 1   Effects of salt concentration, protein concentration, and tissue sources on DNA (ERE)-binding activity. The binding reactions were carried out as described in text. Nuclear extracts from rat uterus or lung treated with EB (10 μg/rat) for 48 hr and with progesterone (500 μg/rat) for 4 hr were incubated with various concentrations of KCl (0, 50, and 100 mM in lanes 1, 2, and 3, respectively) at room temperature and electrophoresed in a 4% polyacrylamide gel plus 5% glycerol for 3 hr in 0.5× TBE buffer. Lane 4, marked "½," was a reaction at 50 mM KCl with the use of half the amount of nuclear extracts from rat uterus. Lane 5 (+) was the same reaction as in lane 2, but with a 500-fold molar excess of cold vit-ERE. Lane 6 (L) was a reaction with the same amount of nuclear extracts from rat lung. L, Lung; F, free probe.

out between pH 7 and 8, although the optimum pH may vary from one protein–DNA system to another. Theoretically, the pH should be close to the physiological pH to maintain the active conformation of the proteins. For stabilizing protein(s) and/or its binding to DNA, dithiothreitol, glycerol, and other factors can be used. The addition of bulk carrier DNA in the binding assay is important, in order to trap nonspecifically binding proteins and permit the detection of a specific protein–DNA interaction (Strauss and Varshavsky, 1984; Carthew *et al.,* 1985; Chodosh, 1988; Revzin, 1989). In our present assay, we have employed poly(dI-dC) as the carrier DNA; 1, 2, and 4 μg of poly(dI-dC) per reaction were tested and gave the same result. Thus, 2 μg of poly(dI-dC) in the binding assay was chosen, a commonly used concentration (Chodosh, 1988). In addition to poly(dI-dC), other nonspecific polynucleotides, such as poly(dG-dC) and poly(dA-dT), and heterologous sequence DNA, such as salmon sperm DNA and *E. coli* DNA, can also be used in the GRA. However, heterologous sequence DNA may have binding sites with varying affinities to the testing sequence and may affect the binding assay. The basic principle for selecting a carrier DNA is that the sequence should bear little similarity to the specific binding site.

Time and Temperature of Incubation

Because protein–DNA interactions are assumed to be thermodynamic processes similar to those in receptor–ligand interactions, the time and temperature of incubation can affect complex formation. It is ideal for an assay to let protein–DNA binding reach a dynamic "steady state." Especially for the quantitation of protein–DNA interactions, one of the fundamental assumptions is that the reaction is at a steady state. Thus, in early experiments an investigation of a time course of a binding assay should be conducted by incubating for various periods prior to gel electrophoresis. The time of equilibrium in the reaction is a function of the association and dissociation rate constants for the protein–DNA interactions (Fried, 1989; Stone *et al.,* 1991). Because the rates of association and dissociation are temperature-dependent processes, it is not surprising that the amount of binding and possibly the time to equilibrium will vary with temperature. It is also inappropriate to run a gel at room temperature if the incubation took place at 4°C for a quantitation assay. For our current ER–ERE interactions, we have found that the binding activity reached equilibrium by 5 min and was constant up to 45 min of incubation at room temperature, in addition to the 15-min preincubation.

Amount of Protein in Assay

The amount of protein in an assay will affect the binding reaction. If the binding is to a biologically relevant site, in the presence of excess DNA, the binding activity should increase in proportion to the concentration of binding protein(s). Thus, it is best to determine the degree of linearity for the binding activity versus protein concentration, and to conduct an assay with a protein concentration known to lie within the range of linearity, in order to obtain the most accurate and reproducible data for an assay. As shown in Fig. 1, the ER–ERE binding activity in the uterus is dependent on protein concentration (see lanes 2 and 4, Fig. 1). We have used 5–20 μg of nuclear proteins for the ER–ERE assay in hypothalamus, uterus, and pituitary and the observed specific binding activity is dependent on the amount of proteins. In other GRAs, as little as 2 μg of nuclear proteins was used and specific binding activity was detected.

Mixing of Sample

Because the binding mixture is viscous (because of glycerol and nuclear extracts) and has a small volume, proper mixing of the sample is important to ensure the accuracy and reproducibility of an assay. Vortexing of the incubation mixture should be avoided, which may disrupt the protein–DNA complex, as suggested by the original description of the GRA (Fried and Crothers, 1981). For our assays the bottom of the Eppendorf tube is tapped, followed by a brief centrifugation (1 second in microcentrifuge at room temperature) to mix and then bring down the solution to the bottom.

Loading Samples and Gel Electrophoresis

Dissociation of complexes during sample loading and subsequent gel electrophoresis can affect the final results of an assay. However, it is interesting to note that complexes capable of dissociating in a few minutes or even seconds in binding buffer can be subjected to electrophoresis for hours without obvious dissociation. Several reasons have been proposed to account for this phenomenon, including increasing affinity of protein for DNA in low ionic strength buffers for electrophoresis, and a "caging effect" of the gel matrix and relatively high protein–DNA concentration while free DNA is moving away during electrophoresis (Fried and Crothers, 1981; Revzin, 1989). In our GRA for ER–ERE interactions, although more than 50% of specific ER–ERE complexes were displaced in 5 min after addition of a 100-fold molar excess of cold probe, the specific complexes had not disappeared after 3 hr of electrophoresis (see Fig. 1). Similar stability has been seen with a different ERE DNA probe (Lauber et al., 1993). Furthermore, there was no difference in the pattern and magnitude of complex formation when loading the gel without running compared to loading the gel while running at high voltage (being careful of electroshock), although it has been suggested that loading the gel while running can minimize the dissociation of complexes during the dead time (while the DNA is entering the gel) (Carey, 1988; Fried, 1989). Some helpful hints can be found in previous reviews and protocols (Carey, 1991; Fried, 1989; Stone et al., 1991) to minimize the dissociation of complexes during loading of the gel and electrophoresis.

Probe

There is no doubt that the selection and preparation of a probe is a crucial factor in a GRA, especially for an oligonucleotide probe and in a quantitative study. If a GRA does not work properly, check the probe first. Several problems may arise from the probe: (a) because differences in a single base of a synthetic oligonucleotide will significantly influence the protein–DNA interactions and functions, mistakes in synthesizing a specific oligonucleotide may cause a GRA to fail; (b) in some protein–DNA interactions, flanking sequences neighboring a specific sequence are important for the interactions (Anolik *et al.*, 1993); therefore it is desirable to include some flanking sequences in the probe although the optimal length is unpredictable so far; (c) unannealed or dissociated single-stranded oligonucleotides may interfere with an assay and the interpretation of results. This may be overcome by the design of hairpin loop oligonucleotides to form a double-stranded probe, which is more stable than probe annealing by reacting two single-stranded oligonucleotides; (d) oligonucleotides may be degraded by nucleases present in cell extracts; and (e) in quantitative studies, the concentration of probe in the assay is critical, therefore any errors in pipetting or loss in the test tubes will affect the results obtained.

Determination of Specificity

One of the most important issues in developing a GRA for crude cell extracts containing mixed protein populations is to determine the specificity of the assay. After selecting appropriate conditions for a GRA, the next phase is to demonstrate that the binding is the one of interest and to determine which complex(es) are specific if multiple complexes are present.

Saturability
Saturability is a minimal requirement for interesting binding in tissues. Any biological tissue contains only a limited amount of a specific binding protein for any given substance and, therefore, if binding is specific for a protein–DNA interaction, by increasing the probe through reasonable concentrations the binding activity should plateau, indicating that all of available binding proteins are occupied by the probe (saturation). If instead the binding is nonspecific, it may not be saturated by increasing the probe but rather will continue to increase in proportion to the amount of probe added. In our ER–ERE GRA, we observed that the top two bands in hypothalamus (see Figs. 2 and 3) and uterus (see Figs. 1 and 3) were saturable by increasing the amount of probe (data not shown), in agreement with the requirement of saturability for a protein–DNA interaction.

Distribution
Specific DNA binding should be associated with the distribution of the binding protein of interest. The DNA binding should be present in tissues in which the protein of interest is known to be present, whereas it should be

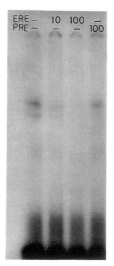

Fig. 2 A competition GRA study shows specific, limited capacity DNA binding to an ERE by nuclear proteins in rat hypothalamus. The experiment was carried out as described in text with 50 mM KCl and electrophoresed in a 4% polyacrylamide gel plus 5% glycerol. Lanes 2 and 3 contained a 10- and 100-fold molar excess of unlabeled vit-ERE, respectively. Lane 4 contained a 100-fold molar excess of unlabeled PRE.

absent in tissues in which the protein of interest is absent. In our ER–ERE GRA, specific complexes were observed in rat hypothalamus (Figs. 2 and 3) and uterus (Figs. 1 and 3), but not in lung (Fig. 1), a tissue known not to contain ER (Curtis and Korach, 1990). A similar conclusion was reached by Lauber *et al.* (1993) comparing hypothalamus with non-estrogen-binding brain regions. On the other hand, the study of the distribution of binding may also lead to the discovery of new loci of binding protein(s) and function, for instance, the β factor (Korner *et al.*, 1989).

Competition Assay

Like enzyme–substrate and receptor–ligand interactions, the protein–DNA interactions also follow the laws for reaching equilibrium in the presence of a competitor, that is,

$$D + P \underset{k_2}{\overset{k_1}{\rightleftharpoons}} DP$$

$$C + P \underset{k_4}{\overset{k_3}{\rightleftharpoons}} CP$$

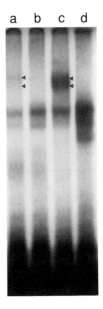

Fig. 3   A supershift assay shows the specificity of ER–ERE interactions. Nuclear extracts from rat hypothalamus (lanes a and b) treated with 17β-estradiol (10 μg/rat) for 1 hr, or from rat uterus (lanes c and d), were preincubated with (lanes a and c) and without (lanes b and d) monoclonal antibody H222 and separated on a 4% polyacrylamide gel with 5% glycerol. The supershifted bands are indicated by arrowheads.

where P represents protein, D is DNA, C is competitor, and k_1 and k_3, and k_2 and k_4, are association and dissociation rate constants for DNA probe–protein and competitor–protein interactions, respectively (Stone *et al.*, 1991; Fried, 1989). Thus, the use of unlabeled competing oligonucleotides in a GRA is the simplest way to approach the specificity of an assay (Carthaw *et al.*, 1985; Stone *et al.*, 1991). Oligonucleotides both related and unrelated to the probe can be employed in a competition assay. The order of addition of protein, competing oligonucleotides, and probe may be not critical for a reversible binding reaction, as long as sufficient time is allowed for full attainment of equilibrium. However, it is usually undesirable to add competing substances last, because this order of addition may underestimate protein occupancy by the competitor, especially for a slowly dissociating interaction. For an irreversible binding reaction, the probe should clearly be added last, preferably after allowing a competitor to approach equilibrium with the protein(s). With various concentrations (a range of two to three orders of magnitude) of competing oligonucleotides, the slope of the competition curve as well as the IC_{50} (the concentration which inhibits 50% of the protein–DNA binding) can be derived, and then the affinity of a competing oligonucleotide to the binding protein can be estimated (Fried, 1989; Stone *et al.*, 1991). We have used the competition assay to assess the specificity of ER–ERE interactions in nuclear extracts from rat hypothalamus (Fig. 2), uterus (lane 5 of Fig. 1), and pituitary (data not shown). The appropriate unlabeled oligonucleotides produced an elimination of the top two bands in the GRA, whereas a closely related but distinct PRE oligonucleotide (Klock *et al.*, 1987) had no significant effect on the binding (Fig. 2). Other ERE-related and unrelated oligonucleotides were also tested and indicated that by this measure our GRA was specific for ERE–protein interactions.

Supershift Assay

Although the competition assay can provide significant information about the affinity and capacity of binding to an oligonucleotide probe, it is unable to tell which protein(s) is interacting with this probe. The use of antibodies in GRA further extends and improves this technology. The addition of a specific antibody against a protein of interest in a GRA may produce either further retardation (supershift) of specific complex(es) (because of the formation of a ternary complex between antibody, protein, and DNA probe) or a blockade of specific complex(es) formation (because of the apparent hindrance of protein–DNA interaction resulting from the interaction between protein and antibody). In contrast, addition of an antibody will have no effect on complex formation, if the protein recognized by the antibody is not involved in the protein–DNA interactions. As shown in Fig. 3, the addition of a specific monoclonal antibody to the human ER, H222 (a gift from G. L. Greene, Ben May Institute, Chicago, IL; see Greene *et al.*, 1984), which has been used in a supershift assay (Klein-Hitpass *et al.*, 1989), in our ERE-binding assay resulted in the emergence of two new complexes with slower mobility and diminished the original top two complexes in both hypothalamus and uterus (compare lanes a and c to lanes b and d in Fig. 3, respectively). For additional controls, the use of a nonspecific monoclonal antibody (from Abbott Laboratories) and an anti-glucocorticoid receptor (GR) monoclonal antibody (from Affinity Bioreagents, NJ) in our GRA did not yield any changes in the complex formation. These results suggest that this ERE GRA detects specific ER–ERE interactions.

Use of Immunoprecipitation in Gel Retardation Assay

An alternative method to characterize the specificity of a protein–DNA interaction in a GRA is the use of immunoprecipitation. After immunoprecipitating a crude protein extract with either a specific or nonspecific control antibody plus a carrier, the samples are used in a GRA. The predicted results would be the reduction or disappearance of specific complex(es) formed between the protein and DNA probe, due to the immunoprecipitation of the specific protein by a specific antibody. A nonspecific control antibody should have no effect on a specific protein–DNA interaction. Furthermore, the addition of the elution from specific precipitates should restore the complex formation. Thus, the specificity of a GRA can be evaluated. We have used a specific polyclonal antibody against the rat ER (a gift from H. Okamura; see Okamura *et al.*, 1992) in our immunoprecipitation GRA to estimate the specificity of the assay. As shown in Fig. 4A and B, immunoprecipitation with this anti-ER antibody in either rat hypothalamus (lane 3 of Fig. 4A) or uterus (lane 3 of Fig. 4B) nuclear extracts significantly diminished the specific band intensity in the GRA. The use of a control antibody in the immunoprecipitation did not result in large changes in hypothalamus (lane 5 of Fig. 4A) or in uterus (lane 5 of Fig. 4B) in the assay. These results further demonstrate the specificity of our GRA.

Shift-Western Blotting

Demczuk *et al.* (1993) have reported the development of a ''shift-Western blotting'' method, which combines GRA with immunoblotting, to identify individual components of protein–DNA complexes containing multiple proteins, their cognate DNA elements, and the ligand when applicable. This approach significantly expands GRA applications in the analysis of protein–DNA as well as ligand interactions. This technique should also be applicable for the determination of specificity of a GRA in crude cell extracts.

Other methods such as denaturation, cross-linking, and excising complex(es), followed by immunoblotting, can also be used in the determination of the specificity of a GRA. The denaturation of cell extracts may destroy the protein and thus eliminate the complex formation, and the denaturation of DNA probe can assess whether a factor binds to single-stranded or double-stranded DNA. Curtis and Korach (1990) have shown that the ER from rat uterus binds to double-stranded, but not single-stranded, ERE elements. We have also observed that the denaturation of nuclear extracts from hypothalamus by heating abolished the binding to ERE probe, suggesting that the conformation of proteins is important for binding. However, whether the protein(s) binding to DNA is necessarily active remains controversial. In studying *trp* repressor Carey (1988, 1991) has stated that only active protein binds to a specific DNA sequence, whereas Cridland *et al.* (1990) reported that selective photochemical treatment of ER in a *Xenopus* liver extract destroyed the ER function but not the DNA binding. The reason for this discrepancy is difficult to discern owing to the difference in the systems

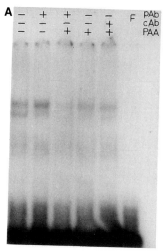

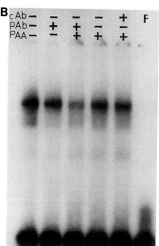

Fɪɢ. 4   The specificity of ER–ERE interactions in rat hypothalamus (A) and uterus (B) assessed by immunoprecipitation experiments with nuclear proteins. Reactions were carried out as described in text and the complexes were separated by a 4% polyacrylamide gel with 5% glycerol. Lanes for both hypothalamus (A) and uterus (B): (1) a reaction without any addition; (2) a reaction with the addition of a specific anti-ER polyclonal antibody (pAb); (3) with the addition of pAb and protein A agarose (PAA); (4) with the addition of PAA alone; (5) with the addition of a nonspecific control antibody (cAb) and PAA; (6) a reaction without nuclear extracts. F, Free probe.

used. Furthermore, accessory factors may be required for ER functions as well as ER binding to DNA (Gronemeyer, 1991).

In summary, multiple approaches have been discussed to assess the specificity of a GRA, a key question in developing a new GRA for crude cell extracts. Each technique has its own advantages and limitations; thus, multiple approaches should be carried out in order to demonstrate the specificity of a new GRA.

DNA-binding Activity versus Functional Activity

Although the GRA is simple, rapid, and sensitive in the study of protein–DNA interactions, it cannot give any information about functional activity. If a study aims to identify functional *cis* elements or *trans*-acting factors, the GRA is not enough and one must employ functional assays. Several functional assays, both *in vitro* and *in vivo* methods, are available, including *in vitro* transcription, transfection, and transgenic and "*in vivo* expression" assays. An *in vivo* expression assay by the use of a defective viral vector (Kaplitt *et al.*, 1991–1993) or liposome technology (Zhu *et al.*, 1993) to deliver a gene into an intact cell in its normal tissue context has advantages in the estimation of promotor function, in the assessment of physiological functions of a gene and in gene therapy, and it will undoubtedly be used more in the future.

Concluding Remarks

The gel retardation assay (GRA) is a sensitive, rapid, simple, and powerful tool for the study of protein–DNA interactions. However, analogous to receptor-binding assays, it is necessary to determine the optimal assay conditions for each protein–DNA interaction. The concentration of salt, the time and temperature of incubation, and other components in the DNA-binding assay are among the factors that can affect an assay, as we illustrated for the ERE–protein assay. A key question for a GRA of crude cell extracts is to demonstrate the specificity of the assay. We have used and discussed multiple methods, including competition, supershift, immunoprecipitation, and shift-Western assays, to characterize the ERE–protein interactions in nuclear extracts from rat neuronal and nonneuronal tissues. Because each method has its advantages and limitations, a combination of methods is the wisest choice. It should be realized that although a GRA carried out under a range of well-chosen conditions can provide a great deal of information about the kinetics and stoichiometry of protein–DNA interactions, it cannot give any information about functional, transcriptional activity. Functional assays, for example, transfection assays, *in vitro* transcription, and transgenic and *in vivo* expression (e.g., viral vector) assays, are required in order to isolate *cis* or *trans* factors in the control of gene expression.

Acknowledgments

We are grateful to Dr. G. L. Greene for providing the H222 antibody, and to Drs. Mona Freidin and Lindsey Grandison for helpful discussions. This study is supported by NIH Grant HD-05751 to D.W.P.

References

Anolik, J. H., Klinge, C. M., and Bambara, R. A. (1993). *Program Abstr. Endocr. Soc. Annu. Meet., 75th* P516.

Brooks, P. J., Funabashi, T., Kleopoulos, S. P., Mobbs, C. V., and Pfaff, D. W. (1993a). *Mol. Brain Res.* **19,** 22–30.

Brooks, P. J., Funabashi, T., Kleopoulos, S. P., Mobbs, C. V., and Pfaff, D. W. (1993b). *Soc. Neurosci. Abstr.* **19,** 483.

Carey, J. (1988). *Proc. Natl. Acad. Sci. U.S.A.* **85,** 975–979.

Carey, J. (1991). "Methods in Enzymology" (R. T. Sauer, ed.). Vol. 208, pp. 103–117. Academic Press, San Diego.

Carthaw, R. W., Chodosh, L. A., and Sharp, P. A. (1985). *Cell* **43,** 439–448.

Chodosh, L. A. (1988). *In* "Current Protocols in Molecular Biology" (F. M. Ausubel, R. Brent, R. E. Kingston, D. D. Moore, J. G. Seidman, J. A. Smith, and K. Struth, eds.), Vol. 2, Suppl. No. 13, pp. 12.2.1–12.2.10. Wiley (Interscience), New York.

Cridland, N. A., Wright, V. E., McKenzie, E. A., and Knowland, J. (1990). *EMBO J.* **9,** 1859–1866.

Curtis, S. W., and Korach, K. S. (1990). *Mol. Endocrinol.* **4,** 276–286.

Demczuk, S., Harbers, M., and Vennstrom, B. (1993). *Proc. Natl. Acad. Sci. U.S.A.* **90,** 2574–2578.

Dignam, J. D., Lebovitz, R. M., and Roeder, R. G. (1983). *Nucleic Acids Res.* **11,** 1475–1489.

Fried, M. G. (1989). *Electrophoresis* **10,** 366–376.

Fried, M. G., and Crothers, D. M. (1981). *Nucleic Acids Res.* **9,** 6505–6525.

Funabashi, T., Brooks, P. J., Mobbs, C. V., and Pfaff, D. W. (1992). *Soc. Neurosci. Abstr.* **18,** 240.

Funabashi, T., Brooks, P. J., Mobbs, C. V., and Pfaff, D. W. (1993). *Mol. Cell. Neurosci.* **4,** 499–509.

Garner, M. M., and Revizin, A. (1981). *Nucleic Acids Res.* **9,** 3047–3060.

Greene, G. L., Sobel, N. B., King, W. J., and Jensen, E. V. (1984). *J. Steroid Biochem.* **20,** 51–56.

Gronemeyer, H. (1991). *Annu. Rev. Genet.* **25**, 89–123.

Kaplitt, M. G., Pfaus, J. G., Kleopoulos, S. P., Hanlon, B. A., Rabkin, S. D., and Pfaff, D. W. (1991). *Mol. Cell. Neurosci.* **2**, 320–330.

Kaplitt, M. G., Lipworth, L., Rabkin, S. D., and Pfaff, D. W. (1992). *Soc. Neurosci. Abstr.* **18**, 588.

Kaplitt, M. G., Rabkin, S. D., and Pfaff, D. W. (1993). *Curr. Top. Neuroendocrinol.* **11**, 169–191.

Klein-Hitpass, L., Tsai, S. Y., Greene, G. L., Clark, J. H., Tsai, M.-J., and O'Malley, B. W. (1989). *Mol. Cell. Biol.* **9**, 43–49.

Klock, G., Strahle, U., and Schutz, G. (1987). *Nature (London)* **329**, 734–736.

Korner, M., Rattner, A., Mauxion, F., Sen, R., and Citri, Y. (1989). *Neuron* **3**, 563–572.

Lauber, A. H., Romano, G. J., Mobbs, C. V., Howells, R. D., and Pfaff, D. W. (1990). *Mol. Brain Res.* **8**, 47–54.

Lauber, A. H., Romano, G. J., and Pfaff, D. W. (1991). *Neuroendocrinology* **53**, 608–613.

Lauber, A. H., Pfaff, D. W., Alroy, I., and Freedman, L. P. (1993). *Soc. Neurosci. Abstr.* **19**, 698.

Lucas, P. C., and Granner, D. K. (1992). *Annu. Rev. Biochem.* **61**, 1131–1173.

Murdoch, F. E., Grunwald, K. A. A., and Gorski, J. (1991). *Biochemistry* **30**, 10838–10844.

Okamura, H., Yamamoto, K., Hayashi, S., Kuroiwa, A., and Muramatsu, M. (1992). *J. Endocrinol.* **135**, 333–341.

Pfaff, D. W. (1980). "Estrogens and Brain Function: Neural Analysis of a Hormone-Controlled Mammalian Reproductive Behavior." Springer-Verlag, New York.

Pfaff, D. W. (1989). *Mol. Neurobiol.* **3**, 135–154.

Pfaff, D. W., Schwartz-Giblin, S., McCarthy, M. M., and Kow, L.-M. (1994). *In* "The Physiology of Reproduction" (E. Knobil and J. Neill, eds.), pp. 107–220. 2nd Ed. Raven, New York.

Pfaus, J., and Pfaff, D. W. (1992). *Horm. Behav.* **26**, 457–473.

Revzin, A. (1989). *BioTechniques* **7**, 346–355.

Romano, G. J., Harlan, R. E., Shivers, B. D., Howells, R. D., and Pfaff, D. W. (1988). *Mol. Endocrinol.* **2**, 1320–1328.

Romano, G. J., Mobbs, C. V., Howells, R. D., and Pfaff, D. W. (1989). *Mol. Brain Res.* **5**, 51–58.

Romano, G. J., Mobbs, C. V., Lauber, A. H., Howells, R. D., and Pfaff, D. W. (1990). *Brain Res.* **536**, 63–68.

Sambrook, J., Fritsch, E. F., and Maniatis, T. (1989). "Molecular Cloning: A Laboratory Manual," 2nd Ed. Cold Spring Harbor Lab. Press, Cold Spring Harbor, New York.

Stone, S. R., Hughes, M. J., and Jost, J.-P. (1991). *BioMethods* **5**, 163–183.

Strauss, F., and Varshavsky, A. (1984). *Cell* **37**, 889–901.

Zhu, N., Liggitt, D., Liu, Y., and Debs, R. (1993). *Science* **261**, 209–211.

[16] Gene Transfection Studies Using Recombinant Steroid Receptors

Klaus Damm

Introduction

The cloning of the human glucocorticoid receptor (hGR) provided the first complete structure of a steroid receptor and revealed a homology to the oncogene v-*erbA* in what was postulated to be the DNA-binding domain of the receptor (1). The cellular homolog of v-*erbA* was subsequently identified as a thyroid hormone receptor, providing the first direct evidence that steroid and thyroid hormones function through receptors with mechanistically similar properties. The isolation of sequences encoding nuclear hormone receptors has since proceeded by many avenues and led to the isolation of cDNAs encoding all the known steroid receptors. Through the availability of the cloned hormone receptor genes, molecular techniques can be employed to characterize the contribution of individual receptor systems to the integrated and complex biological responses inherent in development and homeostasis. The structure and functions of the glucocorticoid receptor have been extensively studied and revealed that the receptor molecule can be conceptually divided into domains that determine DNA-binding, ligand-binding, and transcriptional regulatory properties (1, 2). The central region, responsible for specific DNA sequence recognition, is highly conserved within this receptor gene family and is characterized by invariant cysteine residues forming zinc-binding finger structures. The primary function of this domain is to position the receptor appropriately on target gene sequences in chromatin. The carboxyl-terminal region of the hGR contains the hormone-binding function and is consequently less well conserved with other members of the steroid receptor family. In the absence of ligand this domain inhibits the DNA-binding and activation functions. The amino-terminal portions of the hormone receptors vary in size and show no sequence relatedness. This variability may facilitate functionally distinct tissue- and gene-specific properties and the unique characteristics of the amino-terminal region of each receptor may significantly influence the efficiency with which a receptor can regulate the expression of a particular gene.

This is particularly evident in a comparison of the hGR with the second human corticosteroid receptor, the mineralocorticoid receptor (hMR) (3, 4). Because of the near identity of the DNA-binding domain, the two receptors can interact with the same or closely related target DNA sequences and as a result of extensive sequence conservation in the ligand-binding domain; GRs and MRs are responsive to many of the same ligands. The amino-terminal region, however, displays no significant sequence conservation and exhibits differential effects on the transcriptional regulatory functions of hMR and hGR. The hGR amino terminus contains an important activation function necessary for efficient stimulation of gene transcription. In contrast, the hMR amino terminus does not provide a comparable *trans*-activation function and inhibits the cooperative interaction resulting from the binding of multiple receptor molecules (4). The existence of multiple receptor systems for glucocorticoid hormones raises important issues for the coordinate responses to this stimulus and the use of recombinant nuclear hormone receptors together with the reconstitution of hormone-dependent gene regulation in heterologous cell lines enables an accurate assessment of the functional consequences of ligand–receptor interaction on gene expression.

This chapter covers methodological aspects of the use of cloned nuclear hormone receptors in gene transfection studies, as well as a description of similarities and differences in gene regulation by hMRs and hGRs.

Transfection Strategies

Recipient Cell Lines

When choosing a mammalian host cell line, the species and tissue origin of the cells should be taken into account. Not all types of mammalian cells can be transfected efficiently and some lines of cultured cells may endogenously synthesize high levels of steroid receptors that complicate functional assays. A standard cell line used in many laboratories is the African green monkey cell line CV1. These cells have undetectable levels of endogenous corticosteroid receptors, are easy to handle, and endure the physical stress associated with the transfection procedure. Because of higher transfection efficiencies, strong transcriptional responses, and the neurogenic origin, the human neuro-blastoma cell line SK-N-MC is widely used in the author's laboratory. However, these cells are not as robust as CV1 cells and require higher cell densities for optimal transfection results. Both cell lines can be cultured in Dulbecco's modified Eagle's medium (DMEM) supplemented with fetal calf serum. Phenol red-free medium is required if regulation by certain steroids, especially estrogen and progesterone, is investigated. Even more important is the removal of steroid and thyroid hormones present in the added serum by charcoal and ion-exchange resin "double-stripping."

Vectors Used to Express Nuclear Hormone Receptors in Cell Lines

Several useful eukaryotic expression vectors have been described in the literature and are also available from commercial suppliers. The hormone receptor cDNAs used in the author's laboratory are usually cloned in either the pSG5 vector (5) or derivatives of the Rous sarcoma virus–chloramphenicol acetyltransferase plasmid (2, 3, 6). The eukaryotic expression unit contains a promoter element to mediate transcription of foreign DNA sequences, signals required for efficient polyadenylation of the transcript, and introns with functional splice donor and acceptor sites. Many promoter elements derived from viruses have a broad host range and are constitutively active in a variety of tissues, although considerable quantitative differences are observed among cell types. Widely used enhancer/promoter combinations are derived from the long terminal repeat (LTR) of the Rous sarcoma virus genome (6), the simian virus 40 (SV40) early gene (7), and from human cytomegalovirus (CMV) (8). A number of animal viruses contain DNA sequences that promote high-level replication of the viral genome in the infected cell. Plasmid vectors containing the origin of replication of the SV40 virus replicate to extremely high copy numbers in cells expressing the SV40 large T antigen, such as the CV1-derived COS cells. This system is used for the transient, but abundant, expression of the transfected genes.

Cloned cDNAs almost invariably include the ribosome-binding site and initiation codon found in the natural gene and the efficiency of translation is dependent both on the sequence close to the initiation site and the presence of AUG codons in the leader of an mRNA, which may reduce gene expression. Translation efficiency can be improved by removal of these sequences and the introduction of a consensus ribosome-binding site by site-directed mutagenesis or polymerase chain reaction (PCR).

Hormone-Responsive Promoters and Regulatory Elements

The long terminal repeat of the mouse mammary tumor virus (MMTV) has been widely used as a model system for the study of steroid hormone action (2, 3). This promoter contains several receptor-binding sites or hormone response elements (HREs), enhancer-like sequences that confer glucocorticoid responsiveness via interaction with the GR. Because of the near identity of their DNA-binding domains, mineralocorticoid, progesterone, and androgen receptors also recognize these response elements. A number of hormone-responsive promoters and receptor-binding sites have been described in the literature and can be used to create hormone-regulated transcription units. A comparison of the binding sites resulted in a consensus HRE comprising an inverted repeat of TGTTCT half-sites to which the receptors bind as dimeric molecules with high affinity (1). In the author's laboratory, the regulatory properties of the hGR and hMR are assayed on the MMTV LTR or a synthetic, perfect palindromic response element in conjunction with the thymidine kinase (*tk*) promoter of herpes simplex virus.

Reporter Genes to Monitor Transcriptional Regulatory Properties

Important tools with which to analyze the hormonal regulation of mammalian promoters are reporter genes, which provide convenient assays into the levels of gene expression. A number of different reporter genes with readily detectable activities, such as the *Escherichia coli* genes for neomycin phosphotransferase (*neo*) (7), xanthine–guanine phosphoribosyltransferase (*gpt*) (9), and chloramphenicol acetyltransferase (*cat*) (10), as well as the human growth hormone (11) and α_1-globin genes (12), are frequently used. However, the cDNAs and the assays for the enzymes encoding β-galactosidase (β-GAL) (13) and the firefly (*Photinus pyralis*) luciferase (14) have gained much interest.

The assay for firefly luciferase (LUC) provides substantial advantages over the commonly used CAT assay (14). It has greater sensitivity, is faster and easier to perform, and does not require the use of radioactivity. The luminescent reaction of luciferase is based on the oxidation of beetle luciferin. The enzyme uses ATP to catalyze the formation of an intermediate of luciferin that is subsequently oxidized by O_2. The produced oxyluciferin emits a high-energy photon as its electronic structure converts to the ground state. In the assay, luciferin and ATP are added to cellular extracts and the production of light as a consequence of the oxidation of the substrate is proportional to the quantity of luciferase in the cell extract. The intensity of the luminescent reaction can be measured immediately by the use of a luminometer and the entire procedure from preparing cell extracts to the results requires less than 1 hr.

When measuring the transcriptional regulation of mammalian promoters, it is desirable to cotransfect cells with a second reporter gene construct that serves as an internal control for distinguishing differences in the level of transcription from differences in the efficiency of transfection or in the preparation of the extracts. In the author's laboratory we routinely use the β-GAL enzyme for this purpose (13), which is expressed in the transfected cells from a constitutively active promoter with a broad host range, such as the SV40 early promoter. In physiology, β-GAL hydrolyzes lactose to glucose and galactose and in cell extracts this enzymatic activity can be easily assayed using as substrate the sugar derivative *o*-nitrophenyl-β-D-galactopyranoside 6-phosphate (ONPG). The hydrolyzation of ONPG results in galactose and the chromogenic compound o-nitrophenol. The yellow color emitted from o-nitrophenol is directly proportional to the amount of β-GAL present in the cell extract and can be quantitated in a spectrophotometer set at 420 nm. Extracts of most types of mammalian cells express relatively low levels of endogenous β-GAL activity; however, certain cell types exhibit a high background and in these cases a different reporter gene should be used as an internal control. To calculate the transcriptional induction, the LUC and β-GAL assays are carried out in a constant volume of extract and the results are normalized to a defined level of β-GAL activity.

Transfection Procedure

Many methods have been devised for introducing cloned eukaryotic DNAs into cultured mammalian cells. Popular gene transfection techniques include DEAE-dextran (15) and liposome-mediated DNA transfection (16). The most widely employed technique involves the use of calcium phosphate as a carrier to deliver the DNA (17). Because of its simple protocols, low cost, and relatively high efficiency, transfection mediated by calcium phosphate is the method of choice for experiments that require transient expression of the foreign DNA in large numbers of cells. The electroporation method involving the use of brief, high-voltage electric pulses has been devised for delivering the DNA into cells (18). DNA transfection by electroporation can be extremely efficient and has the additional advantage that it works well with cell lines refractive to other techniques. However, when setting up a variety of different transfection experiments the method is time consuming and requires the handling of large numbers of tissue culture cells. Irrespective of the method used to introduce DNA into cells, the efficiency is determined largely by the cell type and different lines of cultured cells vary greatly in their ability to take up and express exogenously added DNA. Optimal conditions for the particular cell line under study must be defined and it is important to compare the efficiencies of different methods.

Protocols for Transfections and Enzymatic Assays

Calcium Phosphate-Mediated Transfection

Reagents and Solutions

N-2-Hydroxyethylpiperazine-N'-2-ethanesulfonic acid (HEPES)-buffered saline (HBS; $2\times$): 280 mM NaCl, 50 mM HEPES, 1.5 mM

Na_2HPO_4. Adjust the pH to 7.12 with 0.5 N NaOH. Sterilize by passage through a filter (0.22-μm pore size) and store at 4°C

$CaCl_2$ (2.5 M): Filter sterilize and store at room temperature

Tris–ethylenediaminetetraacetic acid (EDTA) buffer (TE, 0.1×): 1 mM Tris-HCl, 0.1 mM EDTA, pH 8.0. Autoclave and store at room temperature

DNA: Supercoiled plasmid DNA is prepared by alkaline lysis and purified by ultracentrifugation in two consecutive cesium chloride–ethidium bromide gradients. Ethidium bromide and cesium chloride are removed by 2-propanol extractions and repeated ethanol precipitations. The DNA pellet is resuspended in sterile 1× TE and quantitated by OD_{260} absorption

1. SK-N-MC cells are maintained in semiconfluent to confluent cultures in DMEM supplemented with 10% fetal calf serum (FCS). CV1 cells require only 5% FCS. Six to 8 hr prior to transfection, harvest exponentially growing cells by trypsinization and replate them at approximately 30–40% confluency (approximately 3–5 × 10^5 cells/60-mm dish). Incubate the cultures for 6 to 8 hr at 37°C in a humidified incubator in an atmosphere of 5% CO_2.

2. For the transfection of a single monolayer of cells, 1 μg of an expression vector encoding the respective steroid receptor, 2.5 μg of the reporter gene, and 1.25 μg of the β-GAL plasmid are used. Carrier DNA, such as pGEM, and 0.1× TE are added to adjust the DNA concentration to 6 μg in 12.5 μl. Transfection experiments are usually performed in quadruplicate (four 60-mm monolayers of cells) and the reaction mixture is prepared as follows: place 50 μl of the DNA mixture in a disposable, clear 5-ml plastic tube, and add 400 μl of 0.1× TE and 50 μl of 2.5 M $CaCl_2$. With gentle mixing on a low-speed vortex, add dropwise 500 μl of the 2× HBS solution. Incubate the mixture for 20 min at room temperature. At the end of the incubation a fine precipitate should have formed. The goal is to avoid the rapid formation of coarse precipitates that result in a decreased transfection efficiency. Several factors such as the amount and quality of DNA and the exact pH of the HBS buffer affect the formation of the precipitate.

3. To resuspend the precipitate, pipette the mixture up and down once. Add 250 μl of the calcium phosphate–DNA solution into the medium in a 60-mm dish. Swirl the dish gently to mix the medium and incubate the transfected cells for 14 to 16 hr at 37°C in a humidified incubator in an atmosphere of 5% CO_2.

4. Remove the medium and wash the monolayer once with prewarmed phosphate-buffered saline (PBS). In contrast to CV1 cells, SK-N-MC cells do not adhere tightly and the PBS should be added slowly to the edges of the dish. Aspirate the PBS and add 4 ml of prewarmed DMEM supplemented

with 10% double-stripped FCS. Add hormone(s) as required and return the cells to the incubator. Cell extracts are prepared 24 hr after the transfection.

Transfection by Electroporation

Reagents and Solutions

E buffer: 50 mM K_2HPO_4, 20 mM potassium acetate. Adjust the pH to 7.35, sterilize by passage through a filter (0.22-μm pore size), and store at 4°C

$MgSO_4$, 1 M: Adjust the pH to 6.7 using 1 N HCl. Filter sterilize

ATP, 0.2 M: Store aliquots at -20°C

Electroporation device: We use the Electroporation system ECM600 (BTX, San Diego, CA). Similar devices are available from other companies, for example, the GenePulser System manufactured by Bio-Rad (Richmond, CA)

1. SK-N-MC cells are maintained in semiconfluent cultures in DMEM supplemented with 10% FCS. Twenty-four hours prior to electroporation, cells are plated at approximately 50% confluency (1–2 × 10^6 cells/100-mm dish).

2. Cells are washed with PBS, trypsinized briefly, harvested in PBS, and pelleted at 1000 g for 3 min at room temperature. Cells are resuspended in 390 μl of E buffer and 10 μl of 1 M $MgSO_4$ per original 100-mm dish.

3. For the transfection of a single monolayer of cells, prepare 5 μg of the receptor expression vector, 5 μg of the β-GAL expression vector, 2.5 μg of reporter gene, and 2.5 μg of carrier DNA. Place the DNA mixture for a duplicate experiment in an Eppendorf tube and add 400 μl of the cell suspension. Incubate for 5 min at room temperature, then transfer to sample cuvette.

4. Discharge a capacitor (charged to 250 V and 500 μF) through the mixture at room temperature. The efficiency of transfection by electroporation is influenced by the strength of the electric field, the length of the electric pulse, and the ionic composition of the buffer. The optimal electric field strength must be determined experimentally by varying the voltage and capacitance settings.

5. Distribute the cell–DNA suspension to two 60-mm dishes and incubate the transfected cells at room temperature for an additional 5 min in an incubator.

6. Add 4 ml of prewarmed DMEM supplemented with 10% double-stripped FCS. Add hormone(s) as required and return the cells to the incubator. Cell extracts are prepared 24 hr after the transfection.

Removal of Steroid and Thyroid Hormones from Serum

Reagents and Solutions

Charcoal, activated (Sigma, St. Louis, MO)

Dextran T70 (Pharmacia, Uppsala, Sweden)

AG 1-X8 ion-exchange resin (Bio-Rad)

Filters: fiberglass prefilters (type GF92; Schleicher & Schuell, Keene, NH) and membrane filters (0.22-μm pore size, Nalgene BTF; Nalge Co., Rochester, NH)

Tris-HCl, 10 mM, pH 7.8

Fetal calf serum (GIBCO-Bethesda Research Laboratories, Gaithersburg, MD)

1. Dextran-treated charcoal is prepared by adding 50 g of activated charcoal powder to 500 ml of 10 mM Tris-HCl pH 7.8. Stir the slurry with a magnetic stirrer for 30 min and centrifuge for 10 min in a table-top centrifuge. Pour off the supernatant, add another 500 ml of 10 mM Tris-HCl, pH 7.8, and repeat this procedure. Pour off the supernatant. Add 5 g of dextran T70 and 400 ml of 10 mM Tris-HCl, pH 7.8. Autoclave and keep at 4°C.

2. Thaw FCS and heat inactivate at 56°C for 30 min.

3. Add 25 g of ion-exchange resin, and put it on a magnetic stirrer at 4°C for 1–2 hr. Remove the resin by centrifugation.

4. Add 10 ml of the dextran-charcoal to 500 ml of FCS. Stir slowly at 4°C for 2 hr or overnight.

5. Centrifuge for 10 min in a table-top centrifuge at 1500 rpm and filter the supernatant through a fiberglass prefilter to remove the charcoal. Sterilize the FCS by passage through a 0.22-μm pore size membrane filter and store aliquots at -20°C.

Preparation of Extracts

Reagents and Solutions

Lysis buffer: 0.1 M KPO$_4$, 1 mM dithiothreitol (DTT); adjust the pH to 7.8. Store at room temperature

1. Aspirate the medium and wash the cells once with PBS (without calcium and magnesium).

2. Add 1 ml of lysis buffer to each 60-mm plate of cells and scrape the cells into an Eppendorf tube with a rubber policeman.

3. Centrifuge the cells at room temperature for 30 sec at 10,000 g.

4. Discard the supernatant and add 50 μl of lysis buffer to the pellet. Vortex to disperse the cells evenly.

5. Freeze-thaw the sample three times in a dry ice–ethanol bath and at 37°C. Make sure the tubes have been marked, preferably on the lid, with ethanol-insoluble ink.

6. Centrifuge the extracts at 4°C for 5 min at 10,000 g.

7. Use the supernatant immediately for enzymatic assays or save at 4°C.

β-Galactosidase Assay

Reagents and Solutions

β-GAL buffer: 60 mM Na$_2$HPO$_4$, 40 mM NaH$_2$PO$_4$, 10 mM KCl, 1 mM MgCl$_2$, 50 mM 2-mercaptoethanol. Store at room temperature

ONPG solution: Prepare a 2-mg/ml solution of o-nitrophenyl-β-D-galactopyranoside 6-phosphate (Sigma) in H$_2$O. Store in aliquots at -20°C

Na$_2$CO$_3$, 1 M: Store at room temperature

Microtiter plates, flat bottom

Microplate reader (e.g., MR5000; Dynatech Co., Burlington, MA)

1. Distribute 10–20 μl of cell extract into the wells. Add 100 μl of assay buffer and start the reaction by adding 25 μl of ONPG solution.

2. Cover the wells and incubate at 37°C until a yellow color appears. Stop the reaction by adding 50 μl of 1 M Na$_2$CO$_3$.

3. Read the optical density (OD) at 420 nm. The reaction is linear between OD$_{420}$ values of 0.1 and 0.8.

Luciferase Assay

Reagents and Solutions

Luciferin: Prepare a 10 mM stock solution by dissolving 10 mg of Luciferin (Boehringer Mannheim, GmbH, Germany) in 2.57 ml of H$_2$O and 1 ml of methanol. Store aliquots at -20°C. Dilute with H$_2$O to 200 μM for the assay procedure

LUC buffer: 100 mM KPO$_4$ (pH 7.8), 5 mM ATP, 10 mM MgCl$_2$. Make fresh each time

Luminometer: Luminometers are sold by a number of vendors, for example, the Lumat LB501 (Berthold Analytical Instruments, Nashua, NH)

1. Add 50 μl of LUC buffer into clear plastic tubes.
2. Add 5–20 μl of cell extract, place the tubes in a luminometer, and inject 100 μl of the diluted luciferin.
3. Read RLU.

Practical Approaches to Assay Gene Regulation by Nuclear Hormone Receptors

The ability of the steroid hormone receptors to regulate the MMTV promoter in the SK-N-MC human neuroblastoma cell line provides an excellent *in vitro* system in which the efficacy of various steroids can be tested. Utilizing this assay the transcriptional responses of hGR and hMR to various steroids can be compared at fixed hormone concentrations. The hGR, for example, exhibits a strong response to the synthetic glucocorticoids RU 28362 and dexamethasone (DEX), as well as to cortisol (CORT) and corticosterone, the natural glucocorticoids (4). Only a weak response to the mineralocorticoid aldosterone (ALDO) is found. The hMR, in contrast, is activated by both mineralocorticoids and glucocorticoids (4). There are, however, several caveats to this type of assay. First, by performing these analyses at one fixed steroid concentration, the differential binding affinities of the hMR and hGR are ignored. Thus, whereas the hMR response to CORT may be optimal at 10 nM, the hGR response is not at a maximum. Therefore, hormone response profiles provide a more accurate comparison of hormone-dependent gene regulation (3, 4). Second, the MMTV promoter has several receptor-binding sites and the hGR in particular has been shown to act cooperatively at several HREs. Data obtained in the author's laboratory demonstrate that the hMR does not display an additive or synergistic increase when multiple binding sites are present and this lack of cooperativity has been localized to the amino terminus (4). The amino termini of several steroid receptors have been implicated in specific regulation of target genes and thus the hMR might have increased transcriptional activity on a set of genes not yet identified. To distinguish between intrinsic *trans*-activation properties of the receptors and the cooperative activity resulting from the binding of multiple receptor molecules, reporter genes containing single binding sites upstream of heterologous promoters, such as the herpes simplex virus thymidine kinase or the simian SV40 early gene promoter, can be used. The absolute transcriptional induction, however, may be drastically reduced, a problem especially significant when assaying MR ligands. To verify weak agonistic effects of hMR ligands, which are hardly detectable by the wild-type hMR owing to its low *trans*-activation potential, the chimeric receptor GMM provides a more sensitive assay system (4, 19). The GMM resembles the hMR in its relative ligand-binding specificity but acts as a much stronger inducer of gene expression owing to the presence of the hGR-derived *trans*-activation domain τ_1 in the amino terminus. Aldosterone, deoxycorticosterone (DOC) and CORT show a clear agonistic activity via the hMR with an induction of LUC activity of $12\times$, $17\times$, and $24\times$, respectively (19). Tetrahydrodeoxycorticosterone (THDOC) and progesterone (PRO) display only a partial agonistic activity of 3.5- and 2-fold, respectively, whereas RU 28362, RU 486, spironolactone, and tetrahydroprogesterone (THP) are inactive as inducers of transcription (19). These agonistic properties were confirmed with the chimeric receptor GMM and revealed a 240-fold induction with THDOC and a 56-fold induction using PRO. Aldosterone, DOC, and CORT resulted in a 1000- to 1200-fold induction. Similar to the data observed with hMR, spironolactone, RU 28362, and RU 486 did not show any activity. This cotransfection assay is particularly useful in determining the antimineralocorticoid properties of steroid compounds. Spironolactone, for example, significantly reduces the ALDO-induced responses of hMR and GMM in a dose-dependent manner with 50% inhibition at 100 nM. This assay also revealed that PRO is even more effective as an ALDO antagonist with an IC_{50} value of 2–3 nM, suggesting that PRO acts as an antimineralocorticoid that is even more potent than spironolactone (19).

Concluding Remarks

Assays measuring binding specificity and affinities of steroid hormones are frequently performed in animal studies. Considerable variations in both parameters have been reported from various species and an extrapolation of pharmacological data is not always possible. The examples described above show that the cotransfection of the human receptors in a human neuronally derived cell line provides a sensitive *in vitro* system suitable for a wide range of problems and modifications of the procedures allow for the analysis of an infinite number of hormone, receptor, and promoter combinations.

Acknowledgments

I would like to thank my colleagues at the Max-Planck-Institute of Psychiatry for critical contributions and discussions to the transfection experiments and Ms. B. Berning for excellent technical assistance. Furthermore, I am indebted to J. Arriza (Seattle, WA) and R. M. Evans (San Diego, CA) for the generous gift of recombinant material.

References

1. R. M. Evans, *Science* **240**, 889 (1988).
2. V. Giguére, S. M. Hollenberg, M. G. Rosenfeld, and R. M. Evans, *Cell* **46**, 645 (1986).
3. J. L. Arriza, R. B. Simerly, L. W. Swanson, and R. M. Evans, *Neuron* **1**, 887 (1988).
4. R. Rupprecht, J. L. Arriza, D. Spengler, J. M. H. M. Reul, R. M. Evans, F. Holsboer, and K. Damm, *Mol. Endocrinol.* **7**, 597 (1993).
5. S. Green, I. Issemann, and E. Sheer, *Nucleic Acids Res.* **16**, 369 (1988).
6. C. M. Gorman, G. T. Merlino, M. C. Willingham, I. Pastan, and B. H. Howard, *Proc. Natl. Acad. Sci. U.S.A.* **79**, 6777 (1982).
7. P. J. Southern and P. Berg, *J. Mol. Appl. Genet.* **1**, 327 (1982).
8. M. Boshart, F. Weber, G. Jahn, K. Dorsch-Häsler, B. Fleckenstein, and W. Schaffner, *Cell* **41**, 521 (1985).
9. R. C. Mulligan and P. Berg, *Science* **209**, 1422 (1980).
10. C. M. Gorman, L. F. Moffat and B. H. Howard, *Mol. Cell. Biol.* **2**, 1044 (1982).
11. R. F. Selden, K. B. Howie, M. F. Rowe, H. M. Goodman, and D. D. Moore, *Mol. Cell. Biol.* **6**, 3137 (1986).
12. P. Mellon, V. Paker, Y. Glutzman, and T. Maniatis, *Cell* **27**, 279 (1981).
13. P. Herbomel, B. Bourachot, and M. Yaniv, *Cell* **39**, 653 (1984).
14. J. R. de Wet, K. V. Wood, M. deLuca, D. R. Helinski, and S. Subramani, *Mol. Cell. Biol.* **7**, 725 (1987).
15. M. A. Lopata, D. W. Cleveland, and B. Sollner-Webb, *Nucleic Acids Res.* **12**, 5707 (1984).
16. D. L. Felger, T. R. Gadek, M. Holm, R. Roman, H. W. Chan, M. Wenz, J. P. Northrop, G. M. Ringold, and M. Danielson, *Proc. Natl. Acad. Sci. U.S.A.* **84**, 7413 (1987).
17. F. L. Graham and A. J. van der Eb, *Virology* **52**, 456 (1973).
18. G. L. Andreason and G. A. Evans, *BioTechniques* **6**, 650 (1988).
19. R. Rupprecht, J. M. H. M. Reul, B. van Steensel, D. Spengler, M. Söder, B. Berning, F. Holsboer, and K. Damm, *Eur. J. Pharmacol. Mol. Pharm. Sect.* **247**, 145 (1993).

[17] Regulation of Neuropeptide Genes: Determination of Responsiveness to Steroids and Identification of Receptors in Brain Nuclei

Joke J. Cox, Sofia Lopes da Silva, Wiljan Hendriks, and J. Peter H. Burbach

Introduction

Steroid hormones control a large number of physiological processes in the nervous system. They are able to alter the peptidergic output of defined neuronal systems. This can be accomplished by the direct effect of steroids on neuropeptide genes. The steroid hormones activate their respective receptors that function as transcription factors. The transcriptional response occurs after binding to a response element on the target gene by the hormone–receptor complex. Therefore, a direct genomic action that alters the expression of the neuropeptide genes is implied.

Analysis of regulation of neuropeptide genes by steroids requires a number of *in vivo* and *in vitro* methods that focus on (a) the effect of *in vivo* steroid treatments on the levels of a neuropeptide mRNA in defined brain regions, (b) the properties of the regulatory elements in neuropeptide genes that underlie their responsiveness to certain steroids, and (c) the presence and nature of steroid hormone receptors in neurons that express the neuropeptide gene. The latter aspect is becoming more and more essential as it is clear that steroid hormones belong to a rapidly expanding superfamily of transcription factors that use common response elements in target genes.

In this chapter the oxytocin gene is taken as an example of a neuropeptide gene that is a target for steroid hormone receptors and related factors. In the brain, the oxytocin gene is abundantly expressed in a subset of magnocellular neurons of the hypothalamus, predominantly localized in the supraoptic (SON) and paraventricular (PVN) nuclei. These neurons send their axons to the posterior lobe of the pituitary gland and form the hypothalamoneurohypophyseal system (HNS). In addition, the oxytocin gene is expressed in several parvocellular neurons.

Previously, we have summarized methods to quantify oxytocin mRNA levels in the SON and PVN of rats (1). These included *in situ* hybridization on brain slices and solution and filter hybridization on microdissected SON- and PVN-containing brain tissue. These methods have been used to assess the responses of the oxytocin gene to treatments with steroid and thyroid hormones and retinoids under basal conditions, and under conditions of altered physiology (e.g., pregnancy, lactation, and hyperosmolality) (2). In this chapter we focus on molecular approaches that define properties of the regulatory region of genes that confer steroid responsiveness and that characterize members of the steroid/thyroid hormone receptor superfamily in brain nuclei.

Determination of Steroid Responsiveness of Genes by Transient Transfection in Cell Lines

For the study of the mechanisms regulating gene expression at the transcriptional level, advantage is taken of the fact that the organization of most genes is modular in nature. Regions mediating regulation of gene transcription can function relatively independently from regions encoding the structure of the gene product. It is thus possible to link regulatory regions of a given gene to another structural gene that acts as a reporter gene. In such a construct, the expression of the reporter gene serves as an indicator for the transcriptional activity of the linked regulatory region (3). A reporter gene should encode a product that is easily assayable, and not normally present in the cell line under investigation. Examples of reporter genes that have been used successfully are those encoding β-galactosidase (β-GAL), chloramphenicol acetyltransferase (CAT), alkaline phosphatase, and firefly luciferase (LUC). Of these reporter genes, the luciferase gene provides the advantage of a simple, quick, sensitive, and inexpensive assay.

Ideally, the regulation of reporter gene expression should be studied in a homologous system, that is, in a cell line that expresses the target gene at abundant levels as well as the steroid receptor. Cell lines derived from tumors of the neural crest (e.g., neuroblastoma, neuroepithelioma, and pheochromocytoma) often express neuropeptide genes that normally occur in the peripheral nervous system [e.g., enkephalin, cholecystokinin (CCK), neuropeptide Y (NPY), vasoactive intestinal peptide (VIP), calcitonin gene related peptide (CGRP), and others] (4). These cells do not always express steroid receptors at high levels. To study the regulatory potency of steroids on neuropeptide genes, it is therefore necessary to introduce promoter–reporter gene constructs together with steroid receptor expression plasmids into cell lines. This cotransfection assay is outlined in Fig. 1. The promoter–reporter gene construct is introduced in the cell line together with a steroid receptor cDNA under the control of a constitutively active viral promoter. A second reporter

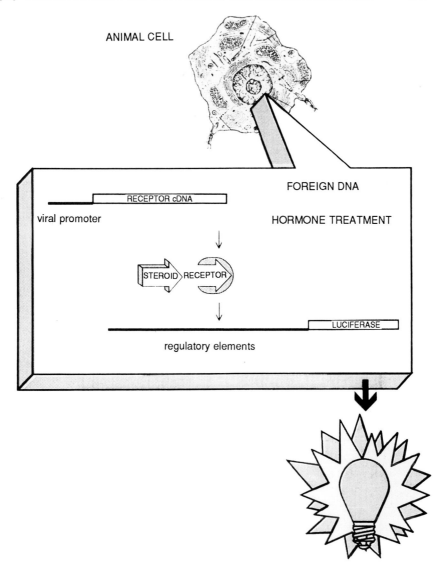

FIG. 1 Principle of the cotransfection assay. A promoter–reporter gene construct, and a steroid receptor cDNA under the control of a constitutively active viral promoter, are introduced into an animal cell. Steroid hormones added to the medium diffuse through the membrane into the cell, where they bind to their receptor. Hormone-activated receptors affect gene transcription by binding to specific regulatory elements in the promoter region, resulting in increased or decreased transcriptional activity that is ultimately measurable through the levels of the enzyme luciferase.

gene driven by a similar promoter is often included to monitor transfection efficiency and to serve as an internal standard (see below). Transfected cells are treated with steroid or vehicle and reporter gene activities are determined.

For many neuropeptide genes, however, no cell lines exist that express these genes (e.g., oxytocin and vasopressin). These genes can be studied for responsiveness to steroid hormones in a heterologous system. In such a system a cell line without expression of the target gene is used to provide a basal transcriptional environment. Often cell lines that are easily propagated and transfected, such as P19 embryo carcinoma cells, CV1 cells, and chinese hamster ovary (CHO) cells, are used and cotransfection is employed. Details of reagents and transfection procedures are given below for P19 embryo carcinoma (EC) cells. The same procedures have been used for other cell types, although optimization of conditions is sometimes required (see also Ref. 5).

Cotransfection in P19 Embryo Carcinoma Cells

Solutions and procedures described here concern the calcium phosphate precipitation method (6) and apply to P19 EC cells. Furthermore a simplified method, based on this procedure, is described for the transfection of JEG3 and Neuro2A cells. For other cell lines, optimal conditions may be slightly different and should be determined empirically (e.g., see Refs. 3 and 6).

Solutions and Materials

Charcoal-stripped fetal calf serum (see also Ref. 3)
1. Add 250 mg of charcoal and 25 mg of Dextran T-70 to 100 ml of fetal calf serum (FCS).
2. Incubate at 56°C for 30 min.
3. Centrifuge at 3000–4000 rpm for 30 min at room temperature.
4. Transfer the supernatant to fresh buckets.
5. Repeat previous steps.
6. Filter the supernatant through a prefilter (type AP; Millipore, Bedford, MA) to remove charcoal.
7. Filter the solution through a 0.45-μm pore size filter (Flowpore, 64-002-04; ICN Biomedicals, Ltd.).
8. Store aliquots at −20°C.

Gelatin (10× Stock)
1. Make a 1% (w/v) solution of gelatin (G-2500; Sigma, St. Louis, MO)

  2. Autoclave for 20 min.
  3. Store at 4°C.
Gelatin-Coated Dishes/Bottles
  1. Dilute the gelatin stock 10 times.
  2. Pipette this solution on the dishes to cover the surface of the dish completely (1.5 to 2 ml per 6-cm dish).
  3. Leave the dishes for 2 hr at room temperature. To hasten this procedure they can be placed in the incubator at 37°C for 1 hr.
  4. Aspirate the dishes.
  5. Wrap the dishes in aluminum foil and store them at room temperature or at 4°C.
HEBS buffer (10× stock)
  1. Combine 80 g of NaCl (final concentration, 1.37 M), 3.7 g of KCl (final concentration, 50 mM), 2.5 g of $Na_2HPO_4 \cdot 12H_2O$ (final concentration, 7 mM), and 50 g of N-2-hydroxyethylpiperazine-N'-2-ethanesulfonic acid (HEPES) (final concentration, 0.21 M).
  2. Add double-distilled water to 975 ml.
  3. Adjust the pH to 7.05 and the volume to 1000 ml.
  4. Autoclave the solution, and store it at 4°C.
  5. Prior to use, filtered glucose solution is added to the buffer (see preparation of 2× HEBS).
HEBS (2×)
  1. Dilute the HEBS 10× stock five times with sterile H_2O.
  2. Add 0.2 ml of a filtered glucose solution (1 g/ml) per 100 ml (final concentration, 10 mM).
  3. Adjust the pH to 7.05–7.08. This step is crucial for the formation of the Ca–DNA complex.
  4. Sterilize the solution by filtration.
Phosphate-buffered saline (PBS) without calcium and magnesium (14200-067 M; GIBCO, Grand Island, NY)
$CaCl_2$ (2.5 M)
Media
  Normal medium: Prepared with non-heat-inactivated FCS and medium with phenol red
  Steroid-depleted medium: Prepared with charcoal-stripped FCS and medium without phenol red. This medium is only used when the effect of hormones, normally present in serum, is to be tested.
Amount of plasmid used per transfection
  Reporter plasmid construct, 6.5 μg
  Expression vector, 1.0 μg
  Expression plasmid as a control for transfection efficiency, 0.5 μg

When the total amount of DNA is less than 8 to 10 μg, carrier DNA (plasmid or herring sperm DNA) should be added in order to obtain a good Ca–DNA precipitate

Lysis buffer

1. Combine 1 ml of Triton X-100 [final concentration, 1% (v/v)], 10 ml of KH_2PO_4/K_2HPO_4, pH 7.6–7.8 (final concentration, 0.1 M), 15 ml of glycerol [final concentration, 15% (v/v)], and 74 ml of H_2O. This mixture can be stored for several months at room temperature

2. Prior to use add 2 μl 1 M dithiothreitol (DTT) (final concentration, 2 mM).

Procedure for Cotransfection Assay

Day 1: Preparation of Cells

Trypsinize the cells and plate them on gelatin-coated dishes (6-cm diameter). A 100% confluent T75 bottle will provide enough cells to plate 30 to 35 six-centimeter dishes. Use normal medium.

Day 2: Transfection

1. Check the 6-cm dishes. They should be about 30–40% confluent.

2. Prepare DNA–$Ca_3(PO_4)_2$ precipitate: use one 15-ml polystyrene tube for each different precipitate. Prepare batches that contain for each plate:

HEBS (2×), 250 μl

H_2O, 240 μl (when the total amount of DNA is diluted in more than 10 μl, adjust this volume)

pRSV-GAL, 0.5 μg

The desired reporter gene(s) and expression vector(s)

3. Drip 27 μl of 2.5 M $CaCl_2$ in this mix for each plate to be transfected.

4. Shake the solution gently and let it rest for 15–30 min. After this period the mix should have an opalescent appearance. If not, the DNA–Ca complex is not formed and the solution should not be used for transfection. The reason for this failure is mostly due to the pH of the 2× HEBS solution, which has not been between the narrow range of 7.05 and 7.08. Furthermore, the quality of the DNA may be important in this respect. CsCl-purified, supercoiled plasmid preparations have performed best in our hands.

5. Aspirate the medium and wash the cells once with PBS and once with 1× HEBS. Slant the dishes and remove all of the HEBS solution.

6. Pipette the DNA–Ca mix dropwise into the dishes and place them for 15–30 min in the incubator. This may cause cell death; therefore it is crucial not to exceed this incubation time and to treat all dishes alike. The exact incubation time depends on the confluency of the dishes; for instance, cells at 30–40% confluency are incubated for 20 min.

7. Add 4 ml of steroid-depleted medium to each dish if the effect of hormones normally present in serum is to be tested. In all other cases replenish the cells with 4 ml normal medium. Leave the cells for 22–26 hr in the incubator.

Day 3: Steroid Treatment

Aspirate the DNA-containing medium from the dishes and replace it with either fresh steroid-depleted medium or fresh normal medium. Add the desired hormone and replace the cells in the incubator for another 22–26 hr.

Day 4: Harvesting of Cells

1. Wash the cells once with PBS.
2. Place the dishes obliquely and remove all of the PBS solution.
3. Add about 400 μl of lysis buffer. Any volume over 250 μl can be used.
4. Leave the dishes at room temperature in a horizontal position for 5 min. Then bang them on the table to facilitate the lysis and place them obliquely for 3 min.
5. Transfer the solution to a microcentrifuge tube (Eppendorf type), which is kept on ice.
6. Spin at maximal speed for 5 min in a microcentrifuge at 4°C.
7. Collect the supernatant which contains CAT, β-galactosidase, and/or luciferase activity.
8. Samples can be stored at −20°C.

Cotransfection of JEG3 and Neuro2A Cells

The major differences with the transfection method described for P1gEC cells reside in the growth rate of the cells and the ease with which the DNA–Ca complex may enter the cell. Both JEG3 and Neuro2A cells grow slowly, but the DNA–Ca complex penetrates the cells easily. This results in the following modification in the "Procedure of the Cotransfection Assay."

The transfection is performed on Day 4 rather than on Day 2, when the cells are about 70% confluent.

After aspiration of the medium, cells are not washed with PBS or 1 × HEBS, but immediately supplied with fresh normal or steroid-depleted medium. The DNA–Ca complex is added to the medium and cells are placed in the incubator for 22–26 hr.

The procedure is continued as described above under "Steroid Treatment" and "Harvesting of Cells."

Luciferase Assay

This assay is performed according to de Wet *et al.* (7).

Solutions

Buffer: Mix 500 μl of 2 M KH_2PO_4/K_2HPO_4, pH 7.6–7.8 (store stock at room temperature), 10 μl of 1 M DTT (store stock at $-20°C$), 200 μl of 100 mM ATP (store stock at $-20°C$), 1 ml of 150 mM $MgSO_4$ (store stock at room temperature), 0.5 ml of 10 mM luciferin (store stock at $-20°C$), and 7.8 ml of H_2O

Procedure

Pipette 100 μl of the sample in a Lumacuvette (Lumac, No. 9200-0; Perstorp Analytical, Oud Beijerland, the Netherlands) and measure the luciferase activity in a Lumac/3m biocounter M 2010A luminometer (Perstorp Analytical). The apparatus injects 100 μl of buffer into each sample. Light production is measured over 2 sec and is expressed as light units.

β-Galactosidase Assay

This assay is adapted from Sambrook *et al.* (8).

Solutions

Mg solution (100×)
   $MgCl_2$, 0.1 M
   2-Mercaptoethanol, 4.5 M
   Sodium phosphate (pH 7.5), 0.1 M
ONPG (1×): 4-mg/ml solution of *o*-nitrophenyl-β-D-galactopyranoside (ONPG) dissolved in sodium phosphate (pH 7.5), 0.1 M
Sodium phosphate (pH 7.5), 0.1 M: Combine 41 ml of 0.2 M Na_2H-$PO_4 \cdot 2H_2O$, 9 ml of 0.2 M $NaH_2PO_4 \cdot 2H_2O$, and 50 ml of H_2O
Na_2CO_3, 1 M
Chloroform

Procedure

1. For each sample to be assayed mix:
   Mg solution (100×), 3 μl
   ONPG (1×), 66 μl
   Sodium phosphate, 101 μl
2. Add 170 μl of this solution to 130-μl sample and mix well.

3. Incubate at 37°C until a yellow color appears.

4. Add 500 μl of 1 M Na_2CO_3 and mix. (The mixture can be stored at −20°C until further analysis.)

5. Add 500 μl of chloroform. Mix well and centrifuge at room temperature for 5 min at maximum speed in a microcentrifuge.

6. Measure the optical density of the supernatant at 420 nm. This is done by pipetting 200 μl of each sample in a microtiter plate. With a microtiter reader the optical density of the samples can be established.

7. As a blank, mock transfected cells are used.

CAT Assay

This assay is performed according to Sambrook *et al.* (8). In this assay [^{14}C]chloramphenicol is converted to butyryl derivatives, which can be separated from the hydrophilic substrate by extraction with an organic solvent.

Solutions

Assay mix: For each sample to be assayed, mix 16.75 μl of 1 *M* Tris-HCl (pH 7.5), 3.75 μl of [^{14}C]chloramphenicol (25,000 dpm/assay, stock: radioactive concentration = 25 μCi/ml, specific activity = 57 mCi/mmol, 175 μCi/mg), 0.45 μl of 25 m*M* ethylenediaminetetraacetic acid (EDTA), 0.15 mg of butyryl-CoA, and 24.05 μl of H_2O
Xylene
Scintillation fluid

Procedure

1. Take one 100-μl sample and add 45 μl of assay mix.

2. Incubate at 37°C for 1 hr. This period can be extended when CAT activity is low.

3. Add an equal volume (145 μl) of xylene and extract the hydrophobic products in the organic phase by vortexing for about 30 sec.

4. Centrifuge in an Eppendorf centrifuge for about 30 min.

5. Collect the (upper) xylene phase in a fresh tube.

6. Repeat the xylene extraction, and pool the two xylene phases.

7. To remove traces of the hydrophilic chloramphenicol from the xylene phase (about 4–5% is retained), the xylene is extracted with 290 μl of water.

8. Transfer the upper xylene phase to a vial for scintillation counting, add 3 ml of scintillation fluid, and count the samples.

Simultaneous Determination of Nuclear Hormone Receptor mRNA in Microdissected Brain Tissue

Various methods are available to determine the presence of steroid hormone receptors in brain nuclei, such as binding assays and immunocytochemical and *in situ* hybridization techniques. Here, we introduce a technique that allows the simultaneous identification of extremely low amounts of receptor mRNA in brain tissue by means of the polymerase chain reaction (PCR). The principle of this technique, which is called dot-blot screening with amplified gene fragments (DOSAGE), is outlined in Fig. 2. The first step is to microdissect the brain nuclei of interest with a small needle according to the method of Palkovits (9). The second step involves the isolation of mRNA and the synthesis of cDNA. This first-strand cDNA is used as the template in the polymerase chain reaction (step 3). To amplify steroid hormone receptor cDNA from the large pool of cDNA derived from brain nuclei, specific primers need to be designed. The PCR product, obtained by PCR amplification with these primers, consists of a mixture of steroid receptor cDNAs. The fourth step concerns the identification of the steroid receptors that are present in this mixture. This is carried out by labeling the entire PCR product and by using this as a probe on a dot blot containing spots of known steroid receptor DNA. Related techniques have been designed and used for analysis of a variety of mRNAs in tissues as small as single cells (10).

Microdissection of Brain Tissue

For microdissection of the SON the following procedure, which is an adaptation of the punch technique of Palkovits (9), is used. Rats are decapitated and the brains are rapidly removed and frozen on dry ice. The tissue is thawed to approximately $-20°C$ and cut with a razor blade in 2-mm slices. A needle with a 1-mm diameter is used to punch a brain nucleus from such a 2-mm slice. This tissue, weighing 2–3 mg, is rapidly transferred to dry ice.

RNA Isolation and cDNA Synthesis

Solutions

RNAzol (Cinna/Biotecx, Friendswood, TX)
Chloroform
2-Propanol
Ethanol (70%)

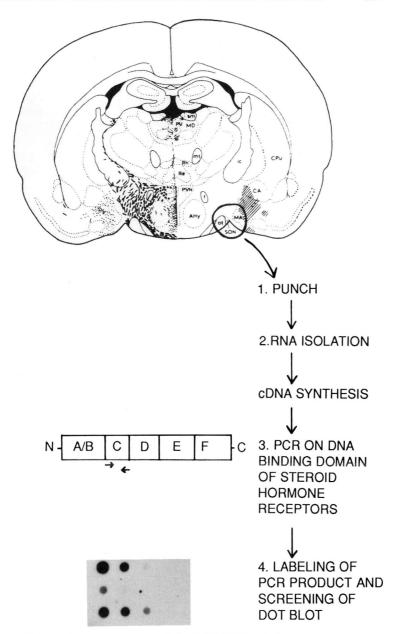

FIG. 2   Schematic representation of the DOSAGE technique (see text).

Procedure

Isolate total RNA from homogenized brain tissue by means of extraction with RNAzol (Cinna/Biotecx), which is based on the methods of Chomczynski and Sacchi (11).

1. Suspend 20 mg of tissue in 400 μl of RNAzol and homogenize it in a 2-ml Eppendorf tube, using a Teflon pestle.

2. Add 40 μl of chloroform, vortex briefly, and incubate on ice for 15 min.

3. Separate phases by centrifugation at 14,000 rpm at 4°C for 20 min and transfer the water phase, containing the RNA, to a fresh tube.

4. Precipitate the RNA with an equal volume of 2-propanol overnight. Precipitate RNA at 14,000 rpm for 20 min at 4°C, and remove excess salt by means of washing with 70% ethanol.

5. Resuspend the RNA pellet, containing approximately 7 μg of RNA, in 50 μl of H$_2$O. A few microliters are used to determine the RNA concentration with the spectrophotometer at 260 nm. Total RNA (2.5 μg) is used for cDNA synthesis. The rest of the RNA is precipitated again and stored at -20°C.

cDNA Synthesis

Solutions

SuperScript reverse transcriptase (RT), a Moloney murine leukemia virus (MMLV–RT) [Bethesda Research Laboratories (BRL), Gaithersburg, MD] (200 U/μl)

Random hexanucleotide primers (0.5 μg/μl)

First-strand cDNA synthesis buffer (BRL) containing 50 mM Tris-HCl (pH 8.3), 75 mM KCl, 3 mM MgCl$_2$

dNTPs (1.25 mM each)

RNase H (Amersham, Little Chalfont, United Kingdom) (15 U/μl)

Procedure

First-strand cDNA is synthesized using SuperScript reverse transcriptase, an MMLV–RT (BRL).

1. Anneal 9 μl, containing 2.5 μg of total RNA, to 2 μl (0.5 μg/μl) of random hexanucleotide primers for 5 min at 70°C and rapidly transfer to ice.

2. Add to this solution:

DTT (BRL) (0.1 M), 2 μl

Superscript RT buffer (BRL) (5×), 4 μl

dNTPs (25 mM each), 1 μl

Superscript RT (200 U/μl), 2 μl

3. Incubate at 37°C for 1.5 hr.
4. Denature at 90°C for 5 min.
5. Add 2 μl of RNase H (15 U/μl).
6. Incubate for another 20 min at 37°C.

Polymerase Chain Reaction Amplification

To amplify steroid hormone receptor mRNAs, oligonucleotide primers need to be designed that anneal to nucleotide sequences that are conserved among steroid hormone receptors. Such conserved regions are found in the DNA-binding domain (12). All receptors that recognize TGACCT motifs in target genes, such as the estrogen, thyroid hormone, and the retinoic acid receptors, share amino acid motifs of high homology in the DNA-binding domain. The DNA-binding domains of receptors recognizing TGTTCT palindromes in target genes, such as the progesterone, glucocorticoid, mineralocorticoid, and androgen receptors, are conserved among this group of receptors, but differ from the DNA-binding domain of the estrogen, thyroid hormone, and retinoic acid receptors. In this protocol degenerated oligonucleotides are used to anneal to steroid hormone receptors that recognize TGACCT motifs in target genes. They are located 5′ and 3′ in the DNA-binding domain. The 5′ primer is

5′ GGA GTC GGT ACC TG(C/T) GA(A/G) GGC TGC AAG GG(T/C) TTC TT 3′

The 3′ primer is

5′ TCC TT(G/C) (G/A/T/C)GC ATG CCC ACT TCG A(A/T/G)G CAC TT 3′

Solutions

Replitherm polymerase buffer: 50 mM KCl, 10 mM Tris-HCl (pH 8.3), 1.5 mM MgCl$_2$, and 0.1% (w/v) gelatin
5′ Primer, 50 pmol
3′ Primer, 50 pmol
dNTPs (10 nmol each)
Replitherm polymerase, 1 U
Mineral oil

Procedure

1. Supplement one-tenth of the cDNA (2 μl) with
   5′ Primer, 50 pmol
   3′ Primer, 50 pmol
   Replitherm polymerase buffer in 23 μl of solution

Boil for 3 min and cool on ice. Add 25 μl of solution containing 10 nmol of each dNTP and 1 U of Replitherm polymerase. Overlay with 50 μl of mineral oil and place in a thermal reactor (Hybaid, Teddington, United Kingdom).

2. Perform PCR: PCR amplification consists of 30 cycles under the following conditions: 1 min of denaturation at 94°C; 1 min of primer annealing at 37°C during the first 15 cycles and 45°C during the following cycles; 2 min of extension at 75°C. In the final cycle, PCR mixtures are incubated at 75°C for 10 min to ensure full extension of all PCR products.

Labeling of Polymerase Chain Reaction Products

The PCR products are separated from the primers on a 1% (w/v) low melting point agarose gel and randomly primed and labeled in gel, using Klenow polymerase (Boehringer GmbH, Mannheim, Germany) and 10 μCi of [α-^{32}P]dCTP. The labeled product is separated from free [α-^{32}P]dCTP by means of precipitation with 0.3 M sodium acetate and a 2.5× vol of ethanol in the presence of carrier DNA.

Preparation of Dot Blots

Solution

SSC (20×)
1. Combine 175 g of NaCl and 88.2 g of sodium citrate.
2. Add double-distilled water to 800 ml.
3. Adjust the pH to 7.0.
4. Adjust the volume to 1000 ml.
5. Sterilize by autoclaving.

Procedure

1. Cloned nuclear hormone receptor cDNAs (e.g., in aliquots of 1, 5, and 25 ng) are dissolved in 50 μl of 6× SSC and denatured for 5 min at 100°C.

2. Before spotting, the nylon membrane is washed with 20× SSC.

3. The samples are cooled to 0°C and spotted with a dot-blot apparatus (Bio-Dot microfiltration apparatus; Bio-Rad, Richmond, CA) on a nylon membrane (Hybond-N; Amersham).

4. The nylon membrane is removed from the dot-blot apparatus, washed with 2× SSC, dried in air, and baked at 85°C for 2 hr.

Hybridization of Dot Blots with Amplified Gene Fragments

Solution

Hybridization mix: Add, per milliliter:
SSC (20×) (final concentration, 3× SSC), 150 μl
Dextran sulfate (50%) (final concentration, 10%, w/v), 200 μl
SDS (10%, w/v) (final concentration, 0.5%, w/v), 50 μl
Denhardt's (50×) (final concentration, 5×) (see Ref. 8), 100 μl
Denatured salmon sperm DNA (final concentration, 50 μg/ml), 50 μg
α-^{32}P-labeled PCR probe (final concentration, 2 × 10^5 cpm/ml, 2 × 10^5 cpm
SSC (0.1×)/0.1% sodium dodecylsulfate (SDS)

Procedure

1. Hybridize dot blots in the hybridization mix at 70°C overnight.
2. Wash the membranes three times in 0.1× SSC/0.1% SDS at 70°C for 20 min.
3. Expose to autoradiographic film (X-Omat AR; Kodak, Rochester, NY) at −80°C for 2 hr, using an intensifying screen.

Concluding Remarks

Cotransfection assays in heterologous expression systems have proved to be essential tools for defining the steroid responsiveness of the oxytocin gene (13), as for many other genes. Figure 3 shows the cotransfection in P19 EC cells of the reporter plasmid pROLUC (having the −363 to +16 upstream region of the rat oxytocin gene cloned in front of the luciferase gene in vector p19LUC (14) and the estrogen receptor expression vector pHEO (having the receptor cDNA under the control of the Rous sarcoma virus (RSV) long terminal repeat in vector pSG5 (15) (from P. Chambon, Unité 184 de l'INSERM, 67085 Strasbourg Cedex, France). In this particular experiment, transfection of pROLUC alone (sample 1) produces a luciferase activity of about 170 light units, over a background of about 10 light units, representing the basal activity of the oxytocin promoter in the P19 EC cell line. Addition of 17β-estradiol does not affect luciferase activity, indicating that the P19 EC cells do not have endogenous estrogen receptors. In cells with endogenous receptors, for example, MCF-7 breast tumor cells, the pROLUC reporter plasmid responds to hormone treatment (16). Cotransfection with pHEO (sample 3) has no effect on luciferase activity, but additional treatment with 17β-estradiol (sample 4) raises luciferase activity about 100-fold. These data illustrate the ligand-dependent transcriptional activity of the estrogen receptor and reveal that the oxytocin gene is estrogen responsive.

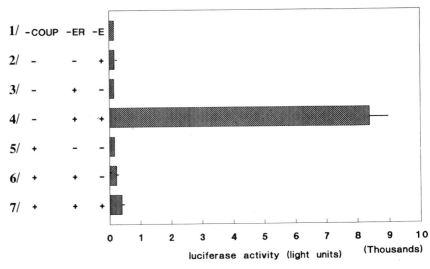

Fig. 3   Stimulation of rat oxytocin promoter activity by 17β-estradiol in transfected P19 EC cells. The $-363/+16$ upstream region of the rat oxytocin gene (pROLUC) was fused to the luciferase gene and cotransfected with pHEO (+ER) and/or pRShCOUP-TF (+COUP). Neither cotransfection with the estrogen receptor or COUP-TF alone, nor with both the receptors, affected luciferase activity. Treatment with 17β-estradiol did not change basal promoter activity. Addition of 17β-estradiol in cells cotransfected with pHEO, however, caused an increase in luciferase activity of about 100-fold. The presence of COUP-TF in the cell prevented the stimulation by the ligand-occupied estrogen receptor. Values are from one experiment performed in triplicate. The standard deviation is indicated.

Cotransfection assays can be further used to test interactions between multiple steroid hormone receptors or other transcription factors. In the same experiment (Fig. 3) it is illustrated that the cotransfected orphan receptor COUP transcription factor I (COUP-TF) can suppress the estrogen response (sample 7) (17). Contransfection of the expression vector pRShCOUP-TF (from S. Tsai, Baylor College of Medicine, Houston, TX) alone (sample 5) or in combination with the estrogen receptor expression vector (sample 6) does not influence the activity of the oxytocin promotor. However, cotransfected COUP-TF prevents the stimulation of promoter activity by 17β-estradiol (sample 7). These data show that COUP-TF interferes with the estrogen responsiveness of the oxytocin gene.

The experiments described above employ cotransfection as an assay to determine the influence of steroid hormone receptors on functional promoter activity. However, they do not reveal the underlying molecular mechanisms. With regard to the oxytocin gene, cotransfection assays have revealed the

TABLE I   Influence of Cotransfection of
pRSV-GAL with pHEO (+ER)
and pRShCOUP-TF (+COUP),
with or without estrogen
treatment, on β-galactosidase
activity[a]

	mean β-Gal	SD
Control	0.339	0.126
ER	0.411	0.049
ER + βE	0.215	0.036
COUP	0.069	0.011
COUP + ER	0.179	0.074
COUP + ER + βE	0.033	0.007

[a] ER, Estrogen receptor; βE, 17β-estradiol; COUP, chicken ovalbumin upstream promotor transcription factor.

stimulatory action of estrogens, thyroid hormone, and retinoic acid as well as the repressive action of COUP-TF. Cotransfection assays can help to locate and define response elements in genes by deleting regions or mutating specific sequences. In this way, elements in the oxytocin gene conferring responses to different steroid hormones have been identified. However, methods with which to analyze the physical interaction between the steroid hormone receptor and the gene are required to characterize the response elements further. In particular, electrophoretic mobility shift assays (EMSAs) and footprint analyses are relatively simple and powerful tools and have been described in detail elsewhere (18, 19).

An important note concerns the use of internal standards that serve to monitor differences in transfection efficiency. In our experiments the RSV promoter appeared not to be inert to different conditions in cotransfection assays. For example, in the experiments described in Fig. 3, cotransfection of pHEO alone increased β-galactosidase activity slightly, but COUP-TF strongly suppressed the β-galactosidase activity. Estrogen treatment of cotransfected cells also reduced β-galactosidase activity significantly (Table I). Within a single experiment β-galactosidase activity appeared to range from 0.033 units (pRShCOUP-TF + pHEO, 17β-estradiol treated) to 0.411 units (pHEO). The effect on β-galactosidase activity appeared to vary from one cell type to the other, and sometimes between experiments. We observed that other widely used viral promoters, simian virus 40 (SV40) and cytomegalovirus (CMV), did not show stable levels of activity. Therefore, internal standards should be used with caution. When condition-dependent variations

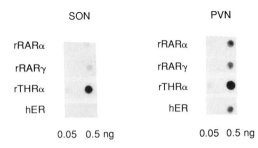

FIG. 4   DOSAGE analysis of SON and PVN samples. cDNA derived from microdissected SON and PVN was used as template in a PCR aimed at the amplification of steroid hormone receptors. The PCR product from the SON and that from the PVN were labeled and used as probes on dot blots containing 0.5 and 0.05 ng of the DNA-binding domains of rRARα, rRARγ, rTHRα, and hER.

occur, the internal standard can be used only to correct for differences in transfection efficiency within groups of identically treated samples, and not to normalize the entire experiment.

In the case of the oxytocin gene, which appears responsive to various members of the steroid/thyroid hormone receptor superfamily, it is essential to know which of them are present in oxytocin-expressing neurons. However, this is difficult to establish by conventional methods, because oxytocinergic cells are few in number, nuclear hormone receptors are expressed at low abundance, and they exist in a superfamily with many members. The DOSAGE technique provides the required high sensitivity and can be applied to many individual factors simultaneously. Sensitivity is provided by PCR amplification of cDNA that can be prepared from very small tissue samples, as small as a single cell (10). Individual members can be detected specifically by hybridization of the labeled PCR population to cloned steroid receptor cDNAs immobilized on a filter.

This approach has been taken to detect nuclear hormone receptors in the SON. By using a PCR primer set directed to the DNA-binding domain of members having the EGCK motif in the P box that recognizes TGACCT *cis* elements in target genes, at least nine different factors have been identified (20). Figure 4 shows the DOSAGE analysis of SON and PVN samples. Spotted on the dot blot are rat retinoic acid receptor α and γ (rRARα and γ), rat thyroid receptor α (rTHRα), and human estrogen receptor (hER). The PCR product derived from the rat SON contains abundant amounts of THRα, a small amount of RAR, and no ER. The rat PVN contains THRα, RAR, as well as ER. When procedures for cDNA synthesis and PCR have been optimized, reproducible results can be obtained in the DOSAGE analysis. However, the value in using DOSAGE as a quantitative assay seems so far to be limited. Owing to the PCR-based amplification a nonlinear bias may be introduced within the population of amplified sequences. The method is semiquantitative at best and can be effectively used to demonstrate the absence or presence of certain factors, or large changes in their relative levels.

References

1. H. H. M. van Tol and J. P. H. Burbach, *in* "Hormone Action, Part K: Neuroendocrine Peptides" (P. M. Conn, ed.), Methods in Enzymology, Vol. 168, p. 398. Academic Press, San Diego, 1989.
2. J. P. H. Burbach and R. A. H. Adan, *Ann. N.Y. Acad. Sci.* **689,** 34 (1993).
3. S. Richard and H. H. Zingg, this series, Vol. 9, p. 324, 1992.
4. M. A. E. Verbeeck, C. L. Mummery, A. Feijen, and J. P. H. Burbach, *Pathobiology* **60,** 127 (1992).
5. K. Damm, this volume [16].
6. A. J. van der Eb and F. L. Graham, *in* "Nucleic Acids," Part I (L. Grossman and K. Moldave, eds.), Methods in Enzymology, Vol. 65, p. 826. Academic Press, San Diego, 1980.
7. J. R. de Wet, K. V. Wood, M. DeLuca, D. R. Helinski, and S. Subramani, *Mol. Cell. Biol.* **7,** 725 (1987).
8. J. Sambrook, E. F. Fritsch, and T. Maniatis, "Molecular Cloning: A Laboratory Manual," 2nd Ed. Cold Spring Harbor Lab. Press, Cold Spring Harbor, New York, 1989.
9. M. Palkovits, *Brain Res.* **59,** 449 (1973).
10. S. M. Nair and J. Eberwine, this volume [19].
11. P. Chomczynski and N. Sacchi, *Anal. Biochem.* **162,** 156 (1987).
12. V. Laudet, C. Hänni, J. Coll, F. Catzeflis, and D. Stéhelin, *EMBO J.* **11,** 1003 (1992).
13. A. J. Van Zonneveld, S. A. Curriden, and D. J. Loskutoff, *Proc. Natl. Acad. Sci. U.S.A.* **85,** 5525 (1988).
14. R. A. H. Adan, J. J. Cox, J. P. van Kats, and J. P. H. Burbach, *J. Biol. Chem.* **267,** 3771 (1992).
15. V. Kumar, S. Green, A. Staub, and P. Chambon, *EMBO J.* **5,** 2231 (1986).
16. J. P. H. Burbach, R. A. H. Adan, H. H. M. Van Tol, M. A. E. Verbeeck, J. F. Axelson, F. W. Van Leeuwen, J. M. Beekman, and G. AB, *J. Neuroendocrinol,* **2,** 633 (1990).
17. J. P. H. Burbach, S. Lopes da Silva, J. J. Cox, R. A. H. Adan, A. J. Cooney, M.-J. Tsai, and S. Y. Tsai, *J. Biol. Chem.* **269** (1994). In press.
18. B. D. Hames and S. J. Higgins, "Gene Transcription: A Practical Approach." IRL Press, Oxford, 1993.
19. G. G. Kneale, "DNA–Protein Interactions: Principles and Protocols." Humana Press Totow, New Jersey, 1994.
20. S. Lopes da Silva, A. M. van Horssen, and J. P. H. Burbach, submitted (1994).

[18] Cloning of Steroid-Responsive mRNAs by Differential Hybridization

Nancy R. Nichols, Jeffrey N. Masters, and Caleb E. Finch

Introduction

Despite the extensive studies on feedback regulation of the hypothalamic–pituitary–adrenal axis by glucocorticoids, little is known about additional target responses of these steroids that regulate cellular functions in the brain, particularly at the level of changes in gene expression. Previously, *in vitro* translation and subsequent two-dimensional gel electrophoresis were used to demonstrate changes in the expression of select poly(A)-containing RNAs in rat hippocampus (1). With this technique, we resolved four mRNA species encoding 50-, 35-, 33-, and 20-kDa polypeptides that responded to acute corticosterone (CORT) treatment (2). To identify the changes in gene activity detected by the two-dimensional gel analysis, and also to increase our sensitivity in selecting responses representing less prevalent mRNAs, we designed a cDNA cloning experiment employing differential screening (±CORT) to isolate mRNAs that are increased or decreased in response to *in vivo* CORT treatment (Fig. 1). Directions for the preparation of reagents and buffers used in this chapter are given in Table I. Sources for particular enzymes and chemicals are specified at their first appearance in the text or in Tables I and II.

In Vivo Glucocorticoid Treatment

Hippocampi from two treatment groups were used to prepare poly(A)-containing RNA for cDNA library construction and screening. In this case, an adrenalectomy (ADX) and high-dose CORT treatment, representing extreme differences in hormonal status, were chosen to maximize changes in gene expression in response to steroid manipulation.

Adult male F344 rats (200–250 g) were multiply housed with a controlled light–dark cycle (lights on 06.00 hr and off at 18.00 hr). Food and water were available *ad libitum*. On day 1, rats were adrenalectomized and maintained on 0.9% (w/v) saline and half were then injected between 09.00 and 10.00 hr on days 2–4 with either CORT [10 mg in oil, subcutaneously (sc)] or 1 ml of Mazola corn oil vehicle. Following decapitation at 6–8 hr after injection

Methods in Neurosciences, Volume 22

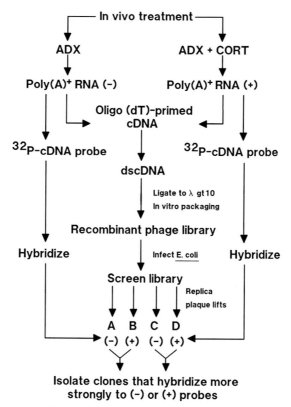

FIG. 1   Flow chart of cDNA library construction and screening. ADX, Adrenalectomy; CORT, corticosterone; poly(A)$^+$, poly(A)-containing; dscDNA, double-stranded cDNA.

on day 4, hippocampi were immediately dissected at 4°C and stored at −80°C until RNA isolation.

Comments

A large number of animals and more than one experiment per treatment group should be combined for the cDNA library and screening, because individual and even cohort variability may give rise to false-positive clones and subsequent unreproducible responses. In inbred strains of rats, these differences may represent effects of stress or handling, rather than genotype. We used two cohorts of 9–10 rats per treatment group for cloning of CORT-

TABLE I Preparation of Reagents and Buffers

Reagent	Amount	Comments
First-strand 5× salts	250 mM Tris, pH 8.3 375 mM KCl 15 mM MgCl$_2$ in deionized and distilled H$_2$O (dd H$_2$O)	Store at 4°C, do not freeze (GIBCO BRL suggestion)
Second-strand 5× salts	94 mM Tris, pH 6.9 453 mM KCl 23 mM MgCl$_2$ 750 μM β-NAD 50 mM (NH$_4$)$_2$SO$_4$ in ddH$_2$O	
5 mM dNTP	5 mM dATP 5 mM dCTP 5 mM dGTP 5 mM dTTP in ddH$_2$O	Individual stocks of 100 mM kept at −70°C
TE and STE	10 mM Tris, pH 7.5 1 mM EDTA For STE, add 150 mM NaCl in ddH$_2$O	Can be made as 10× stock solutions
10× Ligase buffer (−ATP)	660 mM Tris, pH 7.6 66 mM MgCl$_2$ 100 mM DTT in ddH$_2$O	
5× Kinase buffer (−ATP)	300 mM Tris, pH 7.8 50 mM MgCl$_2$ 1 M KCl in ddH$_2$O	
λdil	10 mM Tris, pH 7.5 10 mM MgSO$_4$ in ddH$_2$O	Add a drop of chloroform to plaque plug to inhibit bacterial growth

responsive cDNAs. In addition, careful dissection of brain regions to include only steroid targets, and to avoid heterogeneous tissue sampling (more or less white matter, meninges, vasculature, mass, etc.), should increase the chances of obtaining positive clones.

Poly(A)-Containing RNA Isolation

Poly(A)-containing RNA was prepared from each treatment group for use in cDNA library construction and screening. Total RNA was isolated from pooled samples of 18–20 hippocampi from each group (ADX ± CORT) by a guanidinium thiocyanate/CsCl procedure (3). Poly(A)-containing RNA (4–5 μg) was isolated from 500 μg of each total RNA by two cycles of chromatography on oligo(dT)–cellulose (4). Details of these methods are not given, because other methods of extracting total RNA and preparing poly(A)-containing RNA (1, 5) would be suitable, providing they result in RNA of high integrity.

Comments

Before choosing which pools of poly(A)-containing RNA to use for cDNA library construction and screening, they should be evaluated by RNA blot hybridization (5, 6) with a probe against a known steroid-responding and a nonchanging mRNA (often more difficult to establish *a priori*). Alternatively, the combined technique of *in vitro* translation and two-dimensional gel electrophoresis (1) was used to assess mRNAs for these studies from two experiments. Four reproducible CORT responses against a background of several hundred similar nonchanging translation products assured that the preparations were sufficiently homogeneous, yet contained similar differential mRNA responses (2).

Selection of λgt10 *c*I Insertion Vector

Prior to constructing the cDNA library a cloning vector must be selected and prepared. For this library, we used λgt10 (*imm*[434]*c*I; available from many commercial vendors) and the established methods for using this DNA phage as a cloning vector are described by Huynh *et al.* (7). This insertion vector can accept cDNA fragments ≤7.6 kb, which when inserted into the phage 434 repressor gene (*c*I) at the single *Eco*RI restriction site, generates a *c*I⁻ phage forming a clear plaque; in contrast, a plaque from the noninserted *c*I⁺ is opaque or turbid. Because recombinant plaques are genetically selected on the basis of a clear morphology, it is important to determine that the spontaneous *c*I⁻ level is ≤0.25%, independent of added insert DNA.

Two *Escherichia coli* host strains are required for cloning in λgt10, for example, BNN93 and BNN102 (7). Other host strains, with similar properties for cloning in λgt10, can be obtained from various vendors. Wild-type λgt10 phage are grown for phage DNA preparation, titered, and screened for spontaneous clear plaques on BNN93. BNN102 is used to plate the cDNA library for screening and amplification; after ligation of λgt10 DNA with cDNA inserts and *in vitro* packaging, the turbid plaques of the wild-type phage are suppressed on this strain with a high frequency of lysogeny mutation, leaving only the clear plaques of the recombinant phage for screening. However, before using the λgt10 vector DNA it is important to monitor the following properties (7): (a) that the *in vitro* packaging efficiency is >10⁸ plaque-forming units (PFU)/μg when using high-quality *in vitro* packaging extracts (e.g., Gigapack Gold; Stratagene, La Jolla, CA), (b) that *Eco*RI digestion decreases the packaging efficiency by 10^3, (c) that ligation of *Eco*RI-digested vector DNA gives 5–30% of the packaging efficiency of the undigested vector DNA with no appreciable increase in the percentage of clear plaques over the spontaneous rate, and (d) that the packaging efficiency of vector DNA containing insert is comparable with that of the ligated *Eco*RI-digested vector DNA minus insert.

Comments

Lambda phage vectors have the advantage of high transformation efficiencies due to good commercially available packaging extracts; however, the combination of electroporation of plasmid DNA containing cDNA inserts and certain strains of *E. coli* can now rival phage. More importantly, it is our experience that replica plaque lifts are more reproducible than replica colony lifts; therefore, screening by differential hybridization may have less inherent errors when using a phage library.

Detailed methods for growing and constructing a library in λgt10 are described in Davis *et al.* (8) and Huynh *et al.* (7). Alternatively, the Uni-ZAP XR vector system (Stratagene) has also been used in this laboratory for cDNA library construction and screening by differential hybridization (9). This system combines a λ phage vector with Bluescript plasmid rescue, avoiding the extra steps necessary to subclone restriction fragments from recombinant phage DNA into a plasmid vector for sequencing and making cRNA probes.

Construction of Hippocampal cDNA Library

To construct a rat hippocampal cDNA library containing sequences that increased or decreased in response to CORT treatment, equal amounts of poly(A)-containing RNA from both CORT-treated (+) and ADX (−) groups were pooled as a template for synthesis of double-stranded cDNA (dscDNA). The Gubler and Hoffman (10) modification of the Okayama and Berg (11) protocol was used, which combines oligo(dT)-primed first-strand cDNA synthesis together with RNase H–DNA polymerase I-mediated second-strand synthesis. Specific details of these methods, including fractionation of dscDNAs on a Bio-Gel A-50 (Bio-Rad, Richmond, CA) column, can be found in Huynh *et al.* (7) and in Watson and Jackson (12) and are not further elaborated here.

One of us (J.N.M.) has used a more efficient approach to cloning brain cDNAs. These cloning techniques were specifically developed to better represent the complete mRNA sequences. The original oligo(dT)-primed library was enriched in sequences corresponding to the 3' ends. Random hexamers were used as primers, but other primers can be used. The details of these newer methods are presented, because they are simpler, considerably shorter, and have additional advantages over the previous protocols (see notes and Comments, below).

This protocol is adapted from methods provided by GIBCO BRL (Gaithersburg, MD) (SuperScript RNase H⁻ reverse transcriptase), Promega (Madison, WI) (*Eco*RI adapters), Pharmacia (Piscataway, NJ) (S400 size-select spun column), and through experience gained while making the original oligo(dT)-primed library. This procedure is scaled for 2 µg of poly(A)-containing RNA, but can be adjusted accordingly; as little as 100 ng has been used successfully.

First-Strand cDNA Synthesis

   1. Add the following:

      Poly(A)-containing RNA (1 µg from each treatment group), 2 µg
      Random hexamers (20 ng/µl), 2 µl [or use 20 pmol of primer adapter or
        1 µl of oligo(dT)$_{12-18}$ (500 µg/ml)]
      Deionized and distilled H$_2$O (ddH$_2$O) to 9.5 µl

      Heat to 68° C for 5 min and immediately cool on ice.
   2. Add the following:

      First-strand 5× salts (as supplied with enzyme; Table I), 4 µl
      Dithiothreitol (DTT; 0.1 *M*) (also supplied with enzyme), 2 µl
      dNTP (5 m*M*) (10 nmol of each in final reaction; Table I), 2 µl
      [^{32}P]dATP (or any other dNTP; 800 Ci/mmol), 0.5 µl

      Incubate at 37°C for 2 min.
   3. Add 2 µl of SuperScript (GIBCO BRL, 200 U/µl; use 200 U/µg RNA). Incubate at 37°C for 1 hr (or incubate up to 55°C with some loss of enzyme activity for specific transcripts with extensive secondary structure).
   4. Take 1 µl of synthesized product for acid precipitation (5) to determine the percentage tracer incorporated in order to calculate yield: % incorporation × [4(10 nmol)(10^{-9} mol/nmol)(340 g/mol)(10^6 µg/g)] = µg cDNA. Expect

about 50% conversion of RNA to cDNA (2 μg of RNA generates approximately 1 μg of cDNA).

Second-Strand cDNA Synthesis

1. Add the following:

ddH$_2$O, 86 μl
Second-strand 5× salts (plus NAD) (Table I), 32 μl
dNTP (5 mM), 6 μl
DTT (0.1 M), 6 μl
[^{32}P]dATP, 5 μl
Escherichia coli DNA ligase [10 U/μl (GIBCO BRL); not T4 DNA ligase], 1.5 μl
DNA polymerase I (10 U/μl; GIBCO BRL), 4 μl
RNase H (3 U/μl; GIBCO BRL), 0.5 μl

Incubate at 15°C for 2 hr, then at 68°C for 5 min, followed by cooling on ice.

2. Add 2 μl of T4 DNA polymerase (GIBCO BRL, 5 U/μl). Incubate at 15°C for 10 min.

3. Add 4 μl of 0.5 M ethylenediaminetetraacetic acid (EDTA), pH 8. Take 2 μl for acid precipitation (expect 100% conversion of cDNA to dscDNA), then phenol extract (5) the remainder (1×) and ether extract (1×).

4. Add the following:

Ammonium acetate (4 M) (2 M final concentration), 165 μl
Ethanol, 95–100% (2 vol), 700 μl

Incubate at −70°C for 15 min, then spin in a microfuge for 20 min at 25°C at top speed. *Note:* Because tRNA added as carrier can inhibit packaging reactions, it is not recommended, especially after size selection steps.

5. Remove the supernatant, dry the pellet, and resuspend in 45 μl of TE (Table I).

Linker Addition

Fractionate cDNA on S400 size-select spun column (Pharmacia) before linker–adapter addition.

1. Add the following:

ddH₂O, 50 μl

Ligase buffer (10×, without ATP) (Table I), 10 μl

Run spun column at 400 *g* for 2 min at 4°C in a centrifuge and otherwise according to Pharmacia directions. Collect cDNA in 105 μl.
2. Add the following:

*Eco*RI adapters (10 pmol/μl; Promega), 5 μl (or less)

ATP (100 m*M*), 1.2 μl

T4 DNA ligase (1 U/μl; GIBCO BRL), 4 μl

Incubate at 15°C overnight, then phenol extract (1×). *Note:* By using *Eco*RI adapters, one can avoid having to digest cDNA with *Eco*RI prior to ligation and protecting internal sites by *Eco*RI methylation.

Linker Removal and cDNA Size Selection

For removal of excess linker adapters and to select dscDNAs of ≥500 bp, the linker addition reaction was fractionated on a S400 size-select spun column.

1. Run the S400 spun column as before, except this time in STE (Table I) and collect about 110 μl.
2. Add the following:

Sodium acetate (3 *M*), 10 μl

Ethanol (95–100%), 350 μl

Incubate on ice for 20 min, then spin for 20 min at 25°C in a microfuge at top speed and resuspend in 16 μl of ddH₂O.

Kinase-Ligated cDNA-EcoRI Adapters

1. Add the following:

Kinase buffer (5×) (Table I), 1 μl

ATP (100 μ*M*), 1 μl

T4 polynucleotide kinase (10 U/μl; GIBCO BRL), 2 μl

Incubate at 37°C for 30 min.
2. Add 80 μl of TE. Phenol extract (1×) and ether extract (1×).
3. Add the following:

Sodium acetate (3 M), 10 μl
Ethanol (95–100%), 350 μl

Incubate on ice for 20 min and spin in a microfuge for 20 min at 4°C at top speed. Remove the supernatant, dry the pellet, and resuspend in 10 μl of 1 mM Tris, (pH 7.5)–0.1 mM EDTA. *Note:* Some procedures kinase directly after ligation of adapter followed by size selection. Doing the kinase reaction after size selection avoids having active ligase around during this step.

EcoRI Digestion of Vector

1. Add the following:

λgt10 (10 μg) and ddH$_2$O to 17 μl
*Eco*RI buffer (10×) (provided with enzyme), 2 μl
*Eco*RI (10 U/μl; GIBCO BRL), 1 μl

Incubate at 37°C for 1 hr. Add another 1 μl of *Eco*RI and incubate at 37°C for 2 hr, at 70°C for 10 min, then at 45°C for 30 min (to anneal cohesive ends to form concatamers more readily for optimum packaging efficiency) and place on ice.
2. Add the following:

DTT (200 mM), 1 μl
ATP (100 mM), 0.2 μl

cDNA/Vector Ligation and Packaging

It is recommended that several parallel reactions be set up to test vector ligated alone to ensure that there is not a large percentage of clear plaques. Also, unligated vector should be used to test the completion of *Eco*RI digestion (see Selection of λgt10 *c*I Insertion Vector, above). In addition, it is recommended that two different concentrations of cDNA be tested to obtain the optimal percentage of clear plaques (≤5% of total plaques is optimal in relation to the chances of ligating unrelated inserts into a single vector DNA molecule). Once an optimal ratio of cDNA/vector is found, then the ligation can be scaled up accordingly.

1. Add the following:

λgt10 ($\leq$1 μg) in 2 μl
Tris (0.1 M, pH 7.5), 1 μl, containing 0.2 M MgCl$_2$
ddH$_2$O$_2$, 1 μl
T4 DNA ligase, 0.6 μl

Incubate at 15°C overnight. *Note:* The reaction should be very viscous.
2. Set up an identical reaction with 1 μl of cDNA instead of ddH$_2$O.
3. Set up another reaction with 1 μl of 0.1× cDNA.
4. Package all three reactions with Gigapack Gold *in vitro* packaging extracts (Stratagene) and titer phage at 10^{-4} to 10^{-7} dilutions on BNN102 and BNN93 (7, 9) for determination of the percentage of clear plaques. Store optimally packaged recombinant phage at 4°C in λdil (Table I) for library screening.

Comments

Both cloning protocols gave a cloning efficiency of 5×10^7 recombinants/10 ng dscDNA, which is sufficient to have represented in the library most of the highly complex poly(A)-containing RNA sequences in brain (13). The major difference between the two protocols was in the insert size for randomly picked clones. With cDNA fractionation on a Bio-Gel A-50 column (7), followed by pooling of the fractions that appeared by autoradiography to contain cDNA molecules greater than 0.5 kb, 50% of the recombinants (9 of 18) had inserts smaller than 0.15 kb, and 50% ranged from 0.15 to 3.0 kb. In the newer protocol presented here, all the inserts (10 of 10) ranged from 0.5 to 2.0 kb; therefore the Pharmacia S400 size-select spun columns appear to be more effective in removing small cDNA molecules. In addition, the random-primed library better represents complete mRNA sequences than does the oligo(dT)-primed library, which was enriched in 3' ends.

Screening cDNA Library by Differential Hybridization

The cloned cDNA inserts from the original oligo(dT)-primed library in λgt10 were screened following replica plaque lifts by hybridization with radioactively labeled probes obtained from each poly(A)-containing RNA (±CORT). The scheme for differential hybridization and screening is shown in Fig. 1.

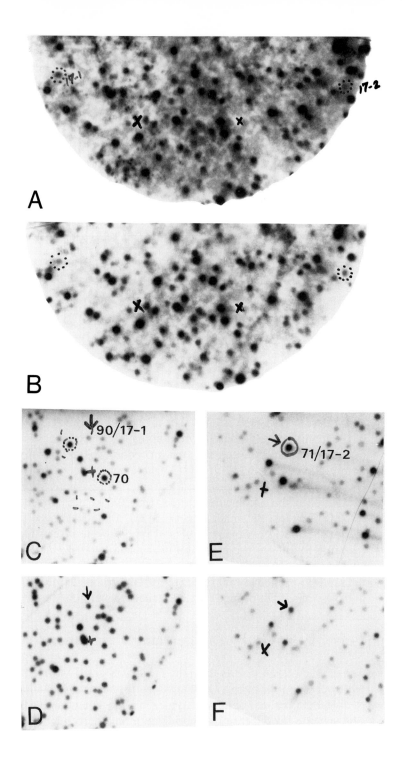

Primary and Secondary Screening of Plaque Lifts

The hippocampal library in λgt10 was plated at a density of approximately 5000 PFU/150-mm plate; 20 sets of quadruplicate nitrocellulose filters (137 mm and labeled A–D in the order of their transfer) were screened on two separate occasions for a total screening of approximately 200,000 clones. Duplicate filters were prepared for each probe. A thorough description of how to perform replica plaque lifts is found in Sambrook *et al.* (5).

In the primary screens shown in Figs. 1 and 2A and B, filters A and C were hybridized with a [^{32}P]cDNA probe transcribed from 0.5 μg of the same hippocampal poly(A)-containing RNA from adrenalectomized rats as that used to construct the library; filters B and D were hybridized wth probe against CORT-treated rats. Probe synthesis with synthetic random hexamer primers and murine reverse transcriptase was done as described in Sambrook *et al.* (5), except RNasin and unlabeled dCTP were omitted (438 pmol of labeled dCTP was added, which is enough to make 0.6 μg of cDNA). Hybridization of 40 filters in a glass dish at 68°C in a shaking water bath was as described in Sambrook *et al.* (5) with the following modifications: prehybridization and hybridization buffers (2 ml/filter) were the same except for probe (1.5 × 10^9 incorporated cpm in 80 ml) and contained 5× SSC (1× SSC is 0.15 *M* NaCl–0.015 *M* sodium citrate), 1% (w/v) sodium dodecyl sulfate (SDS), 0.5% (w/v) nonfat dry milk, 10% (w/v) dextran sulfate, sheared salmon sperm DNA (100 μg/ml), polyadenylic acid (25 μg/ml), polycytidylic

FIG. 2 Plaque-lift filter autoradiographs of A/B pairs compared during primary screening (A and B) and secondary screening (C–F) of two clones. In the primary screen, the A filter (A) was hybridized with the ADX cDNA probe and the B filter (B) was hybridized with the CORT-treated cDNA probe. Plaque 17-1 showed a greater signal on the ADX filter and was a candidate decrease in response to CORT and plaque 17-2 showed a relatively lesser signal on the ADX filter when compared with other nearby background signals and was a candidate increase in response to CORT. An "x" marks the corresponding ink spot on the filter that was used to orient the filters and plates at the time of picking plaque plugs for secondary screening. Phage from plaques 17-1 and 17-2 were rescreened by hybridizing the A filter (C and E) with the CORT-treated probe and the B filter (D and F) with the ADX probe. Clone 90/17-1 (CR90 or transforming growth factor β_1 in Table II) was the isolate (arrow) chosen to rescreen as a decrease in response to CORT treatment. Compare with clone 70, which was a nonchanger by subsequent tertiary screening; at least two different clones are represented on this filter autoradiograph. Clone 71/17-2 (CR71 or glutamine synthetase in Table II) was the isolate (arrow) chosen to rescreen as an increase in response to CORT treatment.

acid (25 μg/ml), and rat DNA (10 μg/ml; C_0t 100). The latter four reagents were mixed and boiled for 5 min before adding to the rest of the buffer. C_0t-fractionated repetitive DNA (available from GIBCO BRL) was added to decrease hybridization to repetitive sequences and background signal. Hybridization was carried out overnight, followed by successive washes in 5× SSC–1% SDS–0.5% nonfat dry milk (1×), then in 2× SSC–1% SDS (2×), then in 0.5× SSC (4× or more if counts are still being removed).

Candidate CORT-responsive cDNA clones that increased and decreased in response to steroid treatment were picked on the basis of differential signals on autoradiographs of A/B and C/D filter pairs. The A/B pair was used for the analysis and the C/D pair was used for confirmation of differential hybridization. The autoradiograph of the A filter was placed on a light box under the appropriate plate of recombinant plaques to line up markers and pick up the candidate plaque with the large end of a Pasteur pipette into λdil for secondary screening.

For secondary screening, <500 clear plaques were plated on 150-mm plates and quadruplicate filters were screened, hybridized and analyzed the same as for primary screening, except A and C filters were hybridized with the probe from CORT-treated rats and the B and D filters with that from adrenalectomized rats. For the secondary screening shown in Fig. 2C–F, 6.8×10^8 incorporated cpm of each probe was added to 250 ml of hybridization buffer containing a total of 130 filters. Filters were exposed to Kodak (Rochester, NY) XAR X-ray film and analyzed for differential signals as described above for the primary screening. A single well-isolated positive plaque was picked with the narrow end of a Pasteur pipette. The phage were then grown in 40-ml liquid cultures or as plate lysates (depending on titers) for isolation of λ phage DNA (5, 8) for tertiary screening. Representative results of secondary screens for a clone that decreased in response to CORT (clone 90/17-1) and one that increased (clone 71/17-2) are shown in Fig. 2C–F. Clone 70 was picked from the secondary screening analysis as a possible increase, but turned out to be a nonchanger on tertiary screening (Fig. 2C–D).

Tertiary Screening of DNA Slot Blots

Clones that maintained a differential response by secondary screening were subjected to a tertiary screening on DNA slot blots. Approximately 10 μg of phage DNA containing inserts was digested with *Eco*RI, then extracted with phenol-chloroform-isoamyl alcohol (25 : 24 : 1; 1×) and ether (1×). Following ethanol precipitation, the DNA pellet was diluted in 300 μl of ddH$_2$O containing 0.01 M EDTA–0.3 M NaOH and incubated at 68°C for 30 min,

TABLE II   Characteristics of CORT-Responsive Clones Isolated by
Differential Hybridization

Clone	Change (CORT/ADX) (-fold)		Size of mRNA (kb)	Sequence identity
	Tertiary screen	RNA blot		
Increases				
CR1, 71	2	3	2.9	Glutamine synthetase
CR3	8	10	2.9	Glycerol-3-phosphate dehydrogenase
CR16	3	4	3.9	Complete sequence/not found
CR36	1.3	2	6.0	Partial sequence/not found
CR62	3	2	2.4	Partial sequence/not found
Decreases				
CR43, 46, 59, 69	0.3	0.4	2.9	Glial fibrillary acidic protein
CR90	0.5	0.7	2.5	Transforming growth factor β_1
No change				
CR8	0.8	0.8	3.1	Proteolipid protein

then placed on ice. The DNA was further diluted with 300 μl of TE, and 100 μl was applied to duplicate nylon membranes (Zetaprobe; Bio-Rad) in triplicate lanes, using a slot-blot apparatus and vacuum application. Following baking in an 80°C oven for 2 hr, replicate blots were hybridized with each cDNA probe, as described above for primary screening, except probes were transcribed from 0.1 μg of poly(A)-containing RNA and blots were hybridized in tubes in 5 ml of buffer containing 10^6 cpm/ml. Under these conditions, the insert DNA is in excess and the probe is limiting. Autoradiographic signals were quantitated by computerized videodensitometry (6) and the change (-fold) was determined by dividing the mean of the three signals from the CORT-treated probe by that of the ADX probe (Table II).

Comments

Eighty-seven plaques showed a differential response to CORT treatment on primary screening and 37 were still positive after secondary screening. Differential slot-blot hybridization revealed that 20 of these were consistently positive, with 9 of them representing increases and 11 representing decreases in response to CORT treatment. Of these 20, four were redundant clones (Table II), 5 contained small inserts and were not characterized further, and 4 have not yet been characterized past tertiary screening (1 increase and 3 decreases). Additional clones were picked as controls that showed no change in response to CORT. From cDNA library construction to tertiary screening, the analysis was completed in 4 weeks.

Other libraries constructed and screened by similar methods in this laboratory include increased cloned responses in rat hippocampus to bilateral and unilateral electrolytic lesions of the entorhinal cortex (9, 14) and increased cloned responses in Alzheimer's diseased hippocampus compared with age-matched controls (15). One unsuccessful differential hybridization cloning experiment did not yield reproducible cloned responses between cohorts of mice in basal forebrain after ovariectomy and estradiol treatment. Consistent with these results, we did not find reproducible changes in mRNA translation products by two-dimensional gel analysis in this model. In retrospect, smaller dissections of the hypothalamus encompassing discrete estrogen targets (e.g., ventral medial hypothalamus) may have resulted in the selection of positive clones.

Analysis of Cloned Corticosterone Responses

Sequence Homology and mRNA Response to Corticosterone

The sizes of inserts of CORT-responsive clones were determined by agarose gel electrophoresis of *Eco*RI-restricted λ clones. The average insert size was 1.5 kb and ranged from 0.7 to 2.8 kb. Many of the 20 clones have been subcloned in a plasmid vector (16) for identification by DNA sequencing and for hybridization analysis of sense and antisense orientation. The *Eco*RI fragments of λ clones were subcloned into either the *Sma*I site of the Bluescribe vector (BS$^+$; Stratagene) by blunt-end ligation or into the *Eco*RI site of the Bluescript vector (BSSK$^+$; Stratagene). Partial DNA sequences of 2–400 nucleotides were determined for the 5' and 3' ends of each insert by a modification of the Sanger chain termination method, using dscDNA templates (17). These sequences were then compared with known genes in GenBank; we are presently using the Internet network to search GenBank at NCBI (National Center of Biotechnology Information; Bethesda, MD), using the Blast program. The identity of the gene encoding the mRNA (if known) is listed in Table II. For many of the clones, the hippocampal mRNA response to CORT was assessed by RNA blot hybridization analysis (6) in an independent cohort of rats and compared to the change (-fold) obtained by tertiary screening. Data on nine clones that have been analyzed to date are summarized in Table II.

Comments

Of the CORT-responsive clones analyzed, one-fourth had inserts that were nearly full length (3 of 11), indicating the completeness of first-strand synthesis. In addition, many clones contained poly(A) tails, as predicted by the nature of oligo(dT) priming. Differential hybridization of 200,000 recombinant clones from a rat hippocampal cDNA library yielded clones for mRNAs that both increased and decreased in response to CORT treatment. In addition, two CORT-responsive mRNAs that were previously resolved by two-dimensional gel electrophoresis of *in vitro* translation products have been identified by hybrid selection with cloned cDNAs: the 50-kDa product is glial fibrillary acidic protein (6) and the 35-kDa product is glycerol-3-phosphate dehydrogenase (18).

We have isolated clones for more prevalent [glutamine synthetase (glutamate–ammonia ligase) and glial fibrillary acidic protein] and less prevalent (glycerol-3-phosphate dehydrogenase, CR16, and transforming growth factor β_1) mRNAs that are also CORT responsive in other brain regions and that exhibit differential anatomical and cellular hybridization patterns in the rodent brain (6, 19). These markers of CORT responsiveness are being utilized to investigate the brain as a glucocorticoid target during aging, stress, and neurodegeneration (20–23).

Concluding Remarks

Cloning by differential hybridization was used to isolate cDNAs for mRNAs that increased or decreased in response to *in vivo* glucocorticoid treatment. Poly(A)-containing RNA was isolated from the hippocampus of adrenalectomized and CORT-treated adrenalectomized rats and pooled to make a dscDNA library in the vector λgt10. About 200,000 recombinant phage, as plaques exhibiting a clear morphology on the *E. coli* host strain, were screened by replica plaque lifts. The filters were then hybridized with [^{32}P]cDNA probes made from poly(A)-containing RNA from either adrenalectomized (−) or CORT-treated adrenalectomized (+) rats and exposed to film. The resulting autoradiographs were compared visually to identify clones (plaques) that hybridized more strongly to (−) or (+) probes, against a background of nonchangers. Following secondary screening by replica plaque lifts and tertiary screening on DNA slot blots, 20 positive clones were obtained, including both increases and decreases in response to steroid treatment. Some of these have been identified by DNA sequencing and comparison to known genes, and others are unknown sequences. These cloned mRNAs represent brain target responses to glucocorticoids, whether direct or indirect, which can be used to monitor changes in the actions of these steroids during diverse physiological and pathophysiological states.

Acknowledgments

This work was supported by ONR Grant NR00014-85-K-0770, The Brookdale Foundation, The John D. and Catherine T. MacArthur Foundation Program in Successful Aging, and NIH Grant AG07909. We thank Nicholas J. Laping for subcloning and sequencing CR71 and Heinz H. Osterburg for performing the hybrid selection experiments.

References

1. J. Poirier and N. R. Nichols, this series, Vol. 7, p. 182, 1991.
2. N. R. Nichols, S. P. Lerner, J. N. Masters, P. C. May, S. L. Millar, and C. E. Finch, *Mol. Endocrinol.* **2,** 284 (1988).
3. B. B. Kaplan, S. L. Bernstein, and A. E. Gioio, *Biochem. J.* **183,** 181 (1979).
4. F. Almaric, C. Merkel, R. Gelford, and G. Attardi, *J. Mol. Biol.* **118,** 1 (1978).
5. J. Sambrook, E. F. Fritsch, and T. Maniatis, "Molecular Cloning: A Laboratory Manual." Cold Spring Harbor Lab. Press, Cold Spring Harbor, New York, 1989.
6. N. R. Nichols, H. H. Osterburg, J. N. Masters, S. L. Millar, and C. E. Finch, *Mol. Brain Res.* **7,** 1 (1990).
7. T. V. Huynh, R. A. Young, and R. W. Davis, *in* "DNA Cloning: A Practical Approach" (D. M. Glover, ed.), Vol. 1, p. 49. IRL Press, Oxford, 1985.
8. R. W. Davis, D. Botstein, and J. R. Roth, "Advanced Bacterial Genetics." Cold Spring Harbor Lab. Press, Cold Spring Harbor, New York, 1980.
9. J. Poirier, M. Hess, P. C. May, and C. E. Finch, *Mol. Brain Res.* **9,** 191 (1991).
10. U. Gubler and B. J. Hoffman, *Gene* **25,** 263 (1983).
11. H. Okayama and P. Berg, *Mol. Cell. Biol.* **2,** 161 (1982).
12. C. J. Watson and J. F. Jackson, *in* "DNA Cloning: A Practical Approach" (D. M. Glover, ed.), Vol. 1, p. 79. IRL Press, Oxford, 1985.
13. B. B. Kaplan and C. E. Finch, *in* "Molecular Approaches to Neurobiology" (I. A. Brown, ed.), p. 71. Academic Press, New York, 1982.
14. J. R. Day, B. H. Min, N. J. Laping, G. M. Martin, III, H. H. Osterburg, and C. E. Finch, *Exp. Neurol.* **117,** 97 (1992).
15. P. C. May, M. Lampert-Etchells, S. A. Johnson, J. Poirier, J. N. Masters, and C. E. Finch, *Neuron* **5,** 831 (1990).
16. K. Struhl, *BioTechniques* **3,** 452 (1985).
17. S. Tabor and C. C. Richardson, *Proc. Natl. Acad. Sci. U.S.A.* **84,** 4767 (1987).
18. N. R. Nichols, J. N. Masters, and C. E. Finch, *Neuroendocrinol.*, in press.

19. N. R. Nichols and C. E. Finch, *Mol. Cell. Neurosci.* **2,** 221 (1991).
20. N. R. Nichols, J. N. Masters, and C. E. Finch, *Brain Res. Bull.* **24,** 659 (1990).
21. N. R. Nichols, N. J. Laping, J. R. Day, and C. E. Finch, *J. Neurosci. Res.* **28,** 134 (1991).
22. N. J. Laping, N. R. Nichols, J. R. Day, and C. E. Finch, *Mol. Brain Res.* **10,** 291 (1991).
23. N. R. Nichols, J. R. Day, N. J. Laping, S. A. Johnson, and C. E. Finch, *Neurobiol. Aging* **14,** 421 (1993).

[19] Molecular Correlates of Corticosterone Action in Hippocampal Subregions

Suresh M. Nair and James H. Eberwine

Introduction

In the mammalian system, glucocorticoids (GCs) modulate metabolic pathways as well as orchestrate physiological responses to various stressful stimuli. These stimuli, in turn, induce secretion of large quantities of GCs that are theorized to play a role in the ability of an organism to adapt to stressful situations. The actions of GCs in the brain are mediated by two types of cytoplasmic receptors, the high-affinity (0.5-nm) type I receptor (also known as the mineralocorticoid receptor, MCR) and the lower affinity (5-nm) type II receptor (glucocorticoid receptor, GCR) (1). Binding of the hormone ligand initiates a chain of events, including dissociation of heat-shock protein (hsp)-90 and other complexed proteins and association of different proteins to bound receptor, dimerization, and translocation to the cell nucleus (not necessarily in that order) (2). In the nucleus, this oligomeric complex binds to *cis*-acting glucocorticoid-response elements (GREs) and modulates transcriptional activity of many genes, leading to altered levels of expression of various cellular proteins.

The hippocampus is a complex neuroanatomical structure that is implicated in learning, memory, development, and maintenance of cognitive functions as well as regulation of the hypothalamic–pituitary–adrenal (HPA) axis. It is thought to transduce the effects of adrenal steroids to the neurons of the paraventricular nucleus (PVN) that express corticotropin-releasing hormone (CRH) and arginine-vasopressin (AVP), which are secretagogues of proopiomelanocortin (POMC). The type I receptor is theorized to play a significant role in this physiological context (3).

Chronic stress, or a paradigm designed to mimic such a state, for example, chronic administration of corticosterone (CORT) to adrenalectomized (ADX) male rats, can cause neuronal cell death in selected subregions of the hippocampus (4). Elevated, yet nontoxic levels of GCs can further potentiate the toxicity of other insults including administration of excitotoxins, antimetabolites, and experimentally induced hypoxia–ischemia (5). These effects of GCs are dependent on the presence of type II receptors, and are dose and time dependent (6, 7). Investigators have also demonstrated a significant inhibition of glucose transport in hippocampal neurons and glia (6–9), and

Methods in Neurosciences, Volume 22

propose that GCs act by compromising energy production in hippocampal neurons, making them more susceptible to metabolic insults (10).

Neuronal cell death due to chronic CORT treatment is likely to be triggered by two distinct, yet temporally overlapping, sets of events. Initial events mediated by MCRs and glucocorticoid receptors (GCRs) specific for CORT would lead to more generalized mechanisms of cell death. Given that these receptors are potent transcriptional modulators, cell death due to chronic CORT treatment is likely to be accompanied by alterations in levels of mRNAs and their corresponding functional proteins. Whereas some changes might be correlative, others may determine cellular susceptibility to this form of cell death. The difference in vulnerability between distinct cell types may arise due to differences in phenotype as well as synaptic and glial connectivity. Thus, characterization of alterations in levels of selected mRNAs is a preliminary step in the elucidation of the molecular mechanisms that underlie cellular vulnerability or increased resistance. It is also possible that morphologically similar cells in each subregion are equally vulnerable to chronic CORT treatment, and variations in cellular responsiveness arise owing to factors such as age of the cell, phase of cell cycle, local connectivity, or differences in ''cellular experience'' or development.

The goal of this study was to characterize concomitant alterations in levels of multiple mRNAs, as well as their temporal development in each subregion of the hippocampus in response to chronic CORT treatment for up to 4 weeks. Poly(A) RNA in tissue sections of total hippocampus and the CA1, CA2, and CA3 subregions of the hippocampus (11) of GC-treated animals was converted into single-strand complementary DNA (cDNA) in the tissue section by a process known as *in situ* transcription (IST) (12). The subregion-specific cDNA populations were then taken through two rounds of linear amplification to generate ^{32}P-labeled antisense RNA (aRNA) probes (13). This procedure resulted in greater than 10^7-fold amplification of the source poly(A) + RNA population. This probe was used to screen slot blots containing plasmid cDNA clones corresponding to various neuronal mRNAs. Because the quantity of DNA loaded on the blots is far in excess of probe, relative autoradiographic intensities of bands that result after appropriate washes are representative of relative mRNA levels in that physiological state. This procedure is called expression profiling (13).

Our intention, therefore, is to characterize diagnostic expression profiles for one or more selected hippocampal subregions that are selectively susceptible to GC-induced cell death in this treatment paradigm, at specific time points. This is a suitable approach to contrast the selective vulnerability of particular subregions, and to elucidate the molecular events that accompany susceptibility as well as increased resistance. This information will enable us to focus on particular subregions for further characterization.

Methods

Treatment Paradigm

Groups of seven adult male Sprague-Dawley rats (average weight, 200 g) are adrenalectomized and administered corticosterone (CORT) (10–15 mg/day) in their drinking water for 1, 2, 3, and 4 weeks (ADC groups). In addition, two groups of both adrenalectomized and sham-adrenalectomized rats receive vehicle [1.5% (v/v) ethanol, 0.9% (w/v) saline] (ADV and SHV groups, respectively). At the end of the treatment period, these animals are decapitated and their brains removed, blocked, and stored at −80°C until further use. Trunk blood taken at this time is used to measure plasma adrenocorticotropin (ACTH) levels, using standard radioimmunoassay (RIA) procedures (14).

Preparation of Sections

Using RNase-free procedures, microscope slides are subbed in 1.5% (w/v) gelatin and used to thaw-mount 20-μm-thick coronal sections through the hippocampus (1). These sections are then postfixed in 4% (v/v) paraformaldehyde, dehydrated, and stored at −80°C until further use.

In Situ Transcription

All procedures are carried out at room temperature unless otherwise specified. Rat hippocampal sections are thawed and dissected so that only the subregion of choice remains on the slide. These were then ringed with rubber cement to form wells. The sections are hybridized for 12–16 hr with oligo (dT)–T7 primer (amplification primer) in 50% (v/v) formamide and 5× SSC (1× SSC is 0.15 M NaCl plus 0.015 M sodium citrate) in a humidified chamber. This primer is an oligonucleotide that has a chain of 24 thymidine bases extended at the 5′ end with the bacteriophage T7 RNA polymerase promoter, and hybridizes to the poly(A) + RNA population from each specific hippocampal subregion (13). Excess primer is washed off in an RNase-free 2× SSC bath. The buffer is exchanged by bathing the sections in IST buffer [50 mM Tris (pH 8.3), 120 mM KCl, and 6 mM MgCl$_2$] for 15 min. This is removed and replaced with IST mix, which has the following composition: IST buffer, 7 mM dithiothreitol (DTT), a 250 μM concentration of each of the dNTPs, RNase inhibitor (0.12 U/μl), sterile RNase-free water, and reverse transcriptase (50 units/section). ^{32}P-Labeled dCTP is added to the IST mix on some sections to monitor specific cDNA synthesis driven by the amplification primer relative to controls (no primer, no enzyme, etc.). This *in situ* reaction is carried out for 90 min in a humidified chamber, using Parafilm to prevent evaporation. The sections are washed for 4 hr in 0.5× SSC to improve the signal-to-noise ratio, and air dried before further analysis. Radioactively labeled sections are apposed to X-ray film to assess the efficiency of specific cDNA synthesis (Fig. 1).

Antisense RNA Amplification: First Round

First-strand cDNA is taken off the sections by trituration with freshly prepared 0.2 N NaOH, 1% (w/v) sodium dodecyl sulfate (SDS). Tissue proteins are precipitated with 5 M potassium acetate, and the cDNA further purified by phenol–chloroform extraction and ethanol precipitation. The single-stranded cDNA is then made double stranded by the Gubler–Hofmann method, the hairpin loop excised, and the cDNA made blunt ended by using standard procedures (15). The cDNA is drop-dialyzed in batches of 5 μl against 50 ml of distilled water for 8 hr to remove any remaining free dNTPs, as they competitively inhibit the enzymatic activity of T7 RNA polymerase, the enzyme used in the next step.

The purified double-stranded cDNA template is incubated with the following reagents: 40 mM Tris buffer (pH 7.5), 7 mM MgCl$_2$, 10 mM NaCl, 2 mM spermidine, 8 mM DTT, 250 μM ATP, GTP, and UTP, 25 μM CTP, 30 μCi of [^{32}P]CTP, 20 units of RNase inhibitor, and 1000 units of T7 RNA polymerase. This reaction is carried out at 37°C for 3.5 hr, and generates ^{32}P-labeled amplified antisense RNA (aRNA). Under optimal conditions, this first round results in 2000-fold amplification of the original poly(A) RNA population (13). A small fraction of the ^{32}P-labeled aRNA is electrophoresed on a denaturing formaldehyde gel along with size markers to assess the size distribution of the first-round aRNA population.

The aRNA amplification procedure is a linear process producing few changes in the relative abundances of individual mRNAs in the population. There are several lines of evidence to suggest this. First, size distribution of first-round aRNA as seen on a denaturing gel is similar to that of the original poly(A)+RNA population. Also, initial R_0T curve studies (not done for this set of experiments) show that the complexity (i.e., the sum total of unique mRNA sequences) of the aRNA population resembles that of the original poly(A)+RNA population. This suggests that most of the mRNA sequences in the original population are retained in the first-round aRNA population.

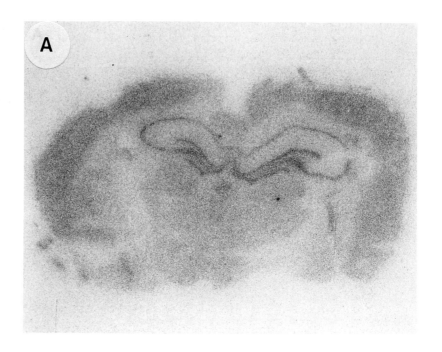

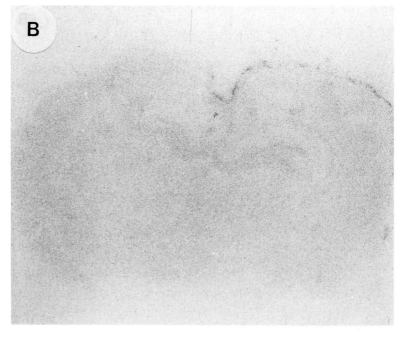

Although the first round of aRNA amplification results in a significant amplification of the source poly(A) + RNA population, other limiting factors such as the small amount of starting material, and the limits of detection of our analysis, make it necessary to take the samples through a second round of amplification. An important consideration at this point is the issue of average length of the aRNA population versus specific activity (radioactive counts incorporated per μg of cDNA template). Although incorporation of ^{32}P-labeled CTP molecules can be increased by decreasing the concentration of "cold" unlabeled CTP relative to ^{32}P-labeled CTP, such a strategy would also tend to decrease the average size of the aRNA population owing to a lowering of substrate concentration to near the K_m of the enzyme. This causes the enzyme to terminate its activity sooner. Incorporation of [^{32}P]CTP also decreases the half-life of the aRNA probe as a result of postsynthetic radiolysis. To prevent a significant decrease in the average size of the aRNA population and still permit radioactive detection of the aRNA, the ratio of "cold" CTP to ^{32}P-labeled CTP is appropriately manipulated.

Antisense RNA Amplification: Second Round

The first-round aRNA population is purified and a mix of random primers is used to initiate corresponding first-strand cDNA synthesis, using reverse transcriptase and dNTPS under appropriate buffer conditions. Because cDNA synthesis occurs in the 5'-to-3' direction, the resulting cDNA population would be in the sense orientation relative to the original poly(A) + RNA population, and most would contain poly(dA) tails. This allows use of the amplification primer to generate double-stranded cDNA that is then blunt ended, dialyzed, and amplified to generate aRNA probes as described for the first round. The cumulative effects of the two rounds of amplification is to amplify the original poly(A) RNA population by greater than 1 millionfold

FIG. 1   *In situ* transcription (IST) in coronal rat brain section (20 μm) using the oligo(dT)–T7 amplification primer (13). The signal was obtained by apposing ^{32}P-labeled sections directly to X-ray film for 20 min at room temperature, and represents total cDNA synthesis (A). Nonspecific cDNA synthesis in the absence of any added amplification primer is presumably primed by endogenous primers present in the tissue (B).

(Fig. 2). This generates enough probe to be used in standard molecular techniques.

The only difference between the first and second rounds of amplification is that the ratio of "cold" unlabeled CTP relative to ^{32}P-labeled CTP is altered to maximize specific activity, without reaching limiting concentrations for probe synthesis (i.e., not less than a final concentration of 10 μM "cold" CTP in the reaction mixture).

For reasons similar to those offered above, the second-round aRNA population is a linearly amplified representation of the original poly(A) RNA population, in which the relative abundances of its component sequences are more or less maintained. Therefore, it is suitable for use as a probe for detection of concomitant changes in levels of many different mRNAs in contrasting physiological states. Probes are generated in the manner described above, from total hippocampus and from CA1, CA2, and CA3 subregions of the longest treated ADC group (4 weeks) relative to vehicle and sham control animals. These were used to examine changes in the pattern of mRNA expression and determine, for each hippocampal subregion if possible, a diagnostic expression profile on chronic CORT treatment for 4 weeks.

Slot-Blot Preparation

Plasmid cDNA clones corresponding to various candidate mRNA sequences are prepared using large-scale (maxipreparation) procedures. Each cDNA clone is linearized using the appropriate restriction enzyme, and stored until further use in 10× SSC buffer. Prior to loading on slot blots, the linearized cDNA in 10× SSC is heat denatured (85°C for 5 min) to remove secondary structure and 1 μg of total DNA is loaded per slot under vacuum. The cDNAs are then fixed to the blots by baking them for 4 hr at 80°C under vacuum.

Probe Addition and Washes

Blots are prehybridized in heat-sealed plastic bags in the following buffer: 50% (v/v) formamide, 5× SSC, 5× Denhardt's reagent, 0.5% (w/v) SDS, salmon sperm DNA (100 μg/ml), and 1 mM sodium pyrophosphate for 8 to 12 hr. ^{32}P-Labeled aRNA probe is heat denatured and allowed to hybridize for 48 hr to the buffered blots (not less than 10^7 counts). The blots are then washed as follows: twice (15 min each) in 2× SSC, 0.1% SDS at 37°C, followed by one wash in 0.2× SSC, 1% SDS at 37°C. If the background is too high, then the latter wash is repeated until the autoradiographic density

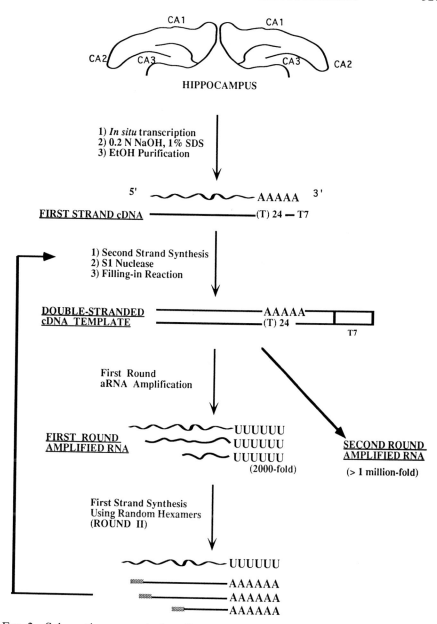

FIG. 2 Schematic representation of two rounds of linear amplification of subregion-specific poly(A) RNA. Under optimal conditions, this results in a greater than 1 million-fold amplification of the poly(A) + RNA population and generates ^{32}P-labeled antisense RNA (aRNA) probes that can be used in standard molecular biological procedures.

of the signal corresponding to 1 μg of plasmid vector on the blots (constituting nonspecific binding of the aRNA probe) is judged to be sufficiently low.

The blots are then air-dried and apposed to X-ray film at $-80°C$.

Data Analysis

Autoradiographic density of bands corresponding to various cDNA clones was measured by densitrometric analysis. The specific signal (above background) of probe bound to each clone was expressed as a percentage of NFL expression for each blot, minimizing variations due to differences in specific activity of probe, and the absolute quantity of probe present. Data analyzed in this manner does not permit absolute quantitation of mRNA levels, but rather gives a picture of relative changes in mRNA levels between the long CORT treatment group (ADC group) and both the ADV and SHV controls. Such an analysis of concomitant expression of multiple candidate mRNAs in contrasting physiological states is called *expression profiling*.

Results

Preliminary results from analyses carried out on hippocampal subregions of the longest treated time point (28 days) relative to adrenalectomized (ADX) vehicle-treated animals and sham controls showed many interesting characteristics. Among the many candidate mRNAs whose steady state levels were examined, the effect of this treatment on GLUR3 mRNA levels was examined across the CA1, CA2, and the CA3 subregions (Fig. 3A). In the CA1 subregion, GLUR3 appeared to be unchanged in the ADX animal relative to the sham-ADX animal, but increased slightly in the CORT-treated animal. In the CA2 subregion, the level of GLUR3 mRNA was threefold higher in the ADX animal relative to the sham-ADX animal and was similar to that seen after CORT treatment for 4 weeks. In the CA3 subregion, there appear to be no significant differences in GLUR3 mRNA levels between the three treatment groups.

The relative levels of mRNAs for GLUR3, GLUR7, and 11-2C (a cDNA encoding a novel sequence, as is explained in Discussion, below) in the CA2 subregion of the CORT-treated animal were examined in relation to corresponding levels in the ADX animal (Fig. 3A). Although GC treatment does not appear to alter mRNA levels for GLUR3 in the CA2 subregion, it extinguishes the signal for GLUR7, and decreases 11-2C signal as well.

Thus, expression of a single mRNA species in different subregions in different physiological states can easily be characterized by this approach.

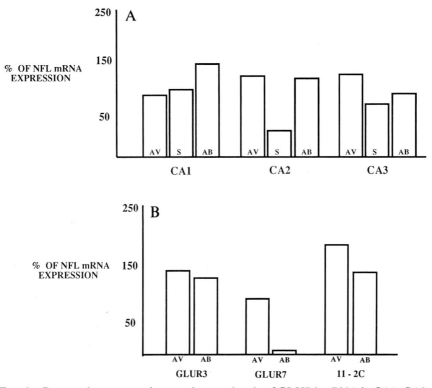

FIG. 3   Bar graphs representing steady state levels of GLUR3 mRNA in CA1, CA2, and CA3 subregions (A). Relative levels of GLUR3, GLUR7, and 11-2C mRNAs in the CA2 subregion are represented in (B). AB, 4-week treatment of male adult adrenalectomized (ADX) rats with 10–15 mg of corticosterone per day; AV, 4-week treatment of male adult ADX rats with vehicle (0.9% saline, 1.5% ethanol); S, 4-week treatment of male adult sham-ADX rats with vehicle.

At the same time, relative abundances of multiple mRNAs in each subregion can also be characterized as a function of physiological state.

Discussion

Glucocorticoids (GCs) including corticosterone (CORT) bind to intracellular receptors such as the mineralocorticoid receptor (MCR) and the glucocorticoid receptor (GCR), leading to formation of oligomeric ligand–receptor complexes that are preferentially localized to the nucleus and act on unique response elements to regulate potently rates of transcription of multiple genes. The resulting gene products very likely influence transcriptional activity of yet other genes, causing a cascading series of cellular and molecular events to occur. In addition to their direct actions on the genome, GCs also act on channel proteins (16,17) as well as specific cell surface receptors (18). Thus GCs are capable of influencing cellular physiology in multiple ways. It is likely that CORT treatment leads to simultaneous changes in levels of multiple proteins, which themselves may lead to other secondary changes. The sum of the alterations in relative levels of one or more proteins to others leads to the observed changes in cellular and tissue physiology that characterizes the physiological state induced by CORT. It is, however, technically difficult to characterize levels of multiple proteins simultaneously. It is easier to characterize alterations in expression of several candidate mRNAs (expression profiling; see Methods), which in turn are suggestive of protein level changes. Conversely, one of the limitations encountered with this approach is that there is no direct *a priori* evidence that changes in all mRNA levels necessarily lead to changes in the levels of their corresponding functional proteins.

Using this approach, the effect of chronic CORT treatment on steady state expression of GLUR3 mRNA was examined in the CA1, CA2, and CA3 subregions. Preliminary evidence indicates that this mRNA is differentially regulated in the CA2 relative to the CA1 and the CA3 subregions. Basal expression of GLUR3 mRNA appears to be relatively lower in the CA2, and adrenalectomy appears to cause an increase in its steady state level that remains unaltered after 4 weeks of CORT treatment.

The analysis of concomitant expression of GLUR3, GLUR7, and 11-2C mRNAs in the CA2 subregion reveals differential regulation of these three mRNAs between ADX and ADX-CORT rats. Corticosterone treatment does not appear to affect GLUR3 mRNA levels in the CA2 subregion, but drastically reduces levels of GLUR7 mRNA, while causing a smaller decrease in 11-2C mRNA levels. These data show that CORT treatment alters the pattern of gene expression in the CA2 subregion. Although it would be imprudent to establish a direct causal relationship on the basis of these data, further evidence generated in this manner along with histological evidence can help direct our attention to a subregion(s) that is selectively responsive to this paradigm of CORT administration.

Although characterization of alterations in levels of multiple mRNAs from selected subregions provides a certain specificity of analysis, it is clear that these subregions are not composed of homogeneous cell populations. There are different cell types within each subregion exhibiting differences in synaptic connectivity and glial associations, age, and morphological phenotype. As a consequence, these cells may have different responses to chronic CORT

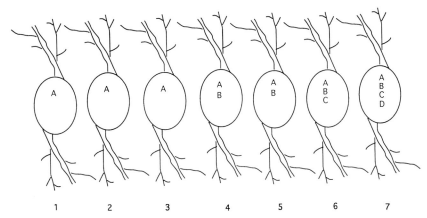

FIG. 4   Schematic depiction of relative abundances of mRNAs, A, B, C, and D in cells 1 through 7. In the composite "tissue" made up of all seven cells, mRNA A occurs at a much higher abundance relative to either mRNA C or D because it is present in all cells, and consequently mRNA A is easier to detect. In cell 7, however, because mRNAs C and D occur at a higher abundance relative to mRNA A than in the composite "tissue," single-cell analysis provides increased sensitivity of detection of alterations in mRNAs C and D.

treatment. Our analysis provides data on average mRNA levels from all of these different cell types, and the magnitude of mRNA level changes in specific cells that are responsive may be masked by the contribution from other nonresponsive cells. Furthermore, glial proliferation in response to neuronal damage or death and consequent overrepresentation of glial mRNAs relative to neuronal mRNAs may alter the resultant expression profiles. For example, CORT-responsive changes in gene expression that are specific for neurons will be underrepresented, whereas the expression of genes that are common to neurons and glia and not necessarily CORT responsive may appear to be increased. Although these questions cannot be easily resolved with the technology as described, this initial set of experiments is necessary to identify the subregion or regions that are particularly susceptible to chronic CORT treatment and to characterize diagnostic expression profiles for the full CORT treatment period of 28 days. A similar characterization for intermediate time points will provide information about the progression of events that lead to the previously described changes.

To characterize alterations in mRNA levels in cells responsive to chronic CORT treatment, single-cell analysis will have to be performed. Figure 4 schematizes why such an approach will increase specificity of analysis. If it is assumed that mRNAs A through D are found in equal abundance in each

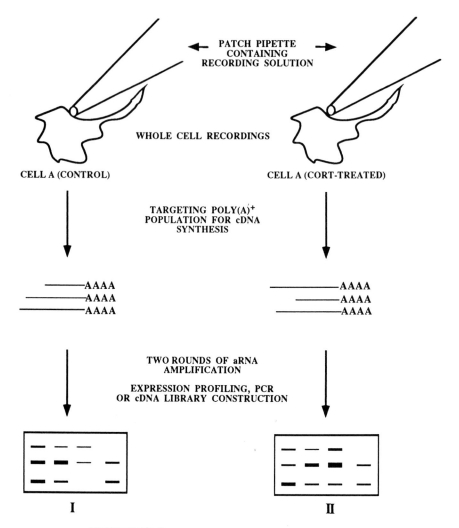

FIG. 5   Single hippocampal neuronal analysis will be used to detect the effects of CORT treatment on whole cell currents as well as levels of multiple mRNAs as follows: a patch pipette electrode (containing both recording solution and reagents necessary for first-strand cDNA synthesis) will be used to measure whole cell currents in acutely dispersed neurons from ADX male rats treated either acutely or chronically with CORT (cell A, CORT treated) relative to acutely dispersed neurons from male ADX rats given vehicle, or control naive neurons (cell A, control). The reagents necessary for first-strand cDNA synthesis will be introduced directly into the cell

expressing cell, then owing to the cellular distribution of these mRNAs it will be easier to detect mRNA A, which is present in all cells, rather than mRNA D, which is present in only one cell. However, using cell 7 by itself as the source of poly(A) RNA will enable characterization of changes in mRNA D as well as the other mRNAs. This occurs because the relative abundance of mRNA D is substantially greater in the single cell when compared to the "tissue sample" composed of cells 1 through 7.

Previously published studies from our laboratory (12, 13) provide evidence that lends credence to this line of reasoning. Poly(A)+RNA from a single acutely dispersed hippocampal neuron was used (after the conclusion of whole cell patch recordings, as explained below) to generate ^{32}P-labeled aRNA probe after two rounds of amplification. This probe was used to screen a hippocampal NG108 cDNA library. Out of several positive clones that were plaque purified and plasmid rescued, 12 were randomly chosen to be sequenced and further characterized. Four of these clones encoded mRNAs, of which three were determined (on the basis of nucleic acid and protein searches) to be previously unidentified sequences. Hybridization of random primed probes generated from one of these novel sequences, 11-2C, to Northern blots containing either total NG108 or rat hippocampal RNAs was not detected even after 1 week of exposure. This implies that this mRNA occurs in NG108 cells and hippocampal tissue at <0.005% abundance. In situ hybridization established that the highest level of 11-2C mRNA was found in the pyramidal cell layer of the hippocampus.

To examine the effects of CORT treatment on individual hippocampal neurons, single-cell analysis will be performed on acutely dispersed neurons from specific rat hippocampal subregions and live slice preparations in which some local synaptic and glial connectivity is maintained. For example, the effect of CORT treatment on single hippocampal neurons could be analyzed in the following paradigm. Neurons (either acutely dispersed or in slice preparations) could be exposed to CORT for short times (minutes, hours, and days) and analyzed relative to naive neurons as follows: a patch pipette electrode backfilled with the reagents necessary for first-strand cDNA synthesis (amplification primer, reverse transcriptase, RNase inhibitor, dNTPs, transcription buffer) will be used to perform whole cell patch recordings (Fig. 5). The tip of the patch pipette will be used to penetrate the cell and

---

via the patch electrode and the resulting cDNA will be processed through two rounds of aRNA amplification. This aRNA can be used to probe slot blots containing candidate cDNA clones, that is, expression profiling (as illustrated by I and II). Template cDNA can also be used to perform PCR analysis for isolation of specific sequences, or to construct cDNA libraries.

introduce the reagents for cDNA synthesis directly into the cell. Next, the cytoplasmic contents will be aspirated into the pipette and incubated at 37°C for 1 hr to generate first-strand cDNA. The samples will then be processed similarly to those described earlier from tissue sections. Such an approach will permit simultaneous electrophysiological and molecular analysis of single neurons (in the case of the slice preparations, synaptic connectivity and glial associations are partially intact) from specific hippocampal subregions. This experimental paradigm could be expanded to include hippocampal neurons from ADX rats chronically treated with CORT for different time periods, combinations of steroids, and so on. Such an approach may help to differentiate initial GCR- and MCR-mediated events in neurons that are specific for CORT from later, more generalized mechanisms of cell death.

Finally, once characteristic expression profiles have been defined for responsive neurons in a certain context, this information can be used to assess the relevance of these changes to neuronal survival. Manipulation of either message levels or proteins independently of the treatment paradigm can be used to modulate or reverse changes in mRNA levels caused by acute or chronic CORT treatment. This will help to realize a better understanding of the role these candidate mRNAs play in the sequence of biochemical events that lead to glucocorticoid-induced cell death.

Acknowledgments

This work was supported by a Merck ADP Fellowship to S.N. and by NIH AG9900 to J.E.

References

1. Funder, J., Feldman, D., and Edelman, I., *Endocrinology* **92**, 994–1004 (1973).
2. Rousseau, G. G., *Biochem. J.* **224**, 1–12 (1984).
3. Akil, H., Kwak, S., Morano, I., Herman, J., Taylor, L., and Watson, S., in "Neurosteroids and Brain Function" (E. Costa and S. M. Paul, eds.), Vol. 8, pp. 31–36. Thieme, New York, 1991.
4. Sapolsky, R., Krey, L., and McEwen, B., *J. Neurosci.* **5**, 1221–1226 (1985).
5. Sapolsky, R. *Progr. Brain Res.* **86**, 13–23 (1990).
6. Virgin, C., Ha, T., Packan, D. R., Tombaugh, G. C., Yang, S. I., Horner, H. C., and Sapolsky, R. M., *Neurochem.* **74**, 1422–1428 (1991).
7. Horner, H., Packan, D., and Sapolsky, R., *Neuroendocrinology* **52**, 57–64 (1990).
8. Phillips, P., Berger, C., and Rottenberg, D. *Neurology* **37**, Suppl. 1, Abstr. 248 (1987).
9. Kadekaro, M., Ito, M., and Gross, P. M. *Neuroendocrinology* **47**, 329–334 (1988).

10. Elliot, E., and Sapolsky, R., *in* "Neurosteroids and Brain Function," (E. Costa and S. M. Paul, eds.), Vol. 8, pp. 47–54. Thieme, New York, 1991.

11. Paxinos, G., and Watson, C., "The Rat Brain in Stereotaxic Coordinates," 2nd Ed. Academic Press, Orlando, Florida, 1986.

12. Tecott, L., Barchas, J., and Eberwine, J., *Science* **240,** 1661–1664 (1988).

13. Eberwine, J., Yeh, H., Miyashiro, K., Cao, Y., Nair, S., Finnell, R., Zettel, M., and Coleman, P., *Proc. Natl. Acad. Sci. U.S.A.* **89,** 3010–3014 (1992).

14. Harlow, E., and Lane, D., "Antibodies: A Laboratory Manual." Cold Spring Harbor Lab. Press, Cold Spring Harbor, New York, 1988.

15. Sambrook, J., Fritsch, E., and Maniatis, T., "Molecular Cloning: A Laboratory Manual," 2nd Ed. Cold Spring Harbor Lab. Press, Cold Spring Harbor, New York, 1989.

16. Souza, E., Goeders, N., and Kuhar, M., *Brain Res.* **381,** 176–181 (1986).

17. Miller, A., Chaptal, C., McEwen, B. S., and Peck, E., *Psychoneuroendocrinology* **3,** 155–164 (1978).

18. Orchinik, M., Murray, T. F., and Moore, F. L., *Science* **252,** 1848–1851 (1991).

[20] Glucocorticoid-Determined Protein Synthesis

Linda A. Dokas

Introduction

Glucocorticoid hormones produce a spectrum of cellular adaptations in neurons, ranging from alterations in enzyme activity to ultrastructural and morphological changes, that are dependent on protein expression. Such changes can be correlated with the period of exposure to glucocorticoids, from rapid alterations that occur within the span of minutes (1) to long-term responses such as the neuronal degeneration that accompanies cumulative exposure to these hormones (2). Although steroids have both receptor-mediated effects on genomic transcription and direct interactions with membrane proteins (3), it is the former that presumably underlie changes in protein expression. Even rapid electrophysiological effects of glucocorticoids, which might be expected to result from membrane-mediated actions, have in some cases been demonstrated to require alterations in protein synthesis (4).

In most instances, the detailed mechanisms by which glucocorticoids regulate gene expression have been defined in molecular systems, utilizing gene constructs expressed in transfected cells. For example, the interactions between glucocorticoid receptors and the AP-1 transcription factors that specify either inhibition or stimulation of gene expression have been characterized in such a manner (5). Although the rigor of such analyses is unquestionable for its ability to characterize gene transcription at the molecular level, it does not provide much information on the types of regulation that govern the posttranscriptional expression and activity of glucocorticoid-determined proteins in intact neurons. Instead, the analysis of protein synthesis in glucocorticoid-modulated neurons can be used to measure more directly the effects of these hormones on protein metabolism and function. When used in combination with other protein-based methodologies, such an approach yields information not only on the translation of glucocorticoid-determined mRNAs, but also on posttranslational factors that regulate the activities of such proteins. An experimental system that adapts itself readily to such an approach is the rat hippocampal slice incubated in the presence of a radiolabeled amino acid following treatment of rats *in vivo* with steroids or agents that modulate endogenous glucocorticoid levels. The aims of this chapter are to describe protocols for the use of such a system, to catalog the ways

Methods in Neurosciences, Volume 22

in which the measurements of protein expression and activity can be combined to understand the multiple levels at which the function of a protein is determined, and to provide an awareness of how such an approach is complementary to analyses of glucocorticoid effects at the nucleic acid level.

Methodology

Administration of Steroid

The use of tissue slices for *in vitro* effects of glucocorticoids is still under discussion because of the risk of receptor loss if conditions required to show functional responses by glucocorticoid manipulation *in vitro* are not strictly observed. The number of intracellular glucocorticoid receptors declines over the course of 2 hr in prepared hippocampal slices (6). With selected conditions of preincubation and steroid exposure, glucocorticoid-mediated effects on neuronal electrophysiology, including those that require protein synthesis and, by inference, a genomic action, can be observed over the course of several hours in hippocampal slices (4, 7). Alternately, a combined *in vivo/in vitro* approach can be used, that is, rats can be exposed to increased levels of steroid for an appropriate amount of time, after which hippocampal slices can be prepared and incubated with a radiolabeled amino acid precursor. Considerable effort in this laboratory has established the fidelity of such an approach, demonstrating that the ability of steroids to alter the physiological state of the rat is reflected at a subsequent level by alteration in the synthesis of glucocorticoid-determined proteins in the rat hippocampus (8–10). The most consistently observed response to elevated corticosterone levels produced either by exogenous injection or physiological stimuli related to the stress response is the synthesis of a hippocampal cytosolic protein with a molecular weight of 35,000. Because synthesis of this protein is enhanced with a short latency as serum corticosterone levels are increased and terminated quickly on reestablishment of basal conditions, it appears to be a valid biological marker of the response of the hippocampus to short-term stress.

Steroids are administered as subcutaneous injections in sesame oil. Sham-treated rats receive sesame oil alone. For a rat of 150–250 g, an injection volume of 1 ml is used. For single or small numbers of repeated injections, this volume poses no problems; however, the total volume of sesame oil given during chronic treatments will not be completely absorbed. The dose of any particular steroid needs to be empirically determined, because the relative potencies of steroids differ, depending on their receptor specificity. For example, as shown in Table I, maximal effects of RU 28362, corticosterone, and aldosterone on the synthesis of the M_r 35,000 hippocampal protein

TABLE I   Effects of Steroids on Hippocampal Protein Synthesis and Glucocorticoid Receptor Binding[a]

| Steroid | Maximal dose (mg) for protein synthesis | K_d for receptor (nM) | | Ref. |
		Type I (MR)	Type II (GR)	
RU 28362	0.05	—[b]	1–2	b, c
Corticosterone	0.5	0.5	2.5–5	c
Aldosterone	2.5	1.5–2.0	29	d

[a] Corticosterone, aldosterone, and RU 28362 (0.005 to 5 mg/animal) were administered to rats as subcutaneous injections in sesame oil. Four hours after injection, hippocampal slices were prepared and the labeling of a cytosolic M_r 35,000 protein was measured with methods as described in text.
[b] T. Hermann, K. Schramm, and R. Ghraf, *J. Steroid Biochem.* **26**, 417 (1987).
[c] J. M. H. M. Reul and E. R. deKloet, *Endocrinology* (*Baltimore*) **117**, 2505 (1985).
[d] E. R. deKloet and J. M. H. M. Reul, *Psychoneuroendocrinology* **12**, 83 (1987).

are consistent with the relative affinities of each steroid for the type II glucocorticoid receptor (GR) (11, 12). To measure accurately the sensitivity of rats to a steroid, adrenalectomized rats should be used so that responses over a dose range are not being measured against that occurring in response to endogenous corticosterone. An additional consideration with regard to the amount of steroid administered is the potency of each in producing peripheral side effects in relation to the total administration time of the steroid, especially if this involves chronic daily injections. Rats will tolerate poorly the more potent glucocorticoid, dexamethasone, when it is given at the same dose as corticosterone over an extended sequence of daily injections, for example.

A standard injection dose used in this laboratory is 5 mg of corticosterone given 4 hr before preparation of hippocampal slices. This dose produces extremely high serum levels of steroid for the first hour after injection, which rapidly fall to stress levels (250–500 ng/ml) and are then maintained for approximately 8 hr (8). Such an injection paradigm serves as a reasonable model for stress delivered over such a period. To standardize the effects of such an injection dose with reference to the circadian rhythm of corticosterone, injections are given before 10:00 AM, so that the peak of exogenous steroid precedes the rise in endogenous steroid (13).

In addition to the administration of exogenous steroid, protein synthesis can be analyzed in hippocampal slices in response to conditions that produce elevation of endogenous corticosterone levels, namely adrenocorticotropin (ACTH) and various physiological stressors. The relationship between these stimuli, levels of serum corticosterone, and effects on hippocampal glucocorticoid-dependent protein synthesis have been determined (9). Serum corticosterone levels rise in a linear manner in response to ACTH injections from 0.25 to 4 units (U)/animal; however, synthesis of the glucocorticoid-dependent M_r 35,000 protein increases in response to the same dose range of ACTH only up to 2 U/100 g of body weight, at which point the serum level of corticosterone is approximately 100 ng/ml. This same relationship is observed regardless of the means by which corticosterone levels are increased. Whether by exogenous administration or by elevation in response to ACTH and stressors, whenever the serum level approaches 100 ng/ml, the synthesis of the M_r 35,000 protein is maximal. This ceiling may be determined by the capacity of the hippocampal pool of glucocorticoid receptors (14).

A number of stressors were compared for their ability both to increase serum corticosterone levels and to increase synthesis of the M_r 35,000 protein (9). Ranked in order of their ability to raise corticosterone levels, the stressors are sham injection, exposure to cold, ether, and immobilization. The same rank order applies to the effect of these treatments on protein synthesis. Elevation of corticosterone levels in response to all stressors is maintained for less than 4 hr because basal levels of steroids are measured at this time after injection. Likewise, the labeling of the M_r 35,000 protein is observed only at the same times.

Preparation and Incubation of Hippocampal Slices

At selected times after injection, the rats are anesthetized by intraperitoneal injection of sodium pentobarbital (5 mg/100 g of body weight). If desired for radioimmunoassay (RIA) analysis, a blood sample is collected by cardiac puncture. Following decapitation, the hippocampus is dissected and placed into a cold solution of Krebs–Ringer bicarbonate (KRB) buffer, pH 7.4, containing 10.8 mM glucose (15). The tissue is sliced into 0.5-mm transverse sections with a McIlwain tissue slicer (Brinkmann Instruments, Inc., Westbury, NY). This thickness of slice has been found to be optimal for incubation studies (16). After sectioning, the slices from one hippocampus are placed into 5 ml of KRB buffer and preincubated in a shaking water bath at 37°C in an atmosphere of 95% O_2/5% CO_2. Fifteen minutes later, the KRB is discarded and replaced with 5 ml of fresh buffer to which is added the radioisotope. [^{35}S]Methionine with a specific activity of approximately 1000 Ci/mmol has been most often used in this laboratory, but a mixture of ^{35}S-labeled methionine and cysteine, which is now available commercially, has also been used as well. The uptake of label into slices and incorporation into protein are linear with respect to isotope concentration up to 100 μCi/incubation and with time for at least 4 hr. Standard conditions used in this laboratory have been 80 μCi of isotope per 3-hr incubation.

Preparation of Subcellular Fractions

Incubations are terminated by the addition of 5 ml of cold 4 mM methionine. Slices are collected by centrifugation at 12,000 g for 10 min at 4°C and the solution above them removed and replaced with 5 ml of KRB and 5 ml of the methionine solution. The centrifugation is repeated. Each set of hippocampal slices is homogenized with a Teflon–glass homogenizer in 1 ml of phosphate-buffered (pH 6.5) 0.32 M sucrose. The whole homogenates are centrifuged at 1000 g for 10 min at 4°C, producing a pellet (P_1) that is resuspended in 4 ml of the same buffer, and a supernatant (S_1). For complete subcellular fractionation, the S_1 fraction is centrifuged at 12,000 g for 20 min at 4°C, producing a mitochondrial–synaptosomal pellet (P_2) and a second supernatant (S_2). The P_2 pellet is resuspended in 450 μl of buffer. When only the cytosolic fraction is required, or for the final step of the sequential centrifugation series, the S_1 or S_2 fraction, respectively, is centrifuged at 100,000 g for 1 hr at 4°C, producing a pellet (P_3, resuspended in 450 μl of the same buffer as for the P_2 fraction) and a final supernatant (S_3), the cytosolic fraction. The overall labeling of these fractions varies, with the highest specific activity [counts per minute (cpm) per milligram of protein] in the S_3 fraction.

Analysis of Proteins

Incorporation of radioactivity into protein is measured by trichloroacetic acid (TCA) precipitation. Triplicate portions (usually 25 μl) of each fraction are added to 1.5 ml of 0.3 N NaOH containing methionine (2 mg/ml) and 80 μl of 30% (v/v) H_2O_2. Samples are incubated at 37°C for 15 min, after which 3 ml of cold 10% TCA is added. Samples are kept at 4°C for a minimum of 2 hr. The reproducibility of triplicate estimates is increased in proportion to the time of precipitation. Each sample is filtered through a Whatman (Clifton, NJ) GF/B glass fiber filter and washed with a total of 16 ml of cold 5% TCA and 4 ml of ethanol. Radioactivity on the filters is measured by scintillation counting. A portion of the filtrate (non-TCA-precipitable radioactivity) may be extracted three times with an equal volume of ether and counted for radioactivity as an estimate of the labeled amino acid pool. No general effect of exogenous corticosterone on total protein synthesis (counts per minute incorporated per milligram of protein) has been seen using the experimental parameters described above. However, analysis of proteins in subcellular fractions by sodium dodecyl sulfate-polyacrylamide gel electrophoresis (SDS-PAGE) demonstrates changes in individual species of protein.

Protein samples are adjusted to contain equal amounts of radioactivity in the volume to be applied to a gel lane and to each is added 0.5 vol of a denaturing solution to produce final concentrations of 0.001% (w/v) bromphenol blue, 5% (v/v) 2-mercaptoethanol, 10% (v/v) glycerol, 2% (w/v) sodium dodecyl sulfate, and 62.5 mM Tris-HCl (pH 6.8). Proteins in the mixtures are separated by discontinuous SDS-PAGE, using a 3% stacking gel and an 11% running gel. Estimates of molecular weights are made in comparison to protein standards run on the same gel. Two-dimensional gel electrophoresis is employed to determine the isoelectric point (IEP) of proteins of interest. As used in this laboratory, this involves isoelectric focusing in the first dimension in a 5% polyacrylamide gel containing ampholines in the pH range of 3.5 to 11, followed by standard SDS-PAGE in the second dimension (10). In the first dimension, both protein samples and the isoelectric focusing gel contain Triton X-100 and urea to maintain the solubility of proteins during the separation. Following staining of either type of gel, in this case with Fast Green, and destaining, they are treated with a fluorographic enhancer (En[3]Hance from Du Pont-New England Nuclear, Boston, MA), using instructions supplied by the company. Gels are dried onto filter paper under vacuum and exposed to Kodak (Rochester, NY) X-Omat film, for an exposure time (based on 10,000 cpm/gel lane) requiring 5 days of exposure. Films are stored at −70°C during the exposure period to minimize background. Quantitation of the effects of steroids on the labeling of protein bands is by densitometric analysis of the autoradiograms.

Variations on the Theme

The basic protocols described above were designed to examine effects of glucocorticoid hormones on hippocampal protein synthesis. But they are equally adaptable for use with other hormones and other brain regions or tissues with simple modifications of the method. Any hormone or compound that acts through one of the steroid hormone–thyroid hormone superfamily of receptors (17) would, in theory, be expected to produce hormone-dependent expression of proteins. For example, the effects of thiouracil-induced hypothyroidism and thyroid hormone replacement on liver and brain protein synthesis have been examined with similar methodology (18).

In thymocytes, glucocorticoid hormones differentially affect the synthesis of heat-shock proteins (19). The experimental protocols described here can be adapted to examine the same possibility in intact tissue slices. Incubation of hippocampal slices with [^{35}S]methionine at 42°C results in predominant labeling of a small number of protein species with the apparent molecular weights of the high molecular weight heat-shock proteins. Hence, labeling under these conditions following *in vivo* treatments can demonstrate the effects of such manipulations on the expression of the major heat-shock proteins. However, because some small mammalian heat-shock proteins do not contain methionine (20), a more appropriate amino acid precursor, such as [^{3}H]leucine, must be chosen to examine synthesis of these proteins.

Protein Synthesis and Other Protein-Based Methodologies

For proteins such as the M_r 35,000 protein, which are responsive to manipulation by glucocorticoid hormones, the use of SDS-PAGE is sufficient to demonstrate and quantitate effects on its synthesis. However, for minor or less responsive proteins, it may be necessary to combine protein labeling with more sensitive means to detect specific proteins. These include two-dimensional gel electrophoresis, immunoblotting, and immunoprecipitation.

When separated on the basis of both molecular weight and isoelectric point, a greater number of glucocorticoid-sensitive proteins become evident, because comigration of proteins in a one-dimensional gel system can mask effects on species with similar molecular weights (10, 21, 22). For example, increased synthesis of an M_r 46,000 protein in response to steroid administration is not prominent on standard SDS-PAGE gels, but becomes more obvious when proteins are separated in two dimensions (10). Use of a broad pH range in the isoelectric focusing dimension allows a survey to be made of a large number of cellular proteins.

Given a glucocorticoid-sensitive protein of a certain molecular weight and isoelectric point, immunoblotting with antibodies to proteins that possess similar properties can provide the minimal requirement to establish identity, that is, comigration between a labeled protein and an immunoreactive band or spot on gels. As a more rigorous test of identity, immunoprecipitation with the antibody can be used to test if it will precipitate a labeled protein from hippocampal subcellular fractions after incubation of slices with [^{35}S]methionine. The labeled protein should be present in the immunoprecipitate prepared from rats in which glucocorticoid levels have been modulated in the same proportion to control preparations as observed in the total subcellular fractions. Both immunoblotting and immunoprecipitation have become standard methods for protein-based laboratories, much as is the case for SDS-PAGE, and therefore detailed protocols will not be presented. Only a brief summary of each method, as used in this laboratory, is given here, but for a more detailed description readers are referred to the original paper in which these methods were used (23).

For immunoblotting, labeled proteins are separated by standard SDS-PAGE, and then electrophoretically transferred to nitrocellulose membrane. Nonspecific binding sites are blocked on the membrane by overnight incubation at 4°C with a solution of dry milk solid and Tween 20. Following incubation with the primary antibody at a predetermined dilution, immunoreactive proteins are visualized with an appropriate alkaline phosphatase-conjugated secondary antibody. For immunoprecipitation, subcellular fractions are dispersed in a detergent-containing buffer and clarified by centrifugation. Aliquots of the supernatant are incubated with the primary antibody and the immunocomplexes are absorbed onto a suspension (5% final concentration) of *Staphylococcus aureus* cells. After centrifugation and washing, immunocomplexes are dissociated by boiling in a sample buffer that prepares the proteins for SDS-PAGE. Following gel electrophoresis, ^{35}S-labeled proteins in the immunoprecipitate are visualized as described above.

Both developmentally (24) and in the adult rat brain (10, 25), examples can be found in which the accumulation of the mRNA for a glucocorticoid-sensitive protein precedes by several days the synthesis of the protein and/or increases in its activity, implying posttranslational regulation. If the glucocorticoid-sensitive protein is an enzyme, the contribution of posttranslational regulation to the expression of the protein can be determined by direct measurement of its activity following manipulation of glucocorticoid levels and comparison of the time course of enzyme activity to the synthesis of the protein as measured by ^{35}S labeling and/or antibody-based analyses.

Experimental Example: Glycerol-3-phosphate Dehydrogenase

A review of the characterization of the M_r 35,000 hippocampal protein provides a detailed example of how protein-based methodology is applied to the identification of glucocorticoid-determined proteins. In conjunction with analysis of the corresponding mRNA species, levels at which the synthesis of the protein are regulated, from availability of mRNA to posttranslational activation, can be identified. That synthesis of this protein reflects availability of mRNA is indicated by the comparable increases in its translation that are observed either when intact hippocampal slices are incubated with [^{35}S]methionine or when mRNA is isolated from the hippocampus and translated *in vitro* following stress or the administration of exogenous steroid to rats (9, 21, 22). However, it should be noted that a different combination of glucocorticoid-sensitive proteins is observed when comparing results obtained with the two experimental models. Although enhanced synthesis of the M_r 35,000 protein is common to both, alterations in the synthesis of several additional proteins are unique to each situation. In slices incubated with [^{35}S]methionine, increased synthesis of only two proteins is apparent, the M_r 35,000 protein with an isoelectric point of 6.6 and an M_r 46,000 protein with an isoelectric point of 6.2. The latter effect is not observed following translation of hippocampal mRNA *in vitro*. Instead, the synthesis of an M_r

50,000 protein (IEP 5.5) is decreased and that of two additional low molecular weight proteins [M_r 33,000 (IEP 6.5) and M_r 20,000 (IEP 6.9)] is increased in response to elevated glucocorticoid levels in the rats from which the mRNA was isolated (21, 22). Such variation may reflect differences in the animal treatment protocols used in each laboratory, but may also indicate the nature of translational control in each system. The *in vitro* translation system may be more efficient in allowing the translation of minor species of mRNA. Alternately, the mechanisms of translational regulation in intact slices may be more complex and reflective of the *in vivo* condition. Thus, there may be advantages in the use of either method.

The properties of the M_r 35,000 protein (sensitivity to type II receptor activation, molecular weight, isoelectric point, chromatographic behavior) suggested that it was glycerol-3-phosphate dehydrogenase (EC 1.1.1.8; GPDH), an oligodendrocyte marker protein that possesses the same molecular characteristics and is steroid inducible (26). This identity was further supported when RNA hybridization with a cRNA probe to GPDH demonstrated that GPDH mRNA levels were responsive to exogenous application of corticosterone and stress, under conditions that corresponded to those that increased synthesis of the M_r 35,000 protein (25). These results imply that the M_r 35,000 protein is GPDH and is translated in proportion to the amount of mRNA that is available. Although the physiological significance of the induction of a glial protein related to intermediary metabolism to stress adaptation remains unclear, increased synthesis of the M_r 35,000 protein is a reliable biological marker of the stress response in the hippocampus.

However, analysis of hippocampal GPDH activity in relation to alteration of glucocorticoid levels demonstrates a lag between translation of the M_r 35,000 protein and enzyme activity. Consistent with previous studies (27), we have found that GPDH activity falls in the hippocampal cytosolic fraction following adrenalectomy. Two to 3 days of corticosterone replacement treatment is required to restore enzyme activity to normal levels (10). Yet, when protein synthesis is examined in hippocampal slices following such treatments, increased synthesis of the M_r 35,000 protein is seen only for a few hours after steroid administration. Possible interpretations of this "uncoupling" between synthesis and activity include the need for an activation step after translation that requires several days, or the existence of several isozymes of GPDH with varying time courses of response. Moreover, administration of exogenous corticosterone to intact rats increases synthesis of the M_r 35,000 protein with no increase in cytosolic GPDH activity. The functional significance of a transient burst of synthesis of the M_r 35,000 protein in intact animals, in which there exists a relatively stable pool of preexisting GPDH, is unknown. Nevertheless, these studies indicate the

TABLE II   Characterization of Glucocorticoid-Determined Proteins

Level of regulation	Method
Genomic transcription	Nuclear transcription (runoff) assay
	Gene construct analysis in transfected cells
Availability of mRNA	RNA hybridization
	In situ hybridization
Translation of mRNA	*In vitro* translation
	Protein synthesis in slices
Protein characterization	1-D and 2-D gel electrophoresis
	Protein purification
	Immunoblotting/immunoprecipitation
Posttranslational activation	Enzyme assays

value of combined analysis of protein synthesis and activity in the definition of such relationships.

Concluding Remarks

Both the advantages and disadvantages of this experimental approach are related to the use of intact hippocampal slices. The major drawback to the use of hippocampal slices is the lack of definition as to the cell type involved. Both neurons and glia have receptors that will allow them to respond to glucocorticoid levels and labeled proteins on gels do not identify themselves as to the cell of origin. However, similar experiments done with cell lines that possess glucocorticoid receptors can be used in comparison with the slice experiments to provide information on this point. Studies on the induction and characterization of GPDH in C6 glioma cells (26), for example, are relevant to the synthesis of the M_r 35,000 protein and to GPDH assays performed with subcellular fractions from the rat brain.

The advantages are that it is a simple, and inexpensive, way to examine protein synthesis in a system that preserves cellular and synaptic structure. The factors that regulate endogenous protein synthesis are intact and the resultant patterns of protein translation and activation are likely to correspond to those that occur *in vivo*. Because the technique involves *in vivo* treatments of rats, changes in protein synthesis reflect the physiological state of the animal. Although this approach does not address the levels of regulation that precede the translation step, they can be combined with more direct measurement of mRNA levels (RNA hybridization and *in situ* hybridization) and transcription (runoff assays and analysis of gene constructs in transfected cells) to provide a more complete description of the effects of glucocorticoids at several molecular levels. Table II provides a concise summary of the combination of methods that can be used to characterize fully the glucocorticoid-determined proteins. Used in combination, these methods provide information on the identity of such proteins and distinguish among the levels at which the synthesis and activity of the protein may be regulated, from alteration in gene transcription to posttranslational activation.

Acknowledgments

The author gratefully acknowledges the experimental work of Drs. L. K. Schlatter and S.-M. Ting, which has laid the experimental basis for the methodological approaches described in this chapter. Research in the author's laboratory has been supported by grants from the American Federation for Aging Research, the VanNess–Thompson Foundation, and the National Institutes of Health (Grant NS 17118).

References

1. M. Joëls and E. R. deKloet, *Proc. Natl. Acad. Sci. U.S.A.* **87,** 4495 (1990).
2. R. Sapolsky, L. C. Krey, and B. S. McEwen, *J. Neurosci.* **5,** 1222 (1985).
3. B. S. McEwen, *Trends Pharmacol. Sci.* **12,** 141 (1991).
4. H. Karst and M. Joëls, *Neurosci. Lett.* **130,** 27 (1991).
5. M. I. Diamond, J. N. Miner, S. K. Yoshinaga, and K. R. Yamamoto, *Science* **249,** 1266 (1990).
6. S. Halpain, T. Spanaier, and B. S. McEwen, *Brain Res. Bull.* **16,** 167 (1986).
7. D. S. Kerr, L. W. Campbell, O. Thibault, and P. W. Landfield, *Proc. Natl. Acad. Sci. U.S.A.* **89,** 8527 (1992).
8. L. K. Schlatter and L. A. Dokas, *Neurosci. Res. Commun.* **1,** 71 (1987).
9. L. K. Schlatter and L. A. Dokas, *J. Neurosci.* **9,** 1134 (1989).
10. L. K. Schlatter, S.-M. Ting, and L. A. Dokas, *Brain Res.* **522,** 215 (1990).
11. E. R. deKloet and J. M. H. M. Reul, *Psychoneuroendocrinology* **12,** 83 (1987).
12. T. Hermann, K. Schramm, and R. Ghraf, *J. Steroid Biochem.* **26,** 417 (1987).
13. J. C. Butte, R. Kakihana, and E. P. Noble, *J. Endocrinol.* **68,** 235 (1968).
14. W. H. Rotsztejn, M. Normand, J. LaLonde, and C. Fortier, *Endocrinology (Baltimore)* **97,** 223 (1975).
15. P. P. Cohen, *in* Manometric Techniques'' (W. W. Umbreit, R. H. Burris, and J. F. Stauffer, eds.), p. 118. Burgess Publishing, Minneapolis, MN 1949.
16. W. D. Lust, T. S. Whittingham, and J. V. Passonneau, *Soc. Neurosci. Abstr.* **8,** 1000 (1982).
17. R. H. Evans, *Science* **240,** 889 (1988).
18. L. A. Dokas, S.-M. Ting, and L. A. Meserve, *Endocr. Soc. Abstr.* No. 657 (1992).

19. R. A. Colbert and D. A. Young, *J. Biol. Chem.* **262,** 9939 (1987).
20. W. V. Welch, *J. Biol. Chem.* **260,** 3058 (1985).
21. N. R. Nichols, S. P. Larner, J. N. Masters, P. C. May, S. L. Millar, and C. E. Finch, *Mol. Endocrinol.* **2,** 284 (1988).
22. N. R. Nichols, J. N. Masters, P. C. May, J. deVellis, and C. E. Finch, *Neuroendocrinology* **49,** 40 (1989).
23. Y.-F. Han, W. Wang, K. K. Schlender, M. Ganjeizadeh, and L. A. Dokas, *J. Neurochem.* **59,** 364 (1992).
24. L. P. Kozak and P. L. Ratner, *J. Biol. Chem.* **255,** 7589 (1980).
25. N. R. Nichols, J. N. Masters, and C. E. Finch, *Brain Res. Bull.* **24,** 659 (1990).
26. J. DeVellis and D. Inglish, *Progr. Brain Res.* **40,** 321 (1973).
27. J. DeVellis and D. Inglish, *J. Neurochem.* **15,** 1061 (1968).

[21] Use of Antisense Oligodeoxynucleotides to Block Gene Expression in Central Nervous System

Margaret M. McCarthy

Introduction

One of the principle challenges in understanding the molecular mechanisms controlling neurological processes is to be able to move beyond the level of description or correlation and arrive at the point of attributing causal relationships between gene activation and cellular responses. The ability to block the translation of individual genes into protein by exploiting the specificity of the genetic code could prove one of the most powerful tools for obtaining this objective. The fact that use of antisense oligonucleotides requires no special equipment or expertise in molecular biology, and that the necessary reagents are commercially available, make this approach all the more attractive.

The use of antisense oligonucleotides *in vivo* has been increasing dramatically and has advanced to the point of phase II trials in humans. However, when employed as a therapeutic agent against cancer or viral infection, a forgiving attitude for nonspecific or "side effects" would be expected. This level of uncertainty may not be acceptable to the nonclinical investigator interested in teasing apart the details of the contribution of a particular gene to complex processes such as development or behavior. Therefore, considerable caution must be used in the exploitation of antisense *in vivo*, including concerns regarding the lack of specificity of antisense under some conditions and the increasing evidence that this approach may not be applicable to all gene products. When beginning an investigation involving the use of antisense deoxynucleotides, the researcher should begin by asking a series of questions.

Nature of the Targeted Protein

Is it encoded by a single gene or does it consist of subunits composing a heterodimer? For example, many rate-limiting enzymes, such as a tryptophan hydroxylase (tryptophan monooxygenase) or glutamate decarboxylase

Methods in Neurosciences, Volume 22

(GAD), are single gene products and have been found amenable to blockage by antisense oligonucleotides (1, 2). In contrast, many neurotransmitter receptors consist of heterodimers with interchangeable subunits. The $GABA_A$ receptor consists of at least 4 separate subunits, with some subunits having variants specified by up to 15 different genes. Presumably many of these subunits are interchangeable, making blockage of the activity of the $GABA_A$ receptor in a meaningful way by antisense oligonucleotides for one subunit unlikely to yield clearly interpretable results.

Is the gene encoding the protein from a multigene family and likely to have homology to other genes in the family? Again, this would be true in the case of the $GABA_A$ receptor, where many of the subunit variants have a high degree of homology. This may also be a problem in attempting to block specifically the translation of steroid receptors, which belong to a supergene family and are characterized by a high degree of overall structural unity (3). Successful use of antisense oligonucleotides against both the estrogen (4) and progesterone (5, 6) receptors have been reported, but the possibility of unintended effects on other steroid receptors was not investigated.

What is the half-life of the protein? Clearly a protein with a half-life measured in days, such as a structural protein, is going to be more difficult to decrease measurably with antisense. Analysis of a kinetic model indicates that frequently translated RNAs producing stable proteins are the most attractive antisense targets (7), provided the half-life of the protein is not exceedingly long. Protein half-life can also be an indicator of specificity as demonstrated by the results of behavioral experiments utilizing antisense to the two separate transcripts for GAD. The half-life of one protein is estimated to be less than 4 hr and antisense to this transcript was effective within 24 hr whereas the second form of GAD has a half-life on the order of 20 to 30 hr and antisense oligonucleotides specific to this form of GAD were not effective until 48 hr postinfusion (2).

Nature of Target mRNA

Approximately how many copies of mRNA are there per cell in a given physiological state? Sometimes this information is available in the literature. For example, the level of vasopressin and oxytocin mRNA in particular brain regions has been found to be consistently high ($>10,000$/cell) and water deprivation will induce a further increase (8). A contrasting situation exists for tryptophan hydroxylase, the rate-limiting enzyme in the serotonin synthetic pathway, in which there are estimated to be approximately 11 mRNA copies per cell in the dorsal raphe nuclei but levels were close to 1000 per cell in the pineal gland (9). When the mRNA of interest is also being examined by *in situ* hybridization using a probe of known specific activity, one can count the number of grains in an area of 1 μm^2 to obtain a rough estimate of the approximate number of copies per cell. If the mRNA copy number per cell is high it may prove difficult to block effectively with antisense oligo.

What is the half-life of the mRNA? This information is often not available. However, if there is reason to believe that the half-life of the mRNA is exceedingly long, as appears to be the case for some neuropeptides such as oxytocin and vasopressin, then this will further complicate attempts to block reliably the synthesis of the peptide. In attempts to block the translation of vasopressin mRNA with daily infusions of antisense oligonucleotides, there was no consistent effect on peptide levels but increasing toxicity in one study (10). Until technology is further developed, highly abundant mRNAs encoding stable proteins may not be accessible to manipulation by antisense.

How large is the population of cells translating the mRNA? Being able to target adequately a sufficient number of the critical synthesizing cells in order to alter physiological responses may be a limiting factor. Direct infusions of labeled oligonucleotide into brain tissue have revealed a restricted area effectively exposed to antisense (11). Intracerebroventricular infusions of antisense, although highly effective, have not been quantified in terms of area of brain directly effected. However, when blocking a subunit of the *N*-methyl-D-aspartate (NMDA) receptor to reduce focal ischemic infarctions, Wahlstedt *et al.* (12) observed in antisense-infused animals a consistent reduction in the effected area throughout its 8 to 10-mm rostral–caudal extent suggesting substantial spread of the infused oligo.

Likely Mechanism of Action of Antisense in the System

Antisense oligodeoxynucleotides that gain access to the cellular cytoplasm are believed to block translation by one of two mechanisms. The antisense oligonucleotides may hybridize to the mRNA and as a result of steric hindrance, or so-called hybridization arrest, prevent the formation of the ribosomal complex and hence translation. Alternatively, the antisense oligonucleotides may hybridize to any region of the mRNA and form a substrate for the nucleases RNase H. In the brain there are two forms of RNase H, a magnesium-dependent and a manganese-dependent form (13), both of which specifically degrade the RNA portion of an RNA : DNA hybrid. The contribution of RNase H activity can greatly enhance the efficacy of an antisense oligonucleotide by degrading the targeted RNA and leaving the DNA oligonucleotide intact to hybridize again. However, there has been some suggestion that RNase H is not a major contributing mechanism to antisense action in the central nervous system (see below).

Designing Antisense Oligonucleotide

Types of Antisense Oligodeoxynucleotides

Antisense DNA is the most commonly used form of antisense both because of its convenience and relatively low price. Antisense DNAs are fairly stable compared to RNA and also can be purchased with additional modifications designed to increase cellular uptake and decrease degradation by endogenous nucleases. The specifics of these modifications have been reviewed in detail (14). Two of the most common modifications consist of replacing an oxygen group on the phosphate–diester backbone with either a methyl group (methyl phosphonate oligonucleotides) or a sulfur group (phosphorothioate oligonucleotides, often referred to as S-oligonucleotides). Experiments designed to address issues of stability of antisense in cells and sera indicated that unmodified oligonucleotides are completely degraded within 30 min in human serum and both methyl phosphonate and phosphorothioate are somewhat degraded in human serum and markedly degraded in 10% (v/v) calf serum within 1 hr of incubation. If various forms of oligonucleotides are exposed to cellular extracts, unmodified DNA is degraded by cytoplasmic and nuclear extract within 1 hr whereas the methyl phosphonates do not appear to be affected. The phosphorothioate oligonucleotides are somewhat degraded by nuclear extract (15, 16). However, it should be borne in mind that these studies are conducted on cellular extracts obtained from cells maintained in culture. By their nature, cells in culture are frequently in or close to a stage of cell division and therefore would be expected to have a high level of nuclease activity, including RNase H. This may not be the case for cells in a particular tissue *in vivo*, as appears to be the case for brain, in which no nucleases are found in the cerebrospinal fluid (CSF) and the activity of RNase H in brain cells appears to be relatively low. For instance, Wahlstedt *et al.* (12) observed a pronounced and significant reduction in the level of binding to the NMDA receptor after intracerebroventricular (icv) antisense oligonucleotide administration to a subunit of the receptor. However, when mRNA for the subunit was quantified, there was no decrease as a result of antisense treatment. This effect would be expected if RNase H was not a contributing mechanism to antisense action in the brain. In other studies, methyl phosphonate antisense oligonucleotide directed against GAD modulated reproductive behavior in a predicted manner (2), yet these oligonucleotides are not substrates for RNase H (17). In contrast, some investigators report a marked (up to 80%) reduction in mRNA levels for the angiotensin I receptor after antisense administration into the brain (R. Sakai, personal communication). Nonetheless, the role of RNase H activity in antisense effects in the brain remains unresolved and may be a contributing factor under some circumstances but not others.

Of particular concern to the neuroscientist may be the apparent toxicity of phosphorothioate-modified oligonucleotides, which have been found to have adverse effects on neurons in culture (12) and when infused directly into brain tissue (2). Many of the problems with modifications of DNA oligonucleotides, in terms of sensitivity to RNase H and toxicity, are being addressed by the creation of copolymers or mixed antisense molecules that combine the use of phosphodiester, methyl phosphonate, and phosphorothioate amidites into one oligonucleotide. In this way the best of each form can be exploited (18, 19). A conservative approach both economically and experimentally is to begin with an unmodified oligonucleotide and explore the advantages of specific modifications as necessary.

There are other forms of antisense oligonucleotides, including antisense RNA and ribozymes. The use of antisense RNAs is relatively uncommon, despite the fact that the hybridization kinetics favor RNA : RNA hybrids over RNA : DNA. However, RNA is more difficult to synthesize and much less stable than DNA, so that it is difficult to maintain its integrity both before and after administration. Continued development of modifications to stabilize and protect RNA antisense may make this a useful tool in the future (20). The discovery of naturally occurring antisense catalytic RNAs that specifically base pair with and cleave target RNA sequences, called ribozymes, has led to the investigation of these molecules as potential therapeutic agents although they have not yet been convincingly employed *in vivo* (21). A detailed discussion of these types of antisense and additional techniques for blocking gene expression is available (22).

Size of Antisense Oligonucleotide

The parameters to be considered in constructing an antisense oligonucleotide for use in the central nervous system are usually similar to those for any other system (23). In general, the oligonucleotide should be no shorter than 13 and no longer than 20 nucleotides. This distinction is based on three major concerns. First, the sequence must be long enough to afford sufficient specificity so that nonrandom associations are minimized. Second, the longer the oligonucleotide, the more likely that it will fold over on itself and sequester part of the molecule, preventing hybridization to the targeted mRNA. Finally, the sequence must be short enough that mismatches in sequence will not be tolerated; in other words, hybridization to nontarget mRNAs will not occur. It has been statistically determined that the shortest sequence that is likely to be unique within the mRNA pool is 13 bases and it has been found that a single base mismatch in a 12-mer will prevent duplex formation, whereas it will only slightly weaken the duplex formed by a 30-mer. However, these encouraging statistics are not supported by analysis of other systems. Many investigators have observed that the dose of antisense oligomer required to destroy a target RNA is frequently toxic and it was suggested and subsequently investigated whether this was the result of destruction of other RNA species as a result of incomplete hybridization (24). Using the *Xenopus* oocyte as a model, these authors found that oligonucleotides do not require perfect complementarity in order to cleave RNA via an RNase H mechanism and that a single mismatch within a 13-mer did not prevent antisense-induced degradation. On the basis of their observations, these authors ruled out the potential for a "therapeutic dose" that would allow only a specific effect. However, an important caveat to this conclusion is that it is based on experiments performed on *Xenopus* oocytes, which are grown at a temperature some 15°C lower than the body temperature of mammals, thus greatly affecting hybridization kinetics (24).

Target on mRNA

Knowledge of whether RNase H activity or hybridization arrest is likely to be the principal mechanism of antisense action is important in guiding the region of the mRNA against which to target the antisense oligonucleotide, and just because a study is being performed in the brain does not mean *a priori* that RNase H is not significant, because the level of enzyme activity has been reported to be elevated in the cerebellum of neonatal rats compared to adults (13). A relatively easy and inexpensive protocol for assaying RNase H activity in tissue is available (25).

When RNase H activity is a contributing factor to antisense action, the oligonucleotide can be made complementary to any portion of the mRNA and be effective. In particular, the oligonucleotide can be directed against the coding region or the 3' noncoding region and remain effective owing to its ability to destabilize the mRNA. When targeting the antisense to the coding region, it is useful to know the secondary structure of the mRNA, as it has been demonstrated that antisense oligonucleotides bind readily to the loop portion of a hairpin structure but have a 100-fold less affinity for the stem portion of this structure (26).

When RNase H activity does not appear to be a major contributing factor to antisense efficacy, then the oligonucleotide must be directed against the 5′ portion of the mRNA. This can include the 5′ cap site, the 5′ noncoding region, or the region encompassing or just downstream of the translation start codon. A useful guiding principle is, when in doubt, go upstream. This is because as a rule, eukaryotic ribosomes search out AUG initiator codons by binding to the cap site and then migrating to the closest (5′-most proximal) AUG codon to begin translation. Once the ribosome complex has been established it appears to be difficult to disrupt mechanically by hybridization to a 15 to 20-mer (27). Further evidence supporting this principle is based on the observation that of several antisense oligonucleotides directed against mRNA for the progesterone receptor, only the construct actually spanning the translation start codon was effective (6). However, others have reported success in the brain using an antisense oligonucleotide just downstream of the start codon (12, 28).

The concern with using antisense oligonucleotides directed to the 5′ portion of the mRNA is the degree of homology to other mRNAs. Analysis of large numbers of vertebrate mRNAs indicates a consensus sequence of 10–12 bases surrounding and including the AUG start codon, referred to as the Kozak consensus sequence (29). Therefore, antisense oligonucleotides designed for this region must be checked carefully against existing sequence databases to avoid significant homology with other mRNAs, and even if significant homology is not found the possibility will remain of having homology to mRNAs not yet sequenced.

The target site on the mRNA can be effectively doubled by using two antisense oligonucleotides in tandem. The closer the two antisense oligonucleotides are on the mRNA, the more effective they are, with maximum effectiveness being obtained by two contiguous oligonucleotides (30).

Last, when technically feasible, it is worth the effort to explore the effectiveness of several antisense constructs against cells in culture before beginning the laborious procedure of *in vivo* administration (6, 12, 28).

Delivery of Antisense Oligonucleotides

Infusions

One of the major challenges in the use of antisense oligonucleotides *in vivo* is to deliver the agent to the target organ with sufficient duration and concentration to block gene expression effectively. One of the more advantageous aspects of antisense infusions in the brain is that a relatively discrete brain area can be targeted or the antisense can be infused directly into the CSF and still be retained around the tissue of interest without dispersion into the general circulation. A discussion of cellular uptake of oligonucleotides in the nervous system is available (11).

When administering antisense oligonucleotides directly into brain tissue, they should be purified of all organic salts and unincorporated bases. If this step has not already been done by the supplier it can be easily accomplished by reversed-phase chromatography using a Waters C_{18} Sep-Pak column, available from Millipore Corporation (Milford, MA). See (31) for a technical description of this technique.

Some of the most spectacular successes to date involving the *in vivo* use of antisense have come from behavioral studies on rats after icv administration of antisense oligonucleotides. Owing to the lack of a specific receptor antagonist, Wahlstedt and co-workers (28) utilized antisense to the neuropeptide Y-Y1 receptor to reduce the level of receptor binding in the brain and to modulate anxiety. Animals bearing chronically implanted cannulas were infused with 50 μg of unmodified antisense or sense oligonucleotide twice daily for 2 days and behavior tested 12 hr after the last infusion. This is a large dose of antisense and an intensive infusion schedule; nonetheless, binding to the receptor was reduced by over 50% and measures of anxiety were reduced by more than 60%. Using the same approach, these investigators also blocked the synthesis of a major component of the NMDA receptor and thereby protected cortical neurons from excitotoxicity (12). These authors report no effects of the antisense treatment on control proteins not expected to change and no effects of scrambled or sense oligonucleotides on the targeted protein.

Direct infusion of antisense oligonucleotides to particular brain regions offers advantages in terms of site specificity and can also be used successfully. For instance, blocking the synthesis of GAD to modulate female reproductive behavior required site specificity because it had been previously determined that GABAergic neurotransmission exerted opposing effects on this behavior in different brain regions (2). Similarly, a desire to address the role played by vasopressive intestinal protein (VIP) in circadian rhythms, specifically in the suprachiasmatic nucleus, required infusion of the antisense directly into this target (32, 33). Alternatively, a target may not be readily accessible to the CSF, as was the case for antisense against c-*fos* infused into the striatum (34).

Another example of successful use of intracerebral administration of antisense oligonucleotides involved the intrahypothalamic infusion of unmodified antisense against estrogen receptor mRNA into 3-day-old rat pups. When behaviorally assayed in adulthood, it was found that this single antisense treatment successfully protected against many of the masculinizing effects of neonatal exposure to testosterone (which is aromatized to estrogen before exerting its effect) in females and also modulated some aspects of brain morphology in normal females (4). A potential contributing factor to the effectiveness of antisense in this paradigm is that neonatal brains have been found to have a higher RNase H activity level than adult brains (13).

Dose and Vehicle

Choosing an appropriate dose requires consideration of the intended frequency of infusion, the estimated abundance of the targeted mRNA, and the possibility of nonspecific effects. In cell culture studies, as concentrations become increasingly high ($>2.5 \ \mu M$), so does the probability of nonspecific reaction with cellular proteins (35).

In initial studies using unmodified oligonucleotides infused directly into the brain tissue, we generally used doses of 500–1000 ng (total when injected bilaterally) with noticeable effect. However, the same dose of phosphorothioate oligonucleotide was twice as effective in modulating a specific behavioral response but may also have increased nonspecific effects to the point of inducing toxicity (2). Therefore as a general rule of thumb it may be appropriate to use 10-fold less of a phosphorothioate oligonucleotide than an unmodified oligonucleotide. In other studies, reproductive behavior has been modified with as little as a single infusion of 400 ng of antisense oligonucleotide to the progesterone receptor (6) and there was no evidence of nonspecific effects in rats after intraventricular infusion of 50 μg of unmodified oligonucleotide twice a day for 2 days (12, 28). In general the dose should be empirically determined for a specific system but with the guiding principal that more is not necessarily better.

Tremendous effort is being put into development of vehicles for or modifications of the oligonucleotides themselves to increase the cellular uptake. Because oligonucleotides are hydrophilic, vehicles and modifications are directed toward reducing the polarity of the molecules and encouraging uptake or diffusion across lipid membranes. This includes the use of polylysine, cholesterol, cationic lipids, liposomes, agents designed as vehicles for transfection, and biotinylated antisense. Most of these have proved effective at getting more oligonucleotide into the cytoplasm and even into the nucleus. However, it is becoming apparent that nonspecific effects of oligonucleotide infusions in the brain are a serious confound and can be explained only as the result of too much oligonucleotide entering the cell. Therefore, at least with regard to the use of antisense in the brain, it does not seem prudent to expend considerable effort on getting more oligonucleotide into the cell until problems with nonspecificity have been eliminated.

In initial experiments involving antisense delivery to the brain we utilized antisense oligonucleotide in a crystalline form by tamping an oligonucleotide : cholesterol mix into the end of an injection cannula that was then placed into a stereotaxically implanted guide cannula aimed at the desired brain structure. The objective was to obtain anatomical specificity combined with slow sustained release of antisense from the cannula tip. However, apparent differences in the rate of oligonucleotide release both within and between animals introduced an unacceptable level of variability into the experimental design and therefore we switched to infusing the oligonucleotide. Extrapolating from neuroendocrine studies in which steroids are dissolved in sesame oil and injected subcutaneously to give a slow continuous release, we used oil as the vehicle for antisense infusion into the brain. However, in lieu of the hydrophilic nature of DNA, the oligonucleotide : oil mixture was heated and sonicated prior to infusion so that the oligonucleotide would be sufficiently suspended. This method of administration has proved effective in several experiments and may have indirectly contributed to cellular uptake of the oligonucleotide (2, 5, 32, 33). Subsequent studies using saline as the vehicle for oligonucleotide infusion into the brain have proved equally effective (12, 28, 36) and if infusions are given frequently (i.e., daily) the oil can induce substantial neuronal damage, making saline the desired vehicle.

Time Course and Frequency of Administration

Additional factors to consider with regard to frequency of infusion are the time course of gene expression of the targeted mRNA. For example, if gene expression is hormonally induced, infusion of antisense for a period 3–6 hr prior to hormone exposure and again 12 to 24 hr afterward would be predicted to bracket the period of increased transcription sufficiently, such that antisense oligonucleotide would be present in abundance at the time the mature mRNA was being produced (36). In the brain we have observed rhodamine-tagged oligonucleotide still present in the cell after 24 hr, but not visible after 48 hr. Presumably the rhodamine would not be retained in the cell on degradation of the oligonucleotide. In a separate experiment, the majority of infused oligonucleotide was recovered intact from brain 5 hr postinfusion; later time points were not measured (11). Thus, when repeated infusions are necessary, a 24-hr interval of administration has proved effective and not overly cumbersome to either the investigator or the animal.

The time course can also be used as an additional control because antisense oligonucleotide would not be expected to have an effect within a short time period. Any action of antisense that was manifested in less than 12 hr should be regarded suspiciously and would require the demonstration that the half-lives of both the protein and the mRNA were sufficiently short to allow for an antisense effect by the presumed mechanisms (see below).

Controls and Nonspecific Effects

Part of what is appealing about the use of antisense oligonucleotides to block protein synthesis is what would seem to be an unprecedented level of specificity based on the genetic code. By exploiting sequence specificity, it should be theoretically possible to block one and only one protein encoded by a particular mRNA. However, this has repeatedly proved not to be the case and the careful use of controls is paramount to any study utilizing antisense oligonucleotides.

Oligonucleotide Controls

The injection of control oligonucleotide that does not encode the gene of interest is essential. This control can be one of three varieties: (a) the sense strand sequence from the same portion of the mRNA, (b) a scrambled oligonucleotide that consists of the same nucleotide composition as the antisense oligonucleotide but in a random order, or (c) a nonsense oligonucleotide that consists of a different nucleotide composition but still lacks complementarity with known mRNAs. Any of these control oligonucleotides is acceptable but the sense strand can create problems with triple-helix formation with DNA if a large quantity gains access to the nucleus, and the G-C content of any nonsense or scrambled oligonucleotides should be similar to that of the antisense oligonucleotides. When using antisense oligonucleotides that span the translation start codon, we have routinely incorporated an antisense start codon and surrounding bases homologous to the Kozak consensus sequence into the scrambled nucleotide control. In this way the remote possibility of a high degree of hybridization to all start codons can be obviated.

Unfortunately, it is not uncommon for any one of these controls to show some level of effect on the dependent variable that is similar to that of the antisense oligonucleotide. The mechanism of these nonspecific effects is poorly understood but it appears to involve a general reduction in protein synthesis, perhaps by disrupting the action of RNase polymerase or by disruption of normal ribosomal functioning (14). Furthermore, using *in vitro* transcription to assess the effectiveness of synthetic S-oligonucleotides, it was found that oligonucleotides longer than 15 bases could inhibit DNA polymerase and RNase H1 in a sequence-independent manner at nanomolar concentrations (37). The inhibition is related to the number of phosphorothioate linkages at the oligomer backbone; if reduced below 15 it seems to lose its inhibitory effect. The effect of S-oligonucleotides on RNase H was biphasic; at low concentrations the S-oligonucleotides hybridize to the target RNA and work as cosubstrates for RNase H, but at high concentrations (when the oligonucleotide is in excess of the substrate) they appear to act as competitive inhibitors of RNase H and actually protect the RNA sequence from degradation (37).

Because of the high probability of nonspecific effects from control oligonucleotide treatment, the incorporation of a vehicle-treated control group or an experimental design that involves a pretest-versus-posttest measure of a dependent variable before and after antisense treatment will help illustrate the degree of any nonspecific effects that are the result of oligonucleotide treatment.

Response Controls

Unfortunately, it is not sufficient merely to utilize treatment controls and then assume all observed effects are the result of a specific reduction of the targeted protein. Additional controls should include measurement of auxiliary proteins that share characteristics with the targeted protein but that should not be affected by the antisense infusion. Furthermore, there are of course a myriad of variables that can influence any one physiological response and this should always be considered in the design of antisense experiments. Despite the incorporation of all the appropriate controls into the experimental design, nonspecific or unexplained effects may still emerge.

Variability

Finally, a consistent and frustrating aspect of the author's experience with the *in vivo* use of antisense has been the high degree of variability of responsiveness. Dependent measures monitored in antisense-treated animals frequently exhibit a standard deviation twice as large as that in control oligonucleotide and vehicle-treated animals and it is not uncommon for the standard error to be close to 50% of the mean. This requires that sample sizes be relatively large to achieve significance. In behavioral studies the author rarely uses fewer than 10 animals and significance levels are often marginal. But more importantly, when examining the data it is clear that some animals show large responses to antisense treatment in the predicted direction, whereas others may exhibit no effect or in fact seem to be responding in the opposite manner from that predicted (or desired!). This individual level of variability may offer some insight into the mechanisms of action of antisense that are particular to its use at the organismal level. For instance, this author has observed a heterogeneous uptake of radiolabeled oligonucleotides by cells in the brain (11), suggesting that populations of cells may respond to antisense differentially. Understanding the sources of this variability may help guide development of antisense tools for the future but for now they remain a source of uncertainty and frustration for the researcher interested in exploiting the exciting new area of antisense technology.

One fact that is becoming increasingly clear is that our understanding of how synthetic oligonucleotides act in biological systems is far from complete. This precise mechanism of action of antisense in blocking protein translation is not known and may vary considerably between different cell types, tissues, and organisms. The potential for nucleotide sequence specific interactions with critical cellular proteins is becoming increasingly evident. Furthermore, the possibility of additional mechanisms not previously considered is hinted at by recent reports of very short latencies of specific antisense effects after infusion into the CNS (38, 39).

Concluding Remarks

When application of antisense therapy appears the optimal approach for investigating the physiological role of a particular gene product, the investigator should begin with a series of practical questions regarding the nature of the protein, its mRNA, and the tissue it is being expressed in. Answers to these questions will help to determine the type, size, and target of the antisense that appear most promising for positive results. In addition, the time course and method of administration, dose, and vehicle are all likely to depend on the specific system under investigation. Last, the investigator must attend with particular care to the appropriate controls, both the oligonucleotide and physiological response controls. Once all of these decisions have been made there is nothing left but to try it, realizing that the result may be positive, negative or, as happens often, without effect. In the latter case there may be no obvious explanation or recourse and it is this last category that most clearly illustrates how much remains unknown regarding the mechanism of action of antisense and the control of gene expression in general. A level of tolerance for uncertainty in these early phases of the use of antisense technology will be crucial to its maintenance as a viable experimental technique in the future.

References

1. M. M. McCarthy, C. MacDonald, and D. Goldman, unpublished data (1994).
2. M. M. McCarthy, D. B. Masters, K. Rimvall, S. Schwartz-Giblin, and D. W. Pfaff, *Brain Res.* **636,** 209–220 (1994).
3. W. Wahli and E. Martinez, *FASEB J.* **5,** 2243–2249 (1991).
4. M. M. McCarthy, E. Schlenker, and D. W. Pfaff, *Endocrinology* (*Baltimore*) **133,** 433–439 (1993).
5. S. Ogawa, U. E. Olazabal, I. S. Parhar, and D. W. Pfaff, *J. Neurosci.* **14,** 1766–1774 (1992).
6. G. Pollio, P. Xue, M. Zanisi, A. Nicolin, and A. Maggi, *Mol. Brain Res.* **19,** 135 (1993).
7. M. Ramanathan, R. MacGregor, and C. A. Hunt, *Antisense Res. Dev.* **3,** 3–18 (1993).
8. J. T. McCabe, M. Kawata, Y. Sano, D. W. Pfaff, and R. A. Desharnais, *Cell. Mol. Neurobiol.* **10,** 59–71 (1990).
9. K. S. Kim, T. C. Wessel, D. M. Stone, C. H. Carver, T. H. Joh, and D. H. Park, *Mol. Brain Res.* **9,** 277–283 (1991).
10. L. M. Flanagan, M. M. McCarthy, P. J. Brooks, D. W. Pfaff, and B. S. McEwen, *Soc. Neurosci. Abstr.* **18,** 205.7 (1992).
11. M. M. McCarthy, P. J. Brooks, J. Pfaus, H. E. Brown, L. M. Flanagan, S. Schwartz-Giblin, and D. W. Pfaff, *Neuroprotocols* **2,** 67–74 (1993).
12. C. Wahlstedt, E. Golanov, S. Yamamoto, F. Yee, H. Ericson, H. Yoo, C. E. Inturrisi, and D. J. Reis, *Nature* (*London*) **363,** 260–263 (1993).
13. Y. Sawai, J. Saito, and K. Tsukada, *Biochim. Biophys. Acta* **630,** 386–391 (1980).
14. E. Uhlman and A. Peyman, *Chem. Rev.* **90,** 544–584 (1990).
15. S. Agrawal, J. Goodchild, M. P. Civeira, A. H. Thornton, P. S. Sarin, and P. C. Zamecnik, *Proc. Natl. Acad. Sci. U.S.A.* **85,** 7079–7083 (1988).
16. S. Akhtar, R. Kole, and R. L. Juliano, *Life Sci.* **49,** 1793–1801 (1991).
17. Walder, R. Y. and Walder, J. A. *Proc. Natl. Acad. Sci. U.S.A.* **85,** 5011–5015 (1988).
18. M. K. Ghosh, K. Ghosh, and J. S. Cohen, *Anti-Cancer Drug Des.* **8,** 15–32 (1993).
19. Q. Zhao, S. Matson, C. J. Herrera, E. Fisher, H. Yu, and A. M. Krieg, *Antisense Res. Dev.* **3,** 53 (1993).
20. Y. Eguchi, T. Itoh, and J.-I. Tomizawa, *Annu. Rev. Biochem.* **60,** 631–652 (1991).
21. L. Chrisey, J. Rossi, and N. Sarver, *Antisense Res. Dev.* **1,** 57–63 (1991).
22. C. Helene and J.-J. Toulme, *Biochim. Biophys. Acta* **1049,** 99–125 (1990).
23. J. J. Toulme, *in* "Antisense RNA and DNA," pp. 175–194. Wiley-Liss, New York, 1992.
24. T. M. Woolf, D. A. Melton, and G. B. Jennings, *Proc. Natl. Acad. Sci. U.S.A.* **89,** 7305 (1992).
25. A. D. Papaphilis and E. F. Kamper, *Clin. Chem. Enzymol. Commun.* **2,** 159–166 (1990).
26. W. F. Lima, B. P. Monia, D. J. Ecker, and S. M. Freier, *Biochemistry* **31,** 12055–12061 (1992).

27. C. Boiziau, N. T. Thuong, and J.-J. Toulme, *Proc. Natl. Acad. Sci. U.S.A.* **89,** 768–772 (1992).
28. C. Wahlstedt, E. M. Pich, G. F. Koob, F. Yee, and M. Heiling, *Science* **259,** 528–531 (1993).
29. M. Kozak, *Nucleic Acids Res.* **15,** 8125–8148 (1987).
30. L. J. Maher, III and B. J. Dolnick, *Nucleic Acids Res.* **16,** 3341–3358 (1988).
31. B. Freie and S. H. Larsen, *BioTechniques* **10,** 420–422 (1991).
32. J. P. Harney, K. Scarbrough, K. Rosewell, and P. M. Wise, *Soc. Neurosci. Abstr.* **19,** 236.15 (1993).
33. K. Scarbrough and P. M. Wise, *Soc. Neurosci. Abstr.* **18,** 6.9 (1992).
34. B. J. Chiasson, M. L. Hooper, P. R. Murphy, and H. A. Robertson, *Eur. J. Pharmacol.* **227,** 451–453 (1992).
35. L. A. Yakubov, E. A. Deeva, V. F. Zarytova, E. M. Ivanova, A. S. Ryte, L. V. Yurchenkko, and V. V. Vlassov, *Proc. Natl. Acad. Sci. U.S.A.* **86,** 6454–6458 (1989).
36. M. M. McCarthy, S. P. Kleopolous, C. V. Mobbs, and D. W. Pfaff, *Neuroendocrinology* **59,** 432–440 (1993).
37. W.-Y. Gao, F.-S. Han, C. Storm, W. Egan, and Y.-C. Cheng, *Mol. Pharmacol.* **41,** 223–229 (1992).
38. D. Mitsushima, D. L. Hei, and E. Terasawa, *Proc. Natl. Acad. Sci. U.S.A.* **91,** 395–399 (1994).
39. I. Neuman, D. W. F. Porter, R. Landgraf, and Q. J. Pittman, *Amer. J. Physiol.,* in press (1994).

Section IV

Cellular Effects of Steroids

[22] Steroid Regulation of Neuronotrophic Activity: Primary Microcultures of Midbrain Raphe and Hippocampus

Efrain C. Azmitia and Xiao Ping Hou

Introduction

Steroids may function as neuronotrophic molecules on a select population of cells from the brain. The steroids may act on neurons or glial cells through either glucocorticoid receptors (GRs) or mineralocoid receptors (MRs). Glucocorticoids regulate the sprouting of serotonergic neurons during early development (1) and adult plasticity (2). Most of the progress made with neuronotrophic factors has utilized tissue culture methods. When brain cells are grown in culture precise control can be maintained over the type and concentration of all external chemicals. In this chapter we describe in detail a method for growing dissociated cells in culture, in order to answer several questions: (a) do steroids trophically interact directly with serotonergic neurons? (b) Do steroids have an indirect effect on serotonergic sprouting mediated by hippocampal cell receptors? (c) Is the trophic effect due to a soluble factor secreted by hippocampal neurons? (d) Is there a difference in the activation of GRs and MRs on any of the above variables?

The development of neurons in culture was first demonstrated in 1911 by Harrison (3). A piece of dorsal root ganglion was placed on clotted blood for observation of growing neurons in a controlled environment. The technique was used to study cell attachment, survival, growth cone movement, and neurite extension. This mainly morphological approach continues to be extremely useful for both anatomical and electrophysiological studies. Biochemical and pharmacological analyses use cell lines or nonneuronal mitotic cells because large numbers of cells from a homogeneous preparation are required.

We have developed a microculture system using primary dissociated cells to accommodate biochemical, pharmacological, and morphological studies (4). A preparation is made of dissociated primary cells from midbrain regions rich in neurons of a known neurotransmitter content. The dissociated cells are plated onto 96-well plates treated with an artificial, positively charged substratum and grown under standard tissue culture techniques. Our automated system can generate thousands of cultures as a means of rapid screening for neuronotrophic and neurotoxic effects of drugs, hormones, and brain extracts (3). The verification that a biochemical change reflexes an alteration in neuronal survival and maturation must be obtained by morphological criteria. To this end, the eight-well chamber slides are utilized in conjunction with immunocytochemistry and/or radioautography.

We present descriptions of brain stem raphe cells grown in a defined medium lacking steroids and compare this to medium containing corticosterone, dexamethasone, or aldosterone. The cultures are used to measure the high-affinity uptake of 5-hydroxytryptamine (5-HT) or the levels of 5-HT in the medium or in the cells. To test if the effects of the steroids are primarily on hippocampal cells, hippocampal cultures are exposed to steroids. Subsequently, serotonergic cells can be grown either in the presence of the hippocampal neurons or in the presence of the hippocampal conditioned medium. High-affinity uptake is used to monitor the growth of serotonergic neurons.

Direct Effect of Steroids on Serotonergic Neurons

To address this issue, raphe midbrain cells are cultured for 24 hr in the presence of fetal bovine serum and then changed to a defined medium containing no steroid. Dexamethasone and aldosterone are added once at concentrations between 10^{-6} and 10^{-11} M. The cultures are allowed to grow for 5 days and then the high-affinity uptake of [^{3}H] 5-HT is measured as an index of the maturation of serotonergic neurons (see Fig. 1).

Cell Preparation

Maximum-barrier (pathogen-free) timed pregnant rats are obtained from Taconic Farms (Germantown, NY) to arrive for use on gestation (day 14). The serotonergic neurons of the brain stem have completed their final mitosis by this time (5) and we can focus on survival and maturational events.

The animals are maintained in a pathogen-free environment if necessary before their use. The rats are anesthetized with CO_2, swabbed with 70% ethanol, and the fetuses removed by cesarian procedure using antiseptic procedures. The entire placenta is transferred to a 150-mm sterile petri dish containing ice-cold D-1 solution [0.8% (w/v) NaCl, 0.04% (w/v) KCl, 0.006% (w/v) $Na_2HPO_4 \cdot 12H_2O$, 0.003% (w/v) KH_2PO_4, 0.5% (w/v) glucose, 0.00012% (v/v) phenol red, 0.0125% (w/v) penicillin G, and 0.02% (v/v) streptomycin] (Sigma Chemical Co., St. Louis, MO). The fetuses are removed from the placenta in a laminar flow hood (model EG-4320, Edgegard Hood; Baker Co., Inc., Sanford, ME) and transferred to fresh D-1 solution. The brains

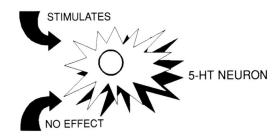

FIG. 1   Direct action of dexamethasone or aldosterone on the growth of serotonergic neurons grown in midbrain raphe dissociated culture. The design of the experiment is to add the steroid and permit time for the neurons to respond. The cells are grown for 5 days in the absence of serum or added steroid in the medium. The selected steroids are tested in a dose–response manner with four cultures assigned to each dose. The results presented in Fig. 2 show that dexamethasone is stimulatory, whereas aldosterone is without effect.

are removed from the skull with two straight-edged watchmaker forceps and placed in a 60-mm petri dish with fresh D-1 solution (5°C).

Two straight-edged No. 11 scalpels (Propper Manufacturing Co., Inc., Long Island City, NY) are used to make a coronal cut at the apex of the mesencephalic flexure and at the pontine flexure. The tegmentum is exposed by cutting dorsally from the cerebral aqueduct and separating the corpus quadrigeminal. Two sagittal cuts are made approximately 0.5–1.0 mm lateral to the midline. The brain pieces from all the fetuses are collected in fresh D-1 solution in a 35-mm petri dish and then transferred to a 5-ml conical tube in about 0.3 ml/10 fetuses. The strips are gently agitated by repeated trituration, using a Pasteur pipette with a fire-polished tip. First a large-bore pipette (>1 mm) and then a small bore pipette (>0.5 mm) is used. The use of trypsin is avoided because it results in a lower yield of viable serotonergic neurons.

The cell suspension is brought up to 1.5 ml/fetus with room temperature (19°C) complete neuronal medium (CNM), consisting of 82.5% (v/v) Eagle's minimum essential medium [MEM (Cat. No. M-0268; Sigma)], 1% (w/v) nonessential amino acids (Cat. No. 320-1140AG; GIBCO-Life Technologies, Inc., Grand Island, NY), 0.5% (w/v) glucose, and 5% (v/v) fetal bovine serum [FBS (Cat. No. F6761; Sigma)]. All solutions except FBS are prepared weekly from powder reagents and sterilized by filtration (Nalgene sterile disposable TC filter unit, 150-ml capacity, 0.2-μm pore size, Cat. No. 155-

0020; Fisher Scientific Co., Pittsburgh, PA). The MEM obtained from GIBCO (Long Island City, NY) contains significantly less calcium chloride than does the MEM from Sigma (St. Louis, MO). The FCS is aliquoted into sterile vials stored at $-70°C$ and thawed immediately before use. The final cell preparation in CNM is allowed to settle for 10 min to precipitate large tissue debris and the clear supernatant is used without centrifugation to avoid damage to the large neurons. The completeness of the dissociation is checked under a microscope before proceeding and the cell density is determined using a hemocytometer (Levy and Levy-Hausser corpuscle counting chamber; Fisher Scientific Co.). Use of a vital stain (trypan blue; Aldrich Chemical Co., Inc., Milwaukee, WI) is used periodically to establish the percentage of viable cells after dissociation (usually greater than 90%).

Preparation of Plates

The 96-well plates (Nunc, Roskilde Denmark; Becton Dickinson and Company, Lincoln Park, NJ) are used for biochemistry and low-power morphological studies, whereas 8-well and 16-well glass slides (Lab-Tek tissue culture chamber/slides; Miles Scientific, Miles Laboratories, Inc., Naperville, IL) are used for high-power morphometric analysis. The two are prepared similarly with a solution of poly(D-lysine) (15 $\mu g/ml$, M_r 70,000–150,000; Sigma) applied to cover the surface the night before (shorter times have been used successfully). The poly(D-lysine) solution is removed, and the wells rinsed once with MEM and filled with CNM before use.

The mesencephalic raphe cells are plated at an initial plating density (IPD) of approximately 1.0×10^6 cells/cm^2 by the addition of 300 μl of the cell suspension to the empty wells (CNM removed from wells immediately before use), using a 1-ml Rainin repipetter with sterile plastic tips (Rainin Co., Woburn, MA). Use of a smaller volume repipetter increases risk of contamination. The solution is briefly agitated before each application. All the trays and slides are filled as quickly as possible without a time break. The trays and slides are covered and kept at room temperature before further solutions are added. Cell attachment is around 90%. The plates are transferred to a CO_2 incubator (model 6300; Napco Incubator) maintained at 36°C with 95% humidity and 5% CO_2/95% air.

Steroid Application

After 24 hr the tissue culture medium is changed to a defined medium containing no serum or steroid. The composition of the medium is insulin (5 $\mu g/ml$), putrescine (10^{-3} M, 20 $\mu l/ml$), sodium selenite (10^{-6} M, 15 $\mu l/ml$), and transferrin (10^{-4} M, 50 μl/ml). In studies to compare the actions of dexamethasone and aldosterone the steroids are dissolved in 95% ethanol, diluted with MEM, and the stock solution filtered (Uniflow, 0.2-μm pore size; Schleicher & Schuell, Inc., Keene, NH). The steroid concentrations are made between 1 mM and 1 pM in serial dilutions in fresh MEM immediately before use. The drugs are applied to the cultures in quadruple from least to greatest concentration in 20-μl aliquots using a 200-μl Rainin repipetter. Controls are placed at the beginning and end, and occasionally in the middle, of our serial dilution experiments.

Uptake Analysis of Serotonergic Neurons

The measure of the high-affinity uptake of [^{3}H]5-HT provides one of the most reliable biochemical estimates of the surface area of a specific neurotransmitter cell (6–9). The uptake is a more reliable indicator of cell number and size than measures of transmitter or enzyme content in the cells or in the medium. In our microcultures of mesencephalic raphe, the amount of [^{3}H]5HT high-affinity uptake is linearly related to the number of plated 5-HT-immunoreactive neurons and to the days in culture.

The cultures are removed from the incubator and the medium is removed. The reaction solution (200 μl) contains 50 nM [^{3}H]5HT (19–27 Ci/mmol; DuPont-New England Nuclear, Boston, MA), and 10^{-5} M pargyline in MEM. The reaction is allowed to proceed for 30 min at 37°C. The concentration of [^{3}H]5HT used has been shown to be selectively retained by serotonergic neurons as demonstrated by radioautographic pharmacology (6, 10), and is five times lower than the concentration needed to detect uptake of 5-HT by astrocytes in culture (11). Nonspecific accumulation and retention of [^{3}H]5HT is calculated by incubating the cultures as described but in the presence of 10^{-5} M fluoxetine (gift of Eli Lilly Co., Indianapolis, IN), which is a recognized specific uptake blocker for serotonergic neurons (12).

After the 30-min incubation period, the medium is carefully and quickly removed and the cultures are washed twice with 0.1 M phosphate buffer (pH 7.4) in a saline solution (0.85%). The cells are air dried before 250 μl of 10% ethanol is added to each culture for 45 min at room temperature. From each well 160 μl is removed and placed in 3 ml of Liquiscint (National Diagnostics, Manville, NJ) and counted for at least 1 min in a Beckman liquid scintillation counter [model LS 1801 (Beckman Instruments, Inc., Somerset, NJ); counting efficiency is 58%]. The specific uptake of [^{3}H]5HT is calculated as the difference between the total and nonspecific accumulation. The specific activity is usually between 1 and 10% of the total uptake (4).

Statistical and Graphic Analysis

The use of microcultures provides a large number of individual wells. Given that the wells contain an equal number of cells from a homogeneous suspension of dissociated cells, then each well is a unit of a normal population and can be treated as such statistically. The counts per minute (cpm) obtained for each well from the scintillation counter are recorded directly into a floppy disk of an IBM PS/2, model 50 computer for descriptive and comparative statistics using a Systat (Systat, Inc., Evanston, IL) software program. The averages, standard deviations (SDs), and standard errors of the mean (SEMs) are obtained for each condition. Analysis of variance and a Tukey *post hoc* analysis are run for each plate separately. Individual t test comparisons for all groups are also generated. Values more than 3 SD from the mean with the inclusion of the value are dropped.

Graphic analysis of the data is produced, using the means and SEMs of the groups on a Sigma Plot (Jandel Scientific, Sausalito, CA) software program. Linear regression and higher order regressions with confidence values are recorded with a hard copy of the graph produced on a plotter (model 7475A; Hewlett-Packard).

Data Presentation

The results of this experiment to determine the effects of dexamethasone and aldosterone are shown in Fig. 2. The results show that dexamethasone has a direct stimulatory action on the development of serotonergic neurons grown in a midbrain raphe culture. A significant effect was apparent at a concentration of 10^{-10} M. In contrast, aldosterone showed an inhibitory response on the serotonergic neurons in the concentration range of 10^{-10} M to 10^{-6} M.

Steroid Action on Hippocampal Conditioned Medium

The hippocampus is known to concentrate circulating adrenal steroids (13). The serotonergic innervation of the hippocampus can also be influenced by circulating adrenal steroids (14). Furthermore, the culturing of fetal hippocampal cells is stimulatory to the development of fetal serotonergic neurons (15). Do glucocorticoids stimulate hippocampal neurons to secrete a soluble trophic factor into the medium? To address this question, fetal hippocampal cultures are grown in the presence of an antimitotic drug to suppress glial

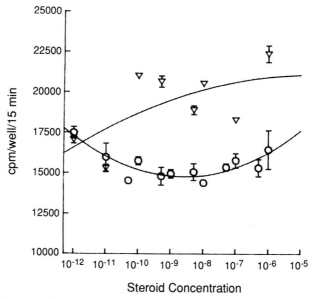

FIG. 2   Effects of dexamethasone (∇) and aldosterone ($\bigcirc$) on the development of the 5-HT reuptake mechanism, which is correlated to total neuronal surface area. It can be seen that dexamethasone is stimulatory at doses between 10^{-10} and 10^{-6} M. Aldosterone is inhibitory at doses between 5×10^{-11} and 10^{-6} M.

proliferation. After 24 hr, the culture medium is removed and replaced with serum/steroid-free medium. These cultures are grown for 2 days and then are treated with steroids (corticosterone, dexamethasone, or aldosterone) for 24 hr and the conditioned medium removed and tested on newly plated fetal raphe cultures (Fig. 3).

Hippocampal Neurons

The steroid effects on the hippocampal neurons and then indirectly on the mesencephalic raphe cells are tested by first growing the hippocampal cells. Previously, we have shown that serotonergic neurons in cocultures of raphe and hippocampal cells grow better than raphe cells alone (16, 17). The fetal hippocampal cells are dissected as described for the raphe cells. In general, hippocampuses are removed from 16 to 18-day-old fetuses because of the problems of removing the meninges from younger tissue (18, 19).

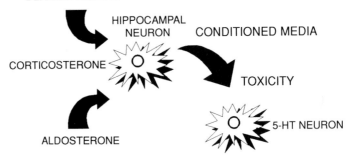

FIG. 3 Design for testing the effects of steroids on the secretion of factors from hippocampal neurons. Hippocampal fetal tissue is grown in serum/steroid-free medium in the presence of an antimitotic to suppress glial growth. The cultures are exposed to steroids in a dose–response manner for 24 hr. The conditioned medium is removed and transferred to raphe cultures grown for 5 days in serum/steroid-free medium. All steroids appeared to produce a release of toxin from the cultured hippocampal neurons.

To remove the hippocampus, the remaining forebrain area is first cut in half along the midline. A pair of pointed watchmaker forceps is then used to cut the fornix. The hippocampus is gently rolled from underneath the cortical surface, using a curved blunt spatula, and separates from the entorhinal cortex ventrally. The meninges and choroid plexus are peeled off the hippocampal tissue. The tissue is dissolved and cell density determined.

Antimitotic Drugs

The fetal central nervous system (CNS) cell mixtures contain both postmitotic neurons and a variety of dividing neurons, glial cells, ependymal cells, and endothelial cells. In short-term cultures (1–5 days), the influence of the dividing cells does not appear to be a problem. In long-term cultures, 5-fluoro-2-deoxyuridine (FdUr, 20 μg/ml) (Cat. No. F-0503; Sigma) and uridine (50 μg/ml) (Cat. No. U-3750; Sigma) are added after 1 day in culture at the first medium change. The antimitotic drug in the CNM is kept for 24–48 hours and then fresh medium is added to the cultures after a single rinse to remove all the FdUr and the cellular debris. Cultures can now be maintained for longer than 2 weeks.

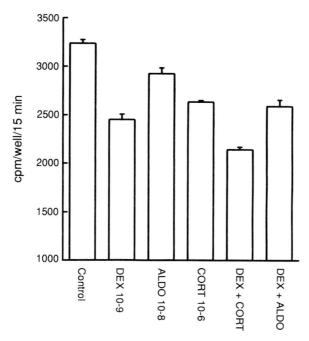

FIG. 4   Effects of hippocampal conditioned medium on midbrain raphe cultures measured as described for Fig. 2. It can be seen that dexamethasone, aldosterone and corticosterone stimulated the release of a 5-HT toxin from the cultured hippocampal neurons. The raphe cultures were immediately transferred to the hippocampal conditioned medium diluted 1:1 with steroid/serum-free medium for 5 days.

Analysis of Medium

The medium at the end of the culture period can be collected and analyzed. Typically, 200 μl of medium is removed and fetal raphe cells are then grown in this medium.

Data Presentation

The effects of hippocampal conditioned medium treated with steroids are shown in Fig. 4. Dexamethasone at a concentration of 10^{-9} M increased the inhibitory effects of medium conditioned by hippocampal cultures. Corticosterone at a concentration of 10^{-6} M produced an inhibition similar to that seen with 10^{-9} M dexamethasone. In contrast, aldosterone at 10^{-8} M pro-

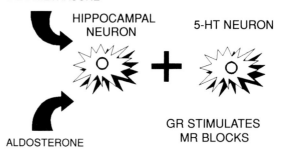

FIG. 5   Design for combining raphe cultures and hippocampal cultures. The hippocampal cultures are exposed to an antimitotic for 24 hr and selected steroids for 24 hr before the addition of the raphe cells. Dexamethasone produced a stimulatory effect on the hippocampal neurons that enhances raphe serotonergic development. Aldosterone was without effect.

duced a smaller decrease. The dexamethasone inhibitory action was increased when combined with corticosterone but not with aldosterone. These results argue against the dexamethasone stimulatory actions seen in Fig. 2 being due to the release of a soluble factor into the medium.

Steroid Action on Hippocampal Neurons

The effects of the steroids could act directly on the hippocampal neuron itself. Hippocampal neurons contain 5-HT$_{1A}$ receptors on their surface and these receptors are influenced by circulating steroids (20, 21). In this experimental design, the hippocampal neurons are grown as described above. When the medium is removed after 4 days to study the conditioned medium, the neurons remaining on the culture plates are used. Fetal midbrain raphe cells are placed on top of the hippocampal neurons and a steroid-free defined medium is added (see above). The coculture is then allowed to grow for an additional 4–5 days (see Fig. 5). The development of serotonergic neurons is measured by high-affinity uptake of [^{3}H]5-HT as described above.

Cocultures

This experimental design is essentially identical to the one described above. However, when the conditioned hippocampal medium with steroids is removed, the hippocampal cells remain attached. At this point the fetal raphe

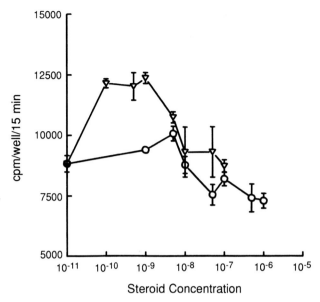

FIG. 6   Effects of dexamethasone (∇) and aldosterone ($\bigcirc$) on hippocampal neurons and their subsequent coculturing with raphe cells. It can be seen that dexamethasone is stimulatory at doses between 10^{-10} and 5×10^{-8} M. Aldosterone is without effect at low dilutions. Both steroids are inhibitory at high, nonphysiological doses between 5×10^{-8} and 10^{-6} M.

cells (as described above) are plated on top of the hippocampal neurons. Two points should be mentioned: first, the hippocampal neurons should not be allowed to dry out after the conditioned medium is removed. In our protocol, the cells are immediately washed with serum/steroid-free medium and a thin layer of solution is left over the hippocampal neurons; second, the number of plated fetal serotonergic neurons should be reduced. Normally the raphe cells are plated at 10^6 cells/cm^2, but in coculture experiments one-third this number is plated (Fig. 6).

Data Presentation

The results show the effects of coculturing raphe neurons with hippocampal cells that had been previously exposed to steroids. The stimulatory effects of target cocultures on serotonergic neurons were increased in hippocampal neurons exposed to dexamethasone at doses between 10^{-10} and 10^{-9} M. A high concentration (10^{-6} M) of dexamethasone showed the opposite effect, an inhibition of raphe serotonergic development. Aldosterone produced an inhibitory effect only at doses greater than 5×10^{-8} M. It should be emphasized that the steroid was given to the hippocampal cells only for 24 hr and the raphe cells were never directly exposed to the steroid. This effect suggest that dexamethasone directly stimulates the neuronotrophic properties of the hippocampal cell with respect to its afferent interactions with serotonergic neurons.

Concluding Remarks

Tissue culture can be used to study the trophic actions of glucocorticoids. In the three approaches outlined in this chapter it was found that dexamethasone, at doses consistent with its actions on the glucocorticoid receptor (GR), can have direct trophic activity on raphe serotonergic neurons, and on their trophic interactions with hippocampal cells. High doses of this steroid, and steroids in general, may produce nonphysiological inhibitory actions. It should be noted that Packan and Sapolsky (22) reported inhibitory effects of dexamethasone at a dose of 10^{-6} M when combined with kainic acid at 10^{-5} M. Our studies demonstrate that low doses of dexamethasone stimulate both raphe and hippocampal neurons.

Tissue culture is an extremely powerful tool for neurobiologists if care is taken to avoid nonspecific effects produced by employing excessive concentrations of hormones or drugs. As a rule, the maximum effect of a hormone in culture is achieved when doses close to the K_d of its receptor are used. The K_d of dexamethasone for the glucocorticoid receptor is 1.5×10^{-9} M (23). At this dose, dexamethasone has trophic activity in culture.

References

1. P. Y. Sze, *Dev. Neurosci.* **3** 217 (1980).
2. F. C. Zhou and E. C. Azmitia, *Brain Res.* **373,** 337 (1986).
3. R. Harrison, *Anat. Rec.* **1,** 116 (1907).
4. E. C. Azmitia, this series, Vol. 2, p. 263, 1990.
5. J. M. Lauder and F. Bloom, *J. Comp. Neurol.* **155,** 469 (1974).
6. E. C. Azmitia and W. F. Marovitz, *J. Histochem. Cytochem.* **28,** 636 (1980).
7. E. C. Azmitia, M. J. Brennan, and D. Quartermain, *Int. J. Neurochem.* **5,** 39 (1983).
8. D. F. Kirskey and T. A. Slotkin, *Br. J. Pharmacol.* **67,** 387 (1979).
9. Y. Nomura, F. Maitoh, and T. Segawa, *Brain Res.* **101,** 305 (1976).

10. G. Doucet, L. Descarries, M. A. Audet, S. Garcia, and B. Berger, *Brain Res.* **441,** 233 (1988).
11. H. K. Kimelberg and D. M. Katz, *Science* **228,** 889 (1985).
12. R. W. Fuller, K. W. Perry, and B. B. Molloy, *Life Sci.* **15,** 1161 (1974).
13. B. S. McEwen, R. E. Zigmond, E. C. Azmitia, and J. M. Weiss, "Biochemistry of Brain and Behavior," p. 123. Plenum, New York, 1970.
14. F. C. Zhou and E. C. Azmitia, *Neurosci. Lett.* **54,** 111 (1985).
15. E. C. Azmitia and P. M. Whitaker-Azmitia, *J. Neurosci.* **20,** 47 (1987).
16. E. C. Azmitia, P. M. Whitaker-Azmitia, and R. Bartus, *Neurobiol. Ageing* **9,** 743 (1988).
17. D. T. Chalmers, S. P. Kwak, A. Mansour, H. Akil, and S. J. Watson, *J. Neurosci.* **13,** 914–923 (1993).
18. G. A. Banker and W. M. Cowan, *Brain Res.* **126,** 397 (1977).
19. M. Segal, *J. Neurophysiol.* **50,** 1249 (1982).
20. B. Liao, B. H. Miesk, and E. C. Azmitia, *Mol. Brain Res.* **328,** (1993).
21. D. T. Chalmers, S. P. Kwak, A. Mansour, H. Akil, and S. J. Watson, *Brain Res.* **561,** 914 (1993).
22. D. R. Packan and R. M. Sapolsky, *Neuroendocrinology* **51,** 613 (1990).
23. E. R. deKloet, *Front. Neuroendocrinol.* **12,** 95 (1991).

[23]   Central Glucocorticoid Receptors and Neuronal Plasticity

Kjell Fuxe, Antonio Cintra, Gerson Chadi,
Jan-Åke Gustafsson, and Luigi F. Agnati

Introduction

It is now well known that there exist large numbers of neuronal and glial cell populations within the central nervous system (CNS) containing nuclear glucocorticoid receptor (GR) immunoreactivity. The highest concentrations are found in stress-sensitive neurons and within many nerve cells of the hippocampal formation (CA1 and CA2 area) and the dentate gyrus (1–3). *In situ* hybridization and immunocytochemical studies have also clearly indicated that embryonic and postnatal GR mRNA and immunoreactivity exist in the rat brain and may become activated by glucocorticoids in fetal and postnatal life (4, 5). Thus, GRs may partly mediate the effects of pre- and postnatal stress on brain neurochemistry and behavioral development. In the present chapter we summarize the evidence, focusing predominantly on our own work, on how GRs may participate in the trophic regulation of the neuronal networks during the life span of rats.

Glucocorticoid Receptors and Growth Factors of Neuronal and Glial Origin

In view of the importance of glia for the survival of nerve cells, it has become important to document further the presence of GR in glial cells, especially in astroglia, within the CNS (2, 3). In Fig. 1, we can now directly demonstrate,

---

Fig. 1   Immunofluorescence histochemistry, showing coexistence of GR (Texas Red) and GFAP [fluorescein isothiocyanate (FITC)] immunoreactivity (A) in the stratum oriens and radiatum of the CA1 area and (B) in the zona reticulata of the substantia nigra (SN) of the male rat. Glucocorticoid receptor immunoreactivity is stronger in the pyramidal nerve cells but distinct nuclear GR immunoreactivity is found in the GFAP-immunoreactive astroglia (arrows). The mouse monoclonal antibody against GR was diluted 1 : 500 and the rabbit antibody against GFAP (Dakopatts, Glostrup, Denmark) was diluted 1 : 500.

Methods in Neurosciences, Volume 22

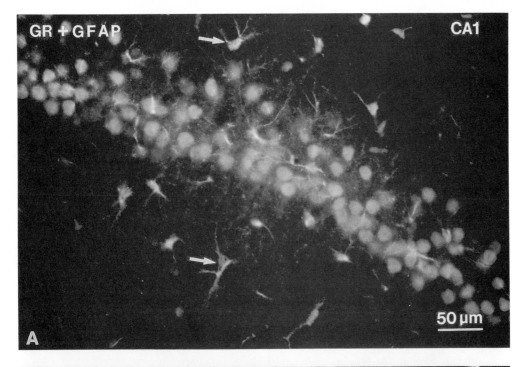

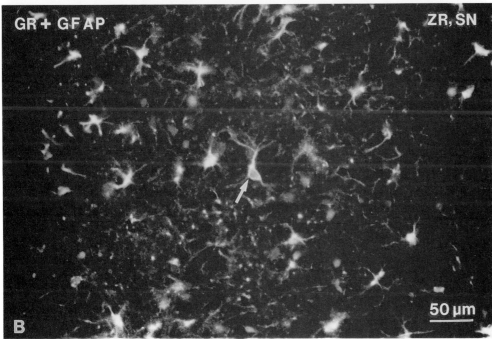

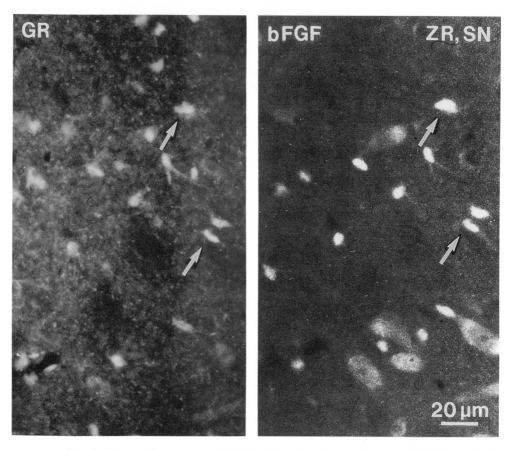

F<small>IG</small>. 2   Immunofluorescence histochemistry showing coexistence of GR (Texas Red) and bFGF (FITC) immunoreactivity in astroglia of the zona reticulata of the substantia nigra in the male rat. Some astroglial cells with coexistence are indicated with arrows. The rabbit bFGF antiserum was raised against bFGF peptide(1–24) [see Chadi *et al.* (6)].

by double immunolabeling procedures using GR and glial fibrillary acidic protein (GFAP) antibodies, that nuclear GR immunoreactivity exists within astroglia of the hippocampal formation and of the zona reticulata of the substantia nigra. The nuclear GR immunoreactivity in glial cells is not as strong as in the nerve cells but is clearly demonstrable within the astroglia. It is of particular interest that GR immunoreactivity can be demonstrated within basic fibroblast growth factor (bFGF)-immunoreactive astroglial cells, as illustrated within the zona reticulata of the substantia nigra (Fig. 2). Thus, the GR may directly influence astroglial trophism within brain regions. We have also demonstrated that corticosterone treatment, in doses that mimic stress serum levels, increases FGF-2 immunoreactivity in astrocytes of the substantia nigra (6). Thus, corticosterone, probably leading to GR activation, may positively influence the trophic state of the substantia nigra via an action of bFGF.

Barbany and Persson (7, 8) have also obtained evidence that glucocorticoids can participate in the regulation of neurotropin mRNA expression in the adult rat representing growth factors mainly of neuronal origin. Thus, 3 days after adrenalectomy a significant reduction in nerve growth factor (NGF), brain derived neurotrophic factor (BDNF), and neurotrophin-3 (NT-3) mRNA levels was observed in the hippocampal formation, effects that were counteracted by treatment with dexamethasone. In a time-course analysis NGF and neurotrophin-3 mRNAs were first demonstrated to be increased and then reduced on treatment with dexamethasone. These results indicate a direct influence of glucocorticoids on neuronal survival via regulation of the formation of various types of growth factors, in this case the neurotrophins. Other experiments demonstrate that adrenalectomy can attenuate kainic acid-elicited increases of mRNAs for neurotrophins and their receptors in the hippocampal formation and cerebral cortex of the rat brain (8). Thus, under conditions of increased activation of glutamate receptor subtypes, glucocorticoids may play an important role in the activation of the neurotrophins within the cortical regions. The ability of hippocampal GRs to enhance voltage-dependent calcium conductance may play a role here (9). However, dexamethasone treatment leads to differential actions on NGF mRNA versus BDNF mRNA increases induced by kainic acid. Thus, dexamethasone potentiated the increase in NGF mRNA but reduced the increase found in BDNF mRNA. Again, these results in the adult rat on neurotrophin gene expression do not clearly support a negative influence of glucocorticoids on neuronal survival. They suggest that GR activation may lead to both positive and negative influences on trophic factors, such as the neurotrophins (see also Refs. 6 and 10). Also, glucocorticoids have been shown to depress activity-dependent expression of BDNF mRNA in neurons of the hippocampus (11) but neurotrophic effects of steroids have been observed on lesion-induced growth in the hippocampus (12).

Finally, Lindholm *et al.* (13) have demonstrated that there exists a differential regulation of nerve growth factor synthesis in astrocytes and neurons by dexamethasone. Both *in vivo* and *in vitro* dexamethasone was found to increase NGF mRNA levels within hippocampal neurons, whereas at least under *in vitro* conditions a reduction in NGF mRNA levels was demonstrated in astrocytes. These results indicate that stress may have a positive influence on neuronal plasticity in the gyrus hippocampus, at least in the adult rat.

Central Glucocorticoid Receptors and Prenatal Development

We have found that during embryonic days 15–22 (E15 to E22) there exists a moderate to strong GR mRNA signal in the neuroepithelia of the medulla oblongata to the telencephalon. A moderate to strong signal was also demonstrated within the neuroendocrine areas of the hypothalamus (paraventricular hypothalamic nucleus and the arcuate nucleus) and within the raphe systems of the medulla oblongata and in the locus coeruleus (5). These results clearly indicate that GRs in embryonic life may participate in the regulation of proliferation events within the neuroepithelia of the brain. In fact, several results indicate that activation of GRs may lead to changes in growth and differentiation during the development (14, 15) involving changes in dendritic development and in maturation (16, 17). On the basis of the present analysis of the prenatal development of GRs, the view is underscored that glucocorticoids can participate in the trophic and functional regulation of the endocrine system in the late fetal period as well as in the regulation of the dendritic branching and synaptogenesis of 5-hydroxytryptamine (5-HT) and noradrenaline neuronal systems, which thus may have consequences for adult life (18–24).

Evidence suggests that prenatal corticosterone affects locomotor activity of the late prepubertal male and female rats (25). In line with these results has been the report that chronic adrenocorticotropin (ACTH) treatment increases striatal dopamine D_2 receptor binding in the developing rat brain (26). An increased interhemispheric coupling of the dopamine systems has also been demonstrated after prenatal stress (27). Thus, taken together, the above experiments and others (for review see Ref. 18) suggest that elevated ACTH/corticosterone levels during the perinatal period could lead to changes in the dopaminergic system at both pre- and postsynaptic sites, which can in turn modify the behavior of the prepubertal, pubertal, and adult rat.

Central Glucocorticoid Receptors and Postnatal Development

Strongly GR-immunoreactive neurons have been demonstrated in the neonatal rat brain by use of GR immunocytochemistry (28). Glucocorticoid receptor immunoreactivity gradually increases within the various cortical areas, including the hippocampal formation, during the postnatal period to reach adult levels at postnatal day 16. Glucocorticoid receptors appear to have a special role in the maturation of the paraventricular and arcuate neurons as well as in the development of 5-HT and noradrenaline neuronal systems. Thus, also in the postnatal period, stress via secretion of adrenocortical hormones may influence the development of distinct neuronal circuits. These results emphasize the possibility that stress-induced secretion of adrenocortical hormones via activating GRs can exert modulatory actions on the maturation of the above-mentioned neurons, which may have long-term consequences for brain functions in adult life (19).

In previous work (29) it was demonstrated that repeated postnatal corticosterone treatment of rat pups leads to a long-term reduction of GR immunoreactivity within the CA1 hippocampal neurons, without any actions on the GR immunoreactivity of the granular cells of the dentate gyrus. These results may reflect a permanent downregulation of GRs within this brain area or a delay in the maturation processes of the hippocampal nerve cells, because the evaluation were made on day 40. It must be emphasized, however, that the mild stress of handling rats produces the same reduction of GR immunoreactivity within the hippocampal neurons of the CA1 area as observed with corticosterone treatment (30, 31). However, maternal deprivation, which must be regarded as a strong stress and involves isolation from the respective mother and litter mates during a daily 5-hr period from postnatal days 1 to 6 (P1–P6), did not produce a downregulation of GR immunoreactivity within the CA1 area. Thus, the extent of the stress is of crucial importance in the determination of the long-term actions of GRs in the hippocampal gyrus (30). Furthermore, the results indicated that mild neonatal stress (pup handling) leads to a long-term modulation of stress-related behaviors as seen from the disinhibition of exploratory behavior in these rats in adulthood (less emotional adult rats). In line with these results, this behavior was also associated with low corticosterone serum levels. However, in contrast, maternal deprivation, representing an intense stress in the postnatal period, failed to elicit such changes. It is unclear to which extent the maintained hippocampal GR immunoreactivity contributed to these results, because the postnatal corticosterone treatment producing a reduction of GR immunoreactivity within the CA1 area also resulted in a reduced responsiveness to stress of the pituitary–adrenal axis (30). It seems likely that under severe stress the neuronal stimuli will dominate in the neuronal and hormonal integrated regulation of gene expression (32, 33). Thus, the protooncogenes, such as c-*fos*, when activated by stress will fully antagonize the actions of activation of GRs either via direct antagonistic protein–protein interaction or via a counteraction at the transcriptional level (34).

In spite of the fact that glucocorticoids in the neonatal period exert an inhibitory influence on growth of brain and on cell proliferation, especially within the cerebellum (35), we have obtained evidence that neonatal corticosterone treatment (10 mg/kg/day on P2, P4, P6, and P8) can to a large extent counteract the monosodium glutamate-induced degeneration of the tuberoinfundibular dopamine neurons and of the arcuate growth hormone-releasing hormone (GHRH) nerve cell populations (36). The mechanisms for this action of corticosterone are unknown but may involve a downregulation of the formation of glutamate receptor subtypes such as the N-methyl-D-aspartate (NMDA) receptor. Alternatively, their regulation may become altered so that they are switched to a desensitized state. These results are also of special interest in the adult because glucocorticoid treatment results in excitotoxic actions on hippocampal neurons at least in terms of enhancement of damage induced by various insults, including brain ischemia (37, 38). The trophic regulation exerted by corticosterone treatment may involve activation of both GRs and MRs (39), which may vary among various neuronal cell populations and also with the age of the animal. Furthermore, the ratio of GRs to MRs in the individual nerve cell may also play a significant role. The results clearly indicate heterogeneities among brain regions in response to GR activation. Thus, it may be emphasized that corticosterone treatment always involves an activation of MRs, although many of them are believed to be tonically activated by corticosterone in view of their high affinity (39, 40). In line with these results, it should be pointed out that adrenalectomy results in hippocampal granular cell degeneration in the rat (41–43) and that aldosterone treatment fully counteracts the adrenalectomy-induced cell death in the dentate gyrus of the adult rat. In contrast, only a partial protection is found with the GR agonist RU 28362 (44). These results underline the existence of a trophic role for MRs and GRs in the control of nerve cell survival in the dentate gyrus. The morphological results suggest the initiation of apoptosis by the adrenalectomy characterized by transformation of the chromatin into distinct bodies. The degeneration therefore represents a phenomenon of programmed cell death (41, 42).

Central Glucocorticoid Receptors and Aging Process

It has been reported that adrenalectomy in adulthood can protect against age-related hippocampal changes (45, 46) and that increases in the corticosterone serum levels can result in neurotoxic effects on hippocampal nerve cells (37, 38). It has therefore been proposed that age-related cell loss in the gyrus hippocampi may be related to overexposure to corticosterone either continuously or via repeated increases as occurs in stress (37, 38, 46, 47). In our own work we have seen substantial strain differences. In the Sprague-Dawley rat a selective reduction of GR immunoreactivity was observed in the CA1 area of the hippocampal formation and in the central amygdaloid nucleus of the aged rat versus many other brain regions (48). Within the CA1 area, also rich in MRs, this reduction in GRs was probably related to a loss of GR-containing neurons, whereas this was not the case in the central amygdaloid nucleus. In all other brain areas analyzed, however, no changes in the GR immunoreactivity were observed. Thus, the activation of the GR alone may not be sufficient to accelerate aging. Instead it may be considered that a combined activation of MRs and GRs in the CA1 area may be mediating the corticosterone cytotoxicity. In contrast, when using stereological procedures (49) to study the Brown Norway rat, we failed to observe any change in the number of GR-immunoreactive nerve cells within the CA1 and CA2 area of the hippocampal formation during the aging process. Thus, in this strain of rats no GR-immunoreactive nerve cells appeared to disappear within the CA1 and CA2 area. Instead, the amount of GR immunoreactivity per cell was reduced during aging, as was also found in the Sprague-Dawley rat. In view of the report of van Eekelen *et al.* (50), showing no changes in the number of GRs in the Brown Norway rat during aging, a reduced intensity of nuclear GR immunoreactivity may reflect a reduced ability of the GR to undergo the conformational changes necessary for translocation and thus activation of the GR response elements. Thus, the results imply that the GR during the aging process may not appropriately regulate nerve cell function either in the Sprague-Dawley rat or in the Brown Norway rat. Such a dysfunction, however, does not seem to be causing the nerve cell death, at least not in the Brown Norway rat. With regard to MR, van Eekelen *et al.* (50) reported a substantial reduction in the number of MRs in all subfields of the hippocampal formation without any associated reduction in the MR mRNA levels, suggesting posttranslational changes in the aged rat. In line with these results we have observed (49) a gradual decline of MR immunoreactivity within the hippocampal gyrus of the Brown Norway rat. The role of MRs and GRs in the aging process of the hippocampal pyramidal neurons remains to be defined, considering also the marked strain differences observed.

Summary

The analysis of GR-immunoreactive nerve cell and glial cell populations in the pre- and postnatal period, in adult life, and in aging emphasizes the involvement of these receptors in the trophic regulation of neuronal networks. The results obtained clearly indicate that GRs may mediate both positive and negative influences on nerve cell survival and function depending on the age of the animal, the nerve cell population analyzed, the brain area analyzed, as well as the possible coexistence with MRs and the degree of glutamate receptor subtype densities. Finally, the effects of glucocorticoids on astroglial trophism in various regions should also be considered. It is concluded that the putative positive actions of glucocorticoids on neuronal plasticity in many nerve cell population have been neglected. Future work will determine the differential trophic effects of glucocorticoids in various nerve cell populations and their interaction with mineralocorticoids in this regulation.

Acknowledgments

This work has been supported by a grant (04X-715) from the Swedish Medical Research Council and by a grant (91/2989) to G.C. from FAPESP (Sao Paulo, Brazil).

References

1. A. Cintra, G. Akner, R. Coveñas, M. de León, A.-C. Wikström, L. F. Agnati, J.-Å. Gustafsson, and K. Fuxe, this volume [9].
2. K. Fuxe, A.-C. Wikström, S. Okret, L. F. Agnati, A. Härfstrand, Z.-Y. Yu, L. Granholm, M. Zoli, W. Vale, and J.-Å. Gustafsson, *Endocrinology (Baltimore)* **177,** 1803 (1985).
3. R. Ahima and R. Harlan, *Neuroscience* **39,** 579 (1990).
4. A. Cintra, R. Coveñas, M. de León, B. Bjelke, J.-Å. Gustafsson, L. F. Agnati, and K. Fuxe, *Neuroprotocols* **1,** 77 (1992).
5. A. Cintra, V. Solfrini, B. Bunnemann, S. Okret, F. Bortolotti, J.-Å. Gustafsson, and K. Fuxe, *Neuroendocrinology* **57,** 1133 (1993).
6. G. Chadi, L. Rosén, A. Cintra, B. Tinner, M. Zoli, R. F. Pettersson, and K. Fuxe, *NeuroReport* **4,** 783 (1993).
7. G. Barbany and H. Persson, *Eur. J. Neurosci.* **4,** 396 (1992).
8. G. Barbany and H. Persson, *Neuroscience* **54,** 909 (1993).
9. D. S. Kerr, W. C. L. Thibault, and P. W. Landfield, *Proc. Natl. Acad. Sci. U.S.A.* **89,** 8527 (1992).
10. K. Fuxe, G. Chadi, L. F. Agnati, B. Tinner, L. Rosén, A. M. Janson, A. Møller, A. Cintra, Y. Cao, M. Goldstein, U. Lindahl, G. David, S. O. Ögren, G. Toffano, A. Baird, and R. F. Pettersson, *in* "Trophic Regulation of the Basal Ganglia Focus on Dopamine Neurons" (K. Fuxe, L. F. Agnati, B. Bjelke, and D. Ottoson, eds.). Elsevier, Oxford, 1993.
11. C. Cosi, P. E. Spoerri, M. C. Comelli, D. Guidolin, and S. D. Skaper, *NeuroReport* **4,** 527 (1993).
12. J. K. Morse, S. T. DeKosky, and S. W. Scheff, *Exp. Neurol.* **118,** 47 (1992).
13. J. Lindholm, E. Castrén, B. Hengerer, F. Zafra, B. Berninger, and H. Thoenen, *Eur. J. Neurosci.* **4,** 404 (1992).
14. M. C. Bohn, *Neuroscience* **5,** 2003 (1980).
15. J. S. Meyer, *Physiol. Rev.* **65,** 432 (1985).
16. M. C. Bohn and J. M. Lauder, *Dev. Neurosci.* **1,** 250 (1978).
17. S. Schapiro, *Gen. Comp. Endocrinol.* **100,** 214 (1968).
18. M. Weinstock, E. Fride, and R. Hertzberg, *Progr. Brain Res.* **73,** 319 (1988).

19. S. E. Alves, H. M. Akbari, E. C. Azmitia, and F. L. Strand, *Peptides* **14,** 379 (1993).
20. W. Rohde, T. Ohkawa, F. Götz, F. Stahl, R. Tönjes, S. Takeshita, S. Arakawa, A. Kambegawa, K. Arai, S. Okinaga, and G. Dörner, *Exp. Clin. Endocrinol.* **94,** 23 (1989).
21. T. A. Slotkin, S. E. Lappi, M. I. Tayyeb, and F. J. Seidler, *Res. Commun. Chem. Pathol. Pharmacol.* **73,** 3 (1991).
22. T. A. Slotkin, S. E. Lappi, E. C. McCook, M. I. Tayyeb, J. P. Eylers, and F. J. Seidler, *Biol. Neonate* **61,** 326 (1992).
23. D. A. V. Peters, *Pharmacol. Biochem. Behav.* **17,** 721 (1982).
24. D. A. V. Peters, *Pharmacol. Biochem. Behav.* **21,** 417 (1984).
25. R. Diaz, S. O. Ögren, M. Blum, and K. Fuxe, *Soc. Neurosci. Abstr.* **19,** 366.6 (1993).
26. C. A. Chiriboga, M. R. Pranzatelli, and D. C. De Vivo, *Brain Dev.* **11,** 197 (1989).
27. E. Fride and M. Weinstock, *Brain Res. Bull.* **18,** 457 (1987).
28. A. Cintra, V. Solfrini, L. F. Agnati, J.-Å. Gustafsson, and K. Fuxe, *NeuroReport* **2,** 85 (1991).
29. M. Zoli, L. F. Agnati, K. Fuxe, F. Ferraguti, G. Biagini, A. Cintra, and J.-Å. Gustafsson, *Acta Physiol. Scand.* **138,** 577 (1990).
30. L. F. Agnati, A. Cintra, G. Biagini, M. Zoli, P. Marama, C. Carani, J.-Å. Gustafsson, and K. Fuxe, *in* "Stress: Neuroendocrine and Molecular Approaches" (R. Kvetnansky, R. McCarty, and J. Axelrod, eds.), p. 751. Gordon & Breach, New York, 1992.
31. G. Biagini, M. Zoli, E. Merlo Pich, P. Marrama, and L. F. Agnati, *in* "Stress and Related Disorders from Adaptation of Dysfunction" (A. R. Genazzani, G. Nappi, F. Petraglia, and E. Martignoni, eds.), p. 139. Parthenon, New Jersey, 1991.
32. R. E. Harlan, *Mol. Neurobiol.* **2,** 183 (1988).
33. H. M. Chao and B. S. McEwen, *Endocrinology* (*Baltimore*) **126,** 3124 (1990).
34. H.-F. Yang-Yen, J.-C. Chambard, Y.-L. Sun, T. Smeal, T. J. Schmidt, J. Drouin, and M. Karin, *Cell* **62,** 1205 (1990).
35. M. C. Bohn, *in* "Neurobehavioral Teratologies of the Nervous System" (J. Yauci, ed.), p. 365. Elsevier, Amsterdam, 1984.
36. M. Zoli, F. Ferraguti, G. Biagini, A. Cintra, K. Fuxe, and I. F. Agnati, *Neurosci. Lett.* **132,** 225 (1991a).
37. R. M. Sapolsky, L. C. Krey, and B. S. McEwen, *Exp. Gerontol.* **18,** 55 (1983).
38. R. M. Sapolsky, L. C. Krey, and B. S. McEwen, *J. Neurosci.* **5,** 1221 (1985).
39. E. R. de Kloet, W. Sutanto, N. Rots, A. van Haarst, D. van den Berg, M. Oitzl, J. A. M. van Eekelen, and D. Voorhuis, *Acta Endocrinol.* (*Copenhagen*) **125,** Suppl. 1, 65 (1991).
40. E. R. de Kloet, A. Ratka, J. M. H. M. Reul, W. Sutanto, and J. A. M. van Eekelen, *Ann. N.Y. Acad. Sci.* **512,** 351 (1987).
41. R. S. Sloviter, A. L. Sollas, E. Dean, and S. Neubort, *J. Comp. Neurol.* **330,** 324 (1993).
42. R. S. Sloviter, E. Dean, and S. Neubort, *J. Comp. Neurol.* **330,** 337 (1993).

43. J. N. Armstrong, D. C. McIntyre, S. Neubort, and R. S. Sloviter, *Hippocampus* in press (1993).
44. C. S. Woolley, E. Gould, R. R. Sakai, R. L. Spencer, and B. S. McEwen, *Brain Res.* **554,** 312 (1991).
45. P. Landfield, *in* "Parkinson's Disease, Vol. 2: Aging and Neuroendocrine Relationships" (C. E. Finch, ed.), pp. 179–199. Plenum, New York, 1978.
46. P. Landfield, R. Baskin, and T. Pitler, *Science* **214,** 581 (1981).
47. P. Landfield, J. Waymire, and G. Lynch, *Science* **202,** 1098 (1978).
48. M. Zoli, F. Ferraguti, J.-Å. Gustafsson, G. Toffano, K. Fuxe, and L. F. Agnati, *Brain Res.* **545,** 199 (1991b).
49. A. Cintra, J. Lindberg, G. Chadi, B. Tinner, A. Møller, J.-Å. Gustafsson, E. R. de Kloet, L. F. Agnati, and K. Fuxe, *Neurochem. Int.* in press (1994).
50. J. A. M. van Eekelen, N. Y. Rots, W. Sutanto, and E. R. de Kloet, *Neurobiol. Ageing* **13,** 159 (1992).

[24] Steroid Action on Neuronal Structure

Catherine S. Woolley and Elizabeth Gould

Introduction

In the late nineteenth century, Santiago Ramón y Cajal speculated that structural changes in the adult central nervous system (CNS) might underlie changes in neuronal function and thus alter behavior. Specifically, he suggested that

> … associations already established among certain groups of cells would be notably reinforced by means of multiplication of small terminal branches of the dendritic appendages and axonal collaterals; but, in addition, some completely new intercellular connections could be established thanks to the new formation of [axonal] collaterals and dendrites. (30)

Ramon y Cajal's ideas developed, at least in part, from a previous suggestion (36) that frequent use of neural connections would result in hypertrophy and increased length of neuronal arborizations, increasing the functional capacity of neurons.

Unfortunately, the idea of structural change within the adult brain remained largely unexplored throughout the early part of the twentieth century as neuroscience focused on a dichotomy between development and adulthood. Development was viewed as a time of great change during which the proliferation and migration of neurons and the growth of dendritic and axonal arborizations occurred. In contrast, the adult brain, having completed these developmental processes, appeared to be structurally stable. This concept of stability in the adult brain was reinforced by (at least) two observations: first, that there were distinct "sensitive periods" for the establishment of certain behavioral patterns (e.g., imprinting, 24, socialization, 32) or for susceptibility to certain forms of mental retardation (e.g., cretinism; 9, 25). The same influences that, when applied during the sensitive period, produced specific changes were relatively ineffective if applied at a later time. Second, a notable lack of regenerative capabilities of adult neural tissue compared to tissue from developing animals (31) further implied that mature neural tissue had lost the capacity for structural modification and was thus relatively rigid.

The concept of a structurally stable adult brain persisted until, in 1964, Bennett and co-workers demonstrated that adult rats exposed to conditions of enriched environmental experience displayed increased cortical weight and cortical acetylcholinesterase (AChE) activity compared to controls housed under conditions of isolation (3). These changes were both qualitatively and quantitatively similar to those observed following differential rearing conditions during development. Subsequently, it was shown that this increase in cortical weight reflected increased thickness of the cortex (7) and increased dendritic field size of cortical pyramidal cells (37). Additional experiments showed that adult rats trained in a maze task also displayed increased branching of cortical pyramidal cells compared to untrained controls (19). Thus, these experiments provided evidence that Ramón y Cajal's speculations were accurate: within previously established neural pathways, the adult brain is capable of use-/experience-dependent modification of connectivity. Remarkably, anatomical changes substantial enough to be visible at the light microscopic level were induced by such subtle, noninvasive manipulations as altered housing conditions or maze training.

The role of steroid hormones in regulation of structural changes in the adult brain was slow to be acknowledged. This was probably largely due to presumptions of a fundamental distinction between organizational and activational effects of steroid hormones (2, 29; see 1 for discussion). Echoing the dichotomy of the anatomists between development and adulthood, this distinction maintains that steroid hormones act during early developmental sensitive periods to permanently organize neural pathways that will be required for future hormone-regulated behaviors (e.g., sexual behavior). In adulthood, these hormones then transiently activate the neural pathways involved in producing certain behaviors. It was generally accepted that the permanence of organizational effects indicates that they were the result of structural modification of neural circuitry. Conversely, the transient nature of activational effects led to the expectation that these changes were restricted to more subtle differences in neuronal biochemistry or electrophysiology. Because activational effects were traditionally not thought to include changes in neuronal structure or circuitry, the possibility that steroid hormones influence the morphology of neurons in the adult brain was largely ignored. This lack of interest is underscored by the observation that even though there were data linking long-term ovarian hormonal state and the effects of enriched versus isolated housing environments on cortical thickness (8, 20, 28), the possibility that steroids rapidly alter neuronal structure was not pursued.

In the 1980s, however, a role for gonadal steroids in the regulation of adult neuronal structure was recognized. These findings indicated that neurons within sexually dimorphic nuclei of the avian song control system and the mammalian spinal cord, which are dependent on gonadal steroids during development, retain their structural sensitivity to these hormones in the adult. In the canary brain, nucleus robustus archistrialis (RA) is highly sexually dimorphic: in the female, RA is smaller than in the male; RA neurons in the female have smaller dendritic trees than in the male. DeVoogd and Nottebohm (6) ovariectomized female canaries at a young age and treated them with testosterone for a period of 4 weeks as adults. Following this treatment, the volume of RA was increased (see also 27) and neurons in the RA had much more extensive dendritic trees than in normal females. Furthermore, these morphological changes were correlated with alterations in behavior: testosterone-treated females were induced to sing as males normally do.

Subsequently, it was shown that dendritic arborization of motoneurons in the spinal nucleus of the bulbocavernosus (SNB) in the adult male rat is dependent on testosterone secretion. The SNB innervates the perineal muscles in the male and, like RA in the canary, is a highly sexually dimorphic nucleus: the SNB in the female rat has many fewer neurons than in the male; female SNB neurons are also much smaller than male SNB neurons. Kurz et al. (23) demonstrated that, in adult male rats castrated for a period of 6 weeks, both the cell body size and dendritic arborization of SNB neurons were substantially decreased; this decrease could be prevented by administration of testosterone. Furthermore, these changes in dendritic morphology were reversible; 4 weeks of testosterone treatment following 6 weeks of castration resulted in dendritic arborization similar to that in a normal male.

In addition, steroid-mediated structural changes in both of these neural regions have been shown to occur naturally. Both the avian song control system and the mammalian SNB are directly involved in reproductive behavior. In some birds and mice the expression of reproductive behavior occurs seasonally, regulated at least in part by seasonal fluctuation of steroid hormone levels. In these cases, structural alterations in the brain (22) or spinal cord (12) parallel changes in hormone levels and occur naturally with the change between breeding and nonbreeding seasons.

While these observations indicate that naturally occurring structural changes in the adult brain do take place over periods of several months, it remained to be explored whether changes in neuronal structure also occur on a much shorter time scale. Clearly, if such structural changes are important mechanisms in behavioral modification, they should occur rapidly. Furthermore, it was not known whether structural effects of steroid hormones occur outside of neuronal populations directly involved in reproductive behavior and are relevant to other aspects of brain function. This chapter focuses on the methods we have used to discover rapid steroid-induced modifications in neuronal structure and survival in the adult hippocampus.

Golgi Impregnation

Background

In 1873, Camillo Golgi published a procedure for the metallic impregnation of neurons and glial cells using potassium dichromate and silver nitrate (13). This technique produced tissue in which a small number of cells were entirely filled with dark precipitate contrasted with a pale yellow background. By allowing visualization of the full extent of axonal and dendritic processes for the first time, Golgi's technique revolutionized study of the nervous system. Previous stains, such as carmine and hematoxylin, demonstrated cell bodies but very little of dendritic processes or axonal fibers. Golgi's technique provided an enormous technical advance: by filling entire cells, this method revealed the overwhelming diversity of cellular structure in the central nervous system and provided the first means to study neuronal structure.

The first application of the Golgi procedure provided an essentially correct morphological description of various neuron types within the cerebellar cortex (13). Later, the Golgi technique provided the first morphological description of glia as a class of cells distinct from neurons (15). Golgi published several modifications of his original technique, including the use of mercuric chloride to produce a precipitate of metallic mercury (15; this version was later modified in 4) and the rapid Golgi method in which tissue is fixed in osmic acid as well as potassium dichromate (14). Over the last 100 years the Golgi techniques have been invaluable in the study of the organization of the nervous system and remain extremely useful in analysis of the development and structure of neurons and glial cells.

Modern versions of the Golgi method are simple techniques: supplies are inexpensive and the procedure requires little specialized equipment and can be completed in as few as 2–3 days. Compared to other more recently developed methods of filling cells to analyze structure, for example, intracellular injection of horseradish peroxidase (HRP) or fluorescent dyes, the Golgi method has the advantage that a greater number of cells per animal can be analyzed. Generally, only one or a few cells per animal can be intracellularly injected. With the Golgi method, however, many cells can be analyzed, increasing the likelihood that the sample mean will accurately reflect the true mean for any particular morphological parameter. Intracellular injections are further limited in that they require specialized equipment to complete.

The primary limitations of the Golgi methods lie in their unpredictability. It is not known why some cells become impregnated and others do not. It remains a possibility that the impregnated cells represent a specialized subpopulation so that sampling these cells may not accurately reflect the true population (although this has never been shown to be the case). Furthermore, because there is no way to determine precisely which or how many cells will become impregnated, an insufficient number of impregnated neurons can require that an experiment be repeated. Finally, the possibility of variation in the degree of impregnation between experiments makes it unacceptable to directly compare morphological data obtained from tissue processed at different times.

Histology

We have used a modified version of the single-section Golgi impregnation technique. In this method, brain tissue is sectioned before impregnation. This provides two advantages: (a) impregnation occurs more rapidly than when larger blocks of tissue are used; and (b) sections can be viewed during the impregnation process in order to determine when impregnation is complete.

For single-section Golgi impregnation, the animals are transcardially perfused with 4.0% (w/v) paraformaldehyde in 0.1 M phosphate buffer with 1.5% (v/v) saturated picric acid, pH 7.4. Following perfusion, the brains are removed from the cranial cavity and postfixed overnight in a solution having the same composition as the perfusate. The brains are then cut into 100-μm sections on a Vibratome into a bath of 3.0% (w/v) potassium dichromate in distilled water and processed as follows:

1. The sections are incubated, lying flat, in 3.0% potassium dichromate overnight.

2. The sections are then rinsed in distilled water, mounted onto plain glass slides (two sections per slide), and a coverslip is glued over the sections at the four corners.

3. The slide assemblies are then incubated in 1.5% (w/v) silver nitrate in distilled water 24–48 hr in the dark at room temperature. The slide assemblies can be periodically removed from the silver nitrate and viewed under a light microscope to monitor the degree of impregnation. To prevent staining of the microscope stage or other surfaces that contact the slide, the silver nitrate solution should be blotted from the back and sides of the slide.

4. When impregnation is complete, the slide assemblies are dismantled, the sections are rinsed in distilled water, dehydrated in graded ethanols, cleared in Americlear, and coverslipped under Permount.

Quantitative Analysis of Golgi-Impregnated Tissue

Quantitative analysis of Golgi-impregnated tissue has been useful for studying the effects of steroid hormones on neuronal structure. We have used Golgi impregnation to evaluate the effects of ovarian steroids on hippocampal dendritic spine density as well as the effects of adrenal steroids on the size and dendritic branching patterns of hippocampal neurons.

Analysis of Golgi-impregnated tissue requires familarity with the cells to be studied. Only a small proportion of cells become impregnated, making it difficult to orient the tissue and identify cells of interest. Furthermore, some cells may be incompletely impregnated or damaged and thus would not be suitable for analysis. It is best to attempt quantitative analysis of Golgi-impregnated tissue only after observation of various Golgi preparations and practice at identifying different cell types. There are three absolute requirements in undertaking quantitative analysis of Golgi-impregnated neurons.

1. All slides containing Golgi-impregnated tissue must be coded prior to analysis and the code should not be broken until the analysis is complete.

2. Only well-impregnated neurons should be analyzed. A well-impregnated neuron will have a black or very dark brown (not red) cell body and a full dendritic tree. Poorly impregnated dendrites may contain short unimpregnated segments or appear jagged or broken.

3. A cell suitable for quantitative analysis must be easily discernible from neighboring impregnated structures that could interfere with analysis.

Dendritic Spine Density

The following is a description of the procedure we have adopted for the determination of dendritic spine density (number of spines per unit length of dendrite) on hippocampal neurons. Generally, we make 3–5 measurements of spine density per dendritic tree (or portion of a dendritic tree) and analyze 6–10 neurons per brain.

1. Neurons should be selected for analysis at a low magnification so that the observer will not be biased by the "spininess" of the cell to be analyzed. Although very spiny neurons often appear to be the most well impregnated, care should be taken not to judge the degree of impregnation of a neuron on the basis of its apparent spine density.

2. On each selected neuron, dendritic segments appropriate for analysis should:

a. be located within a defined range of distances from the cell body, as spine density may vary with distance from the cell body (e.g., 5),

b. remain approximately in one plane of focus so that its length projected in two dimensions using a camera lucida drawing tube will be as accurate as possible, and

c. be of sufficient length (greater than the denominator in the spine density expression).

For each dendritic segment selected, spine density can be measured as follows.

1. The selected segment is traced ($\times 1250$) using a camera lucida drawing tube.
2. All of the dendritic spines visible along that segment are counted.
3. The length of each segment is measured from its camera lucida drawing using an image analysis system [e.g., the Zeiss Interactive Digitizing Analysis System (ZIDAS)].
4. The data can be expressed as number of spines per unit length of dendrite (e.g., 10 μm) and a mean spine density per animal can then be calculated. Reproducible data can be obtained with 5–10 animals per treatment group.

This procedure does not include any attempt to correct for spines hidden either above or beneath the dendritic shaft as is suggested by Feldman and Peters (10). Correction procedures require measurements of dendritic diameter, dendritic spine length, and dendritic spine head diameter (10), which, estimated from light microscopic analysis of Golgi-impregnated tissue, are likely to be inaccurate, reducing the value of the correction. When it is not necessary to estimate the actual density of dendritic spines, but rather to compare relative spine densities, that is, compare spine density between treatment groups, there is little need to correct for hidden spines. It should be noted that a decision not to correct for hidden spines results in spine density values that are likely to underestimate the actual density of dendritic spines. Additionally, the lack of a correction for hidden spines necessitates two further considerations in the selection of dendritic segments for analysis of spine density: (a) segments selected for analysis should all be relatively narrow so that only a small proportion of spines present along the segment will be hidden by the dendritic shaft; (b) segments selected for analysis should all be of a similar diameter, making the effect of hidden spines the same for each treatment group.

Because spine density measurements are dependent on the length of the segment analyzed, changes in the density of dendritic spines can reflect either changes in spine number or changes in dendritic length. For example, an increase in the density of dendritic spines can indicate an increase in the total number of spines or a decrease in the length of the dendrites, that is, the same number of spines on a shortened dendrite would yield an increase in spine density. To address this issue, an analysis of the entire dendritic tree should be undertaken (see below). In cases in which no changes in the number of branch points or the length of dendrites are observed on dendritic trees that show spine density changes, a strong arguement can be made that the density changes reflect alterations in the actual number of spines.

Effects of Ovarian Steroids on Hippocampal Dendritic Spine Density

We have used Golgi-impregnated tissue in order to evaluate the effects of the ovarian steroid hormones estradiol and progesterone on the structure and connectivity of the three principal hippocampal neuron types: CA1 pyramidal cells, CA3 pyramidal cells, and dentate gyrus granule cells (16, 40, 41). Our results indicate that the density of dendritic spines specifically on CA1 pyramidal cells is sensitive to experimental manipulation of estradiol and progesterone levels. Removal of estradiol and progesterone by ovariectomy for 6 days results in an approximately 50% decrease in dendritic spine density. Estradiol treatment protects against this decrease, and progesterone treatment for as few as 5 hours significantly augments the effect of estradiol (Fig. 1). No ovarian steroid sensitivity was observed in CA3 pyramidal cells or dentate gyrus granule cells (16).

The sensitivity of CA1 pyramidal cells to estradiol and progesterone leads to a naturally occurring fluctuation in the density of hippocampal dendritic spines as the levels of these hormones naturally rise and fall during the 5-day estrous cycle (42). Changes in spine density correlate with fluctuating hormone levels and occur quite rapidly. Estradiol and progesterone levels peak on the day of proestrus; spine density values are also highest on proestrus. As soon as 24 hr later during the estrus phase of the cycle, estradiol and progesterone drop to their lowest values and the density of spines decreases by 30%. Spine density then appears to cycle back to high values as estradiol levels gradually rise over the several days of diestrus (Figs. 2 and 3).

A more detailed characterization of the roles of estradiol and progesterone in the regulation of dendritic spine density has indicated that elevated levels of estradiol result in an increase in the density of dendritic spines that is sustained for several days and then declines gradually as estradiol is metabolized over a period of about 1 week. In contrast, progesterone treatment

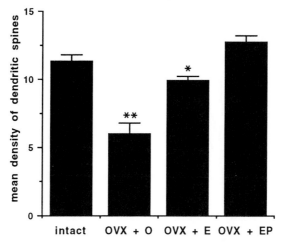

FIG. 1   Mean density of dendritic spines on the apical dendritic tree of CA1 pyramidal cells in female rats that were intact or ovariectomized (OVX) and treated with oil (O), estrogen (E), or estrogen and progesterone (EP). Ovariectomy results in a significant decrease in dendritic spine density compared to all other groups ($p < 0.01$). The OVX + E results are significantly lower than OVX + EP results ($p < 0.05$). (See 16 for details.)

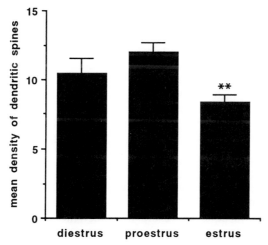

FIG. 2   Mean density of dendritic spines on the apical dendritic tree of CA1 pyramidal cells in female rats in diestrus, proestrus, and estrus. Spine density during estrus is significantly lower than during proestrus ($p < 0.01$). (See 42 for details.)

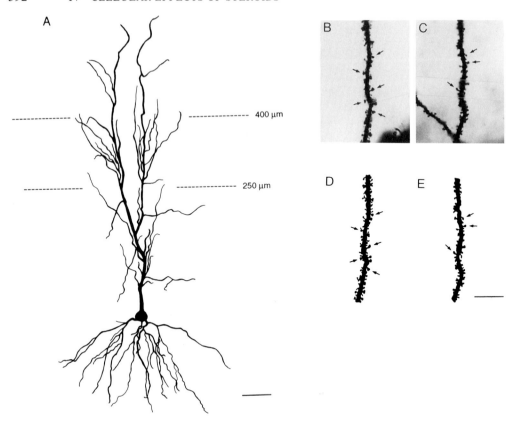

FIG. 3   (A) Camera lucida drawing of a representative CA1 pyramidal cell from the hippocampus of an adult female rat, indicating the distances from the cell body at which apical dendritic spine density measures were taken. (B) Representative photomicrograph of a segment of the apical dendritic tree from a rat during proestrus. Observe the high density of dendritic spines (arrows). (C) Representative segment of the apical dendritic tree from a rat during estrus. Observe the lower density of dendritic spines in (C) compared to (B). (D and E) Camera lucida drawings of the dendritic segments shown in (B) and (C), respectively. (See 41 for details.)

following estradiol has a biphasic effect on spine density: initially, progesterone augments the effect of estradiol, further increasing dendritic spine density, but subsequently results in a much more rapid decrease in spine density than is observed with estradiol alone (Fig. 4). This observation suggested that it is the biphasic effect of progesterone that is responsible for the rapid decrease in spine density that occurs in the shift from the proestrus to estrus phase of the estrous cycle. This role for progesterone was confirmed by

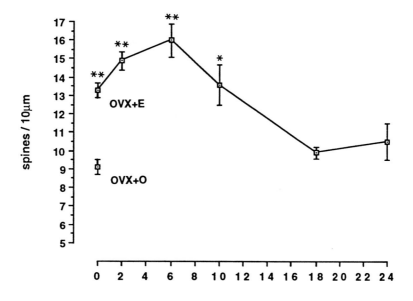

FIG. 4   Mean density of dendritic spines on the apical dendritic tree of CA1 pyramidal cells in female rats that were ovariectomized (OVX) and treated with oil (O) or estrogen (E). The rats that were treated with E were subjected to a time course of progesterone (P) treatment. OVX + E, OVX + EP (2 hr), OVX + EP (6 hr), and OVX + EP (10 hr) were significantly different from OVX + O. (See 41 for details.)

demonstrating that treatment of gonadally intact, cycling rats with the progesterone receptor antagonist RU 486 prevents the proestrus to estrus decrease in spine density (41) (Fig. 5).

To determine whether estradiol-induced changes in spine density are the result of differences in the total number of spines on CA1 pyramidal cells or from an overall expansion and shrinkage of the dendritic tree, we evaluated the effect of the estradiol treatment on CA1 pyramidal cell dendritic arborization. An analysis of both total dendritic length and number of dendritic branch points revealed no differences between ovariectomized and estradiol-treated animals, indicating that estradiol-induced differences in spine density reflect changes in spine number (41).

As dendritic spines are the postsynaptic sites of the vast majority of excitatory input to hippocampal CA1 pyramidal cells, the implication of estradiol and progesterone-induced changes in spine density is that the density of synaptic contacts between CA1 pyramidal cells and their excitatory afferents

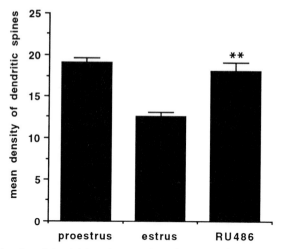

FIG. 5   Mean density of dendritic spines on the apical dendritic tree of CA1 pyramidal cells in female rats in proestrus, estrus, and estrus treated with the progesterone receptor antagonist RU 486. Spine density in estrus following RU 486 is significantly higher than during estrus ($p < 0.01$). (See 41 for details.)

is under the control of these hormones as well. In an electron microscopic study we have found that changes in dendritic spine density of CA1 pyramidal cells that occur with experimental manipulation of estradiol levels or as estradiol and progesterone levels fluctuate across the estrous cycle reflect alterations in synapse density (40).

Estradiol treatment of ovariectomized rats results in a 30% greater density of synapses formed on dendritic spines with no change in the density of synapses formed on the shaft portion of the dendrite (40). Additionally, comparison of synapse density in proestrus rats, in which estradiol and progesterone levels are highest, to synapse density in estrus rats, in which estradiol and progesterone drop to their lowest levels, shows that synapse density fluctuates naturally in parallel with changing hormone levels. Similar to spine density, synapse density drops by as much as 30% in only 24 hr between these two phases of the cycle. As with estradiol-induced changes in synapse density, the naturally occurring fluctuation in synapse density is specific to synapses formed on dendritic spines; we observed no changes in the density of synapses formed on dendritic shafts. These results indicate that light microscopic analysis of dendritic spine density is an effective indicator of synapse density in this system.

Dendritic Arborization

In addition to dendritic spine density, Golgi-impregnated tissue can be used to analyze cell size and dendritic arborization. Cells to be analyzed in this way should not only be thoroughly impregnated and well isolated but should have a cell body in the middle third of the tissue section so that although a portion of the dendrites will be lost in adjacent sections, most of the dendritic tree(s) will be included in the analysis.

In cases in which the cells of interest have heterogeneous morphological characteristics, an equal number of each subtype should be included in the analysis. To avoid analyzing cells of one morphological subtype in one group and those of a different subtype in another group and thus erroneously detecting a treatment difference in the number of branch points and length, an equal number of cells of each type should be analyzed per brain (e.g., 38). Alternatively, morphological subtypes of cells can be analyzed separately. For example, dentate gyrus granule neurons with multiple primary dendrites and those with a single primary dendrite show different branch point numbers and dendritic lengths. In addition, the granule neurons located in the crest of the dentate gyrus have less extensive dendritic trees that those located in the granule cell layers. In a previous study (17), a separate analysis of each of these three cell types was undertaken for each brain.

Cells selected for analysis should be traced (×500) using a camera lucida drawing tube, making certain to indicate clearly each dendritic branch point. From these drawings, the total length of the apical or basal dendritic tree is measured using an image analysis system (e.g., the ZIDAS). Additionally, the number of dendritic branch points can be counted from these drawings. Changes in the pattern of dendritic arborization can also be assessed using a concentric ring analysis (33). In this analysis, a series of concentric rings spaced by equal intervals, for example, 10 μm, is placed over a camera lucida drawing of a cell with the center of the rings on the middle of the cell body. The number of dendritic intersections per ring is counted to determine whether changes in dendritic branching occur in a specific portion of the dendritic tree or generally throughout the tree. Consistent data can be obtained from 5–10 cells per animal, 5–10 animals per treatment group.

It should be noted that the numbers generated from Golgi-impregnated tissue sections reflect only the dendritic length and number of branch points present in a section and thus underrepresent the actual values for entire cells. By underrepresenting the actual values for dendritic parameters, this procedure has the disadvantage that small differences between treatment groups may remain undetected.

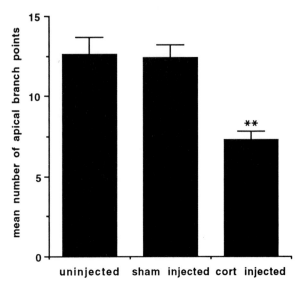

FIG. 6   Mean number of apical branch points of CA3 pyramidal cells in male rats treated with excess corticosterone (cort). Number of branch points following cort injection is significantly lower than that of uninjected and sham injected ($p < 0.01$). (See 39 for details.)

Cross-sectional cell body area can also be determined from camera lucida tracings. Because it can be difficult to delineate the boundary of the cell body versus the initial portions of the dendrites, it is preferable to make these tracings from a higher magnification ($\times 1000$–$\times 1250$) at which the three-dimensional contours of the cell are more apparent.

Effects of Adrenal Steroids on Hippocampal Morphology

We have used Golgi-impregnated tissue to assess the effects of adrenal steroids on neuronal morphology in the hippocampus. Using the single-section Golgi technique, we have shown that manipulations of the circulating levels of adrenal steroids dramatically alter neuronal structure in the hippocampus. Following daily administration of corticosterone for 21 days to adult rats, decreases in the number of dendritic branch points (Fig. 6) and the

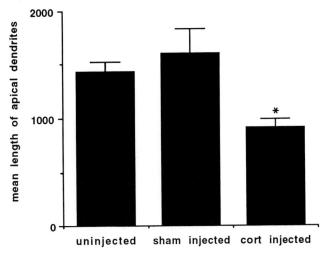

FIG. 7   Mean length of apical dendrites of CA3 pyramidal cells in male rats treated with excess corticosterone (cort). Length of these dendrites is significantly lower following cort injection compared to uninjected and sham injected ($p < 0.05$). (See 39 for details.)

length of dendrites (Fig. 7) of the apical tree of pyramidal cells in the CA3 region were observed (39). In contrast, no changes were observed in the basal dendritic tree of the CA3 region or in CA1 pyramidal neurons or dentate gyrus granule neurons following treatment with excess corticosterone (39). Interestingly, removal of circulating adrenal steroids by adrenalectomy, followed by a 7-day survival, resulted in no change in CA3 pyramidal neurons and small but significant decreases in the number of dendritic branch points (Fig. 8) of granule neurons in the dentate gyrus (17). Decreases in the size of the dendritic tree are suggestive of neuronal atrophy and possibly degeneration.

When using the Golgi technique to examine issues related to cell death and survival, care must be taken to interpret the results accurately. Because the Golgi technique does not impregnate all neurons and the neuronal characteristics that lead to impregnation are unknown, it is possible that impregnated neurons represent a subset of cells, for example, healthier neurons. Moreover, the selection criteria employed during data analysis might prevent detection of evidence for neuronal destruction, as dying cells might appear to be less well impregnated. In instances where neuronal destruction is suspected, adjacent sections should be collected for Nissl staining using cresyl violet (e.g., 17, 39). This point is well illustrated by comparing the effects of high glucocorticoids on pyramidal neurons with those of ADX on

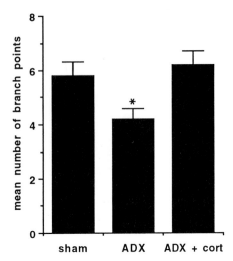

FIG. 8   Mean number of branch points of granule neurons in male rats that were sham operated, adrenalectomized (ADX), or ADX followed by corticosterone (cort) replacement. Number of branch points following adrenalectomy is significantly lower than that of sham operated and ADX + cort ($p < 0.05$). (See 17 for details.)

granule neurons. Daily treatment with a high dose of corticosterone for 21 days resulted in a relatively large ($>40\%$) decrease in the number of CA3 pyramidal cell branch points (39) whereas adrenalectomy (ADX) resulted in a smaller ($<30\%$) decrease in the number of granule cell branch points (17). Moreover, substantial decreases in the length of CA3 pyramidal cell dendrites ($>30\%$) were observed following chronic glucocorticoid administration (39) whereas lesser decreases ($<20\%$) in granule cell dendritic length (Fig. 9) were observed following adrenalectomy (17). Given these results, which suggest a greater degree of neuronal atrophy in CA3 pyramidal cells following corticosterone than in granule neurons following ADX, it is surprising that analysis of Nissl-stained tissue from these same brains revealed massive degeneration in the granule cell layer of adrenalectomized rats (17), whereas a much lesser degree of neuronal atrophy was observed in the pyramidal cell layer of rats treated with high levels of corticosterone (39). Because the results from the Golgi study of granule neurons showed a relatively small decrease in dendritic arborization, Nissl-stained preparations more accurately revealed the effects of ADX on the survival of granule neurons. In contrast, Nissl-stained sections were less revealing than Golgi-impregnated sections when assessing the effects of corticosterone treatment on CA3 pyramidal neurons. The decrease in dendritic branch points and length fol-

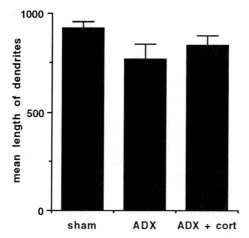

FIG. 9 Mean length of dendrites of granule neurons in male rats that were sham operated, adrenalectomized (ADX), or ADX with corticosterone (cort) replacement. No significant differences were detected. (See 17 for details.)

lowing 3 weeks of corticosterone treatment may signify either the beginning of cell death or dendritic remodeling that is not necessarily a part of the cell death process.

Considerable controversy exists over the histological changes that indicate actual cell death (see 11 for review). With the possible exception of the complete disappearance of a structure or a cell layer, more subtle changes are often difficult to interpret. The following changes, assessed from Nissl-stained preparations, are likely to indicate cell death: (a) The presence of increased numbers of degenerating cells and (b) a concomitant decrease in the number of healthy cells.

It should be noted that detection of degenerating cells requires a familiarity with the morphological characteristics of different types of cell death (see 21 for review). Following ADX, granule neurons undergo apoptosis (34, 35), which is a form of cell death that requires gene expression (21). Apoptotic cells have condensed, darkly stained circular chromatin, no nuclear membrane, and lightly stained or no cytoplasm (e.g., 17, 34, 35). In contrast, following treatment with excess corticosterone, pyramidal cells undergo a form of cell death that more closely resembles necrosis (26). With this form of cell death, the chromatin does not condense; typically the entire cell is more darkly stained and appears to be shriveled (e.g., 39).

Detection of a decrease in the number of healthy cells also requires a familiarity with the cell type of interest. Most brain regions contain numerous cell types. In contrast, the pyramidal cell and granule cell layers of the hippocampus are relatively homogeneous. However, in addition to pyramidal cells and granule cells, these layers contain small populations of inhibitory neurons and some glial cells (18). Because astrocytic reaction is a common characteristic of regions in which neurons are degenerating (e.g., 18), it is important to distinguish between glia and neurons when performing healthy cell counts. This can be accomplished by employing morphological criteria; glial cells are typically smaller and more darkly stained than neurons. However, the observations that reactive glia are typically larger than their resting counterparts (18) and degenerating neurons are usually smaller and more chromophilic than healthy neurons (39) make it difficult to determine with certainty the type of cell counted. In these cases, it is useful to employ immunohistochemical visualization of markers specific to glial cells or neurons combined with Nissl staining. Using cell-specific markers, we have shown that the numbers of mature astrocytes (stained for glial fibrillary acidic protein) and the numbers of radial glia (stained for vimentin) increase in the granule cell layer following ADX (18). Moreover, Nissl-stained degenerating cells that were immunoreactive for the neuronal marker neuron specific enolase, were observed, indicating that granule neurons die following ADX (18). Using a variety of histological techniques, that is, Golgi impregnation, Nissl staining, and cell-specific marker immunohistochemistry, we can conclude with certainty that granule neurons degenerate in response to ADX.

Although the strongest case for cell death can be argued with data showing both decreases in healthy cells and increases in degenerating cells, there are some instances in which this may not be feasible despite the occurrence of massive cell death. Despite a dramatic increase ($>300\times$) in the number of pyknotic cells following ADX with a 7-day survival period (17), a substantial decrease in the number of healthy cells was not observed (18). It was subsequently determined that this small change in healthy cells despite the massive increase in pyknotic cells was probably due to a concomitant increase in the rate of cell birth following ADX (18).

Concluding Remarks

Contrary to the traditional view that hormones influence structure only during development, the results of the studies described above have led us to broaden our view of the range of hormone actions on the adult brain to include rapid, reversible structural changes that are detectable at the light microscopic level.

Acknowledgments

We thank Heather A. Cameron and Bruce S. McEwen for helpful comments on the manuscript.

References

1. A. P. Arnold and S. M. Breedlove, *Horm. Behav.* **19**, 469–498 (1985).
2. F. A. Beach, "Hormones and Behavior." Harper (Hoeber), New York, 1948.
3. E. L. Bennett, M. C. Diamond, D. Krech, and M. K. Rosenszweig, *Science* **146**, 610–619 (1964).
4. W. Cox, *Arch. Mikrosk. Anat.* **37**, 16–21 (1891).
5. N. L. Desmond and W. B. Levy, *Neurosci. Lett.* **54**, 219–224 (1985).
6. T. DeVoogd and F. Nottebohm, *Science* **214**, 202–204 (1981).
7. M. C. Diamond, M. R. Rosenzweig, E. L. Bennett, B. Linder, and L. Lyon, *J. Neurobiol.* **3**, 47–64 (1972).
8. M. C. Diamond, R. E. Johnson, and C. Ingham, *Int. J. Neurosci.* **2**, 171–178 (1971).
9. C. H. Fagge, *Br. Med. J.* **1**, 279 (1871).
10. M. L. Feldman and A. Peters, *J. Comp. Neurol.* **188**, 527–542 (1979).
11. C. E. Finch, *Trends Neurosci.* **16**, 104–110 (1993).
12. N. G. Forger and S. M. Breedlove, *J. Neurobiol.* **18**, 155–165 (1987).
13. C. Golgi, *Gazz. Med. Ital. Lomb.* **6**, 41–56 (1873).
14. C. Golgi, *Arch. Ital. Biol.* **15**, 434–463 (1886).
15. C. Golgi, *Ist. Lomb. Rend.* **12**, 206–212 (1879).
16. E. Gould, C. S. Woolley, M. Frankfurt, and B. S. McEwen, *J. Neurosci.* **10**, 1286–1291 (1990).
17. E. Gould, C. S. Woolley, and B. S. McEwen, *Neuroscience* **37**, 367–375 (1990).
18. E. Gould, H. A. Cameron, D. C. Daniels, C. S. Woolley, and B. S. McEwen, *J. Neurosci.* **12**, 3642–3650 (1992).
19. W. T. Greenough, J. M. Juraska, and F. R. Volkmar, *Behav. Neural Biol.* **26**, 287–297 (1979).
20. W. L. Hamilton, M. C. Diamond, R. E. Johnson, and C. A. Ingham, *Behav. Biol.* **19**, 333–340 (1977).
21. J. F. R. Kerr, A. H. Wyllie, and A. R. K. Currie, *Br. J. Cancer* **26**, 239–257 (1972).
22. J. R. Kirn, R. P. Clower, D. E. Kroodsma, and T. J. DeVoogd, *J. Neurobiol.* **20**, 139–163 (1989).
23. E. M. Kurz, D. R. Sengelaub, and A. P. Arnold, *Science* **232**, 395–398 (1986).
24. K. Lorenz, *J. Ornithol.* **83**, 137–213, 289–413 (1935).
25. F. Lotmar, *Z. Gesamte Neurol. Psychiatr.* **146**, 1–53 (1933).
26. J. N. Masters, C. E. Finch, and R. M. Sapolsky, *Endocrinology* (*Baltimore*) **124**, 3083–3088 (1989).
27. F. Nottebohm, *Brain Res.* **189**, 429–436 (1980).

28. C. T. E. Pappas, M. C. Diamond, and R. E. Johnson, *Behav. Neural Biol.* **26,** 298–310 (1979).

29. C. Phoenix, R. Goy, A. Gerall, and W. Young, *Endocrinology (Baltimore)* **65,** 369–382 (1959).

30. S. Ramon y Cajal (1894), *in* "Cajal on the Cerebral Cortex" (J. Defelipe and E. G. Jones, eds.), Oxford Univ. Press, New York, p. 485.

31. S. Ramon y Cajal, "Degeneration and Regeneration of the Nervous System," Vol. II. Hafner, New York, 1959.

32. J. P. Scott and M. V. Marston, *J. Genet. Psychol.* **77,** 25–60 (1950).

33. D. A. Sholl, "The Organization of the Cerebral Cortex." Methuen, London, 1956.

34. R. S. Sloviter, A. L. Sollas, E. Dean, and S. Neubort, *J. Comp. Neurol.* **330,** 324–336 (1993).

35. R. S. Sloviter, E. Dean, and S. Neubort, *J. Comp. Neurol.* **330,** 337–351 (1993).

36. Tanzi (1893), *in* "Cajal on the Cerebral Cortex" (J. Defelipe and E. G. Jones, eds.), p. 484. Oxford Univ. Press, New York,

37. H. B. M. Uylings, K. Kuypers, M. C. Diamond, and W. A. M. Veltman, *Exp. Neurol.* **62,** 658–677 (1978).

38. Y. Watanabe, E. Gould, and B. S. McEwen, *Brain Res.* **588,** 341–345 (1992).

39. C. S. Woolley, E. Gould, and B. S. McEwen, *Brain Res.* **531,** 225–231 (1990).

40. C. S. Woolley and B. S. McEwen, *J. Neurosci.* **12,** 2549–2554 (1992).

41. C. S. Woolley and B. S. McEwen, *J. Comp. Neurol.* **336,** 293–306 (1993).

42. C. S. Woolley, E. Gould, M. Frankfurt, and B. S. McEwen, *J. Neurosci.* **10,** 4035–4039 (1990).

[25] Electron Microscopic Double and Triple Labeling Immunocytochemistry in Elucidation of Synaptological Interactions between Ovarian Steroid-Sensitive Neurons and Circuits

Csaba Leranth, Frederick Naftolin, Marya Shanabrough, and Tamas L. Horvath

Introduction

Estrogens are the primary steroids involved in the regulation of gonadotropin release. Exposure to estrogen rapidly elicits an inhibition of gonadotropin release, an effect referred to as *negative feedback*. In females, with continued high estrogen levels, this inhibition is reversed by an outflow of gonadotropins from the pituitary gland, the so-called *positive feedback* that initiates ovulation. In males, positive feedback does not occur. This could be due to a direct testosterone-mediated suppression of gonadotropin release or the early androgen exposure-induced differentiation of the neural mechanisms controlling gonadotropin release (1, 2). Early lesion and deafferentation studies of Halasz and Pupp in rats (3) suggested that the site of positive feedback ("cyclic center") is the medial preoptic area (MPOA). This view has been further supported by studies of the local effects of estradiol implants, which showed that the MPOA is the most sensitive to gonadotropin surges induced by estradiol (4). On the other hand, the mediobasal hypothalamus (MBH) seems to be the area most sensitive to negative feedback effects of gonadal steroids (5).

The mechanisms through which gonadal steroids regulate gonadotropin release are not well understood. There are two key observations. First, a substantial number of neurons located in the MPOA and MBH contain estrogen receptors; however, the luteinizing hormone-releasing hormone (LHRH) neurons do not (6). Second, we have demonstrated that treatments resulting in acyclicity in female rats do not result in a decrease in the number, or in an ultrastructural alteration, of LHRH-containing boutons (7). These observations suggest that the feedback effects of estrogen are mediated via estrogen-sensitive neurons or circuits that have either direct or indirect inputs to LHRH neurons.

The morphological delineation of these estrogen-sensitive LHRH-driving circuits is a complex task. Although critically important, such studies have not previously been attempted because of the inability to visualize different transmitters or neuropeptides simultaneously in pre- and postsynaptic structures. To overcome this problem, we have developed and modified several double immunostaining techniques that can be combined with other methods, including degeneration and antero- and retrograde tracer techniques, to elucidate a substantial part of the anatomical basis of gonadal steroid-induced signal transmission to LHRH neurons.

In this chapter, we provide descriptions of these techniques, including fixations, pre- and postembedding double immunostaining procedures, embedding, and even sectioning. Furthermore, we present examples of the applicability of these various techniques in studies including the elucidation of synaptic connections between estrogen-sensitive neurons and circuits controlling gonadotropin release. We hope that these detailed descriptions will aid even the least experienced investigators in the successful application and reproduction of electron microscopy (EM) double immunostaining experiments.

Tissue Preparation

Tissue preparation for electron microscopic (EM) double immunostaining reflects a compromise between the mutually exclusive requirements of structure preservation and deep penetration of antisera or antibodies. Vascular rinsing, fixation, penetration enhancement methods, and even the anesthesia of the animals are especially important for conditions to be met for labeling of two different tissue antigens.

Anesthesia

Although there are no specific studies analyzing the effects of different anesthetics on the outcome of single or double immunostaining, it seems that all of the most commonly used drugs (including ether, methoxyflurane, ketamine, pentobarbital, and chloropent) can be applied with the same satisfactory results. However, to achieve good tissue preservation, the depth of the anesthesia is an important factor; either excessive or marginal levels of anesthesia can result in poor vascular rinsing and perfusion of the fixative.

Rinsing

To remove all the blood from an animal (including cells, which can block small capillaries and give a nonspecific peroxidase reaction, or the serum, which may result in an increased background staining), the vascular system should be rinsed with saline [0.9% (w/v) NaCl in water] prior to fixation. However, the rinsing may contribute to loss of tissue antigens through either anoxic damage or the diffusion of more soluble substances such as mono-amines or γ-aminobutyric acid (GABA) out of the cell prior to cross-linking by the fixative (8). Therefore, the volume of the rinsing solution should be small and the duration of rinsing as short as possible. For example, for a 250- to 300-g rat, no more than 75–80 ml of saline should be used. To avoid the membrane-damaging effect of hypoxia, the saline can be saturated with O_2 by bubbling with Carbogen (CT Air Gas, Bridgeport, CT) (95% O_2 and 5% CO_2). The gas cylinder is connected to a glass cannula, which is placed in the rinsing solution for 20–30 min before perfusion.

Fixation

For normal EM analysis, buffered solutions (pH 7.3, 0.1 mol) containing paraformaldehyde (1.5–4%, v/v) and high concentrations of glutaraldehyde (1.5–2%, v/v) have been known to provide an excellent structure preservation, and the high concentration of glutaraldehyde hardens the tissue so that Vibratome (Lancer, St. Louis, MO) sectioning is greatly facilitated. Unfortunately, higher glutaraldehyde content (greater than 0.2%) can result in a complete loss of immunostaining. On the other hand, visualization of certain tissue antigens, that is, those with antisera generated against an antigen–bovine serum–glutaraldehyde conjugate, requires a fairly high concentration (minimum, 0.8–1%) of glutaraldehyde in the fixative, as do certain immuno-histochemical techniques, for example, immunostaining combined with retrograde horseradish peroxidase (HRP) labeling. In these instances, the Vibratome sections that are cut from the fixed tissue can be treated with a 1% (w/v) sodium borohydride solution (see below) in order to remove unbound aldehydes (9) and, thus, alleviate the penetration problems inherent to using high glutaraldehyde concentrations, while maintaining a good structure preservation.

Although several other fixatives can be used, we recommend the following.

Paraformaldehyde–Glutaraldehyde–Picric Acid Fixative

This fixative contains 4% (w/v) paraformaldehyde, 0.08–0.2% (v/v) glutaraldehyde, and 20% (v/v) saturated picric acid in 0.1 m (pH 7.35) of phosphate buffer (PB). It is perfused via the ascending aorta for a 15- to 20-min period

(150–200 ml for a 200- to 250-g rat). The brains are then removed, the areas of interest dissected, and postfixed for 2–4 hr in the same, but glutaraldehyde-free, ice-cold fixative.

Acrolein Fixative

Acrolein fixative is superb in preserving the antigenicity of all neuropeptides, as well as of tyrosine hydroxylase (tyrosine monooxygenase) (TH), which is the rate-limiting enzyme of dopamine synthesis. However, some tissue antigens, including calcium-binding proteins, specifically parvalbumin, cannot be immunovisualized in acrolein-fixed tissue. There are two routinely used acrolein fixation procedures.

1. After a rapid rinsing (10–50 ml of saline for rats), the animal is perfused with 70 ml of 3.75% (v/v) acrolein in 2% (w/v) paraformaldehyde followed by 200 ml of 2% (w/v) paraformaldehyde. Postfixation for 1–2 hr in the latter greatly improves the ultrastructural morphology.

2. The animal is rapidly rinsed and then perfused with 5% acrolein in PB. This fixative increases the intensity of immunostaining, but sacrifices somewhat the structure preservation. A short postfixation in the same fixative for 20–30 min is recommended.

Remarks

Acrolein is toxic; therefore, preparation of the fixative and perfusion of the animal should be performed under a well-ventilated hood while wearing protective goggles and rubber gloves. To neutralize the acrolein, a 10% (w/v) sodium bisulfite (in water) solution can be used, and should be continuously poured into the chest cavity of the animal during the perfusion.

A complete removal of residual fixatives in the tissue improves the detection of immunoreactivity while reducing nonspecific background labeling. To avoid possible tissue antigen degradation, the postfixation rinsing should be short, but effective. After postfixation, small tissue blocks, no thicker than 2–3 mm, are briefly rinsed in PB, and then 40 to 60-μm sections are cut on a Vibratome. These sections are rinsed three times (10 min each), followed by an incubation for 20–60 min in 1% sodium borohydride in PB. They are then given a final PB rinse (three times, 10 min each).

Penetration Enhancement

Detergents such as Triton X-100, saponin, or protease have been used to permeabilize sections, thus allowing for greater penetration of immunoreagents. However, these substances are normally used for light microscopic preparations only, because even at low concentrations they have an extremely damaging effect on ultrastructural morphology. In addition, Triton X-100 seems to interfere with avidin–biotin binding.

A widely used method for enhancing penetration of immunoreagents in tissues prepared for electron microscopy is the freeze-thaw treatment. After rinsing and sodium borohydride treatment, Vibratome sections are transferred to a vial (10- to 12-mm diameter) containing 0.5–0.8 ml of 10% (w/v) sucrose in PB. The vial is dipped into liquid nitrogen, with constant, slow stirring by hand, until the tissue is frozen; it is then thawed back to room temperature. The sucrose is removed by three 15-min washes in PB.

Cryopreservation of Vibratome Sections

If the sections are to be used later, they can be cryoprotected and stored, even for several years, without losing tissue antigenicity or destroying the ultrastructure. Following rinsing, the freshly cut sections are treated with sodium borohydride, washed, and transferred into a solution containing

PB (0.2 M), 500 ml
Glycerin, 500 ml
Ethylene glycol, 600 ml
Doubly distilled water, 400 ml

The sections are gradually chilled, simply by placing them in a −80°C freezer, where they are also stored until use. A quick freezing in liquid nitrogen is not recommended, as this will completely destroy the ultrastructure.

Double Immunostaining Techniques

Numerous EM double immunostaining techniques have been developed using two contrasting electron-dense labels. Here, we provide the detailed descriptions of just three of the most reliable and widely used. Although our laboratory previously used the diaminobenzidine (DAB)/silver-intensified DAB double immunostaining technique (10, 11), we no longer use or recommend it because of the lack of a consistent positive control for the silver intensification of all of the previously DAB-labeled profiles and the resultant poor structure preservation.

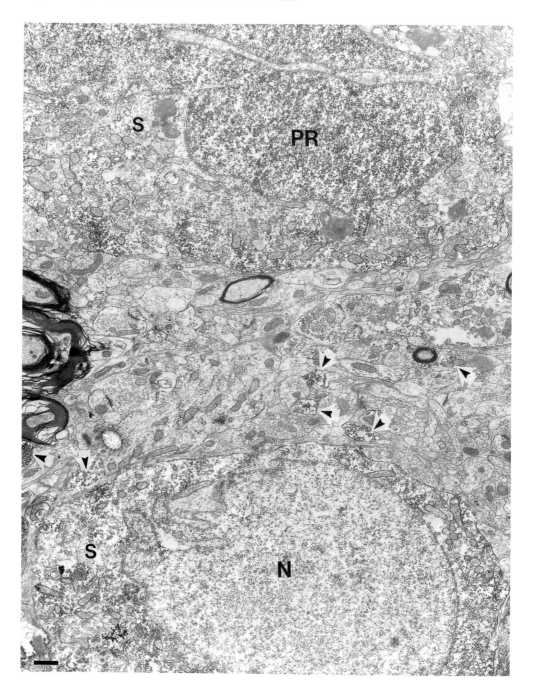

Double Immunostaining Using Two Diaminobenzidine–Immunoperoxidase Reactions

Under certain conditions, when two tissue antigens are localized in different, well-defined compartments of a cell (e.g., nucleus and perikarya), the same immunolabel can be used to visualize both substances. For example, this technique can be applied to visualize nuclear steroid receptor-containing neurons and the transmitter or neuropeptide content of the afferent axon terminals contacting their perikarya, or to determine the transmitter content of these nuclear receptor-containing cells. Electron microscopic immunocytochemical analyses have demonstrated that progesterone receptors (PRs) in the primate and guinea pig hypothalamus are located solely within the cell nucleus (12–14). To determine the transmitter content of these cells, as well as to characterize neurochemically their afferent connections, the following technique has been employed.

In the first immunostaining, the nuclear progesterone receptors were labeled by the avidin–biotin–peroxidase (ABC) method and a DAB reaction (15), using either a monoclonal antibody generated against rabbit uterine PR (M-a-mPRI; Transbio-Sarl, Paris, France), in the case of the monkey tissue, or a second monoclonal antibody generated against human placental PR (M-a-PRII; Transbio-Sarl) for the guinea pig material. The second tissue antigen (either perikaryal or axonal) was visualized using the peroxidase–anti-peroxidase (PAP) technique (16) and a second DAB reaction (Figs. 1 and 2). Control experiments demonstrated the specificity of this double immunostaining procedure. Thus, omitting one of the primary antisera, or using one that was immunoabsorbed, resulted in either nuclear, perikaryal, or axonal immunolabeling only.

These experiments showed that all of the nuclear PR-containing cells in the hypothalami of colchicine-pretreated primates and guinea pigs are GABAergic (12–14), and that some of the PR-containing neurons of the monkey hypothalamus colocalize dopamine (17). Furthermore, using this

FIG. 1 An electron micrograph demonstrating the result of double immunostaining using two immunoperoxidase reactions to visualize immunoreactivity for progesterone receptors (PRs) and glutamic acid decarboxylase (GAD) in the hypothalamus of a colchicine-pretreated African green monkey (*Cercopithecus aethiops*). The soma (S) of both neurons is immunoreactive for GAD, whereas only the nucleus of the neuron (*top*) exhibits immunoreactivity for PR. The nucleus (N) of the other cell does not contain PR. Black and white arrows point to GAD-immunoreactive boutons. Bar: 1 μm.

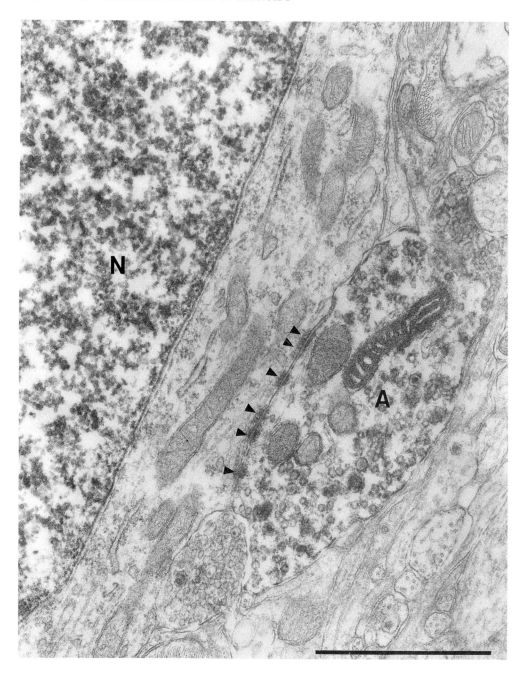

technique, we were able to demonstrate the afferent synaptic connections of these PR-containing cells with several transmitter systems, including proopiomelanocortin-derived peptides (Fig. 2), GABA, catecholamine, neuropeptide Y, substance P, cholecystokinin, and somatostatin, all of which are known to be involved in the regulation of gonadotropin release (12–14, 17).

Double Immunostaining with Immunogold and Immunoperoxidase

The central nervous system, including the neuroendocrine hypothalamus, contains numerous homologous synaptic contacts, e.g., GABA–GABA (18), dopamine–dopamine (19), and even LHRH–LHRH connections (20). Therefore, to avoid misinterpretation of the data, the aforementioned double immunostaining technique using the same DAB chromogen for visualizing different tissue antigens cannot be used in experiments aimed to characterize pre- and postsynaptic profiles neurochemically. However, there are two commonly employed double immunostaining techniques that can be used for this purpose.

In this procedure, one of the tissue antigens is labeled by a DAB chromogen and the other is visualized by gold particles. Instead of gold, ferritin can be substituted (21); however, the size and electron density of the iron core in the ferritin molecule does not allow for an easy recognition of the immunolabeling. The main advantages of the immunoperoxidase/immunogold technique are that (a) the gold particles, in contrast to the diffuse immunoperoxidase reaction, will not cover the synaptic membrane specializations and, thus, the type of synaptic contacts, symmetric or asymmetric, can easily be determined (Fig. 3), and (b) it can be combined with retrograde HRP tracing methods. For example, in a previous study in the rat, we demonstrated that a population of mediobasal hypothalamic (MBH) proopiomelanocortin (POMC)-containing neurons establishes synaptic connection between the MBH and medial preoptic area (MPOA) LHRH neurons (22). It is also known that MBH GABA and catecholamine systems can influence the activity of both MBH POMC-containing cells and MPOA LHRH neurons, thereby indicating that the regulatory action of MBH GABA and dopamine systems

FIG. 2 High-power electron micrograph taken from the ventromedial nucleus of an African green monkey. Double immunostaining using two immunoperoxidase reactions demonstrates a synaptic contact (arrowheads) between an axon terminal (A) immunoreactive for ACTH and the soma of a neuron containing a PR-immunopositive nucleus (N). Bar: 1 μm.

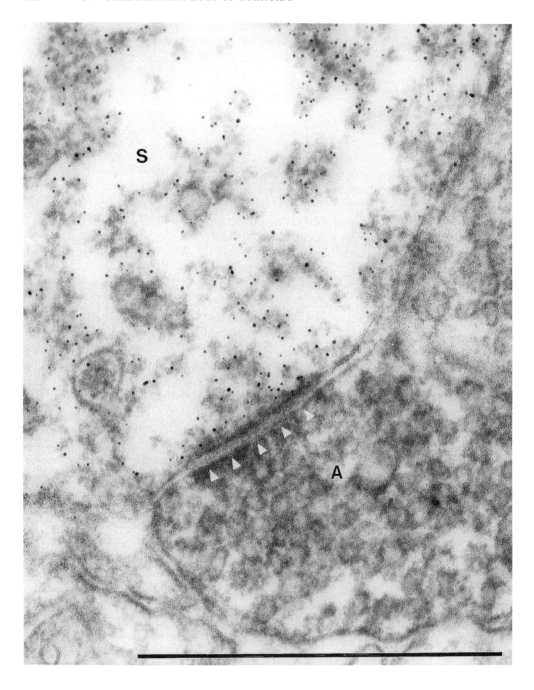

on the LHRH neurons is mediated by projective MBH POMC neurons that contact these neurons. To delineate morphologically such an information flow, an immunogold/immunoperoxidase double immunostaining was carried out on MBH Vibratome sections taken from an animal that received an HRP microinjection into the MPOA. The results of this experiment (23) demonstrated that immunogold-labeled POMC neurons that project to the MPOA (i.e., they also contained retrogradely transported HRP granules) are synaptic targets of immunoperoxidase-labeled GABA (Fig. 4) and catecholamine (Fig. 5) axon terminals.

The major disadvantages of this technique are (a) the poor penetration of the gold-labeled immuno-α-globulins (IgG), (b) the undetectability of gold labeling under the light microscope (thus a correlated light and EM analysis is not possible), and (c) the requirement for a minimum base magnification of $\times 15,000$ for the detection of the gold particles.

Using the immunogold/immunoperoxidase double immunostaining procedure, the two tissue antigens can be labeled either simultaneously, or sequentially. If certain conditions are met, for example, (a) both tissue antigens are very sensitive, (b) the primary antisera are raised in different species, and (c) the second antisera (the gold-conjugated or unconjugated IgG) are raised in the same species, a simultaneous double immunostaining procedure can be carried out. For example, the tissue sections would first be incubated in a solution containing both primary antisera, sheep anti-A tissue antigen and rabbit anti-B tissue antigen, followed by incubation in a mixture of the second antisera, gold-conjugated donkey anti-rabbit IgG and donkey anti-sheep IgG (or biotinylated donkey anti-sheep IgG). Subsequent to this would be an incubation in sheep PAP (or ABC complex) and then, finally, a DAB reaction. This procedure would result in a colloidal gold labeling of tissue antigen B and immunoperoxidase labeling of tissue antigen A.

It is important to note that the dilution of all mixed antisera (both primary and secondary) in simultaneous double staining experiments should be half of that used for single labeling, in order to compensate for the increased dilution produced by the addition of the other antiserum. For example, if the optimal, final dilution for single labeling of a primary antiserum is 1 : 1000, a 1 : 500 dilution should be used in combination with an equal volume of the second primary antiserum.

FIG. 3   High-power magnification of rat MPOA, showing a synaptic contact (white arrowheads) between a TH-immunoreactive axon terminal (A) labeled by an immunoperoxidase reaction and the soma (S) of an LHRH-immunopositive cell labeled with immunogold. Bar: 1 μm.

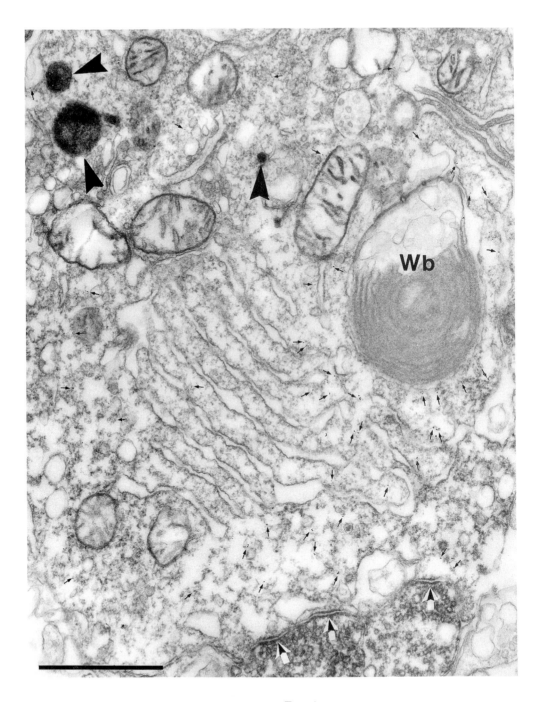

Fig. 4

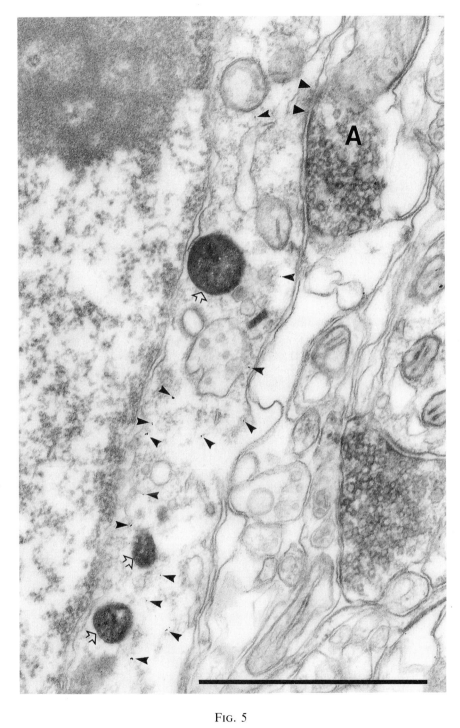

F<small>IG</small>. 5

To avoid the possibility of nonspecific labeling and the use of excessive control experiments, it is more advantageous to perform a sequential double immunostaining procedure. First, the more sensitive tissue antigen is labeled by a PAP or ABC immunostaining method, using DAB as the chromogen, followed by the second immunostaining using the gold-conjugated IgG. The peroxidase product formed by DAB at least partially protects the immuno-globulins of the first labeling series from reaction with the second series (24). Thus, the sequential labeling procedure, with its corresponding controls, is appropriate for double immunostaining using two primary antisera generated in the same species.

Remarks

Instead of IgG-conjugated gold particles (no larger than 15 nm), avidinated gold colloid (particle size, 5–10 nm) can also be used for immunolabeling one of the tissue antigens (Fig. 2). This would be applied after a biotinylated second antiserum, thus adding one more incubation step to the procedure. However, to avoid nonspecific cross-linking, the other tissue antigen should be immunostained by the PAP method rather than the ABC technique.

To increase the penetration depth of the gold-conjugated IgG (or avidinated gold), in either the simultaneous or sequential double immunostaining procedures, a minimum 12-hr (at 4°C) incubation is required in this antiserum.

Double Immunostaining Using Diaminobenzidine and Nickel-Intensified Diaminobenzidine (NiDAB) Reactions

As mentioned previously, the main disadvantage of the immunogold/immu-noperoxidase double immunostaining technique is that the metal particles cannot be seen under the light microscope. The nickel-intensified DAB (NiDAB)/DAB double immunostaining method overcomes this problem. In

---

FIGS. 4 AND 5   Electron micrographs taken from the arcuate nucleus of a rat demonstrate the result of a triple labeling experiment. The retrogradely transported HRP granules (black arrowheads in Fig. 4; open arrows in Fig. 5) injected into the MPOA are visualized with a DAB reaction. Immunoreactivity for β-endorphin is labeled by immunogold (small arrows in Fig. 4; small arrowheads in Fig. 5). TH-immunopositive (Fig. 4) and GABA-immunopositive boutons (Fig. 5) forming synaptic contacts (black-and-white arrows on Fig. 4; large arrowheads on Fig. 5) with the soma of β-endorphin-containing neurons are immunolabeled with an immunoperoxidase reaction. Wb (Fig. 4), whorl body. Bar scales: 1 μm.

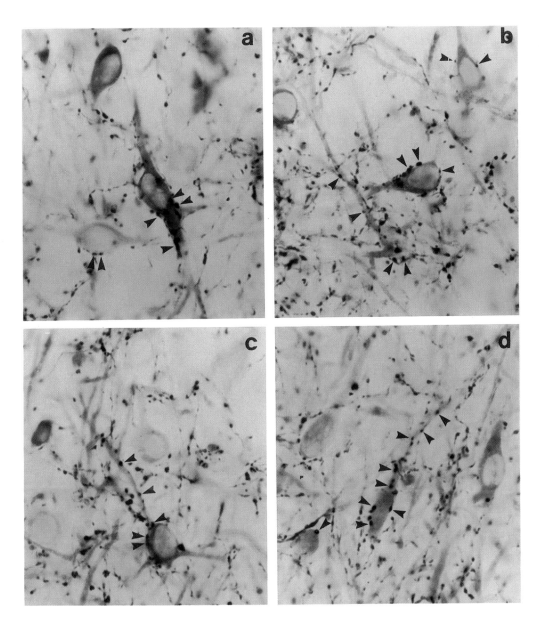

FIG. 6   Color light micrographs of the rat MBH showing the result of a double immunostaining experiment for NPY and β-endorphin using the NiDAB/DAB double immunostaining technique. Immunoreactivity for NPY was labeled by the dark blue to black NiDAB reaction, whereas β-endorphin-immunoreactive neurons were labeled with the brown DAB chromogen. The NPY-immunoreactive boutons (arrowheads) form putative synaptic contacts with the dendrites and somata of β-endorphin-containing neurons. Original magnification × 950.

this double immunostaining procedure, one of the tissue antigens is labeled with a brown DAB reaction product, whereas the other is labeled with a dark blue-to-black Ni-intensified DAB (NiDAB) chromogen (Fig. 6, color plate). The main advantages of this technique are (a) a deep penetration of the two immunolabels, which allows for a quantitative analysis of the putative connections, and (b) its applicability in correlated light and electron microscopic analyses.

The method routinely used in our laboratory is a modification of the technique described by Wouterlood (25), and is as follows.

1. Immunostain the first tissue antigen using the ABC technique, followed by visualization of the tissue-bound peroxidase in NiDAB for a 3- to 10-min incubation. This solution contains

DAB, 15.00 mg
Ammonium chloride, 12.00 mg
Glucose oxidase, 0.12 mg
Nickel ammonium sulfate (0.05 M), 600 μl
β-D-Glucose (10%), 600 μl
Phosphate buffer (pH 7.35, 0.1 M), 40 ml

2. Wash three times (15 min each) in PB.
3. Immunostain the second tissue antigen using the PAP technique.
4. Incubate the sections for 5–10 min in a solution of

DAB, 15.00 mg
Phosphate buffer (pH 7.35, 0.1 M), 30.00 ml
H_2O_2 (0.3%), 165.00 μl

5. After several washes in PB, the sections are wet-mounted onto slides. They can then be analyzed (even with an ×100 oil immersion objective) under the light microscope, and color photographed for documentation of the brown and dark blue-to-black colors of the two separate immunoreactions, and for correlation with the electron microscopic analysis.

6. After osmication, sections are dehydrated, embedded, and serial sections of the previously selected and photographed areas can be examined in the EM (Fig. 7).

Remarks

It is important to monitor the progress of the immunostaining during the NiDAB reaction by frequently examining the sections (every 30 sec after 3 min) under the light microscope, to ensure that they do not overdevelop.

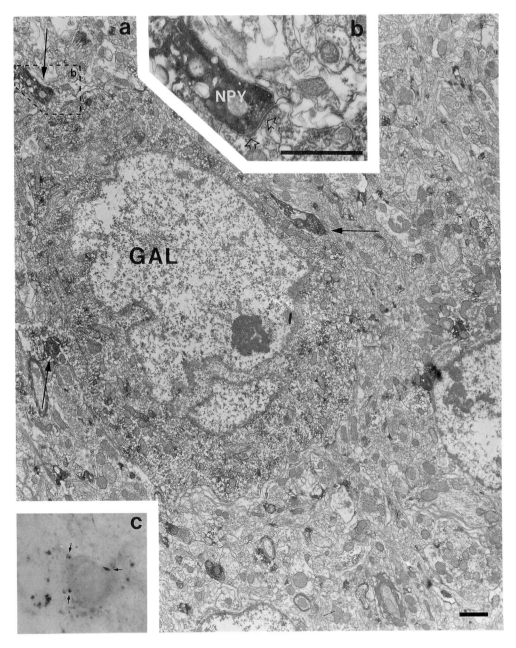

FIG. 7 Light (c) and electron micrographs (a and b) demonstrate the result of a correlated light and electron microscopic double immunostaining experiment using the DAB/NiDAB technique for visualizing neuropeptide Y (NPY) and galanin (GAL)

After 3–4 min of reaction time, the sections can suddenly be overstained, and this can interfere with, if not inhibit, the second immunostaining.

For easier recognition, the NiDAB reaction is always done first, and is normally used to stain smaller profiles (e.g., axons), while the large postsynaptic profiles are labeled second with the light-brown DAB chromogen.

The correlated light and electron microscopic identification and, most importantly, the photodocumentation of the two-color immunolabeled profiles are a precondition to the use of this technique. In the electron microscope, the heavily NiDAB-labeled profiles can be distinguished from the less intensively labeled DAB-stained profiles (Fig. 8) only on ultrathin sections cut from the surface areas of the Vibratome sections. On those sections prepared from deeper regions of the Vibratome sections, the difference between the electron density of the two immunolabels is not obvious.

Further details regarding the applicability of this technique in the elucidation of the synaptic connections of estrogen-sensitive hypothalamic circuits can be found in our previous publications. These studies include the description of (a) the β-endorphin innervation of dopamine cells (26), (b) the neuropeptide Y (NPY) innervation of β-endorphin and dopamine neurons (27), and (c) the NPY innervation of PR-containing dopamine neurons. In this third experiment, using triple labeling immunocytochemistry, the PR-containing nuclei and the NPY-immunoreactive boutons were labeled by an NiDAB chromogen, whereas the TH-immunopositive perikarya were labeled with a DAB reaction (17).

Double Immunostaining for Anterograde Tracer Phaseolus vulgaris Leukoagglutinin and Neuroactive Substances

This technique is useful in determining the extrinsic afferent connections of a population of neurons, the transmitter or neuropeptide content of which can then be determined by a second immunostaining. The advantages of this method over a procedure combining lesion-induced acute axonal degeneration (see below) and immunostaining are that (a) the *Phaseolus vulgaris*

---

in the MBH of a rat. The putative synaptic contacts between the black NPY boutons [arrows in (c)] and the brown GAL-immunopositive neuron were further analyzed under the electron microscope. (a) Same GAL-immunoreactive neuron and NPY-containing boutons (long arrows) seen in (c). (b) High-power magnification of the synaptic contact (open arrows) seen on the boxed area of (a). Bars: 1 μm. Original magnification of (c): ×1250.

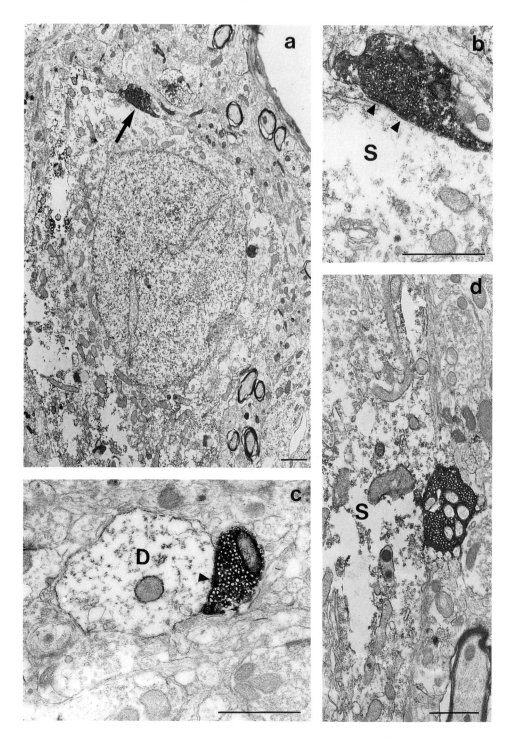

leukoagglutinin (PHAL)-labeled axons can be followed from their cells of origin to their targets; (b) PHAL is picked up only by neurons located in the injection site, thus, PHAL labels only the axons of cells located in the injection site, and not the axons of passage; and (c) PHAL-containing axons and boutons, unlike degenerated boutons, can be seen under the light microscope, thereby allowing for a light microscopic analysis of the putative connections and quick electron microscopic verification of the same contacts.

This technique was employed to determine whether progesterone receptor (PR)-containing GABAergic neurons located in an area between the fornix and ventromedial nucleus of monkeys terminate on MBH LHRH-producing cells. Progesterone reduces the frequency of luteinizing hormone (LH) pulses in a number of species, including primates, as seen by the dramatic slowing of LH pulses during the luteal phase of women and monkeys. This signifies a neural action of progesterone and a reduction in the frequency of hypothalamic gonadotropin-releasing hormone pulse generation. How progesterone regulates gonadotropin release is not known. We have demonstrated that LHRH neurons of guinea pigs and monkeys do not contain PR. However, in the hypothalamus, a large population of GABA cells exhibits nuclear PR immunoreactivity (12–14). Progesterone decreases LHRH neuron activity (28) as does GABA administration (29). In rodents, we have demonstrated GABAergic innervation of LHRH neurons (30). Taken together, these observations suggest that progesterone acts indirectly on the LHRH system, via PR-containing GABAergic neurons. To test this hypothesis in primates, the following experiment was performed.

1. PHAL [2.5% (w/v) in PB, pH 8] was iontophoretically (using a 5.0-μA positive current, 7-sec on/off cycle) delivered to an area between the fornix and ventromedial nucleus (Fig. 9). The duration of the injection via a stereotaxically placed glass micropipette (5- to 10-μm tip diameter) was 30–40 min.

2. Ten days later, the deeply anesthetized animals were sacrificed by a transcardial perfusion of saline (1500 ml) followed by a fixative (1500–2000

FIG. 8 Electron micrographs of ultrathin sections cut from the surface area of a Vibratome section immunostained for PHAL and LHRH, using the DAB/NiDAB double immunostaining procedure. The contrast differences between the NiDAB-labeled PHAL-immunoreactive boutons and the DAB-immunolabeled LHRH-containing somata (S) and dendrite (D) are easily recognized. Arrowheads in (b) and (c) point at synaptic membrane specializations. (b) High-power magnification of the synaptic contact shown in (a) (arrow). Bars: 1 μm.

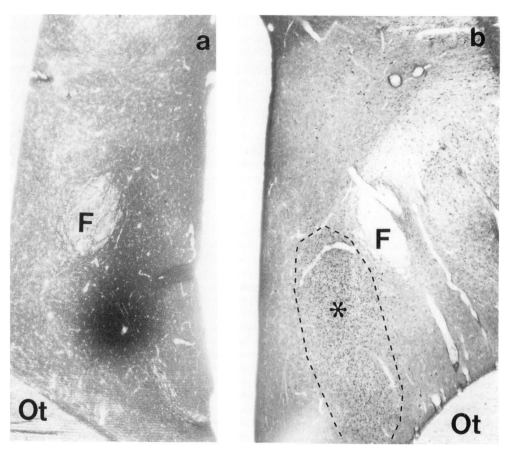

FIG. 9   Light micrographs show the site of a PHAL injection (a) located in an area between the fornix (F) and ventromedial nucleus, and the result of immunostaining for progesterone receptors (b) in the hypothalamus of an African green monkey. Asterisk in (b) labels the center of the PHAL injection. Ot, Optic tract. Original magnification: ×250.

ml) containing 4% (w/v) paraformaldehyde and 0.2% (v/v) glutaraldehyde in 0.1 M PB, pH 7.35.

3. The first immunostaining consisted of visualizing the site of injection and the PHAL-labeled axons and boutons on sodium borohydride-pretreated Vibratome sections (see above) prepared from the hypothalamus.

   a. The sections were incubated for 48 hr (at 4°C) in biotinylated goat anti-PHAL (1 : 200 in PB; Vector Laboratories, Burlingame, CA).

   b. The sections were incubated for 4 hr (at 20°C) in avidin–biotin–peroxidase complex (1:125 in PB; Vector laboratories).

   c. The tissue-bound peroxidase was developed using the dark blue-to-black NiDAB reaction (see above).

4. Immunoreactivity for LHRH was visualized by a second immunostaining using the PAP technique and a final DAB reaction.

5. Light microscopic color photographs were taken of wet-mounted sections exhibiting the putative synaptic contacts between the darkly stained PHAL-labeled fibers originating in the ventromedial nucleus and the light brown LHRH-producing neurons (Fig. 10a and b).

6. Selected Vibratome sections were then postosmicated, dehydrated, and embedded for EM analysis. On serial ultrathin sections, the putative contacts seen under the light microscope were identified, verified, and photographed in the EM.

The results demonstrated that axons originating in an area containing PR-containing GABAergic neurons terminate on MBH LHRH-immunoreactive cells (Fig. 10c). Although 45% of the total population of neurons and practically all of the GABAergic cells of the above-mentioned area contain PR (13), the question of whether all or only a population of the PHAL-labeled boutons contacting LHRH neurons are GABAergic still requires investigation. Performing postembedding immunostaining for GABA on preembedding, PHAL-immunostained material (31) is the best way to approach this question.

Double Immunolabeling Using Pre- and Postembedding Immunostaining

The following experiment was performed on monkeys. Alternate serial ultrathin sections of PHAL-immunolabeled boutons were placed on nickel or gold grids (two or three sections on each grid), and every second grid was immunostained for GABA according to the protocol of Somogyi and Hodgson (32), using a well-characterized antiserum for GABA (33). All steps were carried out on drops of Millipore (Bedford, MA)-filtered solutions (placed on Parafilm) in humid chambers.

1. Incubate for 10 min in 1% (w/v) periodic acid.
2. Rinse in doubly-distilled water.
3. Incubate for 10 min in 2% (w/v) sodium metaperiodate in doubly-distilled water.
4. Rinse in doubly-distilled water.
5. Rinse three times (2 min each) in pH 7.4 Tris-buffered saline (TBS).
6. Incubate for 30 min in 1% (w/v) ovalbumin (in TBS).

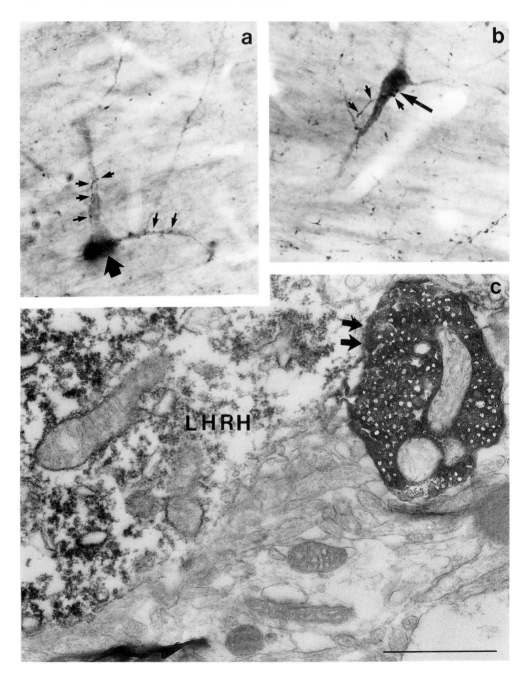

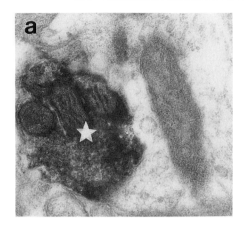
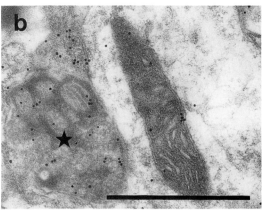

Fig. 11 Serial ultrathin sections of the same PHAL-immunoreactive bouton (star) as in Fig. 10. The section shown in (b), after deosmication, has been postembedding immunostained for GABA, using 20-nm colloidal gold as the immunomarker. Bar: 1 μm.

7. Rinse three times (10 min each) in 1% (v/v) normal goat serum (NGS) in TBS.
8. Incubate for 1–2 hr in rabbit anti-GABA diluted 1:1000 in NGS/TBS.
9. Wash twice (10 min each) in TBS.
10. Rinse for 10 min in 0.05 M Tris buffer (pH 7.5) containing 1% (w/v) bovine serum albumin (BSA) and 0.5% (v/v) Tween 20.
11. 2 hr incubation in gold-conjugated (15 nm) goat anti-rabbit IgG diluted 1:10 in the same buffer.
12. Wash twice (5 min each) in doubly-distilled water.
13. Contrast with saturated uranyl acetate (30 min) and lead citrate (20–30 sec).

The results demonstrated that practically all of the PHAL-containing axon terminals are GABAergic (Fig. 11). Thus, this observation, taken together

Fig. 10 Light micrograph (a and b) and electron micrograph (c) taken from the infundibular nucleus of an African green monkey demonstrate DAB/NiDAB double immunostaining for PHAL (NiDAB) injected into the main location of PR-containing neurons and LHRH cells labeled by a DAB reaction. Small arrows in (a) and (b) indicate the black NiDAB-labeled PHAL-containing fibers contacting the brown LHRH-immunoreactive neurons [arrowhead in (a)]. (c) Electron micrograph of the synaptic contact (arrowheads) indicated in (b) (long arrow). Bar: 1 μm. Original magnification of (a) and (b): ×1250.

with the results of the aforementioned double immunostaining study for PHAL and LHRH, demonstrate that GABAergic PR-containing neurons that occupy an area between the fornix and ventromedial nucleus in the monkey hypothalamus can directly influence (inhibit) the activity of LHRH-producing neurons.

Remarks

This technique has a major limitation in that only those tissue antigens that "survive" the 56–60°C curing required for EM embedding can be postembedding immunostained. The most reliable results have been reported on postembedding immunostaining for GABA on material perfused with fixatives containing a minimum of 0.8% glutaraldehyde (31–33). Finally, all sections selected for postembedding immunostaining should be embedded in Durcupan (ACM Fluka Ronkonkoma, NY).

Double Immunostaining Combined with Degeneration

In many cases, the application of double immunostaining alone cannot answer specific questions. For example, on the basis of pharmacological observations, it was determined that interactions between GABA-, catecholamine-, and LHRH-containing systems of the MPOA play an important role in the control of episodic LH release (34–39). Using a double immunostaining technique we were able to demonstrate the GABAergic innervation of LHRH neurons (21). Specific catecholamine markers (e.g., an antiserum against dopamine), which require high concentrations of glutaraldehyde in the fixative (40), cannot be used in EM double immunostaining. Therefore, to demonstrate catecholamine innervation of GABA and LHRH neurons, an antiserum generated against tyrosine hydroxylase (TH) (which is the rate-limiting enzyme for all types of catecholamine synthesis) had to be used. However, these double immunostaining experiments specify neither the transmitter content (dopamine, noradrenaline, or adrenaline) nor the origin (local or extrinsic) of the TH-immunoreactive boutons terminating on GABA and LHRH neurons in the MPOA.

After surgical or chemical lesions, catecholamine fibers show a unique, acute axonal degeneration. In these boutons, the presence of electron-dense dark bodies, so-called autophagous cytolysosomes, is the first sign of acute degeneration (41). Furthermore, the synaptic vesicles in these degenerating axon terminals, and their transmitter content (even neuropeptides if they are colocalized with catecholamines; see below), are preserved and can be immunovisualized several days after the lesion.

Example 1

These distinctive characteristics of catecholamine fiber degeneration were used to define further the origin and transmitter content of the TH-immunoreactive boutons that terminate on MPOA GABAergic and LHRH-producing cells. After transection of the ascending catecholamine fibers containing epinephrine and noradrenaline, a double immunostaining was performed for TH plus glutamate decarboxylase (GAD) and TH plus LHRH. Immunoreactivity for TH was labeled with immunoperoxidase, whereas GAD- and LHRH-immunoreactive profiles were labeled with immunogold. The results demonstrated that whereas all of the TH-immunoreactive axon terminals contacting LHRH neurons are intact, a large population of TH-containing boutons that innervate GABA cells shows signs of acute degeneration (Fig. 12). Thus, the extrinsic catecholamine fibers, which almost exclusively contain epinephrine in the MPOA, regulate LHRH release via the MPOA GABAergic system. Furthermore, intact TH axons, which represent processes of intrinsic dopamine neurons, can influence the activity of both MPOA GABA and LHRH neurons (42).

Example 2

Similar experimental paradigms were applied in another study to determine whether zona incerta (ZI) dopamine neurons are directly involved in the regulation of LHRH release. First, the anterograde tracer PHAL was iontophoretically administrated (see above) into the ZI. Ten days later, 6-hydroxydopamine (6-OHDA; 1 μg in 0.5 μl of saline containing 0.02% (w/v) L-ascorbic acid) was stereotaxically injected into the same area of desipramine [25 mg/kg, intraperitoneal (ip)]-pretreated animals. Desipramine administration preserves the noradrenergic fibers (43). Two days later, animals were sacrificed and double immunostaining for PHAL and LHRH was performed on Vibratome sections from the site of injection and the MPOA. The PHAL-labeled, degenerated (autophagous cytolysosome-containing) dopamine boutons could not be observed in synaptic contact with LHRH neurons; they were found only in an area more lateral from the main location of the LHRH cells (see Fig. 4 in Ref. 44). These results indicate an indirect involvement of the ZI dopamine system in the regulation of LHRH release.

Example 3

As mentioned previously, not only the TH, but the neuropeptide content of degenerating catecholamine fibers is preserved, and can be immunovisualized 36–48 hr following pathway transection or chemical lesion of the parent

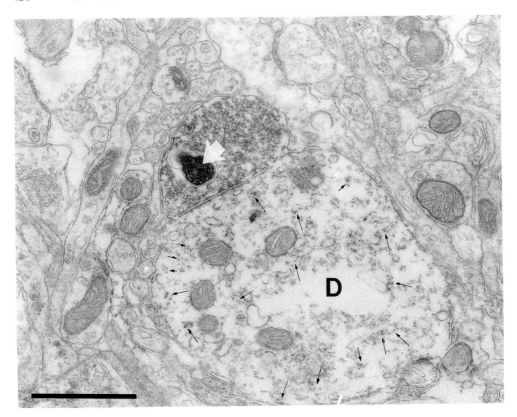

FIG. 12 Electron micrograph shows a synaptic contact between a degenerated, autophagous cytolysosome-containing (white arrowhead), TH-immunoreactive bouton and an immunogold-labeled (arrows), GAD-immunopositive dendrite (D) in the rat MPOA 2 days following a complete transection of the ascending catecholamine fibers. Bar: 1 μm.

neurons (45). This information was put to use in determining whether different neuropeptide-containing dopamine fibers terminate on the same population of lateral septal area neurons.

Both facilitatory and inhibitory dopamine actions have been reported in the estrogen and androgen hormone-sensitive lateral septum. This area receives dopamine fibers from the hypothalamus, as well as from the ventral tegmentum (46). In the hypothalamus of rats, a population of dopamine cells cocontains somatostatin, whereas in a subgroup of ventral tegmental area (VTA) dopamine cells, neurotensin is colocalized (see review in Ref. 47). Because neurotensin increases the firing rate of neurons and attenuates dopamine-induced inhibition (48), one can speculate that the opposite dopamine actions are associated with these two groups of dopamine afferents. An experiment was designed to determine whether the excitatory "neurotensinergic," dopamine axons and the inhibitory, "somatostatinergic" axons terminate on the same population of somatospiny, lateral septal area neurons, which are surrounded by TH-immunoreactive baskets (49). Targeted, stereotaxic microinjections of 6-OHDA were placed into the VTA or into the periventricular area of desipramine-pretreated rats. After a 36- to 48-hr survival time, animals were sacrificed and consecutive Vibratome sections cut from the lateral septal area were immunostained for TH, somatostatin, or neurotensin. The VTA-lesioned animals and the hypothalamus-lesioned rats exhibited degenerated (autophagous cytolysosome-containing) neurotensin- and TH-immunopositive fibers or degenerated somatostatin- and TH-containing boutons, respectively, in contact with lateral septum somatospiny neurons. These data implicate that the hypothalamic "somatostatinergic" dopaminergic afferents can have an inhibitory effect, whereas mesencephalic "neurotensinergic" dopamine afferents exert an excitatory action on the same lateral septal area neurons (47).

Interconnections between Hypothalamic Steroid-Sensitive Circuits and LHRH Neurons

As described above, the absence of meaningful quantities of estrogen receptors in LHRH cells indicates the importance of feedback on estrogen-sensitive neurons that connect with the LHRH neurons. Using the methods described above, a schematic representation of the functional interconnections between different neuronal systems involved in the regulation of gonadotropin release is being composed (Fig. 13). Thus, in the rat, the principal neurons (LHRH) in the regulation of LH secretion are located in the MPOA, and project to the external zone of the median eminence, where they release their product into the portal vessels. At a distance from the MPOA lies the arcuate nucleus (AN), which is the main feedback site and regulator of LHRH production and release. The major link between these two nuclei is the β-endorphin pathway, which probably mediates the effects on LHRH neuronal activity of most neurotransmitter- and neuropeptide-containing systems within this area. This includes the arcuate nucleus dopamine-, GABA (GAD)- (23), galanin- (C. Leranth, F. Naftolin, M. Shanabrough, and T. L. Horvath, unpublished observation), neuropeptide Y- (27), and substance P-containing (C. Leranth, F. Naftolin, M. Shanabrough, and T. L. Horvath, unpublished observation) neurons. The extent of cross-connections between these systems suggests that under various circumstances, one system may exert differ-

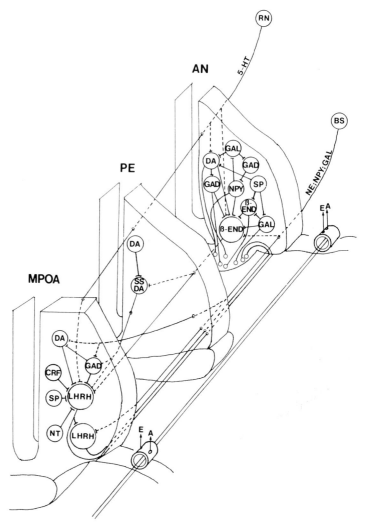

Fig. 13   Schematic representation of the main synaptic interconnections between estrogen-sensitive systems and the LHRH-producing neurons. MPOA, Medial preoptic area; PE, preoptic area; AN, arcuate nucleus; RN, raphe nuclei; CRF, corticotropin hormone-releasing hormone; SS, somatostatin; DA, dopamine; NE, norepinephrine; NT, neurotensin; BS, brain stem; E, circulating estrogens; A, circulating androgens; GAL, galanin; GAD, γ-aminobutyric acid; NPY, neuropeptide Y; SP, substance P; β-END, β-endorphin; 5-HT, serotonin.

ent effects on the LHRH cells by switching its information flow to different target neurons. For example, the arcuate nucleus dopamine cells inhibit the (inhibitory) opiate system via direct synaptic contacts (23) that will result in an ultimate disinhibition on LHRH cells. This has been supported by pharmacological studies utilizing opioids or opioid antagonists showing that the opiate mediation is clearly inhibitory. On the other hand, if the opioid→-dopamine synapses are blocked, dopamine could influence the opiate system indirectly, via inhibition of the GABA neurons (21), which will result in an increase in the opiate inhibition of LHRH release. There are both morphological and pharmacological data supporting this, and also the possibility exists that the other systems of the arcuate nucleus may interact with each other in the same manner. Although more pharmacological observations are needed to investigate this possibility, most of the morphological basis has already been delineated.

From double immunocytochemical methods like those described in this chapter, it has also been determined that in the MPOA (in the vicinity of LHRH neurons), there are direct estrogen–target neural systems (such as dopamine, corticotropin-releasing hormone, substance P, neurotensin, GABA) that innervate the LHRH cells (30,44,50–52). It has been suggested that these neurons play the main role in the positive feedback action of estrogen on LH release, whereas the arcuate nucleus circuits are responsible for the negative feedback. However, such a categorical separation of the two estrogen effects may not be appropriate, because, for example, estrogen-target areas from regions other than the hypothalamus [i.e., dorsal raphe nucleus (serotonin) and brain stem areas (norepinephrine, neuropeptide Y, galanin)] also provide significant inputs on both the arcuate nucleus (23,27,53,54) and the LHRH-secreting neurons (53–56).

Concluding Remarks

Estrogens are the major hormonal mediators in gonadotropin feedback. In the absence of estrogen receptors in luteinizing hormone-releasing hormone (LHRH)-producing neurons, attention has been focused on estrogen-sensitive neurons and circuits that relate to the LHRH neurons. To elucidate the synaptological interconnections between these groups of cells and their functional contacts with LHRH neurons, the development of electron microscopic double immunostaining techniques is needed. This chapter has provided detailed descriptions of anesthesia, fixation, penetration enhancement techniques, and the most frequently used electron microscopic double immunostaining procedures and their combination with retrograde and anterograde tracer techniques, and degeneration. Furthermore, we have furnished an account of the most important synaptic connections between transmitter systems involved in the regulation of gonadotropin release.

Acknowledgment

These studies were supported by NIH Grants HD23830, NS26068, MH 44866 (C.L.), and HD13587 (F.N.)

References

1. Knobil, E. (1980). *Recent Progr. Horm. Res.* **36,** 53–88.
2. MacLusky, N. J., and Naftolin, F. (1981). *Science* **211,** 1294–1303.
3. Halász, B., and Pupp, L. (1965). *Endocrinology (Baltimore)* **77,** 553–573.
4. Goodman, R. L. (1978). *Endocrinology (Baltimore)* **102,** 151–159.
5. Goodman, R. L., and Knobil, E. (1981). *Neuroendocrinology* **32,** 57–63.
6. Shivers, B. D., Harlan, R. E., Morell, J. I., and Pfaff, D. W. (1983). *Nature (London)* **304,** 345–347.
7. Leranth, C., Palkovits, M., MacLusky, N. J., Shanabrough, M., and Naftolin, F. (1986). *Neuroendocrinology* **43,** 526–533.
8. Molin, S.-O., Nygren, H., and Dolonius, L. (1978). *J. Histochem. Cytochem.* **26,** 412–414.
9. Kosaka, T., Nagatsu, I., Wu, J.-I., and Hama, K. (1986). *Neuroscience* **18,** 975–990.
10. Görcs, T., Leranth, C., and MacLusky, N. J. (1986). *J. Histochem. Cytochem.* **34,** 1439–1447.
11. Leranth, C., and Frotscher, M. (1987). *Histochemistry* **86,** 287–290.
12. Brown, T. J., MacLusky, N. J., Leranth, C., Shanabrough, M., and Naftolin, F. (1990). *Mol. Cell. Neurosci.* **1,** 58–77.
13. Leranth, C., Shanabrough, M., and Naftolin, F. (1991). *Neuroendocrinology* **54,** 571–579.
14. Leranth, C., MacLusky, N. J., Brown, T. J., Chen, C. E., Redmond, E. D., and Naftolin, F. (1992). *Neuroendocrinology* **55,** 667–682.
15. Hsu, S. M., Raine, L., and Fayer, H. (1981). *J. Histochem. Cytochem.* **29,** 577–590.
16. Sternberger, L. A., Hardy, P. H., Cuculis, J. J., and Meyers, H. G. (1970). *J. Histochem. Cytochem.* **18,** 315–333.
17. Horvath, T. L., Shanabrough, M., Naftolin, F., and Leranth, C. (1993). *Endocrinology (Baltimore)* **133,** 405–414.
18. Leranth, C., Sakamoto, H., MacLusky, N. J., Shanabrough, M., and Naftolin, F. (1985). *Brain Res.* **331,** 376–381.
19. Leranth, C., Sakamoto, H., MacLusky, N. J., Shanabrough, M., and Naftolin, F. (1985). *Brain Res.* **331,** 371–375.
20. Leranth, C., Segura, L. M. G., Palkovits, M., MacLusky, N. J., Shanabrough, M., and Naftolin, F. (1985). *Brain Res.* **345,** 332–336.

21. Leranth, C., Sakamoto, H., MacLusky, N. J., Shanabrough, M., and Naftolin, F. (1985). *Histochemistry* **82,** 165–168.

22. Leranth, C., MacLusky, N. J., Shanabrough, M., and Naftolin, F. (1988). *Brain Res.* **449,** 167–176.

23. Horvath, T. L., Naftolin, F., and Leranth, C. (1992). *Neuroscience* **51,** 391–399.

24. Sternberger, L. A., and Joseph, S. H. (1979). *J. Histochem. Cytochem.* **27,** 1427–1429.

25. Wouterlood, F. G. (1988). *Histochemistry* **89,** 421–428.

26. Horvath, T. L., Naftolin, F., and Leranth, C. (1992). *Endocrinology* (*Baltimore*) **131,** 1547–1555.

27. Horvath, T. L., Naftolin, F., Kalra, S. P., and Leranth, C. (1992). *Endocrinology* (*Baltimore*) **131,** 2461–2467.

28. Knobil, E., and Hotchkiss, J. (1988). *in* "The Physiology of Reproduction" (E. Knobil and J. Neill, eds.), pp. 1971–1994. Raven, New York.

29. Fuchs, E., Mansky, T., Stock, K., Vijayan, E., and Wuttke, W. (1984). *Neuroendocrinology* **38,** 484–489.

30. Leranth, C., MacLusky, N. J., Sakamoto, H., Shanabrough, M., and Naftolin, F. (1986). *Neuroendocrinology* **40,** 536–539.

31. Freund, T. F., and Antal, M. (1988). *Nature* (*London*) **336,** 170–173.

32. Somogyi, P., and Hodgson, A. J. (1985). *J. Histochem. Cytochem.* **33,** 249–257.

33. Hodgson, A. J., Penke, B., Erdei, A., Chubb, I. V., and Somogyi, P. (1985). *J. Histochem. Cytochem.* **33,** 229–239.

34. Baraclough, C., and Wise, P. (1982). *Endocr. Rev.* **3,** 91–119.

35. Fuchs, E., Mansky, T., Stock, K., Vijayan, E., and Wuttke, W. (1984). *Neuroendocrinology* **38,** 484–489.

36. Honma, K., and Wuttke, W. (1980). *Endocrinology* (*Baltimore*) **106,** 1848–1853.

37. Kalra, S., and Crowley, W. (1982). *Endocrinology* (*Baltimore*) **111,** 1403–1405.

38. Kalra, S., and Simpkins, J. (1981). *Endocrinology* (*Baltimore*) **109,** 776–782.

39. Rance, N., Wise, P., Selmanoff, M., and Barraclough, C. (1981). *Endocrinology* (*Baltimore*) **108,** 1795–1802.

40. Ontaniente, B., Gerffard, M., and Callas, A. (1984). *Neuroscience* **3,** 385–393.

41. Zaborszky, L., Leranth, C., and Palkovits, M. (1979). *Brain Res. Bull.* **4,** 99–117.

42. Leranth, C., MacLusky, N. J., Shanabrough, M., and Naftolin, F. (1985). *Neuroendocrinology* **48,** 591–602.

43. Tabakoff, B., and Ritzmann, R. F. (1977). *J. Pharmacol. Exp. Ther.* **203,** 319–331.

44. Horvath, T. L., Naftolin, F., and Leranth, C. (1993). *J. Neuroendocrinol.* **5,** 71–79.

45. Kagotani, Y., Tsuruo, Y., Hisano, S., Daikoku, S., and Chihara, K. (1989). *Cell Tissue Res.* **257,** 269–278.

46. Lindvall, O., and Stenevi, U. (1978). *Cell Tissue Res.* **190,** 383–407.

47. Jakab, R. L., and Leranth, C. (1993). *Exp. Brain Res.* **92,** 420–430.

48. Stowe, Z. N., and Nemeroff, C. B. (1991). *Neuropeptides* **11,** 95–100.

49. Jakab, R. L., and Leranth, C. (1990). *J. Comp. Neurol.* **302,** 305–321.

50. MacLusky, N. J., Naftolin, F., and Leranth, C. (1988). *Brain Res.* **47,** 391–395.

51. Tsuruo, Y., Kawano, H., Hisano, S., Kagotani, Y., Daikoku, S., Zhang, T., and Yanaihara, N. (1991). *Neuroendocrinology* **53,** 236–245.

52. Hoffman, G. E. (1985). *Peptides* **6,** 439–461.

53. Kiss, J. (1984). *Neurosci. Lett., Suppl. No. 18*, S186.
54. Kiss, J., Leranth, C., and Halász, B. (1984). *Neurosci. Lett.* **44,** 119–124.
55. Merchenthaler, I., Lopez, F., and Negro-Vilar, A. (1990). *Proc. Natl. Acad. Sci. U.S.A.* **87,** 6326–6330.
56. Kalra, S. P., and Crowley, W. R. (1992). *Front. Neuroendocrinol.* **13,** 1–46.

[26] Gene-Mediated Steroid Control of Neuronal Activity

Marian Joëls

Introduction

In the late 1960s McEwen and co-workers (1) showed that corticosteroid hormones, after passing the blood–brain barrier, effectively bind to receptors in the brain. Almost from the beginning, electrophysiological studies were carried out to address the following question: how do corticosteroids, via their intracellular receptors in the brain, affect the characteristic feature of neurons, that is, the transduction of electrical signals? Clearly, the methodological approach was dictated by the technical advances. Initial studies employed multiunit recording in anesthetized animals in combination with peripheral administration of steroids. Later, local application of the steroids became possible with the development of microiontophoretic techniques. A more in-depth study of electrical membrane properties and steroid-induced modulation was feasible only when *in vitro* brain preparations were developed to a level at which they bear relevance to the *in vivo* condition. The latest addition in methods, that is, the use of patch-clamp recording, now allows the examination of steroid actions right up to the level of single ion channels.

Rather than giving a chronological overview of the studies related to steroid actions on electrical properties of neurons, this chapter highlights the methodological considerations that need to be made when studying this subject. The focus is on changes in electrical activity that are accomplished via gene-mediated corticosteroid actions, via the classic intracellular mineralocorticoid and glucocorticoid receptors (MRs and GRs, respectively) in the brain. Investigation of central, gene-mediated actions by sex steroids follows the same principles. For the methodology related to fast steroid-mediated effects, especially those induced by neurosteroids, the reader is referred to [27] in this volume.

When considering the gene-mediated actions of corticosteroid hormones, one may expect some aspects in their mechanism of action to distinguish them from classic transmitters or even from neuropeptides. These aspects emphasize the necessity for a specific experimental approach when investigating the role of corticosteroids in the regulation of neuronal excitability, and are summarized here.

Methods in Neurosciences, Volume 22

1. In general, corticosteroid actions are considered to be conditional, that is, the steroids do not directly change functional processes, but rather restore these processes once they are out of balance. This principle may also hold for corticosteroid actions on electrical activity. If so, little effect of the steroids will be evident when the neurons are at their "normal" resting condition; however, if the neurons are shifted from this condition by activation of ligand- or voltage-induced changes in ion fluxes, steroids will help to restore the temporary disturbance. Such actions can be best investigated if one has control over the ligand- and voltage-gated input to the cell. So far, this can only be fully achieved when recording under *in vitro* conditions. Section II of this chapter reviews the presently available electrophysiological methods, with their specific (dis)advantages when studying corticosteroid actions in the brain.

2. Owing to the genomic mechanism of action, the steroid-induced phenomena will appear after a long delay and they can last for many hours. In principle, this would require recordings that stretch over periods of hours. However, usually recording conditions are not that stable. Therefore, alternative designs are necessary to allow enough time for the induction of genomic effects on the one hand, and to record changes over periods of many hours on the other hand. Section III discusses the timing protocol that can be used when studying steroid effects.

3. Neurons in some parts of the brain, like the hippocampus, contain both MRs and GRs. The two receptor subtypes may evoke entirely different actions. To establish the MR- and/or GR-mediated control of excitability, selective activation of the receptor subtypes is important. This can be achieved when the endogenous ligand is removed (adrenalectomy) and MRs and GRs activated by selectively specific (ant)agonists. Section IV discusses the steroid application routes that are available in combination with the various electrophysiological techniques.

The final part of this chapter briefly alludes to the future possibilities for the investigation of gene-mediated steroid actions on electrical activity in the brain.

Electrophysiological Techniques

Almost every aspect of electrical activity in the brain can be studied, from the electroencephalogram (EEG) to single-channel activity. To choose the most sophisticated technique is tempting, but not always the best choice. For instance, if one is interested in the effect of steroids on electrical activity in relation to certain behavioral situations, recording single-channel activity

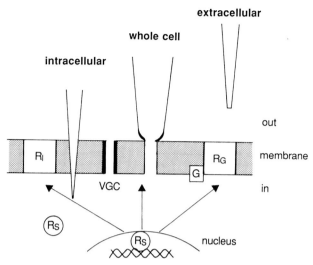

FIG. 1   Activated steroid receptors (R_S) bind to the genome, acting as transcription factors for specific genes. As a result of the altered protein synthesis, properties of the membrane can be changed. These properties comprise the ion fluxes through voltage-gated ion channels (VGC), ion conductances through ionotropic receptors (R_I), and processes linked to transmitter activation of G protein-coupled receptors (R_G). The steroid-induced effects on electrical activity of single neurons can be recorded with an extracellular electrode, an intracellular microelectrode or a patch electrode in the whole cell configuration.

is an unnecessary level of detail, and would require a recording situation (*in vitro*) that precludes behavioral observations. Clearly, the question that needs to be solved determines the options.

In principle, several levels of investigation are possible: activity from the brain as a whole (or from large parts of it), activity from groups of neurons, recording from single cells (see Fig. 1), or at the level of a single ion channel. Table I summarizes briefly which techniques are available. (For further reading on this subject see Refs. 2–5.) The preparations in which these recordings can be carried out are also given. These comprise *in vivo* and *in vitro* preparations, the former used in conjunction with EEG and multiunit recordings, the latter mostly with single-cell and single-channel recordings.

Should corticosteroid actions be preferably studied *in vivo* or *in vitro*? Depending on the level of investigation, sometimes there may be no choice: EEG studies must be carried out *in vivo*, whereas single-channel studies are presently feasible only *in vitro*. However, for other techniques there may be a choice of preparations; e.g., studies of extracellular firing activity can

TABLE I   Techniques Available to Determine Electrical Activity[a]

Source of activity	Technique	Preparation	Agonist application	Advantages	Disadvantages
Whole brain	EEG	*In vivo*	Peripheral, icv	Link to behavior/clinic	Interpretation
Multiunits	Extracellular	*In vivo*	Peripheral, icv, local	Link to behavior	Interpretation
	Extracellular	*In vitro*	Local, perfusion	Control over medium Electrode positioning	Incomplete input
Single units	Extracellular	*In vivo*	Peripheral, icv, local	Link to behavior	Mechanism?
	Extracellular	*In vitro*	Local, perfusion	Control over medium	Mechanism?; incomplete input, poor voltage control
	Intracellular	*In vitro:* slice	Local, perfusion	Control over medium	
	Patch, whole cell	*In vitro:* slice, isolated cells culture	Local, perfusion	Control over extra-/intracellular milieu; Good voltage control	Washout of intracellular components
Single channel	Patch (cell attached, excised)	*In vitro:* slice, isolated cells culture	Local, perfusion	Elucidation of membrane biophysics	Physiological relevance?

[a] Selection of the level of investigation limits the possibilities of the applied techniques, the type of preparation, and agonist application. Specific (dis)advantages for each of the approaches are indicated.

be carried out *in vivo* and *in vitro* (e.g., in slices). Selecting the *in vivo* approach has the advantage that the data are obtained under physiologically more relevant conditions, with intact synaptic connections and "normal" input. However, recording of extracellular activity *in vivo* generally requires the use of anesthetics, so that it is questionable how "normal" the input to the recorded cells really is. This is particularly relevant for the investigation of steroid actions, because the use of most anesthetics is associated with high endogenous levels of corticosteroid hormones. *In vitro* recording certainly has some obvious advantages, such as the easy positioning of the electrodes and the control over the extracellular medium, but here all the variables that are related to "keeping the tissue alive" come into play. Because steroid hormones are implicated in glucose metabolism and viability of neuronal tissue, imposing *in vitro* conditions may be just the sort of "endangerment" that is exacerbated by the corticosteroids. Also, it should be realized that it is unclear what happens to steroid effects once the tissue is cut off from its natural supply of steroid. There are indications that some of the gene-mediated actions induced *in vivo* can extend even to subsequent *in vitro* recording situations; however, how these actions will alter over the *in vitro* recording period (ca. 8 hr) is not clear. The perfect choice of course is to study the question from both an *in vivo* and an *in vitro* approach, but this requires a large investment in time, money, and patience.

Are any of the techniques in Table I particularly suited for the investigation of steroid actions? Again, there is no ready answer to this question. As indicated above, the preference for any technique is mainly dictated by the type of question that needs to be answered. In studying the mechanistic site of putative steroid actions, it may be wise to settle for intracellular current clamp recordings, initially. These recordings give a first insight into which membrane processes are a target for steroid actions. The advantage of current-clamp over voltage clamp recordings is that the current-clamp technique is a far more normal condition for neurons and is therefore more relevant for what the changes in membrane properties could mean in terms of an input–output relationship. Voltage-clamp studies can then be undertaken to elucidate steroid-mediated effects when the cell is at a defined membrane potential.

Alternatively, one may be interested in studying steroid actions in relation to behavioral function, for example, by monitoring electrical activity under stressful conditions, multiunit or (when possible) single-unit recording in freely moving animals is indicated.

In summary, the electrophysiological study of steroid-induced actions in the brain is possible. There is a choice of techniques and preparations, but the appropriateness of each of them depends on the question that needs to be answered.

Timing Protocol

Generally, electrophysiologists have employed recording techniques for the study of rapid and short-lasting changes in electrical properties of neuronal membranes. Most of the presently known neurotransmitter and even neuropeptide actions on membrane conductivity take place within seconds to minutes and are rapidly reversed after washout of the agent. The effects are mediated by ligand-gated ion channels or G protein-coupled receptors, which through a second messenger can alter a membrane conductance.

The mechanism of action for (cortico)steroid hormones is basically different. Thus, the activated steroid–receptor complex binds to the DNA, subsequently altering gene expression and thus protein synthesis. Some of these proteins may be involved in signal transduction, for example, in the synthesis of transmitter receptors, the generation of a second messenger, or the phosphorylation of an ion channel. The resulting changes in electrical membrane properties will therefore be slow at the onset and long lasting. This aspect of timing needs to be taken into consideration, when studying the action of corticosteroids.

To follow the process of activation of steroid receptors, gene transcription, altered protein synthesis, and final change in membrane conductivity in one neuron, stable recording periods of many hours are required. Is this possible? With extracellular recording, stable recordings over periods of hours should, in principle, be feasible. Still, when carried out *in vivo*, one should realize that the neuronal activity also depends on the level of anesthe-

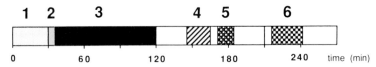

FIG. 2   Timing protocol for *in vitro* evaluation of steroid mediated actions. At the start of the experiment (about 10:00 AM) the rats are subjected to a novel environment (**1**), inducing a mild stress so that influences of a variable background are diminished. After decapitation slices are prepared on ice (**2**); trunk blood is collected for analysis of plasma corticosteroid levels. The slices are kept in a holding chamber at room temperature, for at least 1.5 hr (**3**). Subsequently one slice at a time is transferred to the recording chamber, which is continuously perfused with a warmed (32°C), carbogenated Ringer's solution. Neuronal properties are monitored before steroid application (**4**). Next, steroids are applied (**5**) for 20 min in a known concentration, by the addition to the Ringer's solution. When antagonists are applied, these should be administered for 60 min, starting 20 min before the agonist administration. A delay of 1 to 4 hr after steroid application is introduced to allow enough time for the development of gene-mediated effects. The properties of neurons recorded after this delay are compared with those observed before steroid addition (**6**).

sia, which may very well alter over the course of hours. Also, gradual changes over time in the recording position may not be disastrous in terms of losing the cell, but still may alter the response of the neuron under study.

Intracellular recordings with microelectrodes for hours, even when performed *in vitro*, are difficult to perform. Any slight movement (such as may be induced by the appearance of an air bubble) may interrupt the recording. Also, the tip of the microelectrode may become blocked by debris, altering the recording characteristics of the electrode itself. Similarly, whole cell recordings with patch electrodes are in most cases inappropriate for prolonged recordings: The tip of the electrode is quite large, so that extensive dialysis of the cell content takes place. The resulting changes in, for example, ATP content, which affects ATP-dependent processes, usually result in a considerable decrease in ionic conductances.

Even though recordings that extent over many hours are the preferred approach when examining gene-mediated steroid actions, the yield of such an approach may be extremely low. One may choose to settle for the second-best (and far less frustrating) alternative. The latter makes use of the idea that steroid actions take place anywhere between one to several hours after a brief steroid application *in vitro* (see Fig. 2). If neurons are recorded after such a delay, the ionic conductances or transmitter responses can be compared with the neurons recorded before the *in vitro* steroid perfusion. Only the slowly developing (presumably gene-mediated) effects of the steroids are studied, isolated from possible rapid effects that could be mediated by membrane receptors (see [27] in this volume).

The major drawback to using this approach is that under all circumstances there may be considerable variation in the membrane properties of neurons in a given area. Even in an area with a relatively homogeneous population of neurons, such as the CA1 area of the hippocampus, where most of the neurons are pyramidal cells and few interneurons are encountered, individual variations in shape, size, transmitter receptor distribution, and so on, occur. To rule out the influence of this natural variation, large groups of neurons (>10) should be compared before and after treatment. The results obtained by this "comparison protocol" should be replicated when recording from individual neurons over a period of many hours (even when using a limited number of cells). Also, gradual changes in recording conditions, which may be so trivial as the changing of recording electrode properties over the course of the day, should be taken into account. This means that neurons before (or without) steroid treatment should not be automatically recorded in the first half of the day and steroid-treated groups invariably at the end of the day. This is of particular relevance when testing steroid actions on synaptically evoked responses, because the synaptic connections *in vitro* tend to run down sooner than the intrinsic membrane properties.

Application of Steroids

The *in vitro* application method described above is not the only way to administer steroids. In principle, there are two approaches to study the influences of corticosteroid hormones on electrical activity. First, one can examine how the absence of steroids affects the neuronal properties. Second, one can investigate how application of selective MR and/or GR agonists affects the excitability in the brain. Some aspects of these two approaches are highlighted below.

Because steroids are synthesized outside of the brain, removal of the site of synthesis is relatively easy. Corticosteroid synthesis can also be indirectly prevented by removal of the pituitary. Another approach is to treat animals chronically with MR/GR antagonists. Recently, the possibility of knock-out experiments has come into view.

Regardless of the method used to eliminate the source of the endogenous corticosteroid hormone, the electrophysiological experiment can be started under conditions of a defined state of MR/GR occupation in the brain, that is, with MRs/GRs unoccupied. However, complete removal of corticosteroid hormones from the blood will induce many secondary effects, particularly when several days elapse between the surgery and the experiment. First, there are secondary effects that are the direct result of the removal of the steroids and, therefore, can be restored with steroid replacement. A well-documented example concerns the degeneration of dentate granule cells observed after adrenalectomy (ADX; see [24] in this volume); the electrical properties of neurons receiving input from the dentate gyrus, and of the surviving dentate neurons themselves, may be considerably changed. Another relevant example concerns the upregulation of MRs and GRs after removal of steroids from the blood. Second, there are effects related to hormones other than the steroids, which develop due to the rise in adrenocorticotropin (ACTH) or shortage in noradrenaline associated with ADX. Although the ADX model offers possibilities for studying steroid actions on electrical activity, these side effects must be noted. If possible, it is advisable to perform control studies in sham-operated animals, trying to link the plasma corticosteroid levels to the recorded membrane property. Alternatively, a series of experiments can be performed with variable delay between the surgery and the experiment (e.g., 2 hr, 2 days, and 2 weeks after ADX).

Another approach to establishing a controlled degree of MR/GR occupation is through the administration of selective agonists. There are many good MR and GR(ant)agonists available, some of which need to be metabolized first so that they can be used only *in vivo*. During *in vivo* studies application of the compounds can be achieved by peripheral or intracerebroventricular (icv) injection. Local application of the MR/GR ligands by iontophoresis is possible, but this is meaningful only for gene-mediated steroid effects when combined with prolonged electrophysiological recordings, not a simple undertaking *in vivo*. In principle, the steroid administration can be carried out in animals with an intact hypothalamic–pituitary–adrenal (HPA) axis, but then the MR/GR occupation at the start of the experiment is hard to establish. Also, in animals with an intact HPA axis, the experimental conditions (e.g., anesthesia) may prohibit a proper study of effects mediated by MRs only. Because the endogenous level of the steroid is already high, a considerable degree of the GRs will always be activated.

During *in vitro* studies, steroids can be administered by a brief addition to the medium. Binding studies have shown that a 20-min incubation with 30 nM corticosterone at ca. 30°C should result in considerable occupation of both MRs and GRs. This was confirmed in electrophysiological studies, in which changes in membrane properties were observed for at least 4 hr after such a brief steroid incubation. The steroid actions were at least partly maintained if the tissue was subjected to trituration following the steroid incubation, allowing the use of acutely dissociated cells. Similar observations have been made with aldosterone, the selective GR agonist RU 28362, the GR antagonist RU 38486, and the MR antagonist spironolactone. Some of the antagonists (RU 28318), which need to be metabolized first, were less appropriate for *in vitro* use. Also, the MR antagonist RU 26572, which is best dissolved in polyethylene glycol (PEG), turned out to be less useful, because the vehicle affected electrical activity by itself.

Instead of the steroid application *in vitro*, the animals can be treated *in vivo*, before the *in vitro* preparation procedure is started. Chronic steroid regimes may introduce all kinds of adaptive processes that are complex and hard to unravel. Acute application of steroids, just before the *in vitro* preparation, is less complex; still, it remains difficult to predict to what degree steroid actions established *in vivo* will be maintained *in vitro*.

In short, there are ample possibilities for studying electrical activity in conjunction with changes in MR/GR occupation, perhaps even more so than for many other compounds that are active in the brain. For instance, it is relatively easy to remove steroids altogether from the system, something that is not easily achieved for frequently investigated transmitters such as glutamate or acetylcholine. Furthermore, there are many selective (ant)agonists available for both MRs and GRs, so that either of the receptor populations in the brain can be selectively occupied. Finally, steroids are lipophilic, so that they will readily enter the cells. Because of this, effective doses of the steroids should be around the K_d value of the receptors; again, this is something that is usually not seen for electrophysiological studies with classic transmitters such as acetylcholine or noradrenaline, where the effective dose can be 100 to 1000-fold higher than the K_d value of the receptor.

Future Developments

Research has shown that corticosteroids do affect electrical activity of neurons, for prolonged periods of time. This was shown both *in vivo* and *in vitro* using extracellular and intracellular recording techniques.

With the use of electrophysiology, at least two studies can be made in the near future. First, it will be of imminent importance to record electrical activity under physiologically relevant situations, that is, in freely moving animals. With the current knowledge of putative steroid actions both at the single-cell level and at the behavioral level (see [29] in this volume), it should be possible to establish the link between these two levels of investigation. It will be important to show that all the phenomena described now for *in vitro* single-cell activity indeed develop as predicted under *in vivo* recording conditions.

The second question that can be tackled with the current techniques concerns the steroid receptor-mediated modulation of single ion channel activity. Although the results so far support the fact that steroid hormones can affect ionic conductances, it is not clear whether the modulation takes place at the ion channel itself or at some intermediate step, such as the activation of protein kinases, which, among others, are involved in phosphorylation of the channel.

However, there are other problems that need attention. For instance, little is known about the intracellular mechanism of action for the steroids *in situ*. Most of what we know about the mechanism of action is based on molecular biology studies in transfected cell systems. In these systems, MRs and/ or GRs are expressed together with a promotor–reporter gene construct comprising a hormone responsive element for MRs and GRs. Much as these systems have helped to unravel the hormonal mechanism of action, they may deviate from the *in situ* situation in that (a) only hormone responsive elements for GRs were investigated, (b) the elements mentioned above were overexpressed, and, conversely, (c) not all of the elements relevant for the effect of steroid receptor activation were present. For instance, it is known that transcription factors working on the AP1 complex differentially affect MR- and GR-mediated gene expression. In neurons, these, but also other transcription factors, may be essential in determining the overall effect of the steroid receptor activation. Also, if MRs and GRs compete for the same part of the DNA, their relative distribution may be important for the final action exerted by corticosterone, whereas this may not be so in the artificial expression system.

The answer to this question must be found in future techniques that combine elements from electrophysiology, biochemistry, and molecular biology. It has been shown that, after patch-clamp recording from neurons, the intracellular content of the cell can be aspirated for later evaluation of the mRNA content (see [19] in this volume). This technique is particularly suitable for evaluating the intracellular mechanism of action for steroid hormones. Once the functional changes in ionic conductance or transmitter responses have been established, all elements of the signal transduction pathway involved in the response can be screened by the single-cell amplification method.

Another example of a fruitful combination of techniques involves the single-cell recording of calcium currents and imaging of intracellular calcium concentrations. Thus, it was shown that the amplitude of calcium currents in hippocampal cells is affected by selective MR and GR activation. However, changes in calcium influx do not necessarily result in an overall increase in the intracellular amount of free calcium: binding to calcium-binding proteins, sequestering, and calcium efflux efficacy are some of the mechanisms by which the cell can control its calcium content. In principle, each of these processes can be changed by steroid hormones. Here, combined single-cell calcium imaging and calcium current measurement can give insight into the overall steroid effect on calcium homeostasis, an insight that could never be achieved by either of the techniques alone.

Concluding Remarks

Clearly, although electrophysiological techniques such as single-channel recording will be indispensable in resolving some of the steroid-mediated actions on electrical activity, it is the combination of electrophysiology with a number of other techniques that will be of particular help in establishing the effect of steroids on excitability and the functional relevance of these effects.

References

1. B. S. McEwen, J. M. Weiss, and L. S. Schwartz, Selective retention of corticosterone by limbic structures in rat brain. *Nature* **220,** 911 (1968).
2. N. B. Standen, P. T. A. Gray, and M. J. Whitaker, "Microelectrode Techniques." Company Biologists, Cambridge, England, 1987.
3. J. Chad and H. Wheal, "Cellular Neurobiology: A Practical Approach." Oxford Univ. Press, Oxford, 1991.
4. H. Kettenmann and R. Grantyn, "Practical Electrophysiological Methods." Wiley, New York, 1992.
5. D. I. Wallis, "Electrophysiology: A Practical Approach." Oxford Univ. Press, Oxford, 1993.

[27] Electrophysiological Studies of Neurosteroid Modulation of γ-Aminobutyric Acid Type A Receptor

C. Hill-Venning, D. Belelli, J. A. Peters, and J. J. Lambert

Introduction

It has become apparent that certain steroids may modulate the activity of the nervous system via a direct interaction with several ionotropic neurotransmitter receptors. Of particular interest is the now well-documented ability of several pregnane steroids to modulate allosterically the activity of the γ-aminobutyric acid type A ($GABA_A$) receptor (1). The $GABA_A$ receptor, through the opening of an integral chloride selective ion channel, mediates neuronal inhibition in the central nervous system (CNS). Agents that allosterically enhance the operation of the $GABA_A$ receptor complex, such as depressant barbiturates and benzodiazepine agonists, are associated with anxiolytic, anticonvulsant, and hypnotic activities. Additionally, several structurally diverse intravenous and volatile general anesthetics share the ability to potentiate the action of GABA at the $GABA_A$ receptor, and may thus owe at least a part of their central depressant activity to facilitation of fast inhibitory synaptic transmission.

Early studies demonstrating potent and stereoselective modulation of $GABA_A$ receptor function by pregnane steroids utilized the electrophysiological technique of extracellular recording from slice preparations of the CNS (2,3). Although such studies of population responses to synaptically released transmitter or exogenously applied $GABA_A$ receptor agonists have been, and continue to be, important in documenting the effects of steroids on neurotransmitter systems, they do not provide a direct insight into the molecular mechanisms that underlie such actions. This issue may be pursued using voltage-clamp and single-channel recording techniques applied to single neurons (or paraneurons, as in some of the present studies) maintained in cell culture, or contained within *in vitro* brain slices. To date, studies of steroidal modulation of $GABA_A$ receptors utilizing the above electrophysiological techniques have concentrated, because of their relative simplicity, on cell culture systems. This chapter attempts to summarize some of our experiences in this area. Of course, the electrophysiological approach is only one of several disciplines that have contributed to some understanding of the

Methods in Neurosciences, Volume 22

nature of steroid–GABA$_A$ receptor interactions and the input of radioligand-binding studies with GABA$_A$ receptor-selective probes is described elsewhere in this volume (4). Additionally, space precludes a comprehensive treatment of even electrophysiological studies performed under two-point voltage-clamp conditions. Efforts by this and other laboratories to determine whether steroid regulation of the GABA$_A$ receptor exhibits subunit selectivity are, with the exception of those performed using patch voltage-clamp techniques, excluded. The remainder of such studies, performed on recombinant GABA$_A$ receptors expressed in *Xenopus* oocytes, involve experimental approaches and techniques quite different from those described here. However, many of these are described elsewhere in the series (5, 6).

Choice of Preparation and Recording Technique

Given the widespread distribution of the GABA$_A$ receptor within the central and peripheral nervous systems, many types of neuron are potentially suitable for evaluating the influence of steroids on GABA$_A$ receptor function. However, a number of practical considerations have led our laboratory to concentrate on adult bovine adrenomedullary chromaffin cells, embryonic mouse spinal neurons, and embryonic rat hippocampal neurons maintained in primary cell culture.

In principle at least, potentiation of GABA$_A$ receptor-mediated responses could involve mechanisms unrelated to the receptor itself. In unclamped preparations, in which changes in membrane potential are measured, potentiation might conceivably occur as a consequence of a change in the driving force on chloride ions, contingent on changes in either the resting membrane potential (this is particularly important given the close proximity of the latter to the chloride equilibrium potential in many cells) or the transmembrane distribution of chloride ions. An increase in neuronal input resistance would also produce enhancement. Such possibilities are eliminated if recordings are performed under voltage-clamp conditions with patch electrodes that effectively buffer and define the composition of the cell interior with respect to permeant anions and small molecules. Control of the intracellular concentration of Ca^{2+} may be particularly pertinent, because this divalent cation is known to modulate the affinity of GABA for the GABA$_A$ receptor (7). However, as noted below, dialysis of the cell interior may also perturb biochemical pathways essential to the sustained operation of the GABA$_A$ receptor complex.

Voltage clamp with a single patch electrode can be performed in conjunction with either a conventional patch-clamp amplifier, as in the present studies, or a switch-clamp amplifier. The former allows exceptionally high signal-to-noise recordings to be obtained from relatively small cells, but is not suited to applications where large agonist- or voltage-gated currents are anticipated. A switch clamp may be more appropriate under such circumstances, but at the expense of a reduced signal-to-noise ratio. Neither recording system avoids the potential problem of inadequate space clamp, a lack of uniform voltage control over the membrane surface being clamped, in recordings made from cells possessing extensive processes. In view of the latter, many of our initial studies were performed on paraneuronal bovine adrenomedullary chromaffin cells maintained in cell culture. Under such conditions, many chromaffin cells display a spherical morphology, are of small diameter (8–20 μm), and possess a membrane that lacks extensive infoldings (8). These features provide an electrotonically compact cell with an extremely high input resistance (at least several gigaohms) and low capacitance (several picofarads): characteristics ideal for low noise and fast voltage-clamp recordings with a patch-clamp amplifier. Indeed, with suitable filtering, background noise in whole cell recordings performed on small chromaffin cells is low enough to permit the resolution of single-channel events of relatively long duration (8). The speed and signal quality of the clamp is sufficient to faithfully record rapidly activating voltage-activated conductances such as the sodium current. The geometry and small volume of the chromaffin cell also allow small molecules to diffuse from the patch pipette and reach steady state concentrations intracellularly within only a few minutes (9). This has been exploited in experiments that have attempted to distinguish between extracellular and intramembrane sites of action of the steroids (see below).

The pharmacological profile of the $GABA_A$ receptor population expressed by bovine chromaffin cells appears similar to that documented for centrally located receptors of this class (10). Under the recording conditions employed in our studies, GABA uptake (11) has little influence on responses to exogenously applied agonist as evidenced by the lack of effect of the uptake inhibitor, nipecotic acid (10). Although others (11) have reported chromaffin cells to store and release GABA, the ambient level of GABA present in the recording chamber, which is continually perfused, must be extremely low because "spontaneous" channel openings, when present at all, occur at low frequency. In any event, using high-performance liquid chromatography (HPLC), we have been unable to detect the release of GABA from chromaffin cells. The absence of significant levels of endogenous GABA in such cultures, or an influence of GABA uptake systems on responses to exogenously applied GABA, contrasts with the situation encounted in primary cultures of central neurons and greatly simplifies the interpretation of steroid action. In particular, it is possible to record a direct agonist action of some steroids that cannot be simply attributed to potentiation of "background" GABA. The ability to resolve single-channel events in the whole cell recording mode greatly facilitates the detection of even weak agonist efficacy.

Against the above advantages of the chromaffin cell as a model system are, however, a number of drawbacks. First, cells vary greatly in their sensitivity to GABA (10, 12). Second, the variety of ligand-gated ion channels expressed by the cells is limited; ionotropic receptors for glycine and excitatory amino acids in particular appear to be absent. This limits studies of the pharmacological selectivity of the steroids. Third, whether or not the $GABA_A$ receptors expressed by chromaffin cells are entirely representative of centrally located receptors is unclear. The subunit diversity of this receptor class is known to influence the potency and efficacy of other allosteric modulators, such as the benzodiazepines (13), and it may be unwise to generalize from the results obtained with a single cell type. Finally, we have found it difficult to perform a detailed kinetic analysis of the influence of steroids on single $GABA_A$ receptors in chromaffin cell membranes owing to the presence of multiple, interconverting, conductance states that occur at relatively high frequency (14).

In view of the above, we have additionally examined the influence of steroids on $GABA_A$ receptor-mediated responses in embryonic rat hippocampal neurons and mouse spinal neurons in primary cell culture. Both cell types exhibit prominent responses to GABA and are additionally sensitive to agonists selective for subtypes of glutamate receptor. Spinal neurons were included in our studies because of their robust response to the inhibitory transmitter glycine. These systems permit a detailed evaluation of the pharmacological selectivity of the steroids, both between excitatory and inhibitory amino acid receptors and between receptors that gate a similar anion-selective conductance (i.e., $GABA_A$ and glycine receptors). Additionally, neurons within such cultures establish functional synaptic connections, and the effect of steroids on postsynaptic currents elicited by both evoked and spontaneously released inhibitory and excitatory transmitters may be evaluated. On outside-out membrane patches excised from mouse spinal neurons, $GABA_A$ receptor-gated single-channel currents are reported to exhibit predominantly one conductance state with openings to subconductance levels constituting only a small fraction of the total number of events observed (15). Others have thus found it possible to analyze formally the kinetic behavior of the channel in the absence and presence of steroid modulators (see below). Against these attractions, the ambient level of GABA within cultures of central neurons complicates the interpretation of apparently direct actions of the steroids. Furthermore, when recording whole cell currents, space clamp may be inadequate in cells possessing an extensive neuritic network. To some extent, the latter problem can be minimized by restricting the exogenous application of agonists to the cell soma (by either pressure ejection or ionophoresis), which can be properly voltage clamped. In the following sections, we summarize the protocols that we have used in preparing primary cultures of chromaffin cells, and of spinal and hippocampal neurons.

Cell Culture Techniques

Dissociation and Culture of Hippocampal and Cortical Neurons

Embryonic rat hippocampal neurons are isolated and cultured using minor modifications to the method described by Heuttner and Baughman (16) for culturing neurons of rat visual cortex. Timed rats are killed by cervical dislocation, and embryonic day 17 and 18 (E17–18) embryos are removed into Hanks' balanced salt solution (HBSS; GIBCO-BRL, Gaithersburg, MD) at room temperature (18–22°C). Dissection of the hippocampi is performed in HBSS under a dissecting microscope. The hippocampi are chopped into fragments and incubated in a flask for 60 min at 37°C in 10 ml of an enzyme solution containing 116 mM NaCl, 5.4 mM KCl, 26 mM NaHCO$_3$, 1 mM NaH$_2$PO$_4$, 1.5 mM CaCl$_2$, 1 mM MgSO$_4$, 0.5 mM ethylenediaminetetraacetic acid (EDTA), 25 mM glucose, 1 mM cysteine (Sigma, St. Louis, MO), and papain (20 units/ml; Sigma) (pH 7.4). We accelerate dissociation of the tissue by gentle shaking of the flask every 10–15 min. The tissue fragments are rinsed in 5 ml of HBSS supplemented with bovine serum albumin (BSA; 1 mg/ml) and ovomucoid (1 mg/ml) (both from Sigma) and then transferred into a sterile test tube containing a further 3–4 ml of this solution. The cells are dissociated mechanically by gently triturating the tissue through a fire-polished Pasteur pipette, and the upper layer of dissociated cells is removed and layered onto 5 ml of HBSS containing BSA (10 mg/ml) and ovomucoid (10 mg/ml). The remaining undissociated tissue is further triturated and the resulting top layer of dissociated cells is removed to the BSA–ovomucoid solution. This process is repeated until all the tissue is successfully dissociated. The combined cell suspension is then centrifuged at 100 g for 10 min at room temperature, the supernatant is discarded, and the cells are resuspended in 4 ml of growth medium composed of 88% (v/v) minimal essential medium, 5% (v/v) heat-inactivated fetal calf serum, 5% (v/v) heat-inactivated horse serum, penicillin (50 IU/ml), streptomycin (50 μg/ml), glutamine (2 mM) (all from GIBCO-BRL), and glucose (20 mM). Cells are plated at a density of 1–2 $\times$ 10^5/plate into 35-mm diameter (Falcon Becton Dickinson, Oxford, England) Primaria culture dishes, and incubated in 1.5 ml of culture medium at 37°C in 5% CO$_2$/95% air at 100% relative humidity. Five to 7 days after plating, or when the background nonneuronal cells have reached confluency, the medium is replaced with one including the mitotic inhibitor cytosine arabinoside (10 μM; Sigma) for 48 hr. This is to suppress any further proliferation of nonneuronal cells. Approximately two-thirds of the volume of culture medium is replaced every 5–7 days thereafter, but caution should be exercised during this operation as we find that the cells are sensitive to the removal of used, and addition of fresh, medium. After each exchange, we find there is some cell death, which may be due to the release of glutamate from the neurons. Excess exposure to glutamate can cause neuronal cell injury or loss, and it has been shown that the incubation of N-methyl-D-aspartate (NMDA) antagonists in the medium surrounding tissue-cultured cells reduces the amount of cell death after exposure to glutamate (17). Neurons are used in experiments 10–30 days after plating.

Dissociation and Culture of Spinal Cord Neurons

We prepare cultures of spinal cord neurons from mouse embryos following the method of Ransom et al. (18) with slight modifications. Timed mice are killed by cervical dislocation, and E13 embryos are removed as quickly as possible into HBSS. Each embryo is decapitated, and the spinal cords are dissected out, chopped, and then incubated in a 0.25% (w/v) trypsin (GIBCO-BRL) solution in HBSS, for 30 min at 37°C. One milliliter of culture medium is then added to halt the activity of the trypsin. This medium consists of 78% (v/v) minimum essential medium, 10% (v/v) heat-inactivated fetal calf serum, 10% (v/v) heat-inactivated horse serum, penicillin (50 IU/ml), streptomycin (50 μg/ml), glutamine (2 mM), NaHCO$_3$ (1.5 mg/ml), and glucose (5 mg/ml). The osmolarity of the medium is adjusted to 300 mOsm by the addition of sterile distilled water. After incubation with trypsin, the spinal cord tissue suspension is triturated using fire-polished Pasteur pipettes in the same manner as described for the hippocampal neurons above, and the dissociated cells are transferred to another sterile test tube containing 1 ml of medium. The volume of the cell suspension is then made up to approximately 10 ml by the addition of further medium. Cells are plated at a density of approximately 1×10^6 cells/plate (Nunclon 35-mm culture dishes, GIBCO BRL) and maintained at 37°C and 100% relative humidity in a 10% CO$_2$/90% air atmosphere. After 2–5 days, when the background nonneuronal cells are confluent, the medium is replaced with a maintenance medium that lacks fetal calf serum and is supplemented with the mitotic inhibitors 2-deoxy-5-fluorouridine (15 μg/ml) and uridine (35 μg/ml) (both from Sigma). After 48 hr, this medium is replaced by maintenance medium lacking the inhibitors, and this is refreshed every 7 days. These cells, like the hippocampal neurons, are also sensitive to exchange of medium, and we have found that inclusion of the NMDA antagonist MK-801 (3 μM) in the medium does appear to lessen the degree of cell death. Recordings are made from the neurons 7–30 days after plating.

Dissociation and Culture of Chromaffin Cells

Bovine adrenomedullary chromaffin cells are isolated and cultured as previously described by us (10, 19). Fresh bovine adrenal glands obtained from the local abattoir are immediately trimmed of fat, and the cortex of the gland is punctured several times with a scalpel blade. This allows retrograde perfusion of the gland via the adrenal vein with approximately 10 ml of ice-cold, Ca^{2+}-free Krebs solution (composition: 119 mM NaCl, 4.7 mM KCl, 1.2 mM MgSO$_4$, 2.5 mM NaHCO$_3$, 1.2 mM KH$_2$PO$_4$, 11 mM glucose), supplemented with heparin (10 units/ml). The gland is transported, packed in ice, to the laboratory, and a cut is made around the outer edge of the cortex to just reveal the paler medulla. The gland is then cannulated via the adrenal vein and retrogradely perfused on a modified Langendorff apparatus, which allows the collection and recirculation of the perfusate. Approximately 90 ml of ice-cold Krebs–heparin solution is first passed through the gland and discarded. Krebs solution (approximately 40 ml) containing collagenase (1 mg/ml), BSA (5 mg/ml), and also 1 mM CaCl$_2$ is then recirculated through the gland for 30 min. The temperature and pH of the perfusate are maintained at 37°C and 7.4, respectively, by a water jacket and continuous gassing of the solution with 5% CO$_2$/95% O$_2$. We use Sigma type 1 collagenase, but find there is a great deal of variation between supplied lots. We, therefore, batch test and reserve a batch once we find it is suitable. After collagenase perfusion, the partially digested medulla is removed from the cortex and cut into smaller fragments with a pair of scissors. The fragments are incubated with 8 ml of the above-described Krebs–collagenase solution, gassed with 5% CO$_2$/95% O$_2$, for a further 20 min at 37°C in a shaking water bath. Dissociation of the cells from the tissue fragments is enhanced by trituration of the mixture every 5–10 min. Dissociated cells are isolated by subsequent filtration through gauze (pore size, 80 μm Locker Text, Warrington, England). Usually, a sufficiently large number of dissociated cells is obtained from a single incubation of the tissue with the collagenase solution. However, if there are too few cells, the remaining undissociated tissue can be reincubated with the collagenase solution for a further 20 min, and this process may be repeated three times. The dissociated cell fraction resulting from each incubation is sedimented by centrifugation (100 g, 10 min), and washed twice with 2 ml of BSA (5 mg/ml) in Krebs solution to remove residual collagenase. The cell fractions are combined and layered on a 4-ml cushion of BSA (37.5 mg/ml) in Krebs solution. Following sedimentation under unit gravity for 30–60 min, the cells are resuspended in a growth medium consisting of 87.5% (v/v) medium 199 and 10% (v/v) heat-inactivated fetal calf serum, supplemented with gentamicin (50 μg/ml), penicillin (50 IU/ml), and streptomycin (50 μg/ml) (all from GIBCO-BRL). Cells are plated at a density of approximately 10^5 cells/plate into uncoated 35-mm diameter Nunclon petri dishes, and are cultured at 37°C in 5% CO_2/95% air and 100% relative humidity. Cells are used in electrophysiological studies 1–3 days after plating.

Whole Cell and Single-Channel Recording: Practical Aspects

Patch-clamp recording techniques are now in routine use in a large number of laboratories throughout the world and the methods used in our own studies are doubtless similar to those employed by many other investigators. However, for the convenience of the reader and in the interests of completeness, we include below a description of our practices, limiting the discussion to the electrophysiological methods that we have routinely employed to evaluate the influence of steroids on the function of $GABA_A$ receptors expressed by chromaffin cells and spinal and hippocampal neurons in culture. The methods described are applicable to all three cell types. More comprehensive and detailed accounts of the practical and theoretical aspects of patch clamping can be found in Marty and Neher (20) and Levis and Rae (21).

Experimental Setup

The experimental setup that we use is shown schematically in Fig. 1. The culture dish on which the cells are grown forms the recording chamber, which is placed on the stage of an inverted microscope (Zeiss IM-35, Oberkochen, Germany, or Nikon Diaphot, Tokyo, Japan) equipped with phase-contrast optics. The cells are viewed under ×250–×300 magnification. The recording chamber is continuously perfused at a rate of 3–5 ml/min with an extracellular recording solution (see below), applied from a reservoir via polyethylene tubing and a roller pump (Minipuls 3; Gilson, Villiers Le Bel, France). The bath volume is maintained at approximately 1 ml by aspiration of the recording medium into a trap connected to a vacuum pump. In many of our studies, $GABA_A$ receptor modulators are applied to cells via the perfusate. Although simple, this method does have the disadvantage that all of the cells within a culture are exposed to the drug and equilibration is rather slow (about 2–3 min, including the delay due to the dead space introduced by the polyethylene tubing connecting the solution reservoir and bath). Nonetheless, known bath concentrations of the modulators are eventually achieved

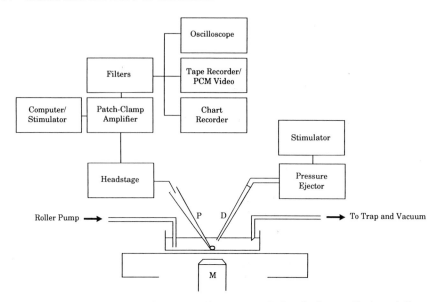

FIG. 1   Schematic diagram of the recording system. P, Patch pipette; D, drug delivery pipette; M, microscope objective.

and, in theory at least, there is no limit on the number of drugs, or drug combinations, that can be tested. In an attempt to reduce the influence of desensitization on agonist-evoked whole cell currents, GABA (10–100 μM) and other agonists are routinely applied locally to cells by ejection from a patch pipette (see below) driven by variable-duration pressure pulses (1.4 × 10^5 Pa), provided by a General Value Corporation Picospritzer II (Fairfield, NJ) driven by a stimulator (Grass S44, Quincy, MA). This method produces reproducible current responses that reach a peak within 100 msec, but suffers from the disadvantage that the concentration of agonist at the receptor plane is unknown. The concentrations of GABA and pulse duration selected do, however, result in responses that are clearly submaximal. In experiments investigating the influence of steroids on the kinetics of single-channel currents elicited by GABA, we have normally applied the compounds via the superfusion system. Increasingly, more elegant methods of drug application employing Y or U tubes or flow pipes driven by stepping motors or piezoelectric translators are employed in the electrophysiological studies of ligand-gated ion channels (e.g., 8, 22, 23). These present the advantages of rapid application of known concentrations of drugs, and should certainly be applicable to future studies of the steroidal modulation of $GABA_A$ receptors.

The positioning of the recording and drug delivery pipettes is accomplished using high-quality micromanipulators. The choice of manipulator will, of course, to some extent be determined by financial considerations, but it is imperative that the instrument exhibit negligible or zero drift over time. In our hands, the manipulators manufactured by Leitz (type M) (Wetzlar, Germany) and Burleigh (PCS 205) (Fishers, NY), although expensive, perform well. The manipulators and microscope are mounted on a vibration isolation table (Wentworth 701, Beds., England). Numerous companies now manufacture several types of patch-clamp amplifier that may employ either traditional resistor feedback or the more recent capacitor headstage technologies. Our experience is limited to the former, and we have routinely used the LM EPC 7 and Axoclamp 1D amplifiers, manufactured by List (Eberstadt, Germany) and Axon Instruments (Foster City, CA), respectively. Simple square-pulse voltage commands are supplied to the amplifier by a stimulator (Grass S48). A personal computer (Dell XP5P60, Limerick, Ireland) running customized software (24) is used to provide more complex voltage jump protocols. Agonist-evoked currents, low-pass filtered at cutoff frequencies of 500 Hz and 1–2 kHz in the case of whole cell and single-channel recordings, respectively, are displayed on a storage oscilloscope (Tektronix 5111A, Beaverton, Oregon) and chart recorder (Gould TA 240, Valley View, OH). The signals are also stored on magnetic tape using either a Racal Store 4DS FM tape recorder (Fullerton, CA) or a PCM 2 A/D VCR adaptor (Medical Systems Corporation, Greenvale, NY) for subsequent computer analysis. In common with many other investigators, we have adopted digital audio tapes as a preferable storage medium, and use a digital tape recorder (DTR 1204; Biologic Scientific Instruments, Claix, France) for this purpose. The programs for the analysis of whole cell and single-channel currents that we have used were designed by J. Dempster (Strathclyde University, Glasgow, Scotland) and are described in detail elsewhere (24).

Composition of Bath and Pipette Solutions

Patch pipettes are filled with a standard solution of the following composition: 140 mM CsCl (or KCl), 2 mM MgCl$_2$, 0.1 mM CaCl$_2$, 1.1 mM ethylene glycol-bis(β-aminoethyl ether)-N,N,N',N'-tetraacetic acid (EGTA), 10 mM N-2-hydroxyethylpiperazine-N'-2-ethanesulfonic acid (HEPES), pH 7.2. Cesium is frequently used as the predominant internal cation in order to suppress membrane K$^+$ currents. For recordings from both hippocampal and spinal neurons, we also include 2 mM Mg-ATP in the standard solution to minimize the run-down of GABA-evoked currents (25). We have found that inclusion of Mg-ATP alone in the pipette solution is insufficient to prevent such rundown in chromaffin cells. Instead, a solution specifically designed to sustain GABA$_A$ receptor activity (26) is currently employed in such experiments [composition: 140 mM CsCl (or KCl), 2 mM MgCl$_2$, 10 mM HEPES, 1 mM

BaCl$_2$, 2 mM Mg-ATP, 10 mM 1,2-bis(2-aminophenoxy) ethane-N,N,N',N'-tetraacetic acid (BAPTA)]. This solution reduces the rate of run-down of GABA currents, but in our hands, high-resistance seals appear to form less frequently when using this saline. The external bathing solution is of the following composition: 140 mM NaCl, 2.8 mM KCl, 2 mM MgCl$_2$, 1 mM CaCl$_2$, and 10 mM HEPES-NaOH, pH 7.2. For both hippocampal and spinal neurons, 300 nM tetrodotoxin (TTX) may be included in the external solution to suppress synaptic activity if desired.

Electrode Construction

Electrodes for whole cell and single-channel recording are fabricated from glass capillary tubing [either soft Kimble (Toledo, OH) R-6 soda lime or medium Corning (Corning, NY) 7052 borosilicate glasses] using a Narishige (Tokyo, Japan) PB 7 vertical two-stage microelectrode puller. The borosilicate glass, owing to its lower dielectric loss factor, is preferable for lower noise recordings of channel activity in excised membrane patches. The tip of the patch pipette, viewed under ×250 magnification, is routinely fire polished using a home-built microforge, which consists essentially of a variable power source that heats a glass bead located at the apex of a V-shaped length of 0.2-mm diameter platinum–iridium wire. Electrodes for whole cell recording should have the lowest resistances (typically 1–2 MΩ or so) compatible with gigaseal formation in order to minimize the series resistance (see below) that can result in inadequate control of the membrane potential in whole cell recordings performed with a patch-clamp amplifier. However, it should be noted that the selection of low-resistance pipettes may exacerbate the time-dependent run-down of agonist-evoked currents (see above). Electrodes with resistances in the range of 5–10 MΩ have proved suitable for single-channel recording from outside-out membrane patches excised from either chromaffin cells or hippocampal neurons. It is desirable to coat the electrode shank, as close to the tip as is possible, with Sylgard silicone elastomer 184 (Dow Corning, Midland, MI), which is subsequently cured by passing the electrode briefly through the heated coil of the puller. This reduces the capacitance of the pipette when immersed in recording solution, and helps to prevent the latter from creeping up the shank with the subsequent introduction of noise. Electrode tips are filled by negative pressure (the glasses employed lack a filament), and are then back-filled with an internal solution prefiltered through Millipore (Bedford, MA) filter (0.22-μm pore size) to remove particulate matter that might otherwise interfere with seal formation. Electrodes should be filled with the minimal volume of electrolyte that reliably contacts the Ag/AgCl wire or AgCl pellet of the electrode holder, and should not, under any circumstance, extend into the holder itself, which if wetted will add substantial noise to the recording. In our experience, it is not necessary to filter the extracellular solution routinely.

Forming Gigaseals and Obtaining Whole Cell Recordings

The formation of the gigaseal between the patch pipette and cell membrane is monitored by supplying a repetitive square-voltage command of brief duration to the pipette. The resulting pipette current, from Ohm's law, is inversely proportional to the electrode resistance. Some patch-clamp amplifiers (e.g., Axopatch 1D) have built-in oscillators that supply such signals, whereas others (e.g., List LM EPC-7) require the use of an ancillary stimulator. While advancing the electrode through the air/extracellular medium interface and toward the cell surface, slight positive pressure (supplied by mouth) is applied to the pipette via polyethylene tubing connected to the suction port of the electrode holder. This prevents debris from contaminating the pipette tip. Prior to seal formation, junction potentials arising at various points within the recording system are offset to 0 mV, using the junction potential null control of the patch-clamp amplifier. It is important to appreciate that any liquid junction potential arising at the interface between extracellular and internal solutions at the pipette tip is also offset by this procedure, but will "reappear" both in patch and whole cell recording configurations, introducing an error in the observed holding potential. The origin of liquid junction potentials, and their measurement and influence on holding potential, are beyond the scope of the present account, and a more specialized and advanced review is recommended (27). As the pipette tip contacts the cell membrane, its resistance should increase. It is our practice to advance the electrode further until the resistance at least doubles. At this time, relief of positive pressure may be sufficient for gigaseal formation but, more frequently, the application of slight negative pressure is required. Seal formation may occur dramatically as the sudden disappearance of all but the capacitive current transients associated with the imposed voltage steps, or may develop more gradually as a progressive decline toward zero of the amplitude of the square-current response to such steps. In stubborn cases, alternating pulses of positive and negative pressure, or the application of a negative potential to the pipette, may sometimes coax the development of a seal. The seal resistance, measured at high gain from the current response to brief hyperpolarizing voltage steps (i.e., positively directed pipette potentials) should be 10 GΩ or higher. Before proceeding to establish a whole cell recording, the amplifier is switched into voltage-clamp mode, the steady holding potential adjusted to a value not dissimilar to that anticipated for the cell resting potential (routinely 60 mV, pipette negative in our experiments). Finally, the current transients associated with the step changes in potential and the charging/discharging of the pipette capacitance are electronically neutralized, using the appropriate capacitance cancellation circuitry of the amplifier.

It is common practice to rupture the cell membrane within the pipette, and thus establish electrical continuity between the pipette and cell interior, by the application of negative pressure to the pipette. However, other methods, such as the inclusion of polyene antibiotics (e.g., nystatin, amphotericin B) within the pipette solution, which have the effect of permeabilizing the membrane (28), or the brief application of voltages that exceed the dielectric constant of the membrane (known colloquially as "zapping"), are alternative methods of obtaining a whole cell recording. Permeabilization of the membrane as opposed to its rupture has the advantage, or disadvantage, dependent on the aims of the experiment, of minimally perturbing intracellular biochemical events. Successful establishment of the whole cell recording configuration is heralded by an increase in the current noise in the recording and the reappearance of capacitive transients to step changes in pipette potential. Such changes, which develop over several minutes when using the permeabilization technique, but suddenly when the patch is disrupted by negative pressure or zapping, derive from the increase in membrane area associated with the whole cell.

Irrespective of the method chosen to initiate whole cell recording, when using a patch-clamp amplifier it is important that a low-resistance pathway (the series resistance, R_S) is formed between the pipette and cell interiors. This is verified by estimating the value of R_S as follows. A capacitive transient associated with a known voltage command (V_p) that evokes no active current is captured on the oscilloscope screen. The peak of the capacitive current (I_C), measured with all filters bypassed, can then be used to estimate R_S roughly from the relationship $R_S = V_p/I_C$. Strictly, the transient should be fitted with an exponential function to derive this information. Alternatively, and more quickly, faith may be placed in the precalibrated "G-series" and "slow capacitance cancellation" controls of the patch-clamp amplifier, which, when adjusted interactively, nullify the transient and provide estimates of both R_S and the capacitance (C) of the whole cell. The effective series resistance ($R_{S,eff}$) can be reduced electronically by series resistance compensation, but in our experience greater than 80% compensation is rarely possible. Voltage commands are not applied to the cell throughout a recording, but it is important to periodically recheck R_S. Clogging of the pipette tip with cytoplasmic components or resealing of the membrane are not uncommon and can impact greatly on the quality of the data. Gentle suction often remedies the problem. What constitutes a tolerable R_S depends on the nature of the experiment, the electrical characteristics of the cell (and the attitude of the investigator). Too high a series resistance poses two potential problems: (a) inadequate control of membrane voltage when large currents flow through R_S and (b) the inability of the cell membrane potential to rapidly follow a step change in command potential.

Formation of Outside-Out Membrane Patches

Outside-out membrane patches are formed by gradually withdrawing the patch pipette from the cell after a whole cell recording has been established. This has the effect of drawing out a strand of membrane that eventually breaks and often reseals such that its extracellular surface faces into the recording chamber. Agonists are then applied to the patch, either by bath application or pressure ejection, and provided that the appropriate receptor is present, single-channel activity is recorded. Prior to attempting to form an outside-out patch, any repetitive voltage commands or R_S compensation are switched off, and the holding potential is set to 0 mV with the amplifier still in voltage-clamp mode. An alternative procedure is to switch to the constant current mode of recording during the process of excising the patch. Successfully formed patches have a resistance greater than several gigaohms and do not show bursts of irregularly shaped currents on polarization.

Applications of Techniques in Investigating Steroid Modulation of GABA$_A$ Receptors

Utilizing whole cell recording techniques applied to bovine chromaffin cells, we have demonstrated unequivocally that the anesthetic steroids interact with the GABA$_A$ receptor to enhance the GABA-evoked whole cell current (14, 19). Such effects were evident at very low (30 nM) extracellular concentrations of the steroid and, like the central depressant effects, were stereoselective (19, 29) (Fig. 2). The functional GABA$_A$ receptor is presumed to be composed of five subunits (13, 30), and molecular biology has revealed a diversity of such subunits (13, 31). Hence, although much of the pharmacology of the bovine chromaffin cell is common to the GABA$_A$ receptors of central neurons and mRNAs encoding some of the neuronal GABA$_A$ subunits are also evident in these adrenal cells, the possibility existed that the neurosteroid effect we had reported was atypical. However, similar observations have been made in rodent hippocampal and spinal neurons (32, 33). These data demonstrate a specific effect of the neurosteroid on the GABA$_A$ receptor, but reveal little of the molecular mechanism(s) underlying this effect. A clue

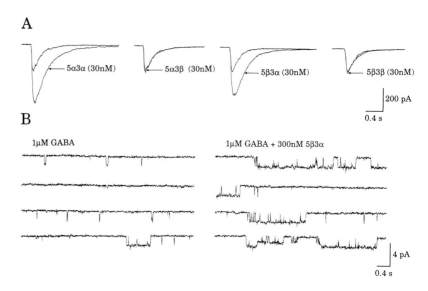

FIG. 2   Steroid modulation of GABA-evoked whole cell and single-channel currents. (A) Whole cell currents evoked by pressure applied GABA (100 μM, 20 msec, 1.4×10^5 Pa) recorded under voltage clamp from a bovine chromaffin cell held at -60 mV. The bath application of either 5α-pregnan-3α-ol-20-one (5α3α; 30 nM) or 5β-pregnan-3α-ol-20-one (5β3α; 30 nM) enhances both the amplitude and duration of the inward current response. Identical concentrations of the corresponding 3β-ol isomers (i.e., 5α3β and 5β3β) produce no significant effect. Steroids were applied via the superfusate for 5 min. Each pair of traces represents the computer-generated average of four responses to GABA recorded in the presence (arrowed) and absence of the steroid. Currents were recorded with a symmetrical transmembrane distribution of Cl ions and were low-pass filtered at a cutoff frequency of 500 Hz. (B) Single-channel currents elicited by GABA (1 μM) in the presence (right) and absence (left) of 5β-pregnan-3α-ol-20-one (300 nM). Recordings were made from an outside-out membrane patch excised from a bovine chromaffin cell. Single-channel currents were recorded at a holding potential of -60 mV, and low-pass filtered at a cutoff frequency of 1 kHz. Drugs were applied via the superfusate. The traces are consecutive.

comes from studies utilizing whole cell clamp techniques on embryonic rat hippocampal neurones maintained in cell culture (32). This preparation develops functional synapses *in vitro*. Therefore, it is possible to stimulate a "presynaptic" neuron (with locally applied glutamate) and to record the resultant GABA-mediated inhibitory postsynaptic current (IPSC). By the judicious choice of holding potential and the selection of suitable pipette and extracellular salines, these synaptic currents can be recorded in the absence of glutamate-mediated excitatory postsynaptic currents (EPSCs). Under these conditions, the steroidal anesthetic alphaxalone has little influence on the IPSC amplitude or rise time, but greatly prolongs their decay (32). As the time constant (τ) of decay of the IPSC approximates to the mean channel open time, one interpretation of this observation is that the steroid acts to prolong the open time of the chloride channel activated by GABA.

This view is supported by the influence of alphaxalone on GABA-evoked current noise recorded with whole cell clamp from rat spinal neurons and subsequently quantified by fluctuation analysis (33). A detailed description of this approach is beyond the scope of this chapter, but may be found in Stevens (34) and Anderson and Stevens (35). Briefly, the application of GABA results in the activation of a number of associated chloride channels. For a steady state, nondesensitizing, whole cell current, the number of open channels fluctuates about a mean value. The variance of this current around the mean will clearly be determined by the single-channel current amplitude. Information about the mean open time of the channels can be obtained by investigating the frequency composition of this variance. For example, if the GABA channels have a brief open time of 2 msec, then relatively high-frequency fluctuations occur, whereas if the open time is much longer, say 40 msec, then the current variance is associated with lower frequency components. Therefore, by investigating the frequency dependence of the current variance, an estimation of the mean channel open time can be made. Utilizing this approach, Barker and colleagues (33) demonstrated that alphaxalone had little influence on the GABA single-channel conductance, but shifted the current variance to lower frequencies consistent with a prolongation of the GABA channel open time. This approach clearly uses a mathematical treatment of a population response to determine the properties of individual channels. One of the assumptions made is that the channels activated by GABA are of a unitary conductance. This assumption now seems unlikely.

Although still useful, this approach has now largely been replaced by patch-clamp methods, which allow the activity of single ion channels to be monitored directly. When applied to outside-out membrane patches made from bovine chromaffin cells, GABA elicits single-channel currents that exhibit a number of conductance states (12, 14). Furthermore, switching between conductance states within openings is evident. This complexity prevents a formal kinetic analysis of the action of neurosteroids on single GABA$_A$ receptor channels. Notwithstanding this limitation, visual inspection of GABA-activated channels (see Fig. 2) reveals a dramatic change in the kinetic behavior induced by the steroid (14). The GABA-activated channels of mouse spinal neurons also exhibit multiple conductance states, and switching within such states. However, single-channel recordings from outside-out patches reveal these events to be dominated by a main conductance state of 28 pS (15). Analysis of recordings in which only this conductance is evident demonstrates that GABA activates channels exhibiting three kinetically distinct open states. Steroids, such as androsterone and 5β-pregnan-3α-ol-20-one, prolong the mean channel open time by changing the proportion of open states to favor the longer openings. Hence, the neurosteroids do not prolong all GABA channel openings per se, but influence the kinetic behavior of the receptor such that transitions to a naturally occurring long open state are now favored. A similar mechanism has been reported for the barbiturate modulation of GABA$_A$ receptors, although the neurosteroids were additionally reported to increase the single-channel opening frequency (15). However, it is now evident that neurosteroids, in the absence of GABA, can directly activate the GABA$_A$ receptor complex and this may further complicate kinetic modeling (14). This direct activation will now be considered further.

At concentrations ($\geq$300 nM) generally greater than those required for the threshold enhancement of GABA-evoked currents, alphaxalone and other anesthetic steroids directly activate the GABA$_A$ receptor (14, 19) (Fig. 3). On isolated outside-out membrane patches these steroids evoke single channels with a conductance and reversal potential similar to those induced by GABA. Furthermore, steroid-induced single-channel or whole cell currents are blocked by the GABA$_A$ receptor antagonists bicuculline and picrotoxin, and are enhanced by diazepam and phenobarbitone (14, 19). Collectively, these observations are consistant with the steroid acting to activate the GABA$_A$ receptor directly. It has been suggested that, in cultures of central neurons, this apparent direct effect might be produced by the neurosteroid merely enhancing the action of background GABA and certainly functional GABAergic synapses in these cultures are evident. However, direct effects are clearly present on bovine chromaffin cell outside-out membrane patches spatially isolated from neighboring cells (19). Furthermore, in our experience, these adrenal cells do not store or release detectable amounts of GABA (see also 11).

The relatively high potency and stereoselectivity of both the modulatory and direct effect suggest a specific interaction of the steroid with the GABA$_A$ receptor protein. However, because of their lipophilic nature, the local membrane concentration of the neurosteroid will greatly exceed that present in the aqueous phase. Furthermore, the active steroids are known to perturb membrane lipid ordering. In an attempt to distinguish between potential protein and membrane sites of action of the steroids, we have included relatively high concentrations of either alphaxalone or 5β-pregnan-3α-ol-20-one within the patch pipette solution in experiments performed on bovine chromaffin cells (36; see Fig. 4). Empirical determinations performed by Pusch and Neher (9) on the same cell type, using recording conditions closely resembling our own, indicate that molecules with molecular weights approxi-

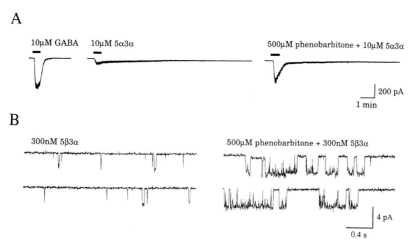

FIG. 3   Direct agonist actions of pregnane steroids. (A) Whole cell currents recorded from a bovine chromaffin cell in response to bath-applied GABA (10 μM), 5α-pregnan-3α-ol-20-one (5α3α; 10 μM), and the latter in the presence of phenobarbitone (500 μM). Note the large enhancement of the steroid-induced current by phenobarbitone. Currents are unaveraged responses to single applications of the agents recorded at a holding potential of -60 mV. Data were low-pass filtered at a cutoff frequency of 500 Hz. (B) Single-channel currents evoked by bath-applied 5β-pregnan-3α-ol-20-one (5β3α; 300 nM) recorded from an outside-out membrane patch excised from a bovine chromaffin cell. Note the prominent increase in the opening probability of the steroid-activated channels in the presence of phenobarbitone (500 μM). Channel currents were recorded at a holding potential of -100 mV and low-pass filtered at a cutoff frequency of 1 kHz.

mating those of the steroids diffuse from the pipette and rapidly reach a steady state concentration within the cytoplasm. On the assumption that the exterior and interior faces of the membrane do not exhibit marked differences with regard to the partitioning of steroids, the inclusion of such compounds within the patch pipette should result in their accumulation within the cell membrane. Significantly, such steroid-loaded chromaffin cells show no obvious increase in spontaneous channel activity, yet respond to extracellularly applied steroid with a large increase in noise indicative of GABA$_A$ receptor activation. Moreover, low concentrations of extracellularly applied steroid produce enhancement of GABA-evoked current responses in both control and steroid-loaded cells, and there is no significant difference in the magnitude of this effect between the two experimental conditions. Collectively, such data suggest the site of steroid action to be extracellularly located, the most probable candidate being the GABA$_A$ receptor protein it-

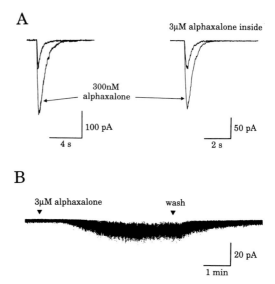

FIG. 4   The site of steroid action is extracellularly located. (A) Whole cell currents elicited by pressure-applied GABA (100 μM, 20 msec, 1.4×10^5 Pa) from a control bovine chromaffin cell (left), and a cell dialyzed with a patch pipette solution containing 3 μM alphaxalone (right). Note that under both experimental conditions, the extracellular application of a relatively low concentration (300 nM) of alphaxalone produces a large enhancement in the amplitude and duration of the inward current evoked by GABA. Each current trace illustrated is the computer-generated average of four responses to GABA that were recorded at a holding potential of -60 mV and low-pass filtered at a cutoff frequency of 500 Hz. (B) High-gain whole cell recording from a bovine chromaffin cell dialyzed with a pipette solution containing 3 μM alphaxalone. In the absence of extracellularly applied drug, channel activity (apparent as "noise" in the recording) was of low frequency and was similar to that observed in similarly sized cells dialyzed with an alphaxalone-free solution. Extracellularly applied alphaxalone (3 μM), by contrast, elicits a pronounced increase in membrane current fluctuation. Data were obtained under the recording conditions described in (A).

self. If this is correct, it may eventually prove possible to identify specific subunits of the receptor that confer, or at least modulate, steroid sensitivity. However, in one study performed on recombinant GABA$_A$ receptors expressed in a mammalian cell line (37), no evidence for a subunit-selective action of the steroids was found. Contrary to results obtained with native receptors (14, 15, 33), an increase in the frequency of single-channel events was reported to dominate steroid-induced potentiation at the recombinant receptors. Clearly, this is an area that warrants further investigation.

The pharmacological selectivity of the pregnane steroids has been evaluated in whole cell recordings performed on chromaffin cells and central neurons (19, 36). In the former system, alphaxalone was found to block inward current responses mediated by "neuronal-type" nicotinic receptors, but only at a concentration 300-fold higher than that necessary for enhancement of GABA-evoked responses. Moreover, the blockade of the nicotinic response was mimicked by the behaviorally inactive 3β-ol isomer of alphaxalone (betaxalone) and gave no indication of stereoselectivity (19). In experiments performed on voltage-clamped central neurons, alphaxalone (10 μM) was found to be devoid of either potentiating or inhibitory effects at the ($\pm$)-α-amino-3-hydroxy-5-methylisoxazole-4-propionic acid(AMPA)/kainate or NMDA classes of glutamate receptor or the strychnine-sensitive glycine receptor (36). Thus, of the ligand-gated ion channels that might logically be suspected as mediators of the behavioral effects of the pregnane steroids, only the GABA$_A$ receptor has so far been identified as a target.

Concluding Remarks

The rapid induction of anesthesia by some steroids (38) is consistent with a nongenomic action for such compounds. Some 40 years after these observations, radioligand binding and extracellular recording techniques, applied to membrane homogenate and brain slice preparations, respectively, identified the GABA$_A$ receptor as a likely locus of steroid action (2, 3). The application of patch- and voltage-clamp techniques to mammalian neurons and paraneurons grown in cell culture, confirmed this suggestion and further demonstrated that the enhancement of GABA$_A$ receptor function results from the steroid dramatically changing the GABA channel kinetics (14, 15).

The utilization of whole cell clamp techniques has now been extended to include brain slice preparations (39). Hence, it is now possible to investigate the actions of steroids on mature functional GABAergic synapses under voltage clamp (40). Finally, the application of the techniques of molecular biology, together with the electrophysiological approach described here, represents a new and powerful approach by which to probe the mechanism of neurosteroid action (37). Indeed, the electrophysiological studies of steroid-insensitive GABA receptors (41) combined with site-directed mutagenesis may define the locus and nature of the neurosteroid-binding site on the GABA$_A$ receptor.

Acknowledgment

This work was supported in part by grants from the MRC and SHERT to J.J.L.

References

1. M. A. Simmonds, *Semin. Neurosci.* **3,** 231 (1991).
2. C. N. Scholfield, *Pfluegers Arch.* **383,** 249 (1980).
3. N. L. Harrison and M. A. Simmonds, *Brain Res.* **323,** 287 (1984).
4. L. D. McCauley and K. W. Gee, this volume [13].
5. M. B. Boyle and L. K. Kaczmarek, this series, Vol. 4, p. 157, 1991.
6. D. Bertrand, E. Cooper, S. Valera, D. Rungger, and M. Ballivet, this series, Vol. 4, p. 174, 1991.
7. M. Inoue, Y. Oomura, T. Yakushigi, and N. Akaike, *Nature (London)* **324,** 156 (1986).
8. E. M. Fenwick, A. Marty, and E. Neher, *J. Physiol. (London)* **311,** 577 (1982).
9. M. Pusch and E. Neher, *Pfluegers Arch.* **411,** 204 (1988).
10. J. A. Peters, J. J. Lambert, and G. A. Cottrell, *Pfluegers Arch.* **415,** 95 (1989).
11. Y. Kataoka, Y. Gutman, A. Guidotti, P. Panula, J. Wrobleski, D. Cosenza-Murphy, J. Y. Wu, and E. Costa, *Proc. Natl. Acad. Sci. U.S.A.* **81,** 3218 (1984).
12. J. Bormann and D. E. Clapham, *Proc. Natl. Acad. Sci. U.S.A.* **82,** 2168 (1984).
13. C. I. Regan, R. M. McKernan, K. Wafford, and P. J. Whiting, *Biochem. Soc. Trans.* **21,** 622 (1993).
14. H. Callachan, G. A. Cottrell, N. Y. Hather, J. J. Lambert, J. M. Nooney, and J. A. Peters, *Proc. R. Soc. London, Ser. B* **231,** 359 (1987).
15. R. E. Twyman and R. L. MacDonald, *J. Physiol. (London)* **456,** 215 (1992).
16. J. E. Heuttner and R. W. Baughman, *J. Neurosci.* **6,** 3044 (1986).
17. D. W. Choi, J.-Y. Koh, and S. Peters, *J. Neurosci.* **8,** 185 (1988).
18. B. R. Ransom, E. Neale, M. Henkart, P. N. Bullock, and P. G. Nelson, *J. Neurophysiol.* **40,** 1132 (1977).
19. G. A. Cottrell, J. J. Lambert, and J. A. Peters, *Br. J. Pharmacol.* **90,** 491 (1987).
20. A. Marty and E. Neher, *in* "Single Channel Recording" (B. Sakmann and E. Neher, eds.), pp. 107–122. Plenum, New York, 1983.
21. R. A. Levis and J. L. Rae, "Methods in Enzymology" (B. Rudy and L. Iverson, eds.), Vol. 207, p. 14. Academic Press, San Diego, 1992.
22. N. Akaike, M. Inoue, and O. A. Krishtal, *J. Physiol. (London)* **379,** 171 (1986).
23. C. Frank, H. Hatt, and J. Dudel, *Neurosci. Lett.* **77,** 199 (1987).
24. J. Dempster, "Computer Analysis of Electrophysiological Signals." Academic Press, San Diego, 1993.
25. A. Stelzer, A. R. Kay, and R. K. S. Wong, *Science* **241,** 339 (1988).
26. Q. X. Chen, A. Stelzer, A. R. Kay, and R. K. S. Wong, *J. Physiol. (London)* **420,** 207 (1990).
27. P. H. Barry and J. W. Lynch, *J. Membr. Biol.* **121,** 101 (1991).
28. S. J. Korn, A. Marty, J. A. Connor, and R. Horn, this series, Vol. 4, p. 364, 1991.

29. J. A. Peters, E. F. Kirkness, H. Callachan, J. J. Lambert, and A. J. Turner, *Br. J. Pharmacol.* **94,** 1257 (1988).
30. W. Sieghart, *Trends Pharmacol. Sci.* **13,** 446 (1992).
31. H. Lüddens and W. Wisden, *Trends Pharmacol. Sci.* **12,** 49 (1991).
32. N. L. Harrison, S. Vicini, and J. L. Barker, *J. Neurosci.* **7,** 604 (1987).
33. J. L. Barker, N. L. Harrrison, G. D. Lange, and D. G. Owen, *J. Physiol. (London)* **386,** 485 (1987).
34. C. F. Stevens, *in* "Membranes, Channels, and Noise" (R. S. Eisenberg, M. Frank, and C. F. Stevens, eds.), pp. 1–20. Plenum, New York, 1981.
35. C. R. Anderson and C. F. Stevens, *J. Physiol. (London)* **235,** 665 (1973).
36. J. J. Lambert, J. A. Peters, N. C. Sturgess, and T. G. Hales, *in* "Steroids and Neuronal Activity" (D. Chadwick and K. Widdows, eds.), pp. 56–82. Wiley, Chichester, England, 1990.
37. G. Puia, M. R. Santi, S. Vicini, D. B. Prichett, R. H. Purdy, S. M. Paul, and P. H. Seeburg, *Neuron* **4,** 759 (1990).
38. Selye, H. (1941). *Proc. Soc. Exp. Biol. Med.* **46,** 116.
39. Edwards, F. A., Konnerth, A., Sakmann, B., and Takahashi, T. (1989). *Pfluegers Arch.* **414,** 600.
40. Edwards, F. A., Konnerth, A., and Sakmann, B. (1990). *J. Physiol. (London)* **430,** 213.
41. Chen, R., Belelli, D., Lambert, J. J., Peters, J. A., Reyes, A., and Lan, N. C. (1994). *Proc. Natl. Acad. Sci. U.S.A.* **91,** 6069.

[28] *In Vitro* Approaches to Studying Glucocorticoid Effects on Gene Expression in Neurons and Glia

Martha Churchill Bohn

Introduction

Glucocorticoid hormones are important regulatory molecules that influence many aspects of development and aging of the nervous system. In addition, they play roles in modulating responses to stress, learning and adaptation, mood, and levels of sensory detection (for reviews see 1–7). These pleiotropic effects are mediated primarily through intracellular receptors that are expressed widely throughout the brain in many classes of neurons and glial cells (8–11). In addition, many neurons, particularly those in the hippocampal formation, and apparently some glial cells, coexpress glucocorticoid and the mineralocorticoid receptors (GRs and MRs, respectively) (12–16). These two receptors apparently compete for the same glucocorticoid ligands, but with different affinities (17). The functional significance of coexpression of these receptors in a single cell is not known; however, electrophysiological and behavioral studies suggest that they may mediate antagonistic effects (2, 18).

Although the widespread effects of glucocorticoids on the brain are well documented, little is known about their effects at the molecular level in defined types of neurons and glia. Furthermore, in cells that coexpress GRs and MRs, it is not known whether these act as transcriptional regulators on the same or different sets of genes. This is particularly important to understanding glucocorticoid effects on the hippocampus, which is remarkable for its high levels of expression of both receptors and for the paucity of molecular data on the effects of glucocorticoids on gene expression in this brain region.

Our goal has been to establish and characterize *in vitro* systems of glial and neuronal cells in which glucocorticoid effects on gene expression in defined cell types might be studied more directly. This has involved establishing cultures of relatively pure hippocampal neurons and cerebral astrocytes that are grown under serum-free, defined conditions. The cells are then characterized for expression of GRs and/or MRs and their corresponding mRNAs using immunocytochemistry and an RNase protection assay, respec-

Methods in Neurosciences, Volume 22

tively. Glucocorticoid effects on gene expression are studied in the cultures through the use of two-dimensional (2D) gel electrophoresis to study steady state levels of metabolically labeled whole cell and secreted proteins, and the levels of corresponding mRNAs are studied by running 2D gels of ^{35}S-labeled *in vitro* translation products of poly(A)$^+$ RNA.

Immunocytochemical Localization of Glucocorticoid and Mineralocorticoid Receptors in Hippocampal Neurons and Cerebral Astrocytes

We have used light and confocal microscopy to characterize the expression and subcellular localization of GRs and MRs in cultured neurons and glial cells (8, 13). In addition, electron microscopy has been used to study subcellular localization of GR in neurons *in situ* (19, 20). Relatively pure cultures of hippocampal neurons and cerebral astrocytes were prepared using previously published procedures with minor modifications (8, 21–23). Cells are grown in defined, serum-free medium without the addition of steroids to the medium, except transiently to study possible translocation of receptors, as discussed below. Cells are stained for immunoreactivity to GRs or MRs by using a biotin-conjugated secondary antibody and streptavidin conjugated to fluoresceinisothiocyanate (FITC).

The majority of hippocampal neurons and astrocytes in culture contain GR immunoreactivity and MR immunoreactivity (Fig. 1) (8). Because some steroid receptors are known to translocate to the nucleus in the presence of ligand, whereas other receptors appear to be primarily localized in the nucleus even in the absence of ligand (24, 25), we have studied the localization of GRs and MRs in the presence and absence of glucocorticoid. In all types of glial cells studied, including astrocytes, oligodendrocytes, C6 glioma, and schwannoma, GR immunoreactivity is primarily cytoplasmic in the absence of steroid and nuclear in the presence of glucocorticoid (8, 13). In contrast, the localization of MR immunoreactivity in astrocytes is not affected by the presence or absence of corticosterone or aldosterone, but is generally distributed over both cytoplasmic and nuclear compartments (13).

In hippocampal neurons, the pattern of MR staining observed is similar to that in astrocytes, that is, the presence of ligand has no obvious effect on its distribution. Neurons have high levels of MR immunoreactivity over both cytoplasm and nucleus. Examination of these neurons with confocal microscopy shows a particularly intense perinuclear staining. In contrast, in the majority of pyramidal-shaped neurons in culture, GR immunoreactivity is primarily nuclear in the presence of corticosterone and cytoplasmic in its

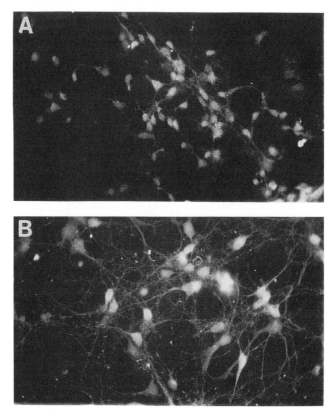

Fig. 1   Immunofluorescence showing (A) GR immunoreactivity and (B) MR immunoreactivity in dissociated hippocampal neurons grown in astrocyte conditioned medium
lacking steroids and fixed (A) after treatment with corticosterone (10^{-6} M, 30 min)
or (B) without corticosterone. Note that GR immunoreactivity is mainly nuclear in
the presence of ligand. In contrast, MR immunoreactivity is distributed throughout
cells, although nuclei are heavily stained even in the absence of ligand.

absence. Particularly notable is the high level of GR immunoreactivity in
neuronal processes in the absence of steroid as observed with confocal
microscopy (13). A similar localization is observed at the electron microscopic level in tissue sections (19, 20). In many smaller, bipolar-shaped
neurons, however, no apparent translocation of GR is observed, with GR
immunoreactivity present in both nucleus and cytoplasm (13). It is not clear
whether these neurons are as yet too immature to express the translocation
mechanism, or whether the GR, like the MR, does not translocate in
some neurons.

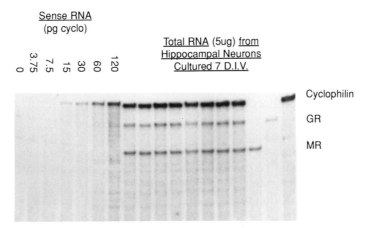

FIG. 2 RNase protection assay for GR mRNA and MR mRNA, as well as the internal control cyclophilin mRNA, in 5 μg of total RNA prepared from dissociated hippocampal neurons. On the left of the gel are varying amounts of cyclophilin sense RNA for the purpose of determining absolute amounts of mRNAs and normalizing among experiments. D.I.V., days *in vitro*. (Adapted reference 26 by permission of Elsevier Science Publishers.)

Expression of Glucocorticoid and Mineralocorticoid Receptor mRNAs in Cultured Neurons and Astrocytes

To strengthen the immunocytochemical data with molecular data and to establish an assay for studying factors that influence the expression of GRs and MRs at the level of mRNA, an RNase protection assay was established for the simultaneous measurement in the same tube of mRNAs for GRs, MRs, and cyclophilin, an internal control mRNA not regulated by glucocorticoids (Fig. 2) (26). Using this assay, we found that protected bands corresponding to GR mRNA and MR mRNA are present in both neurons and astrocytes (27). However, the level of expression in astrocytes was found to be markedly lower than that in neurons. Both messages are easily detected in 2.5 μg of total RNA from neuronal cultures, whereas 1 μg of poly(A)$^+$ RNA is needed for detection in astrocytes. Interestingly, higher levels of MR mRNA than GR mRNA are expressed by cultured hippocampal neurons (27). This is similar to the expression observed in hippocampus *in vivo* (26). However, in astrocytes, the level of MR mRNA is significantly lower than that of GR mRNA (27). The characterization of GR and MR expression in these *in vitro* systems combined with this sensitive assay for mRNA levels provides an opportunity for investigating more directly the role of various extracellular factors in regulation of GR and MR expression.

Effects of Glucocorticoids on Gene Expression, Using Two-Dimensional Gel Electrophoresis

To begin to identify novel genes that are regulated by glucocorticoids in the brain, we have been studying glucocorticoid-induced alterations in the steady state levels of proteins and *in vitro* translation products of poly(A)$^+$ RNA. Using separation of ^{35}S-labeled proteins by 2D gel electrophoresis, we have observed that addition of glucocorticoids to cultured type 1 cerebral astrocytes significantly changes the steady state levels of about 5% of the total cellular proteins, as well as the secreted proteins (M. C. Bohn *et al.*, unpublished observations, 1988). In addition, the pattern of secreted proteins is markedly altered from that secreted by control astrocytes. Because many of these changes probably represent secondary responses, rather than primary inductions, we have extended this observation to the mRNA level and have looked at mRNA levels after short treatments with corticosterone in the presence or absence of cycloheximide to inhibit protein synthesis. A study of ^{35}S-labeled *in vitro* translation products from astrocyte poly(A)$^+$ RNA after separation by ultrahigh-resolution giant 2D gel electrophoresis reveals 12 *in vitro* translation products, likely representing 10 mRNA species that are regulated by corticosterone (28). Eleven products are significantly increased, and one decreased, most within 3 hr of corticosterone treatment. Because addition of cycloheximide does not prevent these changes, they probably represent alterations in transcription. However, other mechanisms, such as changes in the rate of mRNA degradation, cannot be excluded. Two of these proteins have been identified as glucocortin and glutamine synthetase (glutamate–ammonia ligase) (28), whereas the others remain to be identified. Because large amounts of protein can be loaded on giant gels, future studies will permit the isolation and sequencing of these unknown proteins.

Applications of *in Vitro* Systems

Future use of these neuronal cultures in which GRs and MRs are expressed will simplify studying certain cellular and molecular aspects of glucocorticoid effects in the brain. The action of various agents, such as steroids and neurotransmitters, on steroid receptor expression in specific cell types can more easily be studied. The methods developed for studying glucocorticoid mRNAs in astrocytes, as discussed above, can be applied to hippocampal neurons, or immortalized hippocampal neurons. The identification of proteins secreted by astrocytes whose levels are affected by glucocorticoids might be studied for their effects on cultured neurons with the idea that some glucocorticoid effects on neurons might be indirectly mediated by astrocytes. Finally, these systems are amenable to the study of glucocorticoid-regulated genes by other molecular techniques, such as subtractive hybridization or the differential gene display method (29).

Procedures

Immunofluorescence of Glucocorticoid and Mineralocorticoid Receptor Immunoreactivity in Cultured Neurons and Astrocytes

Dissociated cultures of hippocampal neurons are prepared from embryonic day 17.5–18 (E17.5–18) rat hippocampus according to previously published methods of Banker and Cowan (22). Neurons are plated on poly-D-lysine-coated coverslips placed in the bottoms of 24-well culture plates in modified Eagle's minimal medium (MEM) (Cat. No. #320-1200AJ; GIBCO, Grand Island, NY) in the presence of 10% (v/v) horse serum for 4 hr to permit attachment. Cultures are then washed and switched to serum-free MEM without phenol red, but containing supplements and N1 components (36) lacking progesterone, as shown below. The neurons also can be grown in astrocyte conditioned medium. This is prepared by incubating confluent type 1 astrocytes in the same medium described above for 24 hr. The medium is then collected, centrifuged to remove cellular debris, and frozen at −80°C.

MEM-Hipp Medium
See (23) for details.

   MEM powder for 1 liter (Cat. No. 410-1700; GIBCO)

Add per liter:

D-Glucose	4 g/liter
L-Glutamine	292 mg
Sodium bicarbonate	2.2 g
HEPES	2.83 g
Sodium pyruvate	110 mg
KCl	1.1 g

N1 Components
See (36) for details.

Insulin	5 μg/ml
Transferrin	5 μg/ml
Selenite	$3 \times 10^{-8} M$
Putrescine	$1 \times 10^{-4} M$
Bovine serum albumin (BSA; fatty acid free)	1 mg/ml
(A-0281; Sigma, St. Louis, MO)	

Purified type 1 astrocytes are prepared according to the methods of McCarthy and DeVellis (21). After other cell types have been removed, the astrocytes are subcultured on poly-D-lysine-coated coverslips and grown to the desired cell density in Dulbecco's MEM (DMEM; Cat. No. 432-2100; GIBCO) supplemented with 10% (v/v) fetal bovine serum. At least 24 hr prior to staining, or labeling with [^{35}S]methionine, as discussed below, cells are thoroughly washed and switched to defined, serum-free medium DMEM without phenol red plus N1 components, as above, lacking steroids.

Prior to staining for GR or MR immunoreactivity, some cultures are treated for varying times, for example, 30 min–2 hr, with dexamethasone, corticosterone, or aldosterone (10^{-6}–$10^{-8} M$). Note that careful handling of the coverslips for the following steps is critical. All solutions are added to the sides of the wells containing the coverslips so as not to disturb or dislodge the cells. Cultures are washed in phosphate-buffered saline (PBS) and fixed on ice for 30 min following the addition of ice-cold 4% (w/v) paraformaldehyde in 0.1 M sodium phosphate (pH 7.4). Cultures are washed three times in PBS and then permealized with 0.3% (v/v) Triton X-100 (T-6878; Sigma) in PBS containing normal serum (horse or goat), as appropriate, for 15 min at room temperature. After rinsing the cells with 0.1% (v/v) Triton X-100 in PBS, they are incubated at 4°C for 16–48 hr in primary antibody diluted in 0.3% (v/v) Triton X-100, 1% (w/v) BSA, and 1% (v/v) normal serum (horse or goat, as appropriate). For GR immunoreactivity, the mouse monoclonal antibody Bugr-2 (Cat. No. MA1-510; Affinity BioReagents, Neshanic Station, NJ) is used at a dilution of 1:250. For MR immunoreactivity, rabbit polyclonal antiserum MR-2 or MR-4, previously purified by immunoadsorption, is used at 1:100. Note that the optimal dilution of each batch of antiserum needs to be determined by staining with decreasing dilutions of each. Following incubation with the primary antibody, cells are washed with 0.1% Triton X-100 in PBS containing 0.25% BSA and incubated with an affinity-purified, biotinylated secondary antibody (horse anti-mouse or goat anti-rabbit, as appropriate; Vector Laboratories, Burlingame, CA) diluted 1:250 in PBS containing 0.1% Triton X-100, 1% BSA and 1% normal serum for 90 min at room temperature. Following several washes, cells are incubated with fluorescein isothiocyanate (FITC)-conjugated avidin (cell sorter grade, Cat. No. A-2011; Vector Laboratories) used at 1:200 in PBS with 0.1% Triton X-100 for 30 min at room temperature. The FITC-stained cells are washed several times, with the final wash lacking Triton X-100, and coverslipped in glycerin. If desired, nuclei can be stained with propidium iodide for viewing by confocal microscopy. In this case, cells are washed following the FITC incubation, incubated for 20 min at room temperature in RNase (10 mg/ml in PBS) (Cat. No. R-5882; Sigma) diluted 1:200, washed with PBS, and incubated with propidium iodide at 0.1 μg/ml (stock solution, 50 μg/ml in PBS, Cat. No. P-5264; Sigma).

RNase Protection Assay for Glucocorticoid and Mineralocorticoid Receptor mRNA

RNA

Total RNA is prepared from cultured cells following washing with PBS using a modification of the method of Chirgwin *et al.* (30). Cells are lysed for 5 min in ice-cold 4 M guanidine thiocyanate in 25 mM sodium acetate (pH 6.0) containing 1% 2-mercaptoethanol (2 ml/100-mm plate), and then passed through a 22-gauge needle. RNA is purified by ultracentrifugation through a 5.7 M CsCl cushion in an SW-60 rotor (Beckman, Fullerton, CA) at 35,000 rpm for 24 hr at 20°C.

Probes

Three antisense probes and one sense probe are prepared for each assay. The antisense probes are (a) a 670-bp cyclophilin probe (31), (b) a 510-bp probe of the amino terminus of GR generated from a plasmid subcloned from pXR14 (32), and (c) a 420-bp probe of the 3' terminus of rat MR generated from plasmid prMREH (12). A sense cyclophilin probe is used to enable the determination of absolute amounts of mRNAs, as well as to make quantitative comparisons among assays run on different days. Varying levels of cyclophilin sense RNA are run in each assay as determined by the specific activity of tritium-labeled RNA as described in detail in the Ambion RPA kit manual (Ambion, Austin, TX). Antisense probes are labeled using the Riboprobe Gemini System II Kit (Promega, Madison, WI) and [^{32}P]UTP. After incubation with RQ1 DNase, the labeled cRNA transcripts are precipitated in the presence of 50 μg of carrier yeast tRNA with ammonium acetate–ethanol. The pellets are dried in a Speed Vac (Savant, Hicksville, NY) and resuspended in 8 μl of loading buffer (solution E in Ambion RPA kit; Cat. No.

1400), boiled for 3–4 min, and purified by electrophoresis through a 6% (w/v) polyacrylamide gel using a Protean II xi cell system (Bio-Rad, Richmond, CA) at 450 V for 2–2.5 hr. The locations of the specific cRNAs are located by exposure to X-Omat film (Kodak, Rochester, NY), excised from the gel, and placed in 300 μl of probe elution buffer (solution F, Ambion RPA kit) at 37°C overnight.

Protection Assay and Quantitation

A quantitative RNase protection assay is performed using the Ambion RPA kit (Cat. No. 1400) according to manufacturer directions with minor modifications. The three antisense probes (3×10^4 cpm/sample of cyclophilin and 3×10^5 cpm/sample each of GR and MR) are added to 5 μg of total RNA. Several standards of sense cyclophilin ranging from 0 to 240 pg of RNA are run separately, with only the antisense cyclophilin probe added. The samples are precipitated with ammonium acetate–ethanol, resuspended in 20 μl of solution A (Ambion RPA kit), boiled at 90°C for 3–4 min, transferred to 45°C, and left overnight. The following day, 200 μl of RNase digestion buffer (solution B) containing 100 units of RNase T1 (Boehringer Mannheim, Indianapolis, IN) is added to each sample and incubated for 1 hr at 30°C. Then, 10 μl of proteinase K (solution D) and 10 μl of 20% (w/v) SDS (solution D2) are added and the samples incubated for 15 min at 37°C. The protected RNA hybrids are extracted with phenol–chloroform–isoamyl alcohol (25:24:1) and precipitated with ammonium acetate–ethanol. The pellets are dried, resuspended in loading buffer (solution E, Ambion RPA kit manual), and purified by electrophoresis through a 6% polyacrylamide gel. After fixation, protected bands are visualized on preflashed Kodak X-Omat film at −80°C using an intensifying screen. Following scanning with a Molecular Dynamics 110A laser scanner (Molecular Dynamics, Sunnyvale, CA), band densities are determined using Quantity One (Protein Databases, Inc., Huntington Station, NY) software run on a Sun Sparc workstation. The bands are corrected for lane differences using the internal cyclophilin band and for differences in the probe sizes. The absolute amount of mRNA is determined by comparison to the bands of cyclophilin sense standards run on the same gel.

Two-Dimensional Gel Analysis of in Vitro Translation Products

Poly(A)$^+$ RNA is isolated from total RNA by adsorption twice to oligo(dT) as described by Kingston (33); it is then washed twice with 0.5 ml of 80% ethanol and dissolved in diethyl pyrocarbonate (DEPC; Sigma)-treated water

(1 μg/ml). This is translated *in vitro* in the presence of [^{35}S]methionine according to a previously published method, with modifications (28,34). An aliquot of poly(A)$^+$ RNA (2.5 μg) is transferred to a clean tube and dissolved in 4 μl of 7 *M* methylmercuric hydroxide (prepared in DEPC water) at room temperature for 10 min. To this, 46 μl of translation mix (see below) is added, and the reaction allowed to proceed at 30°C for 60 min. A sample of translation mix containing no RNA is included to determine background incorporation.

To determine the degree of incorporation of label into protein, 2 μl of the translated mixture is transferred to 1 ml of 1 *M* NaOH–1.5% (v/v) hydrogen peroxide–BSA (0.5 mg/ml) in a 12-ml polystyrene tube and heated to 37°C for 10 min. Four milliliters of cold 20% (w/v) trichloroacetic acid (TCA) containing 2% (w/v) casamino acids is added, vortexed, and left at 4°C for 30 min. The precipitated proteins are collected on glass fiber filters premoistened with 1–2 ml of the TCA–casamino acid solution. A 12-sample Millipore (Bedford, MA) filtration apparatus is convenient for this step. The tubes are washed twice with 4 ml of 8% TCA and added to the filters. The filters are then washed with 1–2 ml of acetone, removed, allowed to dry on blotting paper, and counted.

The remaining translated mixture is prepared for 2 D gel electrophoresis by transferring it to an airfuge tube, washing the reaction tube with 25 μl of water, and adding this to the tube. Polysomes are sedimented by centrifuging at 100,000 *g* for 15–20 min at room temperature. The supernatant is carefully transferred to a 1.5-ml Eppendorf tube, quick-frozen on dry ice, and lyophilized. The pellet is then dissolved in 150 μl of lysis buffer for running 2D gels according to previously published methods (35).

Translation mix
    DEPC water, 21.4 μl
    Amino acid mix minus methionine (Promega), 5.7 μl
    [^{35}S]Methionine/cysteine (Tran^{35}S-label, Cat. No. 51006; ICN, Costa
       Mesa, CA), 28.5 μl
    RNasin (40 units/μl; Promega), 7 μl
    Rabbit reticulocyte lysate (Promega), 200 μl (one vial)
Lysis buffer
    Urea (9.5 M)
    Nonidet P-40 (NP-40), 2% (v/v)
    Ampholytes (1.6% pH range 5–8, 0.4% pH range 3.5–10%; Pharmacia,
       Piscataway, NJ), 2% (w/v)
    Mercaptoethanol, 5% (v/v)

Metabolic Labeling of Cultured Cells for Study of Total and Secreted Proteins by Two-Dimensional Gel Electrophoresis

Cells in 35-mm culture dishes are washed three times with 1 ml of methionine-free medium [L-methionine-free DMEM (Cat. No. 320-1970; GIBCO) supplemented with 0.1% (v/v) fetal bovine serum (FBS) and 4 mM L-glutamine]. Cells are then labeled in 0.5 ml of methionine-free medium with 200 μCi of [^{35}S]methionine/cysteine (Tran^{35}S-label, Cat. No. 51006; ICN) for the desired time (usually 0.5–3 hr) at 37°C. The supernatant containing the labeled secreted proteins is carefully pipetted into tubes containing 2.5 μl of 200 mM phenylmethylsulfonyl fluoride (PMSF; Sigma) and placed on ice. The labeled cells are rinsed three times with ice-cold PBS containing 1 mM MgCl$_2$, lysed with 200 μl of lysis buffer for 3–4 min, transferred to tubes, and stored at −80°C. The supernatants are then centrifuged at 12,000 g for 5 min at 4°C to remove cellular debris, dialyzed overnight against 1 mM ethylenediaminetetraacetic acid (EDTA), lyophilized, dissolved in 150 μl of lysis buffer, and stored at −80°C.

Prior to running 2D gels, each sample is assayed for incorporation of label into protein. Whatman (Clifton, NJ) filters (Cat. No. 1003323) are labeled with the sample number using a pencil and a straight pin is placed through the middle of the filter. A sample (2 μl) is spotted onto the filter and the filter floated on ice-cold 10% TCA containing 1% methionine for approximately 45 min, during which time the filters sink. The filters are then washed twice with 5% TCA lacking methionine and four times with 99% ethanol for 15 min each. The filters are allowed to air dry under a light bulb and counted in 8 ml of scintillation fluid.

Concluding Remarks

One approach toward dissecting the complex actions of glucocorticoids on the brain is to turn to *in vitro* systems of identified neurons or glial cells grown under defined conditions. For such systems to be useful, however, it is necessary to characterize glucocorticoid and mineralocorticoid receptor expression in cells grown *in vitro*. This chapter reviews methods for studying expression of GR- and MR-immunoreactive proteins and mRNAs in cultured hippocampal neurons and cerebral type 1 astrocytes. Methods have also been described for labeling and preparing cells to study glucocorticoid effects on secreted and total cellular proteins, as well as proteins translated *in vitro* from poly(A)$^+$ RNA by 2D gel electrophoresis.

Acknowledgment

The author wishes to thank Mrs. Rita Giuliano and Mr. Syed Hussain for their technical assistance in setting up the procedures reviewed. The writing of this chapter was supported in part by the National Institutes of Health, NS20832.

References

1. M. C. Bohn, *in* "Neurobehavioral Teratology" (J. Yanai, ed.), p. 365. Elsevier, Amsterdam, 1984.
2. E. R. De Kloet, *Front. Neuroendocrinol.* **12,** 95 (1991).
3. A. J. Doupe and P. H. Patterson, *in* "Adrenal Actions on Brain" (D. Ganta and D. Pfaff, eds.), p. 23. Springer-Verlag, Berlin, 1982.
4. B. S. McEwen, E. R. de Kloet, and W. Rostene, *Physiol. Rev.* **66,** 1121 (1986).
5. R. M. Sapolsky, *Progr. Brain Res.* **86,** 13 (1990).
6. J. S. Meyer, *Physiol. Rev.* **65,** 946 (1985).
7. M. C. Bohn, *in* "Autonomic–Endocrine Interactions" (K. Unsicker, ed.), Autonomic Nervous System Series. Harwood, Chur, Switzerland, 1994. In press.
8. U. Vielkind, A. Walencewicz, J. Levine, and M. C. Bohn, *J. Neurosci. Res.* **27,** 360 (1990).
9. J. A. M. Van Eekelen, J. Z. Kiss, H. M. Westphal, and E. R. De Kloet, *Brain Res.* **436,** 120 (1987).
10. K. Fuxe, A. C. Wikström, S. Okret, L. F. Agnati, A. Hårfstrand, Z.-Y. Yu, L. Granholm, M. Zoli, W. Vale, and J.-Å. Gustafsson, *Endocrinology (Baltimore)* **117,** 1803 (1985).
11. K. Fuxe, A. Cintra, L. F. Agnati, A. Hårfstrand, A.-C. Wikström, S. Okret, *et al.*, *J. Steroid Biochem.* **27,** 159 (1987).
12. J. L. Arriza, L. W. Swanson, and R. M. Evans, *Neuron* **1,** 887 (1988).
13. D. L. Barker, R. Giuliano, Z. Krozowski, and M. C. Bohn, *Soc. Neurosci. Abstr.* **18,** 478 (1992).
14. J. P. Herman, P. D. Patel, H. Akil, and S. J. Watson, *Mol. Endocrinol.* **3,** 1886 (1989).
15. J. A. M. Van Eekelen, W. Jiang, E. R. DeKloet, and M. C. Bohn, *J. Neurosci. Res.* **21,** 88 (1988).
16. M. C. Bohn, E. Howard, U. Vielkind, and Z. Krozowski, *J. Steroid Biochem.* **40,** 105 (1991).
17. E. R. De Kloet, A. Ratka, J. M. H. M. Reul, W. Sutanto, and J. A. M. Van Eekelen, *Ann. N.Y. Acad. Sci.* **521,** 351 (1987).
18. M. Joëls and E. R. de Kloet, *Proc. Natl. Acad. Sci. U.S.A.* **87,** 4495 (1990).
19. Z. Liposits, R. M. Uht, R. W. Harrrison, F. P. Gibbs, W. K. Paull, and M. C. Bohn, *Histochemistry* **87,** 407 (1987).
20. Z. Liposits and M. C. Bohn, *J. Neurosci. Res.* **35,** 14 (1993).
21. K. D. McCarthy and J. DeVellis, *J. Cell Biol.* **85,** 890 (1980).
22. G. A. Banker and W. M. Cowan, *Brain Res.* **126,** 397 (1977).

23. M. P. Mattson and S. B. Kater, *Brain Res.* **490,** 110 (1989).
24. K. R. Yamamoto, *Annu. Rev. Genet.* **19,** 209 (1985).
25. W. V. Welshons, M. E. Lieberman, and J. Gorski, *Nature* (*London*) **307,** 747 (1984).
26. M. C. Bohn, D. Dean, S. Hussain, and R. Giuliano, *Dev. Brain Res.* **77,** 157 (1994).
27. M. C. Bohn, S. Hussain, R. Giuliano, M. K. O'Banion, and D. Dean, *Soc. Neurosci. Abstr.* **18,** 479 (1992).
28. M. K. O'Banion, D. M. Young, and M. C. Bohn, *Mol. Brain Res.* **22,** 57 (1994).
29. P. Liang and A. B. Pardee, *Science* **257,** 967 (1992).
30. J. M. Chirgwin, A. E. Przybyla, R. J. MacDonald, and W. J. Rutter, *Biochemistry* **18,** 5294 (1979).
31. P. E. Danielson, S. Forss-Petter, M. A. Brow, L. Calavetta, J. Douglass, R. J. Milner, and J. G. Sutcliffe, *DNA* **7,** 261 (1988).
32. R. Miesfeld, S. Okret, A.-C. Wikström, O. Wrange, J.-Å. Gustafsson, and K. R. Yamamoto, *Nature* (*London*) **312,** 779 (1984).
33. R. E. Kingston, *in* "Current Protocols in Molecular Biology" (F. M. Ausubel, R. Brent, R. E. Kingston, D. D. Moore, J. G. Seidman, J. A. Smith, and K. Struhl, eds.), p. 4. Greene Publ. Assoc. and Wiley (Interscience), New York, 1989.
34. R. A. Colbert and D. A. Young, *J. Biol. Chem.* **261,** 14733 (1986).
35. D. A. Young, B. P. Voris, E. V. Maytin, and R. A. Colbert, *in* Enzyme Structure," Part I (C. Hirs and S. Timasheff, eds.), Methods in Enzymology, Vol. 91, p. 190. Academic Press, New York, 1983.
36. J. E. Bottenstein, *in* "Methods for Serum-Free Cultures of Neuronal and Lymphoid Cells," p. 3. A. R. Liss, Inc., New York, 1984.

Section V

Steroid Effects on Integrated Systems

[29] Behavioral Approaches to Study Function of Corticosteroids in Brain

Melly Silvana Oitzl

Introduction

Evolution of species is the result of improving adaptation to changes in environmental conditions. The higher forms of life can exist only in a relatively narrow range of physical and chemical factors that are determined by the genetically encoded morphological and metabolic properties of the organism. This implicates the presence of a mechanism ensuring homeostasis of the system. A set of reference signals represents an optimal input from internal and external signals. Any deviation is continuously evaluated as reducing or increasing the gap between the reference set and the actual signal. The static forms of reference sets (genetically controlled systems) are supplemented by ever-changing dynamic adjustments of the organism to a variety of reference signals from its internal and external environment, that is, motivation, context, and learning. Behavior is the final output of these interactions and represents, at the same time, a mean of the organism to control its input. Behavioral responses are thought to be represented in a set of (hierarchically) organized feedback systems: (a) the genetically determined species-specific responses; behavior determined by (b) the state of the organism (motivation), (c) the context (environment), and (d) learning, the most plastic ability of the organism to adapt its reactions to specific environmental requirements. The capacity to organize and synchronize the various systems is largely ascribed to the hippocampus (1). Hormones appear to be critically involved in organizing these dynamic processes of adjustment of different biological systems.

Ever since their discovery, corticosteroids have been tightly linked to stress and adaptation, a system of homeostasis (2, 3). Stress responses occur following a wide variety of physiological and psychological stimuli. These stimuli influence the activity of the hypothalamic–pituitary–adrenal axis, which in turn is modulated by corticosteroids. The demonstration of two types of intracellular receptors in the brain, mineralocorticoid receptors (MRs or type I) and glucocorticoid receptors (GRs or type II), which both bind corticosterone (rat), and their special location in the hippocampus require a detailed analysis of the behavior controlled by the action of corticosteroids via these two receptor types. In this chapter two of the behavioral tasks and the automated method for behavioral observation used in our laboratory to identify the role of corticosteroids and their receptor-mediated effects are described. This also includes a brief description of the underlying rationale to facilitate the understanding of specific adjustment of the experimental designs.

Automated Computerized Registration of Behavior

The use of computer imaging techniques is rapidly increasing in neuroscience. In behavioral observations freely moving animals are monitored by video-cameras connected to a video digitizer and a computer. Automated systems for tracking the moving pattern have more and more become a standard technique. In the two behavioral tasks described below, the motion of the rat (swimming and locomotion) and its distribution in time and space were monitored by such an image analysis system (4) (Video Tracking & Motion Analysis System, Noldus Information Technology, Wageningen, The Netherlands). This automated computerized registration and analysis of behavior, described below, allows a detailed description of underlying psychological processes.

Behavioral Analysis

Hardware
The hardware consists of a video digitizer (frame grabber) connected to an IBM compatible 80386 computer. Any standard video signal, either from a camera or a video recorder, may be connected to the digitizer. For displaying the output an RGB monitor is required. In our setup a CCD video camera (Sony; light sensitivity, 3 lx) was used. It is fixed at the center above the behavioral apparatus.

Software
Because the flexibility of the system is essential for the study of animal behavior under various environmental and experimental conditions, routines for subtracting background, thresholding, and defining objects of interest are used. Data acquisition and data analysis are divided into different subroutines allowing different analysis procedures on the same data.

A picture of the empty apparatus (e.g., water maze, open field) is digitized and after introducing the animal in the apparatus, another picture is taken. These two pictures are subtracted from each other (object remains). Irrelevant "objects" (e.g., reflection of the water, urine, sawdust) are removed via a threshold operation. These imaging commands are performed in every cycle of the program. The sampling rate depends largely on the processor frequency of the computer (i.e., the minimal sampling rate), but can also be adjusted to the specified experimental condition. We used a sampling rate of 2/sec in all tasks.

In every cycle of the routine the analysis module uses the last position and the present position of the animal for assessing changes in position (distance), reflecting locomotor activity and presence of the animal in a certain area of the test cage. An important feature is that all kind of variables, such as areas of interest, time periods, and even sampling rate can be computed after data collection has taken place. Motion patterns can be printed. Examples are given in Figs. 2B and 3B.

General Preconsiderations for Behavioral Experiments on Effects of Corticosteroids

Owing to the differential affinity of corticosterone to its two receptors (MR high and GR low affinity), their differential distribution in the brain, and proposed time of action in the stress response [sensitivity (MR) and offset (GR) of the stress response], it is advisable to keep to strictly controlled schedules: (a) the circadian rhythm of hormones, time-related effects of drug treatment, and interactive effects (drug × period of the circadian rhythm) will influence the behavior of the animal. We run behavioral tests mainly during the 12-hr light-on period (7:00–19:00 hr), the behaviorally less active time of rats. It is the period of low, resting (morning levels; occupation mainly of MRs), and slowly increasing to peak levels (evening levels; significant occupation of GRs) of circulating corticosterone. Testing takes place either at the trough (within the first 4 hr after light-on period) or the peak (starting 2 to 3 hr before light-off period) of circulating corticosteroids. Also, the duration of the light-on/light-off period is critical; (b) days before the behavioral observations start, rats should be brought to the experimental room or an adjacent chamber to avoid the stress response to transportation; (c) handling of the animals should start several days before they are subjected to the behavioral tasks.

Corticosteroid Receptors, Information Processing, and Spatial Behavior

The processing of a complex set of stimuli, that is, internal and/or external signals, comparing with a set of reference signals, and the consequent "choice" of the (behavioral) output have been ascribed mainly to one structure of the limbic system, the hippocampus (1, 5). Under circumstances in which the difference between actual and reference signals becomes too big and attempts to restore homeostasis fail, the organism will experience stress. Information processing and behavioral adaptation (restoring homeostasis) are closely linked to the function of corticosteroids and the hippocampus.

Corticosteroids are important regulators of the function of the hippocampal formation. This is emphasized by the high density and colocalization of MRs and GRs in hippocampal neurons, the stabilizing effect of these receptors on the excitability of these neurons [(3); see Joëls in this book]. Moreover, MRs are thought to be involved in the regulation of the onset, whereas GRs are involved in the suppression of the stress response. It is generally accepted that the hippocampus provides an essential contribution to learning and memory and plays a key role in the reactivity of the animals to novelty (5, 6). Thus, behavior specifically related to hippocampal functioning (requiring a spatiotemporal coordination) represents the mean to elucidate the role of corticosteroids in information processing. Two spatial tasks, the water maze and open field, were chosen to unravel the role of corticosteroid receptor-mediated effects.

The behavioral tasks are described followed by an example demonstrating the involvement of corticosteroids and their specific MR- and GR-mediated behavioral aspects. In example 1, the functionality of the corticosteroid system was altered by the pharmacological approach. The central mineralocorticoid receptors were blocked by intracerebroventricular (icv) injection of a specific MR and GR antagonist. Rats were tested in the water maze. In example 2 the endocrinological approach was used. Adrenalectomized rats (ADX) and ADX rats substituted with corticosterone were observed in the open field with object. In both situations, rats were tested in the morning. Mineralocorticoid receptor-mediated effects were expected to be reflected in the response of an animal to a new environment: the search and exploratory strategy used, that is, the state- and context-related information processing, and GR-mediated effects should be reflected in processes of learning and memory. These differential effects of corticosteroids mediated by MRs and GRs have been demonstrated (7, 8).

Spatial Navigation Learning

A spatial navigation task, the so-called Morris water maze (9, 10) has proved useful for characterizing the neurochemical basis of spatial learning (11) and the role of the neuroanatomical systems (e.g., septum and hippocampus) required for spatial behavior (12). The procedure consists of (a) a series of training trials, in which the rat must locate a hidden platform, followed by

(b) a free swim trial without platform. Processes of learning and memory and behavioral strategies can be assessed. The flexibility of the various components of the tasks allows an easy adjustment to the appropriate experimental design.

Apparatus

The water maze consists of a large circular pool (90 to 250 cm in diameter) filled with warm water ($26 \pm 1°C$). A platform, usually a cylinder, is hidden below the water surface, at a fixed position, equidistant from the side and the center of the pool. For automated recording, a black pool with clear water and a black platform provide a very dark background excellently suited for detecting white animals (e.g., Wistar and Sprague-Dawley rats). Furthermore, it allows tracking of the animal while it dives. Using colored rats, like Brown Norway rats, requires the addition of chalk or milk to the water to increase the contrast between animal and background; white rats must be marked with a black spot on their head. The illumination of the experimental setup should be indirect, preferably in an dimly lit experimental chamber. The testing room contains numerous extramaze cues. All stimuli of the experimental chamber must remain in a fixed position to allow the rat an optimal spatial orientation. Fill the pool with fresh water every day.

Training to Find the Platform: Acquisition, Consolidation, Retrieval of Spatial Information

Different start positions, usually four, are equally spaced around the perimeter of the pool. They are used in a random sequence. A trial begins by placing the rat into the water facing the wall of the pool at one of the starting points. If the rat fails to find the platform within 120 sec it is placed onto it and remains there for 30 sec. Thereafter, it is placed for 15 sec (short intertrial interval) into a waiting cage to run the computer-program for the next trial. After the last trial the rat is dried under a red light heating lamp. Usually blocks of four trials are run on consecutive days. The number of training trials per day, their intertrial interval, and the number of training days can be varied.

Analysis and Interpretation of Training Trials The time needed (latency) to find and swim the distance to the platform, as well as the swimming speed (distance divided by latency), can be expressed per trial and per training block.

The rat learns the location of the submerged platform with the help of extramaze cues. The degree of spatial learning (acquisition, consolidation, and retention) will be reflected in the decrease of latency and swim distance to the platform over trials. The number of routes to reach the platform is unlimited. Different so-called search/escape strategies are possible. At successive periods of spatial learning rats can use two strategies. The first strategy, exploration, occurs mainly during the acquisition trials. It allows the animal to explore the maze in an efficient way and to search in all areas of the pool for a possibility to escape. The second strategy, goal-directed escape, reflects the retention of spatial information gained during acquisition. With progressive training, rats will locate the platform directly, independent of the start position. When using several training days, there will always be some animals that will perform a series of "direct to the platform escapes" and then start to explore the pool again for other escape possibilities.

One of the advantages of the image-analysis system is the concurrent estimation of the distance swum. This measure allows one to control for the motor ability of the animal. For example, (a) long latencies and short distances swum to find the platform indicate that the animal has difficulties in the control of motion without concurrent learning deficits; (b) long distances swum and no difference in latencies to find the platform are indicative of a different search/escape strategy. Therefore, it is advised to use both latencies and swim distances to provide a more differentiated picture of the treatment effects.

Furthermore, the performance during the first training trial of each daily block is of special interest: it will reflect long-term memory (the strength of consolidation and retrieval of spatial information) and the strategy the rat uses to solve the task.

Statistics   Statistical significance of the data is generally tested by an analysis of variance, followed by *post hoc* tests on specified trials.

Free Swim Trial (probe trial)

At a certain time after training (hours to days), the platform is removed from the pool, the rat swims for 60 sec, and the swim pattern of the animal is analyzed with respect to the former location of the platform.

Analysis and Interpretation of Free Swim Trial   Time spent and distance swum in the four quadrants of the pool (platform, adjacent left and right, and opposite to the platform) and the total distance swum are analyzed. Distances swum per quadrant in relation to the total distance can also be estimated as well as the number of times the rat crossed the exact location of the platform (accuracy).

The rat will show a certain swim pattern in relation to the previous location of the platform. The performance of the animal reflects several aspects: (a) sensory–motor capacity, (b) the strategy the animal uses to locate the platform and the tendency to switch between strategies, and (c) memory (consolidation and retrieval of information, and long-term and short-term memory). Even with the knowledge that the differentiation between strategy and memory is difficult in the case of impaired performance, it is important to estimate the relative contribution of these aspects to the performance. For example: the earlier in the process of spatial learning and the closer in time to the block of training trials, the more the swim pattern of the animal will reflect the search strategy used to locate the platform. If the latency to find the platform during the training trials is comparable between treatment groups, a different swim pattern during the free swim trial will indicate a different search strategy.

The term *transfer trial*, which is widely used in the literature, refers to memory processes only, and should be avoided.

Statistics Although others use parametric tests, it is advised to use nonparametric statistics (distribution of data—cutoff latency of 60 sec; percentages). The distribution of time and distance over the four quadrants within groups can be tested by a Friedman analysis of variance. Differences for a certain quadrant between groups can be tested by a Kruskal–Wallis analysis of variance (more than two groups) or a Mann–Whitney U-test (two groups).

Times of Treatment to Influence Information Processing
Animals can be treated at specially chosen times related to their performance and underlying information processing (13). Surgical and/or pharmacological intervention before training might influence acquisition, consolidation, and retrieval of information; immediately after training it will alter processes of consolidation.

Example 1: Mineralocorticoid and Glucocorticoid Receptor-Mediated Effects of Corticosterone on Spatial Navigation of Rats in Water Maze
We chose a training procedure consisting of a block of four trials with a short (15 sec) intertrial interval, given on two consecutive days. Male Wistar rats received an icv injection into the lateral ventricle (a) posttraining: immediately after their fourth trial on day 1 [100 ng of RU 38486 (GR antagonist) per microliter, or 1 μl of vehicle] and (b) before training: 30 to 45 min before their first swim trial on day 2 [100 ng of RU 28318 (MR antagonist) per microliter, or 1 μl of vehicle (artificial cerebrospinal fluid)]. The free swim trial was run 90 to 120 min after the fourth training trial on day 2. The time of injection was chosen for the following reasons: (a) posttraining: influence on learning (consolidation); (b) before training: influence on acquisition or

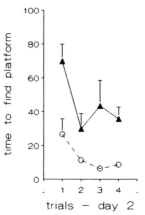

FIG. 1   Preventing the action of GR-mediated effects of corticosterone directly after the fourth trial on day 1 (posttraining treatment with a GR antagonist, ▲; vehicle, ○) impairs spatial navigation on day 2. During all trials rats need more time to find the platform (mean in seconds + SEM).

retrieval and the search pattern in the free swim trial, and the possible differentiation between learning and context-dependent behavior.

Result

Preventing the GR-mediated action of corticosteroids after the training on day 1 resulted in an impaired performance 24 hr later (i.e., longer latencies to find the platform; Fig. 1). Rats also swam longer distances to locate the platform (not shown). In addition, blockade of MRs did not influence the ability of the rats in acquiring the task. The training trials on day 2 were comparable for both groups: latency and distance swum (Fig. 2A) decreased in a similar time course over trials. However, the swim pattern of the free swim trial was remarkably changed in rats treated with the MR antagonist (Fig. 2B): they swam with accuracy to the location of the platform. However, thereafter they focused their search for escape possibilities to other locations, whereas vehicle-treated rats searched most intensely at the former location of the platform.

Exploration of Novel Spaces

Novelty is both an effective stimulus for exploratory behavior and fear (5). The exploratory activity of an animal in a novel environment is the overall result of the features of the environment and the internal state of the animal.

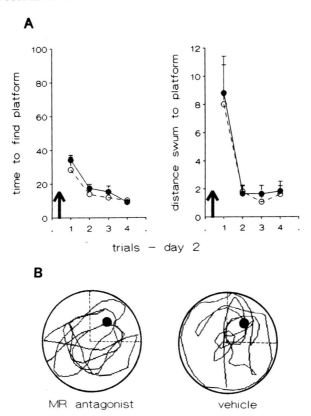

FIG. 2   Blocking the action of corticosterone before the training on day 2 (pretraining treatment is indicated by the arrow; ●, MR antagonist; ○, vehicle) does not influence (A) the time (in seconds) and distance (in meters) swum to find the platform (mean + SEM). However, the swim pattern of the free swim trial (i.e., no platform is present) shows that rats treated with the MR antagonist display an altered search strategy. The rats spent less time searching in the former platform quadrant (mean percentage of time ± SEM: 27.8 ± 1.2 versus 35.4 ± 2.3 in vehicle controls). Two representative samples of the swim pattern are given (B). The filled circle indicates the previous location of the platform.

The response in a novel situation will be completely different from the one in a familiar environment. The "detection" of novelty, that is, result of the comparison of the input signal and the reference set, will lead to the inhibition of ongoing behavior and initiate exploration. The terms *reactive locomotor activity* or *behavioral reactivity* (14) were chosen to represent the sum of horizontal movements of an animal and the time the animal spends in certain areas of a novel environment. We assess behavioral reactivity in a large open field with an object in its center. This open field task was developed in our laboratory.

Open Field Apparatus

A black, square board (125 × 125 cm), 35 cm above the floor, is situated in a dimly lit experimental chamber (200 × 200 × 300 cm). A black object (7.5 × 7.5 × 8 cm) is located at the center of the plate. White rats give a perfect contrast to the background, whereas brown rats can be easily marked with a self-adhesive white sticker on their back (or use a less dark background). Experimental conditions [dim light (50 lx), indirect illumination] are created in such a way that rats can display a high level and differential distribution of reactive locomotor activity. Bright illumination is known to suppress their locomotor activity and increases fear-related behavior.

Procedure

Behavioral testing starts by taking the rat from its home cage and placing it immediately into one of the four corners of the plate, head pointing to the center. Choose the corners in a counterbalanced order because rats have a tendency to spent more time in the corner area where they have been introduced to the open field. Start the computer program, which is preset to stop at a certain time. Usually, we observe the animals for 10 min. Then take the animal back to its home cage and clean the plate and object with an 1% (v/v) acetic acid solution (to remove or spread odor) before running the next animal. To keep environmental conditions comparable for all animals, also clean the apparatus before testing the first rat.

Analysis and Interpretation of Data

For analyzing the behavioral pattern, the area of the open field is subdivided into areas of interest. Time samples of any length can be chosen. We subdivided the open field into a center (one-ninth of the plate, 42 × 42 cm) and peripheral area. The time or percentage of time the rat spent and the distance walked in these two areas, the number of entries into the areas, as well as the total distance walked are calculated. We observe the animals for 10 min, divided into five periods of 2 min each. The latter provides a more detailed picture of the distribution of the behavioral pattern in time. For example, general locomotor activity (total distance traveled) decreases over time, that is, habituation, whereas the distance traveled in the center (object area) remains comparable over time or shows an increase. Time spent in certain areas can show a comparable course.

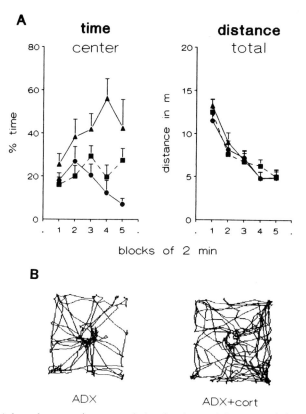

FIG. 3 (A) Adrenalectomy increases behavioral reactivity toward the object in the center of the open field, expressed by the percentage of time (mean ± SEM) the animal spends in the center. If ADX (▲) rats were injected with a low dose of corticosterone (ADX + cort, ●) before the test, their behavior was comparable to sham-ADX (■) animals. The total distance walked did not differ between the groups. An example of the walking pattern of an ADX and an ADX + cort rat is given in (B).

Data provide information on the state of activity of the rat and the differential distribution of exploration. They reflect possible approach and avoidance tendencies, and differentiate between general locomotor activity and stimulus-directed behavior.

Statistics

Statistical significance can be tested by an analysis of variance, followed by a *post hoc* test.

Example 2: Effect of Adrenalectomy on Behavioral Reactivity toward Object in Open Field

One day before behavioral testing, rats were bilaterally adrenalectomized (ADX) or sham-operated (sham-ADX). One hour before the open field test, ADX animals received an injection of corticosterone [50 μg/kg subcutaneously (sc), sufficient to occupy mainly MRs] or vehicle (1 ml/kg). Sham-ADX rats received vehicle only. Animals were observed in the open field with the object in its center for 10 min.

Results

Adrenalectomy changed the pattern of the animals in the time spent and distance walked in the center—close to the object—and in the peripheral area of the open field. Corticosterone normalized the behavior of ADX animals, which is indicative of an MR-mediated effect (Fig. 3). The total distance walked was comparable between the groups. It is the distribution of exploratory activity in space and time and not the general locomotor activity that was affected by the absence of endogenous corticosterone.

Concluding Remarks

To use behavior as a means to elucidate mechanisms and processes of homeostatic control, behavioral methodology must be embedded in experimental settings incorporating the knowledge derived from other disciplines of neural sciences. By this approach we demonstrated that corticosteroids represent a hormonal signal that influences the processing of information in an organizing and synchronizing fashion. Moreover, the experimental designs allowed the differentiation between MR- and GR-mediated actions. It is reasonable to assume that information processing at the level related to the state of the organism and the context is largely under the influence of MR-mediated effects; GR-mediated actions of corticosteroids will incorporate learning processes. It is the integrative design of an experiment, and not the behavioral task itself (which can easily be automated), that will help in our understanding of the mechanisms underlying behavior.

Acknowledgment

This research was supported by the Dutch Foundation of Advanced Research, NWO program-grant 900-546-092.

References

1. J. A. Gray, "The Neuropsychology of Anxiety." Oxford Univ. Press, (Clarendon), London, 1982.
2. B. S. McEwen, E. R. de Kloet, and W. Rostene, *Physiol. Rev.* **66,** 1121 (1986).
3. E. R. de Kloet, *Front. Neuroendocrinol.* **12,** 95 (1991).
4. B. M. Spruit, T. Hol, and J. Rousseau, *Physiol. Behav.* **51,** 747 (1992).
5. J. A. Gray, "The Psychology of Fear and Stress." Cambridge Univ. Press, London, 1987.
6. V. Chan Palay and C. Kohler, "The Hippocampus: New Vistas." Alan R. Liss, New York, 1989.
7. M. S. Oitzl and E. R. de Kloet, *Behav. Neurosci.* **106,** 62 (1992).
8. M. S. Oitzl, M. Fluttert, and E. R. de Kloet, *Eur. J. Neurosci.* **6,** 1072 (1994).
9. R. G. M. Morris, *Learn. Motiv.* **12,** 239 (1981).
10. R. G. M. Morris, *J. Neurosci. Methods* **11,** 47 (1984).
11. R. K. McNamara and R. W. Skelton, *Brain Res. Rev.* **18,** 33 (1993).
12. R. Brandeis, Y. Brandys, and S. Yehuda, *Int. J. Neurosci.* **48,** 29 (1989).
13. J. L. McGaugh, *Annu. Rev. Psychol.* **34,** 297 (1983).
14. C. Gentsch, M. Lichtensteiner, and H. Feer, *Experientia* **47,** 998 (1991).

[30] Adrenocorticosteroids and Cardiovascular Regulation: Methods for Surgery and Blood Pressure Measurements

Elise P. Gómez Sánchez

Introduction

Mineralocorticoids in excess produce sodium retention, potassium wasting, increase in salt appetite, and hypertension. The elevation in blood pressure appears to be mediated by actions of these steroids in the central nervous system (CNS), as well as in the kidney and the vascular smooth muscle (1, 2). Both type I (mineralocorticoid) and type II (glucocorticoid) receptors are found in discrete areas of the brain in overlapping patterns. Type I receptors are concentrated in the hippocampus, amygdala, lateral septum, and hypothalamus, particularly in the periventricular regions, areas known to be or suspected of being important in the regulation of adrenocorticotropin (ACTH) release, arousal, fluid and fluid osmolality equilibrium, and the maintenance of normal blood pressure. Ablation of the anteroventral third ventricle (AV3V) area of the anterior hypothalamus, the area postrema, or the central, but not peripheral, sympathetic system, prevents the development of mineralocorticoid–salt hypertension in the rat. The AV3V area in the rat appears to mediate the central pressor effects of angiotensin II (AngII) and integrates information about fluid and electrolyte homeostasis and baroreceptor input.

Experimental mineralocorticoid hypertension is induced by the systemic administration of an excess of mineralocorticoid, usually deoxycorticosterone acetate (DOCA), or aldosterone in several species, including humans, dogs, pigs, sheep, and rats. The development of hypertension is often accelerated by removing one kidney and increasing salt consumption. To study further the role of the CNS in mineralocorticoid hypertension, steroids, their antagonists, and antagonists of possible intracellular effector mechanisms of mineralocorticoid action have been infused into the brain at doses smaller than those required for a systemic effect. Such studies assume that adrenocorticosteroids equilibrate across the blood–barrier, whether they are administered intracerebroventricularly (icv) or systemically, and that icv infusion produces greater concentrations in the brain than in the rest of the body owing to dilution by the systemic circulation. The icv infusion of aldosterone

Methods in Neurosciences, Volume 22

in amounts that have no effect on blood pressure when infused systemically produces an elevation of blood pressure similar both in amplitude and time of onset to the subcutaneous (sc) infusion of much larger amounts in both rats and dogs (3, 4). The hypertension is dose dependent and reversible after 5 weeks of significant blood pressure elevation. The concomitant icv infusion of prorenone, a mineralocorticoid antagonist, blocks the pressor effect of aldosterone. The acute icv injection of a bolus of aldosterone also elevates the blood pressure of rats (5).

The polydipsia and polyuria associated with systemic mineralocorticoid excess occurs with the sc, but not the icv, doses of aldosterone. The icv infusion of RU 28318, a mineralocorticoid receptor antagonist, at a dose that is without effect when infused sc, prevents hypertension but not saline polydipsia produced by the sc administration of aldosterone (2) or DOCA (6). Corticosterone binds to the type I receptor with an affinity similar to that of aldosterone. However, its icv infusion or that of a selective type II receptor agonist does not elevate the blood pressure; instead, the concomitant icv infusion of corticosterone blocked the hypertensive effect of aldosterone (2).

These and similar studies suggest that the mineralocorticoid, or type I, receptors in the brain have a role in the genesis of clinical mineralocorticoid hypertension. The purpose of this chapter is to describe the chronic intracerebroventricular infusion methods used to complement the receptor and lesioning studies. The design of such a study must consider the acute sensitivity of the hypothalamic–pituitary–adrenal axis and the catecholaminergic system to stress from any source. In the ideal situation the administration of test agents and the measurement of blood pressure should not, in themselves, alter the blood pressure.

Rat Models of Mineralocorticoid Hypertension

There are several models of mineralocorticoid hypertension in rats: systemic mineralocorticoid–salt excess, of which DOCA–salt is the prototype, central mineralocorticoid excess, and adrenal regeneration hypertension. The outbred Sprague-Dawley rat is commonly used and can be expected to develop hypertension in a consistent fashion. The reduction of renal mass and salt loading is not absolutely necessary for mineralocorticoid excess models, but they greatly accelerate and enhance the development of hypertension and thus facilitate the studies (7). For mineralocorticoid-salt hypertension, aldosterone (0.5 or 1 μg/hr) is delivered sc by miniosmotic pump (Alza Corp, Palo Alto, CA) or by sustained release pellets (Innovative Research of America, Toledo, OH) and 0.9% (w/v) NaCl plus 0.15% (w/v) KCl is substituted for water to drink. The KCl prevents the potassium depletion produced by systemic mineralocorticoid excess, which produces cardiovascular and metabolic consequences (8). An excess of DOCA in implanted silastic strips or pellets containing 100–200 mg/kg and lasting several weeks or injected sc in oil (6–10 mg twice a week) is also effective.

For central mineralocorticoid hypertension aldosterone (5 or 10 ng/hr) is infused icv and the rats drink either 0.9% (w/v) NaCl or 0.9% (w/v) 0.15% (w/v) KCl, if there is a group receiving sc aldosterone.

For adrenal regeneration hypertension, rats undergo a right nephrectomy and adrenalectomy, left adrenal enucleation, and drink 0.9% (w/v) NaCl.

Doses of corticosteroids, their antagonists, and other agents to be studied are chosen so that they do not cause overt clinical illness or distress, which could independently alter blood pressure and the humoral environment, particularly that of the adrenocorticosteroids and catecholamines. When possible, the stress of repeated injections or tethering for chronic infusions is avoided by using implanted miniosmotic pumps, sustained release pellets, or silastic implants. Solutions are sterilized by filtration. Invasive procedures are done aseptically with the minimum trauma possible. Although rats are noted for their hardiness, an animal struggling to survive a traumatic surgery or sepsis is a poor model for the study of adrenocortical function or blood pressure control. As isolation represents a social stress for rats, whenever possible the animals are housed in stable groups of three or four. Metabolic studies in which urine samples from a single animal requires isolation are a problem that can be mitigated by acclimatization to the cage. Notwithstanding, under optimum conditions it takes 3–4 days for 24-hr urinary corticosterone values to stabilize (9).

Preparing Lateral Ventricular Cannulas and Catheters

Lateral ventricular (icv) cannulas are constructed of 20-gauge stainless steel tubing cut in 1.5-cm lengths and the ends ground smooth. The cannula is placed in a perpendicular 18-gauge hole milled to a depth of exactly 4.5 mm into a block of lexan, then bent at a 90° angle precisely 4.5 mm from one end. Generally a 4- to 5-cm PE50 catheter suffices to connect the cannulas to miniosmotic pumps. To vary the infusate over the pumping period, a length of PE50 tubing adequate to contain the total infusate for the pumping period is coiled on a glass rod, leaving 3–4 cm straight at the cannula end. The rod and coil are held in hot, then cold, water to compact the tubing and avoid kinks. When the catheter is filled, the different infusates are separated by minute air bubbles to prevent mixing of the solutions. This method has been used to avoid interactions between infusates and the pump interior, but this is seldom a problem. The cannula–catheter assembly is washed under pressure with distilled water, followed by air. Clean, dry assemblies are gas sterilized along with labeled resealable containers in which filled pump assemblies will be stored in preparation for implantation.

Infusion Pumps and Solutions

For most experiments the 14-day miniosmotic pump, which delivers 0..45–0.5 μl/hr, is chosen. Although this size may require several pump changes for a long experiment, the longer duration pump is large enough to cause discomfort for the rat in restraining tubes and feeding tubes of some metabolism cages. Because each pump lot differs slightly in pumping rate, the same lot is used for a given experiment. The artificial cerebrospinal fluid provides Na^+ (150 mEq/liter), K^+ (3 mEq/liter), Ca^{2+} (2.5 mEq/liter), Mg^{2+} (1.6 mEq/liter), HCO^{3-} (25 mEq/liter), PO^{3-}_4 (0.5 mEq/liter), and Cl^- (135 mEq/liter). For 500 ml of solution, add, in order, to distilled water: 3.653 g of NaCl, 93.0 mg of KCl, 11.3 mg of KH_2PO_4, 69.37 mg of $CaCl_2$, and 81.3 mg of hydrated $MgCl_2$; dissolve, then add 1.05 g of $NaHCO_3$. Cover and leave out overnight to equilibrate with the atmosphere, then adjust the pH to 7.4 with HCl. Filter the solution into a sterile container and store refrigerated. Stock solutions of steroids and antagonists are prepared in distilled propylene glycol, then stored at $-70°C$. The propylene glycol comprises no more than 4% of an infusion solution, depending on the solubility and concentration of the steroid. Propylene glycol is added to all of the infusion solutions, including the controls, in an equal amount. The mineralocorticoid antagonist RU 28318 (Roussel Uclaf) is a potassium salt, the lactone form being too insoluble. Potassium gluconate, used to provide the control animals with an equivalent amount of K^+, and the RU 28318 are dissolved in 140 mEq of NaCl, rather than in the artificial cerebrospinal fluid (CSF), because of solubility problems. Infusates are prepared in their final form and sterilized by filtering through sterile filters (0.2-μm pore size) immediately before filling and implanting the pumps. When there is to be a delay of more than 1 hr between filling and implantation, the filled pumps are stored in sterile sealed containers. Because of the risk of contamination and of early and variable activation of the pumps even by atmospheric water, long-term storage of filled pumps is not recommended.

Anesthesia and Surgery

Isoflurane or halothane anesthesia is delivered in an open system with an inhalant anesthesia machine. The expense of inhalent compared to injectable anesthetic is justified by the safe, rapid induction and recovery times and flexible anesthesia duration. Postoperative morbidity, hypothermia, and involuntary movements, which are particularly relevant for animals already trained and midway through an arduous experiment, are greatly reduced.

A step-down adaptor allows the connection of commercial intravenous (iv) tubing to the mixed gas port, bypassing the CO_2 scrubber, which in turn connects to a multiport stopcock manifold and short iv tubings connected to syringe barrels serving as anesthetic masks. This allows for one or several animals to be anesthesized at the same time without wasting anesthesia. To scrub the anesthetic gasses and vent the fumes from the construction of the acrylic skull caps away from the surgery site, a small activated charcoal air filter (available at most drug stores) is modified by covering the intake opening with a rigid plastic sheet into which a hole is cut to tightly fit a 4-in. diameter flexible clothes dryer hose. The free end of the hose is placed near the heads of the patients.

Innovar Vet [fentanyl citrate (0.4 mg/ml) plus droperidol (20 mg/ml)], at a dose of 0.03 ml/100 g body weight, intraperitoneal (ip), is used as a premedication for procedures in which the skull is trephined. Food and water are not withheld from rats before surgery, even adrenalectomy or nephrectomy, because rats usually cease eating and drinking after the lights go on in the morning and because rats do not vomit. Salivation is not a problem with halothane or isoflurane, nor have we documented a problem with bradycardia when the vagus nerve is manipulated during catheterization of the carotid artery, and therefore atropine premedication is not routine. Antibiotics are not routinely administered.

The animals are placed on a warm water blanket during surgery and recovery and are watched throughout for respiratory depression. Recovery from inhalent anesthesia is complete within 5 min; however, the rats may succumb to respiratory depression and postural anoxia for an hour after this dose of Innovar Vet. Therefore they must be watched and repositioned if necessary. The sedative/narcotic is worth the risk because it significantly shortens surgical recovery. The morning after icv surgery, those animals receiving the analgesia are better groomed and, judging from their postsurgical weights, have started eating and drinking sooner than those receiving surgical anesthesia alone.

Expediency and statistically valid protocols usually require the surgical preparation of many animals at the same time. Notwithstanding, the value of aseptic procedures cannot be overemphasized. Although rats are amazingly resilient animals that can survive surgical insults and septic conditions other animals cannot, such conditions constitute unacceptable interferences with the very parameters to be studied, blood pressure and the homeostatic mechanisms controlling it. Standard aseptic surgical practices are followed. Easy and inexpensive practices that help prevent contamination when surgery on many animals must be done over several hours include the following: (a) segregating the instruments used to open and handle the skin from those used for intraabdominal or pump-handling tasks, (b) wiping instrument tips between use on each animal with sterile one-use-only bits of paper towel and dipping them in a container of ethanol (do not allow wet drapes to form a wick from contaminated surfaces), and (c) using paper towels cut in thirds and sterilized in the instrument pack as individual disposable surgical drapes. Hot bead sterilizers that sterilize instrument tips in less than a minute are useful if a great many consecutive surgeries are contemplated. It is critical that the miniosmotic pumps be kept sterile. They constitute a significant foreign body that the rats have trouble clearing of contamination if they become infected.

Lateral Ventricular Cannulation

For lateral cerebroventricular (icv) cannulation, the top of the head is carefully shaved to avoid scraping or nicking the ears and inciting postsurgical self-trauma. A 1- to 1.5-cm skin incision is made down the midline between the ears and the occiput, and the periosteum pushed away from the cap site with a cotton-tipped swab. The opening in the skull for the cannula is placed 1.5 mm caudal to the bregma and 1.0 mm lateral to the sagittal suture. In the beginning a small piece of a clear plastic ruler is used until the surgeon can find the trephine site by sight. The skull is hand-drilled with an 18-gauge disposable needle with care not to penetrate past the dura. The depth of the cannula is predetermined by the position of the right angle made in the cannula as described above. Shallow tapholes forming an equilateral triangle with the cannula hole are started with the trephine needle and two 000, 3/32-in. stainless steel self-tapping screws (Smal Parts, Inc., Miami, FL) are anchored in, but not through, the skull. A 1- to 1.5-cm skin incision is made at least 1 cm caudal to where the pump will ultimately lie along the back. If the animal has undergone a concomitant nephrectomy and/or adrenalectomy, the same incision is used. After the trephine hole and screws have been placed, an Allis forceps is tunneled under the skin forward from the caudal incision, a fresh drape is placed over the anesthesia mask and head of the rat, and the pump end of the cannula and pump assembly are dragged back until the cannula is over the opening in the skull. The assembly is rotated so that the cannula is perpendicular with the skull when released, reducing the tendency for pendular motion once the cannula is placed into the brain. The cannula is then pushed into the trephine hole until the horizontal portion is flush to the skull. A cap of dental acrylic (Lang, Chicago, IL) is formed to encompass the screws and cannula, including the junction with the catheter, to anchor the cannula securely. It is important that the cap be smooth and as small as possible but still encompass the critical parts, especially if the rats will be using a metabolic cage with a tunnel feeder. The skin on the head is closed with 4-0 monofilament prolene suture. If the rats remove the sutures before the wound heals, one should suspect that the cap was too rough, the wound was contaminated intraoperatively, or the sutures were placed too tightly. The position of the cannula within the lateral ventricle must be checked postmortem by dye infusion or visual inspection of the cannula tract if dye is contraindicated. With a little practice icv cannula placement by this method becomes quite accurate and rapid compared to the use of stereotaxic device. In addition, when a large number of animals are involved, it is more difficult to maintain aseptic conditions between rats if a stereotaxic device is required.

Although a straight catheter or one with only a few coils is easy to pull under the skin, a long coil tends to drag and fold on itself. A 1-ml syringe case with its end cut off and heat polished is used to tunnel under the skin, then the pump and coils are passed through. Flank and back skin incisions are closed with wound clips or prolene suture.

When osmotic pumps connected to catheters are changed, the end of the catheter stretched by the pump modulator is removed to ensure a tight fit for the new pump. If the change is complicated by a seroma or abscess, the capsule is incised to its full axial length, drained and flushed thoroughly with saline, and the new pump placed in a different location. A mixture of penicillin G procaine and benzathine at 30,000 units each per kilogram is given sc for an infection.

Nephrectomies, Adrenalectomies, and Adrenal Enucleations

Nephrectomies, adrenalectomies, and adrenal enucleations are performed through a flank incision just caudal to the ribs and ventral to the epaxial muscles. Chromic gut (4-0) is used to close the peritoneum and abdominal muscles, and wound clips or prolene suture is used to close the skin incisions. For the removal of a kidney, the capsule is gently stripped and a titanium hemoclip or 4-0 silk suture is used to ligate the renal pedicle before severing it. When only one kidney is to be removed, the right kidney is taken because the compensatory enlargement of the remaining right kidney has been found occasionally to cause venous compromise owing to its close apposition to the major vessels, whereas the left kidney has adequate space for enlargement.

Adrenalectomy does not require ligation in rats; the vessels are briefly clamped before cutting or tearing. Care must be taken not to avulse part of the capsule containing glomerulosa cells, which will regenerate. Adrenalectomized rats are given 0.9% NaCl to drink. Exogenous glucocorticoids have not been required after adrenalectomy, perhaps because healthy, well-acclimatized rats are used and anesthetic, surgical, and environmental stress are minimized. In long-term adrenalectomy studies one must always consider the possibility of hypertrophy and/or hyperplasia of adrenal rests, particularly in male rats. Apparently it is not uncommon for adrenal cells to be dragged caudally during embryogenesis by the migrating testes and deposited along the tunica vaginalis. These cells, presumably in response to high levels of ACTH after adrenalectomy, soon become able to provide significant amounts of adrenocorticosteroids that can confound studies using "adrenalectomized" rats. If adrenalectomized animals are not to be used within 5–7 days of adrenalectomy, females are used, if possible, and the completeness of adrenalectomy verified by corticosterone measurements.

Adrenal regeneration hypertension entails the removal of the right kidney and right adrenal gland and enucleation of the left adrenal gland. Adrenal enucleation involves gently securing the adrenal gland with nontraumatic forceps, incising the capsule about one-third the width and depth of the gland (a No. 11 blade or 18-gauge needle used as a blade works well), then gently pressing out and removing the nucleus comprising the medulla, fasciculata, and reticularis.

Choice of Method to Measure Blood Pressure

Clinical effects of corticosteroids on the blood pressure are fairly chronic, therefore most studies in this area require the continuous or repeated measurement of the blood pressure over several days or weeks. The choice between direct and indirect methods is dictated by the size and temperament of the subject, the tolerance for imprecision, the likelihood that the method itself will alter the blood pressure, and the cost in both labor and equipment. Although humans, dogs, pigs, and sheep have been used in the study of the effects of adrenal steroids on the blood pressure, rats are the most extensively studied model. The small size of the rat can present a problem for blood pressure measurements. The direct measurement of blood pressure with indwelling fluid-filled catheters and minimum-displacement pressure transducers or catheter-tipped transducers offers more information about the cardiovascular system than indirect plethysmographic methods, which generally are limited to giving an estimate of systolic pressure in small subjects. Recording blood pressure from an unstressed animal may be cited as a reason to use direct pressure measurements, however, chronic stressors inherent in these methods are also significant, and include frequent catheter flushing, restraint systems to prevent the animal from traumatizing the equipment or themselves, isolation so that other animals do not interfere, chronic irritation produced by equipment, infection, and vascular damage due to the size of the catheter required for high-fidelity recording relative to the size of the rat's arteries. The advent of miniaturized telemetry systems using catheter-tipped pressure transducers, which are less subject to drift during their implantation life, circumvent many of these problems, particularly that of the stress of restraint and/or tethering required for both direct and indirect systems. Unfortunately, the cost and the frequency of equipment failure and repair of telemetry for rats is still prohibitive for most laboratories (10).

Indirect Tail Plethysmography

We have chosen an indirect tail plethysmographic method using a photoelectric pulse transducer in trained unheated rats for most of the rat blood pressure monitoring in our laboratory (IITC, Wood Hills, CA). An automatic cuff inflater is extremely helpful because it produces consistent cuff inflation and deflation intervals and maximum pressure, 220 mmHg, so that the rat becomes accustomed to the procedure and is neither surprised by erratic inflation/deflation rates nor hurt by overinflation. Heating the rat to vasodilate the tail as required by some plethysmographic equipment produces stress, which can differ from one rat to the next, and has been shown to have different effects on the blood pressure in different rat strains (11, 12). In addition, the relationship between the core blood pressure and that of the heat-vasodilated tail depends on the vascular tone and volume status of the animal, parameters that are modulated by mineralocorticoids (11–14). The most important determinant of reliable indirect blood pressure measurements is adequate training of the rat, the vigor of which depends in great part on the strain and, to some extent, sex of the rat. Sprague-Dawley rats are relatively calm, easily trained rats compared to the Fischer, Wistar-Furth, and some SHR rats (12). Females in general tend to remain more alert, even after training; well-trained males often fall asleep in restraining tubes. Blood pressures should be measured at the same time of day, by the same person, in the same quiet place. Rats generally are calmest in the morning, perhaps because it is the beginning of their major sleep cycle. While our upper threshhold for hearing is about 20 kHz, rats hear and communicate at 10–70 kHz. Loud ultrasonic noise, which we cannot perceive, has adverse effects on rodents, including the induction of hypertension. Sources for such noise, which include computer monitors, running water, vacuum hoses, and components of electronic blood pressure recording systems, are difficult to eliminate but must be minimized (15).

A 20-min restraint period for a naive rat elevates corticosterone and cate-cholamines severalfold and reduces total body oxygen consumption by one-third (11). Although proper training can mitigate this effect, prolonged re-straint can also facilitate the stress response on the next exposure. We have found that short, frequent training periods, starting with the rat running through the restraining tube and working up to the requisite 15–20 minutes slowly, so that the individual never becomes "hysterical" or goes into a hyperalert "freeze" stance, works best. Although our machine can accom-modate 12 animals wearing individual tail cuffs, we generally work with 3 or 4 trained animals at a time to avoid excessive duration of restraint. A rat is allowed a few minutes in the tube before 5–10 cuff inflation cycles are performed. If five consecutive pressures differ by less than 5 mmHg, the average is taken and the procedure is done on the next animal. If the pressures are erratic, we go on to the next animal to give the first time to calm. After a period of no less than 10 min from the recording of the first "good" average, the pressure cycle is repeated. If the second average of five consecutive pressures differs from the first by no more than 5 mmHg, the average is recorded as the pressure of the day. A refractory rat is returned to its cage to calm while the pressures of other animals are taken, then tried again. Restraining an animal longer than 15 or 20 min because it is restless is counterproductive (16). Pressures and body weight are measured twice a week. Weight loss or failure to grow normally is often the first indication of a problem that, if not resolved, could jeopardize the experiment.

Direct Blood Pressure

Direct blood pressures are measured on occasion. Because anesthetics, even cocktails designed to minimize cardiovascular effects, elevate corticosteroids and catecholamines, most of our direct measurements are made in conscious rats (17). One of the primary problems in successfully miniaturizing direct methods used in humans and larger subjects is the size and stiffness of the catheter required for high-fidelity recording relative to the size of the vessels. To obtain good pulsatile readings of the blood pressure acutely, a relatively large, stiff catheter can be threaded into the carotid artery or aorta through the femoral artery or the stump of a renal artery after the kidney is removed. However, tying off the artery around the catheter promotes the rapid over-growth of the tip with fibrous tissue, sometimes within days, making a one-way valve through which fluid can be flushed, but none withdrawn, nor pressure measured. Occlusion of the femoral artery is unacceptable for chronic studies in rats because of severe ischemia, resulting in significant distress and, not uncommonly, self-mutilation and loss of the limb. The use of a softer catheter material, such as silastic, appears to decrease vascular damage and the rate of fibrosing of the tip, but fidelity of the signal is lost due to the compliance of the catheter. The advent of less thrombogenic catheter materials, polymer-complexed heparin, and miniaturized catheter-tipped pressure transducers has mitigated the problem somewhat.

There are probably as many catheter constructs for rat arterial cannulation as there are laboratories measuring chronic arterial blood pressure. We generally place chronic catheters into the descending aorta just cranial to the iliac bifurcation. For long-term catheters in mice and rats, the portion of the catheter to be inside the vessel is generally made of silastic, 0.012- or 0.02-in. inner diameter, (Dow Corning, Midland, MI) swedged onto 10- to 20-cm PE10 or PE50 (Clay-Adams, Parsippany, NJ). The end of the silastic is softened by dipping it into hexane or toluene so that the PE can be pushed into it for about 2 mm. The silastic portion is trimmed without a bevel to leave 0.7–1 cm. Alternatively, we pull a gentle taper of about 0.5 cm with an inner diameter of about 0.02 in. (0.05 cm) in the end of the PE50 while rolling it in a stream of hot air. The taper facilitates placement and, more important, allows more flow around the intralumenal part of the catheter, while maintaining fairly good fidelity. A fine wire such as that used to clean pen stylets is inserted into the catheter and a bleb is made on the outside of the catheter, just distal to the taper, by heating the area, then pushing it together. The bleb forms a convenient nonslip anchor for a suture to secure the catheter to the retroabdominal fascia. The distal end of the PE50 is coiled tightly around a glass stirring rod, then held in hot water for a minute, then cooled in a beaker of cold water, leaving 5–10 cm straight, depending on the size of the animal. It is easier to coil enough catheter for two catheters so that one can hold onto the straight portions of the two. A single coil is made 1 cm from the point at which the catheter exits the vessel, providing reserve to allow stretch of the catether and a convenient suture point to anchor it further. The finished catheter is treated with TDMAC–heparin (Polysciences, Inc., Warrington, PA), then gas sterilized. Note that intraoperative trimming of the catheter will remove the TDMAC–heparin coating on that edge.

Procedures for the placement and use of catheters require aseptic technique and sterile components and fluids. Sepsis, besides being a source of stress and cardiovascular compromise, directs the attention of the rat to the catheter. A 3-mm segment of iv tubing is used as a cuff to kink the proximal 3 cm of the heparinized (100 U/ml) saline-filled catheter to prevent back flow. The catheter is introduced into the aorta just cranial to the iliac bifurcation through a slit made with a microsurgical blade or 25-gauge needle so that the catheter is a tight fit. Stiffening the silastic tip in iced saline facilitates its insertion. Cyanoacrylate glue can be used to plug small leaks around the catheter. A

4-0 cardiovascular silk suture is tied where the silastic cuffs the PE or at the bleb in the PE, then secured to the retroperitoneal fascia. The distal portion of the catheter is tunneled through the lower quadrant of the abdominal wall and skin to exit between the scapulae with a sailor's needle. The cuff producing the kink is removed to allow flushing of the catheter, with care not to overheparinize the animal, and the catheter is accommodated in the lower abdomen to avoid tension on the catheter in the vessel. The distal coils, except for one, remain under the skin. A titanium hemoclip is used to kink it 3 mm from the end and another is used to secure the end to the skin. The rat can then be returned to its cage without a jacket. To take the blood pressure or a blood sample, one coil of catheter is gently pulled out of the skin, the catheter crimped with an iv tubing-clad hemostat, the section wiped with povidone or alcohol, the hemoclipped portion of the catheter cut and discarded, and a Luer stub fitted. The contents of the catheter are withdrawn, the volume noted and discarded, and heparinized saline of the volume of the catheter plus 0.03 ml infused; then the Luer stub is connected to a pressure transducer. If a blood sample is to be taken, the volume of the sample is replaced with saline or lactated Ringer's before filling the catheter with heparinized saline. Blood sampling can be done in a well-trained rat in its home cage during the day, while it normally sleeps, with minimum disturbance. The same restraining tube with a slot cut out along the long axis is used for direct as well as for indirect blood pressure measurements. To store the catheter, another catheter volume of heparinized saline is infused and the catheter crimped with a new hemoclip and secured to the skin. The advantage of this method is that the animals seem oblivious to the catheter and are not chronically stressed by jackets and swivel devices. The disadvantage is that the design allows only short blood pressure monitoring periods under confined conditions. The number of times a catheter is to be used determines the length of catheter and number of coils needed. An alternative is to load a 1-ml syringe with the exact amount of fluid desired to fill the catheter; then, with the syringe firmly in the Luer stub and the plunger maximally inserted, the syringe is cut with a shears or rongeur just above the rubber portion, leaving the end as a disposable plug. This type of catheter end must be protected by a jacket. For short experiments in which significant growth is not anticipated, a low-cost alternative to mesh jackets can be fashioned from 2-in. duct tape. A strip the length of the girth of the torso plus 4 cm is cut and laid sticky side up. Another strip of tape 4 cm shorter and folded to be 1.5 cm more narrow is centered, sticky side down, onto the first. Ample leg holes 0.5 cm apart are cut in the center. While ensuring that the skin and fur are comfortably in place, the jacket is placed on the yet anesthetized rat so that it closes in back and the tape along the edges of the jacket are pressed against the fur. The two sticky ends of the tape are folded and trimmed to encompass and support the catheter, Luer stub, and occluder. The catheter envelope is folded over and secured to the jacket with a short piece of tape with a folded edge to ease removal. This jacket is secure enough to support a swivel system.

Cardiac Puncture

Cardiac puncture under isoflurane anesthesia is a convenient method for obtaining blood from rats and mice when the acute stress of anesthesia is not contraindicated. The entire anesthetic time, including induction, is rarely more than 3 min, and the animal is ambulatory in less than 5 min. A syringe and 25- or 26-gauge needle is usually used, but a 22-gauge needle and 3-ml vacutainer tube have also been used safely. Once the procedure is learned, drawing the sample is so quick that the anesthetic is turned off as soon as the animal is recumbent, leaving the mask in place to deliver oxygen. The needle is placed in the apex of the angle made by the xyphoid process and the left ribs, at a 45° angle, along a plane parallel with the midline. The needle is advanced carefully until the heart beat is felt, then thrust into the left ventricle. If no blood can be withdrawn even after rotating the needle, the needle is retracted to the level of the chest wall before changing the angle to avoid lacerating the heart or lungs. Usually 0.5 ml of blood is drawn, but 3 ml can be obtained from 250-g rats with no overt clinical sequelae.

Concluding Remarks

In measuring many natural phenomena the very parameters one seeks to quantitate may be altered. This is certainly the case when studying the central control of the blood pressure. This chapter addresses problems inherent in chronic systemic and central infusions and measurement of blood pressure, methods that complement, but will never supplant, others, for example lesioning, pharmacological manipulations of neurotransmitters, or anatomical localization of receptors and biochemical markers, studies that have been described in detail in other chapters of this book and elsewhere. Blood pressure or behavioral changes are terminal effects that can be modified not only by extraneous variables, including method of measurement, but also by unforeseen changes secondary to experimental manipulations. Because the perfect experimental design is rare or nonexistent, it is crucial that several approaches with different potential problems be used to examine the central nervous system with its myriad interactions and complexities.

References

1. D. F. Bohr, *Hypertension* **3,** II-160 (1981).
2. E. P. Gomez-Sanchez, *Am. J. Hypertens.* **4,** 374 (1991).
3. E. P. Gomez-Sanchez, *Endocrinology (Baltimore)* **118,** 819 (1986).
4. Y. Kageyama and E. L. Bravo, *Hypertension* **11,** 750 (1980).
5. D. T. W. M. van den Berg, E. R. De Kloet, H. H. van Dijken, and W. de Jong, *Endocrinology (Baltimore)* **126,** 118 (1990).
6. P. Janiak and M. J. Brody, *FASEB.* **2,** 5719 (1988) (Abstr.).
7. E. P. Gomez-Sanchez, *J. Hypertens.* **6,** 437 (1988).
8. J. Mitchell, W. D. Ling, and D. F. Bohr, *J. Hypertens.* **2,** 473 (1984).
9. E. P. Gomez-Sanchez and C. E. Gomez-Sanchez, *Steroids* **56,** 451 (1991).
10. C. Guiol, C. Ledoussal, and J. M. Surge, *J. Pharmacol. Toxicol. Methods* **28,** 99 (1992).
11. G. M. Walsh, *Proc. Workshop Blood Pressure Meas. Hypertens. Anim. Models* p. 1. U.S. Dep. Health, Educ., Welfare, NIH, Bethesda, Maryland, 1977.
12. A. U. Ferrari, A. Daffonchio, F. Albergati, P. Bertoli, and G. Mancia, *J. Hypertens.* **8,** 909 (1990).
13. R. D. Buñag and J. Butterfield, *Hypertension* **4,** 898 (1982).
14. K. C. Kregel and C. V. Gisolfi, *J. Appl. Physiol.* **68,** 1220 (1990).
15. G. Clough, "New Developments in Biosciences: Their Implications for Laboratory Animal Science," p. 239. Nijhoff, Dordrecht, Netherlands, 1988.
16. M. Kreutz, D. Hellhammer, R. Murison, H. Vetter, U. Krause, and H. Lehnert, *Acta Physiol. Scand.* **145,** 59 (1992).
17. D. R. Gross, "Animal Models in Cardiovascular Research." Nijhoff, Dordrecht, Netherlands, 1985.

[31] Steroids and Central Regulation of Immune Response

Bernd Schöbitz

Introduction

Glucocorticoids are thought to be the principal mediators by which the central nervous system (CNS) regulates the immune system via the hypothalamic–pituitary–adrenal (HPA) axis. The antiinflammatory effects of these steroids are well documented owing to their wide use in pharmacotherapy (1). They range from the treatment of urticaria to the suppression of transplant rejections. At the cellular level glucocorticoids affect many immune parameters such as the number of circulating mononuclear cells and neutrophils. Moreover, the adrenocortical steroids decrease antibody formation and suppress vascular leak, cell recruitment, and leukocyte diapedesis; and they inhibit the release of lysosomal enzymes. The molecular mechanisms underlying these pleiotropic effects are still poorly understood. Our knowledge of the immune suppressive effect of glucocorticoids stems from studies with synthetic derivatives of naturally occurring steroids used at pharmacological doses. It is, however, not known whether physiologically relevant alterations in the concentrations of endogenous glucocorticoids significantly affect the immune response.

There is increasing evidence that the inhibition of the biosynthesis of proinflammatory cytokines might be the major target for glucocorticoid action in the immune system (2, 3). This hypothesis explains a number of immune suppressive effects of these steroids because cytokines play a major role as soluble mediators in the host defense against infection through their involvements in the immediate, antigen-independent stimulation of the immune system and the delayed antigen-specific immune activation of lymphocytes. The most typical examples of multifunctional cytokines are interleukin 1 (IL-1), IL-6, and tumor necrosis factor (TNF) (4, 5). A potent and well-characterized activator of these cytokines is the bacterial endotoxin lipopolysaccharide (LPS). Glucocorticoids suppress proinflammatory cytokine actions, and they control the expression of interleukins after their induction with proinflammatory stimuli such as LPS (2, 3). Accordingly, the genes encoding IL-1, IL-6, and TNF contain glucocorticoid-responsive elements in their promoters. The interaction between cytokines and glucocorticoids is bidirectional because IL-1, IL-6, and TNF activate the HPA axis and raise

Methods in Neurosciences, Volume 22

the concentrations of circulating corticosteroids (6). This closed feedback loop provides an interesting model with which to study the communication between the CNS and the immune system.

Control of Cytokine Gene Expression by Glucocorticoids

Gene expression is controlled at the level of transcription, translation, and by changes in the stability of mRNA and protein. Additional mechanisms such as alterations in the splicing and nuclear transport of the heterologous nuclear RNA have also been observed. However, some studies have shown that these mechanisms usually point in the same direction when glucocorticoids regulate cytokine expression. For example, when activated macrophages are treated with dexamethasone a decrease in the transcription rate of IL-1β mRNA is accompanied by a reduced stability of the message (3).

Determination of Cytokines by Bioassays

Cytokines can be determined in extracellular fluids, cell lysates, tissue, homogenates, and supernatants of cultured cells. The cytokine content can be quantified by enzyme-linked immunoassays and radioimmunoassays, using monoclonal antibodies. The results should be interpreted with caution because the extent to which these antibodies detect precursors, degradation products, and multimeric forms of cytokines is not always well defined. Moreover, a number of antibodies is available commercially for determination of human and murine interleukins, but not for interleukins from other species. These difficulties are in part overcome in bioassays based on cell lines that specifically respond to particular cytokines with proliferation, death, or immune reactions, for example, antibody formation. Some examples are given below. In many cases the sensitivity of bioassays exceeds that of immunoassays.

Interleukin 1 and IL-6 induce mitogenesis of D10.G4.1 mouse helper T cells and B9 hybridoma cells (7), respectively. In contrast, cell death triggered in the murine fibrosarcoma cell line WEHI-164 is a measure of the TNF concentration (8). It is not possible to distinguish between the α and β forms of IL-1 and TNF as they bind to the same receptor. For this reason, and owing to the putative influence of unknown factors in the samples that may interact with the responsive cells, the application of an independent method (e.g., Northern blot analysis) is recommended.

The appropriate collection of samples is of paramount importance to exclude artifacts caused by partial degradation of cytokines. Usually, the preparation of homogenized tissue bears more problems than analyses of supernatants from cultured cells and plasma or serum. When quantifying intracellular cytokines the use of detergents should be avoided as they can interfere with the bioassay. Tissue samples can be frozen in liquid nitrogen. Brain structures or other tissues with little protease activity (such as muscle) can be homogenized by 20 strokes (Polytron, Kinematica, Switzerland) in sterile, ice-cold phosphate-buffered saline (PBS) at a 5:1 ratio of buffer volume and tissue weight. The homogenate is centrifuged at 10,000 g for 15 min at 4°C to remove cell debris. Sterile filtration of the supernatants is required to avoid microbial contamination during subsequent cell culture. An aliquot should be assayed for the total protein content, for example, according to the method of Lowry. Analyses of lymphoid tissues including spleen, liver, and bone marrow require the addition of the following protease inhibitors to PBS: antipain, leupeptin, aprotinin, pepstatin A (each 1 μg/ml), and phenylmethylsulfonyl fluoride (PMSF) (2 mM). The effect of this mixture on the bioassay should be tested.

Interleukin 1 and IL-6 are determined by their property to induce in a dose-dependent manner the proliferation of D10.G4.1 helper T cells and B9 hybridoma cells, respectively. The induction of DNA synthesis in these cells is quantified by measurements of the incorporation rate of [^{3}H]thymidine into acid-precipitable material. Because this method is labor intensive and produces high disposal costs, nonradioactive methods based on the determination of the number of viable cells by colorimetric assays have been developed (9). In metabolically active cells the cell number is directly proportional to the reducing capacity of dehydrogenase enzymes. The hydrophilic tetrazolium compound MTS [3-(4,5-dimethylthiazol-2-yl)-5-(3-carboxymethoxyphenyl)-2-(4-sulfophenyl)-2H-tetrazolium] in combination with the electron-coupling reagent phenazine methosulfate (PMS) has been established as a novel substrate for dehydrogenases because this dye does not require addition of organic solvents for solubilization. MTS is reduced to a formazan product that can be measured by its absorbance at 490 nm.

1. Samples containing interleukins and a standard of human recombinant IL-1β and IL-6 are titrated by serial twofold dilutions (standard concentration between 2 ng/ml and 2 pg/ml) with RPMI-1640 medium [supplemented with 5% (v/v) fetal bovine serum and 50 μM 2-mercaptoethanol (ME)]. Sample volumes of 50 μl are pipetted to a 96-well titer plate.

2. B9 or D10.G4.1 cells, which have been cultured for 2 days in the absence of IL-1 and IL-6, are washed three times with PBS. The cell density can be determined by trypan blue exclusion, using a hemocytometer.

3. The cell density is adjusted to 100 cells/μl of supplemented RPMI-1640, and a volume of 50 μl is added to the titer plate.

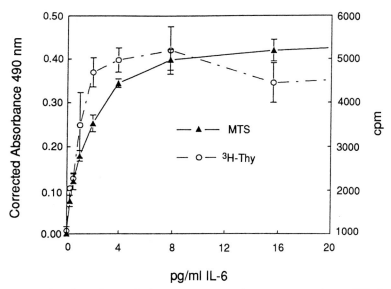

FIG. 1 Proliferation of B9 cells in response to various concentrations of IL-6 measured with a colorimetric proliferation assay based on the determination of the dehydrogenase activity in viable cells (CellTiter 96 AQ; Promega Corp., Madison, WI) (▲) and the [³H]thymidine incorporation assay (○). Details of the procedure are described in text (with permit from the Promega Technical Bulletin 169, Promega Corp., Madison, WI).

4. The plate is incubated for 48 to 72 hr at 37°C in a humidified 5% CO_2 atmosphere.

5. The enzymatic reaction is started with 50 μl of a combined solution of MTS (in Dulbecco's phosphate-buffered saline, pH 6.0) and PMS (pure). The final concentration of MTS is 333 μg/ml and that of PMS is 25 μM.

6. After 4 hr the absorbance at 490 nm is recorded, using an enzyme-linked immunosorbent assay (ELISA) plate reader. The background absorbance due to spontaneous reduction of MTS is typically between 0.1 and 0.2. The value is subtracted from all other absorbances, and the corrected absorbance is plotted versus the concentration of the IL-1 and IL-6 standard.

The concentration of IL-1 and IL-6 in the samples (expressed in U/ml) is determined from this standard curve. One unit per milliliter is defined as the growth factor concentration necessary to give one-half the maximal response at the plateau of the plot [50% effective dose (ED_{50})] (Fig. 1).

The physiological function of TNF in killing tumor cell lines is used to determine the TNF content. For this purpose the WEHI-164 fibrosarcoma cell line has been applied widely (8). Cell death can be determined as the ^{51}Cr release from $^{51}CrO_4$-labeled cells or by the decrease in the [^3H]thymidine incorporation rate. These radioactive techniques can now be replaced by measurements of the stable cytosolic enzyme lactate dehydrogenase (LDH) released into the culture supernatant by nonviable cells (10). The enzyme reduces lactate to pyruvate and produces NADH from NAD. This reaction is coupled to the diaphorase reaction, which reduces the tetrazolium salt INT [p-iodonitrotetrazolium violet; 2-(4-iodophenyl)-3-(4-nitrophenyl)-5-phenyltetrazolium chloride] to a red formazan product. The resulting absorbance at 490 nm is proportional to the amount of LDH and to the number of lysed cells. The TNF content in the sample is equivalent to the 50% lethal dose (LD_{50}), which is the concentration of the cytokine that gives 50% cell killing or LDH release. Three factors contribute to background absorbances: (a) phenol red in the culture medium, (b) LDH in the culture medium, and (c) spontaneous death of WEHI-164 cells. This requires recordings of pure medium and WEHI cells as controls. The procedure for TNF determinations follows the above-mentioned protocol for IL-1 and IL-6, with the exception that 2000 cells/well (50 μl) are incubated with samples (50 μl) overnight. Fifty microliters of a substrate solution containing INT, NAD, lactate, and diaphorase is added and the titer plate is incubated at room temperature for 30 min. The reaction is stopped by the addition of a stop solution. As a positive control pure LDH is used. One hundred percent of cell lysis is determined by the addition of lysis solution to WEHI cells.

When animals are subjected to treatments with various xenobiotics it should be taken into account that tissue samples may contain significant amounts of the compound used for the treatment of the animals. This may influence the results of the bioassay *in vitro* and thus the accuracy of the analysis. For example, the administration of high doses of dexamethasone in parallel with LPS to rats does not inhibit the synthesis of IL-6 *in vivo* because the glucocorticoid effect becomes manifest only when the animals are pretreated with the steroid for at least 30 min. However, dexamethasone, which is present in the tissue samples, may well block the proliferation of B9 cells *in vitro*. Thus, as an additional control the bioassay should be carried out in the presence of various concentrations of dexamethasone.

Determination of Cytokine mRNAs

The mRNAs of particular cytokines can be determined by several techniques. Northern blot hybridization is widely used for this purpose because mRNA levels can be determined reliably, the method is highly specific, and the sensitivity is sufficient to allow the detection of cytokine mRNAs after induction by inflammatory stimuli. The detection limit of the method is lower when poly(A) RNA rather than total RNA is examined (11). Moreover, Northern blot analyses will reveal information on the size of the investigated mRNA. A complete analysis consists of the following steps: (a) extraction of total RNA and, if required, preparation of poly(A) RNA, (b) agarose electrophoresis and blotting of the RNA, (c) hybridization, and (d) quantitation of the signal.

A rapid and simple method for RNA extraction developed by Chomczynski and Sacchi (12) can be applied to cultured cells or tissue samples. The isolation of RNA from whole tissues requires particular care during homogenization, which is carried out in a denaturing solution [4 M guanidinium isothiocyanate, 25 mM sodium-citrate, 0.5% (w/v) sarcosyl, 0.1 mM 2-mercaptoethanol; henceforth called solution D]. When cells are disrupted the intracellular release of lysosomal RNases causes a rapid breakdown of RNA. Lymphoid tissues in particular contain large amounts of RNases. Therefore, solutions and samples should be cooled on ice. A rapid homogenization of tissues is achieved with a Polytron. The tissues can be frozen in liquid nitrogen before homogenization and stored at −70°C. Frozen samples should be reduced to pieces of less than 5-mm diameter at −70°C before they are defrosted in solution D with a Polytron.

1. For the extraction of RNA a volume of 500 μl of the mixture of homogenized tissue and solution D is combined with 50 μl of sodium acetate, pH 4, 500 μl of phenol, and 100 μl of chloroform–isoamyl alcohol (49:1). The mixture is kept on ice for 15 min.

2. After centrifugation (15,000 g, 10 min, 4°C) the upper phase is recovered and mixed with 500 μl of 2-propanol to precipitate the RNA at −20°C for 1 hr.

3. After centrifugation (15,000 g, 10 min, 4°C) the pellet is redissolved in solution D and precipitated as described above.

4. The pellet is then washed with 75% ethanol followed by absolute ethanol, dried in a Speed-Vac (Savant, Hicksville, NY), and dissolved at 65°C (15 min) in diethyl carbonate-treated, autoclaved water containing 0.5% (w/v) sodium dodecyl sulfate (SDS).

5. The concentration of RNA is determined by ultraviolet (UV) spectrophotometry at 260 nm. The absorbance coefficient of pure RNA is 25 liter g^{-1} cm^{-1} (11).

RNA (20 μg/sample) is separated by electrophoresis with agarose gels [0.8 to 1.5% (w/v) agarose depending of the size of the mRNA] that contain

2.2 mM formaldehyde in electrophoresis buffer (11). Ethidium bromide is added to visualize the RNA under UV light at 312 nm (11). The dye can be pipetted directly to the samples before they are put onto test gels to check whether the RNA is intact and present in equal amounts in the samples. Gels containing RNA for subsequent blotting and hybridizations should not contain ethidium bromide because the dye can disturb the hybridization, and it is recommended that the filters be stained (after blotting the RNA) with methylene blue instead (11). For maximal efficiency the RNA is blotted to nylon filters with 10× SSC (1× SSC is 15 mM sodium citrate, 150 mM NaCl); the buffer is soaked vertically through the gel using pressure or vacuum (11). Subsequently, RNA is irreversibly cross-linked to the membrane by UV irradiation (1.5 J/cm^2 for wet filters). The filters can be used for several hybridizations with different probes.

Hybridizations can be performed with synthetic oligonucleotides, with cDNA, or with cRNA probes. The use of cRNA probes will increase the sensitivity of the method because RNA–RNA hybridizations are more efficient than DNA–RNA hybridizations. This is because cRNA probes can be labeled to higher specific radioactivities and because there is no competition between hybridization and reassociation of the probe in the buffer. However, some cRNA probes have a greater tendency to produce artifacts during the hybridization. For instance, we have observed that a rat IL-6 cRNA probe, but not the corresponding cDNA (13), can bind avidly to ribosomal RNA and other RNA species on the filter. This nonspecific signal could not be stripped, even under stringent conditions. Despite the development of nonradioactive methods the use of ^{32}P-labeled nucleic acids is still preferable because of the high sensitivity. Complementary DNA is best labeled with the multiprime reaction. Complementary RNA can be labeled with an *in vitro* transcription by RNA polymerase. These protocols are not presented here because commercial kits are available from a number of suppliers (see also Ref. 11). The removal of nonincorporated nucleotides from the labeling reaction by gel filtration on Sephadex G-50 columns will markedly reduce the background signal. Prehybridization is performed in a buffer containing 50% (v/v) formamide, 5× SSPE [1× SSPE is 150 mM NaCl, 10 mM NaH$_2$PO$_4$, 1.25 mM ethylenediaminetetraacetic acid (EDTA)], 2% (w/v) bovine serum albumin, 2% (w/v) polyvinypyrrolidone 40, 2% (w/v) Ficoll 400, denatured salmon sperm DNA (100 μg/ml), and 0.1% (w/v) SDS for 1 to 8 hr at 42°C. A volume of 0.1 ml of the buffer per square centimeter of the membrane is used. After prehybridization 10^6 disintegrations per minute (dpm) of the labeled probe is added to 1 ml of hybridization buffer. Complementary DNA probes must be denatured at 95°C for 5 min and chilled on ice prior to use. The filters are incubated overnight at 42°C in sealed plastic bags or in rotating glass vials, using a hybridization oven. Subsequently, the radioactive buffer is discarded, and the filters are washed twice in $2\times$ SSC at room temperature, once at $60°C$ in $2\times$ SSC, and finally at $60°C$ in $0.2\times$ SSC. Because the blots should not be dried they are sealed in plastic bags before exposure to X-ray films.

The intensity of the bands of the mRNAs on the films can be determined by scanning the absorbance with a densitometer or with an image analysis system. The signal should be corrected for unavoidable differences in the amount of RNA in the samples. For this purpose a second hybridization is performed after stripping the probe from the filter (10). The control hybridization is performed with cDNA or cRNA probes that encode proteins that have a constant level of gene expression. Usually glycerol-3-phosphate dehydrogenase (GAPDH) or actin fulfill this criterion. Alternatively, an internal control can be applied for normalization of the signal. For this purpose, a subclone, which contains a partial sequence of the cytokine mRNA to be determined, must be prepared. This deleted mRNA is then transcribed by RNA polymerase *in vitro*. When a defined amount of this RNA in the picogram range is pipetted to the samples prior to the separation of the RNA on agarose gels, the synthetic RNA will give an additional band on the blot that then serves as an internal standard. With this approach the relative intensity of the signal of a particular mRNA can be transformed into absolute amounts expressed as weight units or copy numbers (14).

We studied the effect of endogenous corticosterone on LPS-induced IL-6 expression in brain and peripheral tissues (15). Samples were analyzed 3 hr after the induction of IL-6 because peak levels of IL-6 and its mRNA occur between 2 and 4 hr after an intraperitoneal injection of LPS (100 μg/kg, *Escherichia coli* endotoxin serotype 0111:B4). Rats were divided into three groups: sham-operated controls, adrenalectomized animals, and rats that were pretreated with corticosterone (10 mg/kg, subcutaneously) 1 hr before the LPS administration. RNA was isolated from hippocampus, hypothalamus, pituitary gland, adrenals, and spleen. By using Northern blot analyses with a rat IL-6 cDNA (13), the IL-6 mRNA was quantified. The results are shown in Fig. 2A and B. Adrenalectomy or corticosterone pretreatment did not affect the IL-6 mRNA content in hippocampus and hypothalamus (Fig. 2B). However, IL-6 mRNA was further increased in spleen and pituitary of adrenalectomized animals (Fig. 2A and B). The IL-6 induction caused by LPS was completely blocked in spleen, adrenals, and pituitary gland after corticosterone pretreatment (Fig. 2A and B).

These findings suggest that the expression of IL-6, which is a reliable marker of the inflammatory response, is tightly controlled by endogenous corticosterone at physiological concentrations.

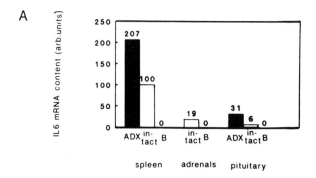

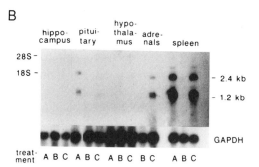

FIG. 2   Effect of corticosterone on IL-6 induction. Lipopolysaccharide (from *E. coli*, 100 µg/kg, intraperitoneal) was injected into adrenalectomized animals (black bars), sham-operated controls (open bars), and rats treated with corticosterone (10 mg/kg) subcutaneously 1 hr before the injection of LPS. After 3 hr the animals were decapitated and RNA was isolated. Interleukin 6 mRNA was analyzed by Northern blot analyses. Values are given on top of the bars. [From Schöbitz *et al.* (15), with permission.] (A) Relative IL-6 mRNA levels in spleen, adrenals, and pituitary gland of rats 3 hr after the injection of LPS. (B) Northern blot analysis of IL-6 mRNA rats. Lane A, adrenalectomized animals; lane B, animals pretreated with corticosterone, lane C, sham-operated controls. Copyright © 1993, The Endocrine Society.

Determination of Cytokine mRNA Transcription

Transcription of RNA cannot be studied in whole animals or intact cells because the isolation of nuclei is a prerequisite to quantify the rate of transcription of individual genes by nuclear run-on assays (16). However, this method provides a system that is the closest to the intact cell. The observed biosynthetic activity reflects the completion of RNA chains that are initiated for transcription before the nuclei were prepared. When nuclei are isolated from whole tissues that contain different cell types the heterogeneity of these nuclei must be taken into account. The removal of blood for the preparation of nuclei from liver is achieved by perfusion of the organ with an isotonic buffer (0.14 M NaCl, 10 mM Tris-HCl, pH 8.0). The tissue is then cut into small pieces and rinsed in the perfusion buffer.

1. Tissue pieces are homogenized with a suitable homogenizer [e.g., Dounce (Wheaton, Millville, NJ) homogenizer for soft tissue] in buffer A [0.32 M sucrose, 3.0 mM $CaCl_2$, 2.0 mM magnesium acetate, 0.1 mM EDTA, 0.1% (v/v) Triton X-100, 1.0 mM dithiothreitol (DTT), 10.0 mM Tris-HCl, 0.1 mM PMSF, 1 mM ethylene glycol-bis(β-aminoethyl ether)-N,N,N',N'-tetraacetic acid (EGTA), 1 mM spermidine, pH 8.0]. A volume of 10 ml/g tissue is used.

2. The suspension is filtered rapidly through several layers of cheesecloth to remove connective tissue, and the filtrate is homogenized using a homogenizer with a tight pestle.

3. The homogenate is diluted with 1 to 2 vol of buffer B (2.0 M sucrose, 5.0 mM magnesium acetate, 0.1 mM EDTA, 1.0 mM DTT, 10.0 mM Tris-HCl, 0.1 mM PMSF, 1 mM EGTA, 1 mM spermidine, pH 8.0).

4. The mixture is layered over a cushion of buffer B and centrifuged at 30,000 g for 45 min at 4°C.

5. The pellet containing the nuclei is resuspended in storage buffer [25% (v/v) glycerol, 5.0 mM magnesium acetate, 0.1 mM EDTA, 5.0 mM DTT, 50 mM Tris-HCl, pH 8.0].

6. The density of the suspension is determined after crystal violet (70 μg/ml in 8.3 mM sucrose) staining in a hemocytometer. Aliquots can be frozen and stored in liquid nitrogen.

7. Cell nuclei (10^7) in a volume of 100 μl are mixed with 10 μl H_2O and 40 μl of a solution containing 0.6 M KCl and 12.5 mM magnesium acetate.

8. The reaction is started with 50 μl of nucleotide buffer [2.0 mM ATP, 2.0 mM GTP, 2.0 mM CTP, 0.2 mM UTP, RNasin (2 U/μl), 6.0 M [α-^{32}P]UTP (400 Ci/mmol)]. The mixture is incubated for 30 min at 25°C.

9. Samples of 5 μl are taken before the start of the reaction, and after 10, 20, and 30 min, and spotted on glass fiber membranes.

10. The filter is washed with 50 ml of ice-cold trichloroacetic acid (TCA) and 5 ml of ethanol, and the radioactivity is determined in a scintillation counter. The time–response curve serves to judge the quality of the nuclei and the general transcription rate. In a typical reaction the amount of radioactivity incorporated into TCA-precipitable material will increase 20- to 100-fold during the 30-min period.

11. The reaction is stopped after 30 min with a solution of 1% (w/v) SDS, 10 mM EDTA, pH 7.0. Subsequently, 200 μl of sodium acetate, 2 ml of phenol, and 400 μl of chloroform are added and kept on ice for 15 min.

12. After centrifugation at 10,000 g for 10 min at 4°C, the aqueous phase containing the RNA is removed. The RNA is recovered by the addition of 200 μl of 0.4 mM sodium acetate and 5 ml of ethanol and centrifugation at 10,000 g for 10 min at 4°C.

13. The pellet is washed with 70% ethanol and dissolved in 450 μl of 0.5% (w/v) SDS, 0.3 M sodium acetate.

14. The RNA is precipitated again with 1 ml of ethanol, centrifuged, and dissolved in 50 μl of 0.5% SDS.

15. Five microliters of the solution is diluted (1:10), and 5 μl of the dilution is mixed with 2 ml of 10% (v/v) TCA and kept on ice for 15 min. The suspension is spotted on a glass fiber filter, washed twice with 30 ml of ice-cold 10% (v/v) TCA containing 1% (w/v) sodium pyrophosphate, and once with 5 ml of ethanol.

16. The radioactivity on the filter is determined in a scintillation counter. The value reflects the total amount of RNA that is used for the hybridization.

The radioactive RNA is then hybridized to filter-bound plasmids encoding the sequence of particular cytokines. The corresponding plasmids are linearized with restriction enzymes. Denatured, cleaved plasmid DNA (5 μg) is diluted with 20× SSC and spotted on nylon filters, using a minifold. The DNA is fixed by UV irradiation as described above. Vector DNA is used as a control to correct for nonspecific binding. Hybridization and washing conditions are identical to those applied to the Northern blot analyses. To quantify the amount of RNA the radioactivity bound to the dots can be determined using a densitometer after exposure of the filters to X-ray films. Alternatively, dots can be punched out for determination of the radioactivity by liquid scintillation counting. These values are corrected by subtracting the background signal. The relative transcription rates are expressed in parts per million (ppm) by division with the values of the total amount of RNA (expressed in dpm) that was present in the incubation.

Determination of Cytokine mRNA Stability

Several methods have been developed to determine the half-life of particular mRNAs. These techniques cannot be applied to whole tissues because isolated cells must be incubated with transcriptional inhibitors or, in pulse–chase experiments, with nucleotides. For this reason the stability of cytokine mRNAs can be analyzed in cell culture only. However, to induce the corresponding cytokine mRNA these cells can be stimulated either *in vitro* or *in vivo*. The latter approach requires cell preparation after the treatment of experimental animals or humans. Subsequent steps are performed *ex vivo*. In many cases monocytes or macrophages have been used because these cells are able to produce large amounts of IL-1, IL-6, and TNF mRNA.

When comparing the half-life of a particular mRNA determined with different methods, the absolute values may differ (17). However, it is possible to determine accurately relative changes in mRNA stability evoked by different treatment of cells or experimental animals. The most convenient method is based on measurements of mRNA decay when *de novo* synthesis of RNA is inhibited. Actinomycin D and 5,6-dichloro-1β-ribofuranosylbenzimidazole have been widely used as transcriptional inhibitors. Depending on the cell type, the concentrations of these drugs vary in the range of several grams per milliliter for both compounds. The concentration used should inhibit the incorporation of [^{3}H]uridine into TCA-precipitable material, but it should not produce cytotoxic effects within the culture period that is necessary for the analysis. The viability of cells can be judged by trypan blue exclusion. Because the half-life of cytokine mRNAs is usually less than 2 hr, culture periods are no longer than a few hours. This brief period does not produce alterations in the amount of total RNA, owing to the large stability of ribosomal RNA (which constitutes more than 80% of the total amount of RNA). Therefore, the quality of ribosomal RNA (visualized by agarose gel electrophoresis) and the amount of total cellular RNA can be used to judge the quality of cells in culture when RNA synthesis is inhibited. Cells are harvested at short intervals after addition of the inhibitors. From these samples total RNA is prepared as described above. The determination of the mRNA should be done by Northern blot analysis to show that the cytokine mRNA is intact and that, under the conditions that have been used, the hybridization signal is specific. In subsequent experiments dot-blot analyses can be performed instead. A second hybridization can be performed with a cDNA encoding a stable RNA, for example, 18S ribosomal RNA or actin mRNA, to normalize for the amounts loaded on the filter. If the total cellular RNA content changes during the culture period the total RNA amount should be determined to correct the decay of the examined mRNA by the decay of total RNA. The orcinol method can be applied for this purpose (11).

Concluding Remarks

Communication between the CNS and the immune system can be studied in a closed feedback loop comprising the interaction between cytokines and glucocorticoids. The inhibition of the biosynthesis of proinflammatory cytokines is thought to be the major target for the immune suppressive effects of exogenous and endogenous glucocorticoids. Cell biological and molecular biological methods, including cytokine bioassays and cytokine mRNA quantitation, and stability and transcription assays, can be carried out to study this interaction. These methods, applied *in vitro* and *in vivo*, will provide valuable physiological and pharmacological data on the mode of action of glucocorticoids.

References

1. A. G. Gilman, T. W. Rall, A. S. Nies, and P. Taylor, "Goodman and Gilman's Pharmacological Basis of Therapeutics," 8th Ed. Pergamon, New York, 1990.
2. H. N. Baybutt and F. Holsboer, *Endocrinology (Baltimore)* **127**, 476 (1990).
3. S. W. Lee, A. P. Tsou, H. Chan, J. Thomas, K. Petrie, E. M. Eugui, and A. C. Allison, *Proc. Natl. Acad. Sci. U.S.A.* **85**, 1204 (1988).
4. S. Akira, T. Hirano, T. Taga, and T. Kishimoto, *FASEB J.* **4**, 2860 (1990).
5. C. A. Dinarello, *Blood* **77**, 1627 (1991).
6. H. O. Besedovsky and A. del Ray, *Front. in Neuroendocrinol.* **13**, 61 (1992).
7. M. Helle, L. Bouije, and L. A. Aarden, *Eur. J. Immunol.* **18**, 1535 (1988).
8. T. Espevik and J. Nissen-Mayer, *J. Immunol. Methods* **95**, 99 (1986).
9. T. Mosmann, *J. Immunol. Methods* **65**, 55 (1983).
10. C. Korzeniewski and D. M. Callewaert, *J. Immunol. Methods* **64**, 313 (1983).
11. J. Sambrook, E. F. Fritsch, and T. Maniatis, "Molecular Cloning: A Laboratory Manual." Cold Spring Harbor Lab. Press, Cold Spring Harbor, New York, 1989.
12. P. Chomczynski and N. Sacchi, *Anal. Biochem.* **162**, 156 (1987).
13. W. Northemann, T. A. Braciak, M. Hattori, F. Lee, and G. H. Fey, *J. Biol. Chem.* **264**, 16072 (1989).
14. D. C. DuBois, R. R. Almon, and W. J. Jusko, *Anal. Biochem.* **210**, 140 (1993).
15. B. Schöbitz, M. van den Dobbelsteen, W. Sutanto, F. Holsboer, and E. R. de Kloet, *Endocrinology (Baltimore)* **132**, 1569 (1993).
16. B. D. Hames and S. J. Higgins, "Transcription and Translation: A Practical Approach." IRL Press, Oxford, 1986.
17. S. Harrold, C. Genovese, B. Kobrin, S. L. Morrison, and C. Milcarek, *Anal. Biochem.* **198**, 19 (1991).

Perspectives

[32] Steroid Hormone Effects on Brain: Novel Insights Connecting Cellular and Molecular Features of Brain Cells to Behavior

Bruce S. McEwen

Introduction

Steroid hormones continue to provide novel insights into a variety of biological phenomena, just as they have during the past 100 years, from cellular and molecular events to the understanding of how behavior is developmentally programmed and regulated. The paper by Berthold in 1849 is often cited as the first true experiment in the field of endocrinology, because it involved castration and transplantation of testes in the rooster; and it was also the first experiment in behavioral endocrinology because it demonstrated a role for testicular secretions in controlling sexual, aggressive, and crowing behavior of the rooster (1).

Not only endocrinology, but also behavioral neuroscience, owes its beginnings at least in part to this study, because steroid hormones have provided important lesions regarding brain mechanisms underlying behavior and behavioral plasticity. This includes the important findings regarding the process of sexual differentiation of the brain and the differences between the permanent developmental actions of steroids early in life and the reversible actions they exert on the mature brain that affect synaptic connectivity and synaptic transmission. There is also the realization that parts of the adult brain are capable of considerable plasticity, including synaptic turnover and neurogenesis, that is regulated by steroid hormones.

The study of steroid hormones is playing an important role in the development of cellular and molecular neurobiology. The characterization of intracellular steroid hormone receptors that bind to DNA in the 1960s provided the first concrete examples of *trans*-acting gene regulatory proteins in eukaryotic cells; and the study of transcription regulators has taken off from this origin (2, 3). The implications of transcriptional regulation of brain function encompass not only the information provided by the external hormonal signals but also the additional input of transcriptional regulation by second messengers derived from neuronal activity, which act often in combination with steroid receptors via so-called composite response elements (2, 3).

Methods in Neurosciences, Volume 22

The study of steroid receptors in brain tissue has provided additional methods beyond autoradiography and cell fractionation procedures to isolate nuclei and study cytosol receptor binding ([6] in this volume). Immunocyto-chemistry at the light and electron microscoptic level ([9]–[11] and [25] in this volume) and *in situ* hybridization histochemistry ([12] in this volume) provide new tools for measuring steroid receptors that do not require removal of the hormone-secreting tissue from the animal or injection of radioactive steroid. And quantitative *in vitro* autoradiography ([8] in this volume) has provided new possibilities for receptor localization not requiring huge quanti-ties of expensive isotope injected into animals. Moreover, techniques of molecular biology have provided the ability to study structure–activity rela-tionships via modified genes inserted into cells ([14] and [16] in this volume), to inhibit gene expression via use of antisense oligonucleotides ([21] in this volume), and to study DNA–protein interactions in search of regulatory sites in cloned genes and their promotor regions ([15] in this volume).

Besides these important methodological considerations, the study of ste-roid hormone actions in the brain has revealed some important aspects of hormone action that are not so evident in other tissues. In particular, the nongenomic actions of steroids are especially suited to the neurotransmitter mechanisms of brain function and to investigations using electrophysiology ([7], [13], and [27] in this volume); and the intriguing "neurosteroids," formed by brain tissue from cholesterol, raise important questions about the sources and compartmentation of steroids ([3] in this volume). The following over-view considers some of these topics, using as a starting point chapters in this volume.

Gene Expression in Brain in Light of Composite Response Elements

At first, the control of gene expression by steroids was believed to be a relatively straightforward process involving steroid response elements in the promotor regions of a subset of genes. When these genes are expressed in a cell containing steroid receptors, then the result was expected to be hormonal regulation of the gene product. Although this scenario does apply quite frequently, there are additional complications related to the fact that most genes are regulated by more than one *trans*-acting factor (2, 3). There are cell-type specific transcription factors, as well as factors such as c-*fos* that are turned on by neuronal activity, and other factors such as cyclic 3'5' AMP response element binding protein (CREB) that are stimulated by phos-phorylation arising from cyclic AMP or calcium-stimulated protein kinases.

Genes that are regulated by so-called composite response elements, de-pending on the binding of more than one *trans*-acting factor, add another

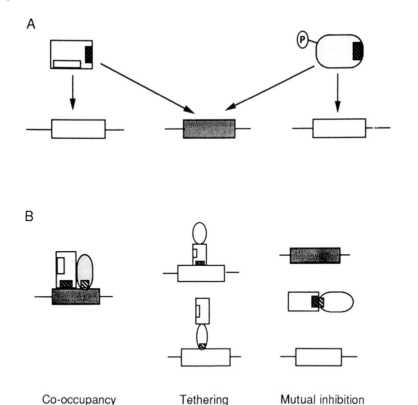

FIG. 1 Diagram of simple and composite response elements involved in the regulation of transcription of genes by steroid hormones, immediate early genes, and second messenger-regulated DNA-binding proteins.

dimension to the complexity of understanding regulation of transcription in brain cells. The fact that the brain uses a multiplicity of second messengers, as well as the so-called immediate early genes, means that multiple events converge to produce a final transcriptional regulatory event (4–6).

Genomic and Nongenomic Effects of Steroids on Brain and Other Tissues

If this were not complicated enough, the fact that steroid hormones produce rapid actions on excitable membranes via nongenomic means adds yet another level of complexity to the way in which hormonal signals regulate neural

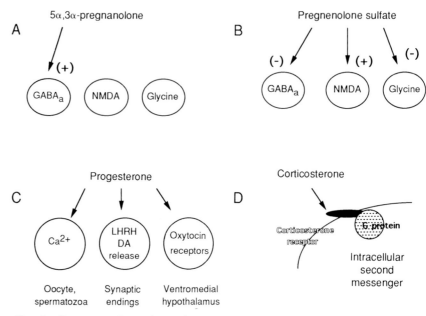

FIG. 2   Spectrum of membrane-based actions of steroids on excitable tissues.

activity. Identification of nongenomic effects of steroids actually predates the genomic receptors, in that the work of Hans Selye around 1940 provided the first clear evidence regarding the anesthetic potency of progesterone and derivatives of progesterone (7).

Nongenomic effects have been most thoroughly studied for the γ-aminobutyric acid type A (GABA$_A$) receptor, and there is a recognized receptor site that regulates opening of the chloride channel along with GABA, benzodiazepine agonists, and agents that interact with the so-called barbiturate site ([13] in this volume). Potentially important anxiolytic as well as anesthetic and sedative–hypnotic effects have been recognized for steroids acting as pharmacological agents ([13] in this volume). In contrast to the pharmacological effects of these steroids and their therapeutic implications, the nature of physiological effects of *in vivo*-generated metabolites of progesterone and deoxycorticosterone is under intensive investigation but is still somewhat of a mystery. Progress in the study of lordosis behavior in the hamster indicates the involvement of GABA$_A$-active steroids along with genomic mechanisms (8).

A study on the actions of A ring-reduced progesterone and deoxycorticosterone metabolites in intact cells in culture has revealed another layer of complexity in interpreting their effects (9). In this study, progestin receptors were transfected into neuroblastoma cells along with a reporter gene responsive to the progesterone receptor; tetrahydrometabolites of the two parent steroids were able to activate the reporter gene, but apparently after oxidation to the 5 α-dihydro form (9). Thus any tetrahydro steroid that is not sterically hindered and subject to back conversion to the dihydro derivative may be able to activate an intracellular progestin, glucocorticoid, or mineralocorticoid receptor.

Nongenomic actions of other steroids such as estradiol and glucocorticoids have also been revealed by electrophysiological means and from the study of binding (10). However, binding has been the most difficult aspect of nongenomic receptors to demonstrate; nevertheless success with the newt, *Taricha granulosa*, indicates that this is possible and that conditions must be sought to obtain binding in other species ([7] in this volume).

Besides the clearly nongenomic effects of certain steroids, there are actions of steroid hormones on cellular excitability that are sometimes fast enough to create confusion as to whether they are genomic or nongenomic. This is true of the rapid actions of adrenal steroids on neuronal activity in hippocampal neurons ([26] in this volume). Other, slower actions of steroid hormones affect the excitability of muscle via effects on ion channels (11, 12). The specific examples of estrogen and glucocorticoid induction of the mRNA for a form of the potassium channel illustrate these type of effects (13, 14).

Yet other effects of steroid hormones involve a bewildering combination of genomic effects and interactions of these effects with neurotransmitter systems, making the distinction between genomic and nongenomic effects almost meaningless. One such example deals with estrogen effects in the CA1 region of the rat hippocampus involving induction of new synaptic contacts ([24] in this volume) as well as direct and indirect estrogen effects on electrical excitability (15) and on the induction of N-methyl-D-aspartate (NMDA) receptors (16), glutamate decarboxylase mRNA (17), and GABA_A receptors (18) in the CA1 region. One of the peculiar features of these events is that only the GABA-containing interneurons within the CA1 cell body regions have been shown to contain intracellular estrogen receptors (19), making it necessary to postulate an indirect effect of estrogens on the NMDA receptors and synapse formation occurring in the CA1 pyramidal neurons. Another unusual aspect of the CA1 response to estradiol is that blockers of NMDA receptors prevent estrogen induction of new synapses on CA1 pyramidal neurons (20). It will be necessary to determine what the primary effect of estradiol is, and the pathway—genomic or membrane action—through which estradiol produces its synapse-inducing stimulus before we will begin to understand how this interesting effect comes about.

Neurosteroids: Brain as a Source of Steroids

Neuroactive steroids that work via nongenomic mechanisms are often called "neurosteroids" as a short-hand reference. The term "neurosteroid" really means a steroid generated in the brain from cholesterol ([3] in this volume). The principal neurosteroids are pregnenolone and its sulfate and dehydroepiandrosterone (DHEA) and its sulfated form. Both of these steroids have actions at micromolar concentrations that potentiate excitatory amino acid actions on nerve cells, but the concentrations required for these effects raise serious doubts as to their physiological significance.

The generation of neurosteroids is also a fascinating topic, and the presence in brain of cytochrome *P*-450 enzymes is an important indicator of this capability ([4] in this volume). Glial cells play a role, and the mitochondrial "peripheral-type" benzodiazepine receptor has a key role in the side-chain cleavage of cholesterol to generate pregnenolone (21). Studies indicate that pregnenolone and DHEA formation can be regulated by drugs that alter the "peripheral-type" benzodiazepine receptor and that a pool of rapidly turning over neurosteroid results as well as a more stable pool. The compartmentation and functions of these different pools may hold secrets as to the role of neurosteroids in the brain, because attempts to deplete the total brain pool of neurosteroid by adrenalectomy and castration have proved ineffective (22). It is difficult to demonstrate the importance of a natural substance unless you can get rid of it and then replace it.

Neurosteroids are reputed to be convertible in small quantities to steroids such as progesterone (22). However, it is doubtful that the brain is a major source of hormonal steroids of the types secreted by the adrenals and gonads because, if it were, then there would be no science of behavioral endocrinology and Berthold (1) would never have been able to make the landmark observations as to the dependence of certain behaviors on the gonads.

An exception to the minor role of the brain in providing steroids to the rest of the body is the role of the avian brain of at least some species in making estrogens from circulating androgens; these estrogens are released into the blood and provide estrogen throughout the body (23).

Steroid–Carrier Proteins and Blood–Brain Barrier

A special feature of the brain compared to other tissues is the "blood–brain" barrier, a collection of mechanisms that limit penetration of substances into the brain, or at least pump them out when they enter ([1] in this volume). Steroid hormones gain access to the brain, it is thought, because of their lipophilic nature. Indeed, early studies of uptake of radiolabeled steroids

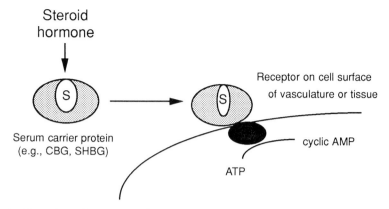

Fig. 3   Function of serum carrier proteins as ligands for receptors on certain target cells.

revealed how readily and in high concentrations steroids such as progesterone and estradiol entered the brain and how poorly corticosteroids such as dexamethasone were able to enter it (24). An application of this information has shown, for example, that doses of dexamethasone that result in blood levels typical of a dexamethasone suppression test (DST) used in diagnosis of depressive illness reach primarily the pituitary gland and show limited penetration of the brain (25).

Naturally occurring adrenal steroids bind to carrier proteins, such as corticosteroid-binding globulin (CBG), α-fetoprotein (AFP) and sex hormone-binding globulin (SHBG). These proteins play an important role in limiting the access of steroids to the brain, as for example is shown by the important role of AFP in protecting newborn rat pups from natural estrogens in the mother's milk; estrogens such as diethylstilbestrol and moxestrol, which do not bind to AFP, gain access to neural estrogen receptors and defeminize the brain (26). Carrier proteins such as CBG and SHBG have also been shown to bind to their own receptors on certain cells and result in activation of transmembrane events, including the formation of cyclic AMP (27, 28) (see Fig. 3). It is not clear whether this type of signaling occurs in brain capillaries or in the ependymal lining of the ventricles, but this is a possibility.

Importance of Steroid Metabolism and Compartmentation

In addition to generating the neurosteroids pregnenolone and DHEA from cholesterol, the brain transforms adrenal and gonadal steroids in important ways. The presence and distribution of cytochrome *P*-450 enzymes in brain provide important insights into the types of transformations of steroids and where they occur ([4] in this volume; see also Mellon and DeScheppers (29)]. The ability of the brain to take up and transform steroids via redox mechanisms is also important for the design of novel delivery systems ([2] in this volume).

In the case of aromatization of testosterone or conversion of testosterone to 5α-dihydrotestosterone, the conversion results in a steroid hormone that acts via genomic receptors. In the case of the conversion of progesterone to allopregnanolone or of deoxycorticosterone (DOC) to tetrahydro-DOC, the product interacts with the $GABA_A$ receptor to facilitate chloride ion movement (30).

For other conversions, the result is not so clear. For the enzyme 11β-hydroxysteroid dehydrogenase (11-HSD), the product is not known to be active on any genomic or nongenomic receptor ([5] in this volume). In the kidney, this mechanism is an important means of ensuring that cortisol in humans and corticosterone in rats do not act as mineralocorticoids (31). In brain, the role of 11-HSD is less clear ([5] in this volume). It may function as a device to reduce the potency of circulating glucocorticoids, but the level of 11-HSD is such that it is not nearly as potent a mechanism as in the kidney; in fact, low doses of corticosterone are not impeded in their access to mineralocorticoid receptors in hippocampus (32), site of one of the highest 11-HSD activity in brain (33), the way they are in kidney (31).

The two steps in the generation of tetrahydro derivatives of various steroids appear to be compartmentalized, in that neurons may be responsible for the 5α reduction and glia for the 3α-hydroxysteroid dehydrogenase step (34).

Another type of compartmentation is important, namely, the regionalization of the aromatase enzyme complex in the brain, which results in providing estradiol to a limited population of estrogen receptors (Fig. 4) (35). This is particularly important because estrogen receptors are present in both sexes and estradiol is an important hormone in the male, mediating some of the actions of testosterone via aromatization. Note in Fig. 4 the different pattern of estrogen receptor occupancy in males with circulating testosterone compared to females with circulating estradiol. In contrast, the 5α reduction of testosterone is widespread enough throughout the brain that it does not provide a limitation to the access of 5α dihydrotestosterone to androgen receptors (35).

Developmental Actions of Steroids

Aromatization and 5α reduction of testosterone play essential roles in generating steroids that bind to receptors during early development and produce sexual differentiation of the brain (36). Studies of sexual differentiation of

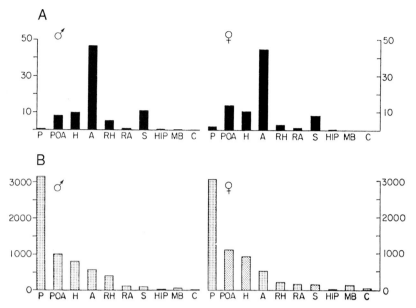

FIG. 4 Role of testosterone metabolism in providing steroid to estrogen receptors in male and female rat forebrain. Estradiol radioactivity (fmol/mg DNA) recovered in cell nuclei from different brain regions 2 hr after [³H]testosterone or [³H]estradiol administration. (A) 17β-Estradiol as a testosterone metabolite (5.7 μg/kg); (B) 17β-estradiol infused directly (2.7 μg/kg). P, Pituitary; POA, medial preoptic area; H, basomedial hypothalamus; A, corticomedial amygdala; RH, rest of hypothalamus; RA, rest of amygdala; S, septum; HIP, hippocampus; MB, midbrain; C, parietal cortex. The y axis shows radioactive steroid concentration in fmoles/mg DNA. [From Lieberburg and McEwen (35), with permission. Copyright © 1977, The Endocrine Society.]

the brain have revealed that it consists not only of permanent differences in the survival and neuroanatomical connectivity of neurons in a wide range of brain regions, but also of differences in responsivity of mature neurons to circulating steroid hormones. For example, estradiol has some effects on the male brain that differ from those on the female brain, as well as many actions that are similar between the sexes. One striking example is the induction by estradiol of synapses in the ventromedial hypothalamus (VMN) of female rats, an event that occurs cyclically during the estrous cycle of the adult female rat (37). In contrast, estradiol fails to induce synapses in the male VMN even though it contains estrogen receptors and even though those receptors mediate equal levels of oxytocin receptor induction in the male and female VMN (38, 39). Interestingly, the male VMN has a larger number of synapses compared to the female, which appear to be stably wired in place by the developmental actions of testosterone (40).

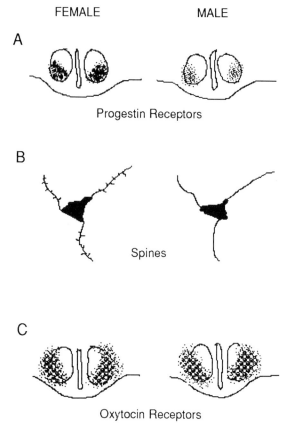

FEMALE   MALE

A

Progestin Receptors

B

Spines

C

Oxytocin Receptors

FIG. 5   Summary of sex differences in the ventromedial nuclei (VMN) of the hypothalamus in relation to three changes induced by estradiol. (A) Estradiol induces
more progestin receptors in the female VMN than in the male VMN; (B) estradiol
induces more spines, and therefore new synapses, on dendrites in the female VMN
than in the male VMN; (C) estradiol induces the same number of oxytocin receptors
in the female and male VMN. [From McEwen (42), with permission.]

Ironically, those actions of testosterone during early development that
suppress female sexual behavior and alter properties of the VMN most
likely represent actions mediated by aromatization and thus by the estrogen
receptor (41). Thus, whereas synapse formation directed by testosterone
(via estradiol) during development is permanent, that which occurs in adult
life under the direction of estradiol and progesterone is cyclic, as long as
the hypothalamus has escaped the developmental actions of testosterone
(Fig. 5) (42).

In contrast to sexual differentiation, far less is known about the actions of other steroids during development. Studies of the role of glucocorticoids in neuronal development in the hippocampus reveal a negative controlling role of adrenal steroids acting via type I receptors to limit cell division and cell death in the developing dentate gyrus (43). It is noteworthy that dentate gyrus development is also influenced, directly or indirectly, by testosterone and by thyroid hormone, because male rats have a thicker dentate gyrus and neonatal hyperthyroidism increases dentate gyrus size as well as CA3 pyramidal neuron branching and synaptic connectivity (44, 45).

The actions of thyroid hormone, another member of the steroid–thyroid hormone superfamily, are fascinating because, like the developmental actions of testosterone and estradiol, they are qualitatively different from the actions they produce on mature neurons. Thyroid hormone, for example, causes a reduction in hippocampal pyramidal neuron spine density when it is given to a young adult rat, whereas it produces the hypertrophy of CA3 neurons alluded to above when it is given in early postnatal life (45). Thus the uses to which hormonal signals are put change as the cells that respond to them differentiate and mature.

Structural Changes in Brain Not Confined to Development

We have noted that whereas steroid receptors remain the same, the uses to which they are put often change depending on the developmental stage. In particular, some of the developmental actions of gonadal steroids are qualitatively different from the actions that the same hormones exert in adult life. However, not all effects of steroid hormone change qualitatively between early development and adult life. Perhaps the most striking example involves the dentate gyrus, which continues to undergo neurogenesis and neuronal turnover in adult life (43). And this process continues, both during early development and in the mature brain, to be regulated by adrenal steroids. It is conceivable that the functional role of this continuing plasticity of the dentate gyrus is to provide a means of enlarging and then reducing the size and functional capability of the hippocampus. This may be particularly important in species, such as the vole, in which males use spatial information during the mating season; and indeed, males have a seasonally larger hippocampus than females (46).

One of the most frequent misconceptions of the difference between hormone actions in development and adult life is that structural changes in brain are confined to early development and that what hormones do in adult life is to regulate synaptic transmission in stable circuits. The example of hormonally regulated cell turnover in the adult dentate gyrus is one illustration of structural plasticity in the adult brain. Another is synaptogenesis. We have noted the cyclic synaptogenesis controlled by estradiol in the VMN of the adult female rat, which is undoubtedly involved in the cyclic hormonal control of feminine sexual behavior. Another example of cyclic synaptic change occurs in the CA1 region of the hippocampus. In this region, it has been possible to study hormonal control of synaptogenesis, involving sequential actions of estradiol followed by progesterone ([24] in this volume). One of the most striking findings is that progesterone acts as the signal that terminates the synaptogenesis and promotes destruction of the newly formed synapses (47). Thus, the decline in synapse density between the proestrus peak and the estrus trough, less than 24 hr later, is blocked by Ru 486, an antagonist for intracellular progestin receptors (47).

Conclusions: Connecting Cellular and Molecular Aspects of Steroid Hormone Action to Behavior

Steroid hormone effects on the adult brain are diverse at the cellular level, and one of the primary reasons for their diversity is that they are ultimately concerned with regulating integrative brain function and behavior. Besides regulating structural changes in adult as well as developing brains, there are many other cellular and molecular aspects of steroid hormone actions on neurons and glial cells. Each of these effects is interesting to study in its own right because of what it tells us about the function of brain cells and the integration of genomic and nongenomic mechanisms of action at the cellular and molecular level.

At the same time, it is an article of faith that each of these effects fits into a larger picture of the role of the brain in integrating physiological responses and regulating behavior. The challenge is the following: how do we find out the role of each effect in integrated brain function? In other words, how can we appreciate the diversity and understand underlying molecular mechanisms and then apply this knowledge to elucidating the role of these individual events in behavior? This section addresses these issues.

Cellular and Molecular Analysis of Mechanism

At the cellular level, steroid hormone effects on brain cell function include regulation of synapse formation, dendritic branching and cell turnover ([24]

in this volume; see above discussion), growth responses ([22], [23] in this volume), excitability ([26] in this volume; see also above discussion), neuropeptide formation ([17] in this volume), and expression of neurotransmitter receptors (48–52).

Regulation of neurotransmitter receptor expression and neuropeptide gene expression appears to represent relatively straightforward examples of steroid hormone action at simple or composite response elements (2) (Fig. 1), and methods for studying them are now being developed ([15]–[17] in this volume). The other cellular events are likely to be more complicated, or at least there are fewer ideas concerning the types of genes that are regulated by steroids and that are rate limiting for these processes.

The following are some questions that remain to be answered: (a) are trophic effects of steroids explained by the regulation of neurotrophin production, as they appear to be in the elegant studies of Toran-Allerand et al. (53)? (b) Does structural plasticity involve steroid effects on expression of key structural proteins such as the cytoskeletal tau protein [(54); see (55) for review] or cell death genes (43)? (c) Are the rapid, but apparently genomic, steroid effects on neuronal excitability ([26] in this volume) explained by rapid regulation of ion channel expression or expression of protein kinases and protein phosphatases that regulate ion channel and receptor activity via phosphorylation and dephosphorylation?

Considering that many of the genes regulated by steroid hormones may not be known *a priori*, what are the means for finding them? Cloning of steroid-responsive mRNAs has been successful in identifying a number of glucocorticoid-responsive genes in neurons and glial cells, including some gene products not previously recognized ([18] in this volume). However, progress is slow and many genes that are identified are already known. Another important strategy with single-cell resolution is the generation of cDNAs at the single-cell level, followed by polymerase chain reaction (PCR) amplification; this approach has the advantage of focusing on regulatory events within single cells or defined cell populations, and, in principle, it also represents a means of generating cDNAs for screening and identification of novel steroid-regulated genes ([19] in this volume). The new methods of differential display (56) may be useful in accelerating and otherwise improving the identification of novel genes.

At the level of protein formation and identification, the labeling and identification of steroid-regulated proteins in hippocampal slices has freed the investigator from complexities of *in vivo* labeling and has facilitated identification of one stress-regulated protein ([20] in this volume). Going beyond slices to cells in culture, the elegant procedure of two-dimensional gel analysis allows for the identification and monitoring of hormone effects on a large range of cellular proteins ([28] in this volume), whereas transfection with recombinant steroid receptors and target gene constructs has freed the investigator from the constraints of always using only what differentiated cells themselves provide ([14] and [16] in this volume).

At the nongenomic level, there are also daunting challenges to experimental skill, as well as opportunities for future progress in this fascinating area. Binding assays have not been successful in identifying cell surface steroid receptors, but, where they have been effective, they have revealed possible coupling to second-messenger systems via G proteins, which may be useful in illuminating the cellular mechanism and links to neuronal excitability, as well as providing opportunities for cloning of the receptors themselves ([7] in this volume). When even one membrane receptor is cloned, it will be interesting to determine whether there is any homology with the intracellular receptors, in view of data that suggest that intracellular steroid receptors may have associations with dendrites and membrane elements and possibly even synaptic endings ([11] in this volume) (57). The $GABA_A$–benzodiazepine receptor complex has, of course, already been cloned; here, the problem is to identify the structure–activity relationships for the many combinations of subunits of this receptor system and to recognize possible subtype specificity for A ring-reduced steroid metabolites ([13], [27] in this volume). Studies showing the ability of A ring-reduced metabolites of aldosterone to affect salt appetite in the brain and renal sodium flux (58, 59) make this an interesting possibility.

With regard to the capacity of the brain to make "neurosteroids," one of the most puzzling aspects is the cellular compartmentalization and function of these substances ([3] in this volume). Are they conjugated, or otherwise covalently bonded to cellular structures, and what role do they play in maintaining the electrical "insulation" of nerve cells, on the one hand, and the response to damage on the other hand? This latter possibility is brought into focus by the fact that the mitochondrial "peripheral-type" benzodiazepine receptor, which is involved in the cleavage of cholesterol side chain to generate pregnenolone (21), is markedly upregulated by neural damage (60).

Integration of Mechanisms with Physiology and Behavior

The greatest challenge of all is integrating what is known about the individual cellular effects of steroid hormones and identifying their ultimate contributions to integrated brain function and behavior. Steroid hormones regulate a number of important aspects of brain function and coordinate them with events elsewhere in the body. Sex hormones regulate and coordinate reproduction and reproductive behavior (61–63); glucocorticoids modulate the sleep–waking pattern of rest and cognitive activity ([29] in this volume), along with food-seeking behavior and metabolism (64, 65) and cytokine responses to damage and infection ([31] in this volume), mineralocorticoids, along with glucocorticoids, regulate blood pressure ([30] in this volume) and the intake and retention of salt and body fluids (66–68).

It is clear in each of these examples from what is already known at a mechanistic level that both genomic and nongenomic mechanisms are involved, and that neurotransmitters and circulating hormones cooperate in various ways to bring about the behavioral and physiological changes; furthermore, many mechanisms known to operate in the periphery apparently also operate in the brain, for example, regulation of cytokine production ([31] in this volume). But how do we find out if a given gene product plays a role in brain function and behavior? Two approaches are available based on genetic engineering. One is the use of antisense for specific genes ([21] in this volume), and the application of this technique has already led to fascinating results concerning, for example, the role of estrogen receptors in brain sexual differentiation (69). Another approach is the "knock-out" strategy, or the variant of introducing a gene expressing antisense RNA as a means of neutralizing sense mRNA expression (70). This latter approach is very powerful, but so powerful that the entire organism is affected in ways that are sometimes difficult to unravel, as exemplified by the results of introducing an antisense-producing gene for the glucocorticoid receptor into mice (70).

A final and most important aspect of steroid hormone action is the developmental effects that have been alluded to throughout this chapter. Sexual differentiation is only one example of these effects and thyroid hormone actions during development are another (see above). Equally important are the long-term effects of early experience, especially stress, in which adrenal steroids and thyroid hormone may play an important role (71). Like the impact of sexual differentiation on adult brain responses to the environment (see above), one of the most intriguing aspects of early experience is how it influences the degree to which "wear and tear" during adult life affects the neural response to stress and ultimately how rapidly the brain ages [see (70–72) for discussion of these issues].

Thus, circulating steroid hormones have a profound influence on all aspects of brain function, from conception until death (73), and the chapters in this volume have documented the tremendous progress and promise that studies of hormone action in the nervous system have for further progress, as well as some of the ways that this progress contributes to the advance of the neurosciences.

Acknowledgments

Research cited in this review is supported by NIH Grants NS07080, MH41256, and MH43787. The author wishes to thank many laboratory colleagues, past and present, for their contributions to the findings and ideas discussed above.

References

1. A. A. Berthold, *Arch. Anat. Physiol. Wiss. Med.* **16,** 42 (1849).
2. J. Miner and K. Yamamoto, *TIBS* **16,** 423 (1991).
3. J. Miner, M. Diamond, and K. Yamamoto, *Cell Growth and Differentiation* **2,** 525 (1991).
4. M. Sheng and M. E. Greenburg, *Neuron* **4,** 477 (1990).
5. T. E. Meyer and J. F. Habener, *Endocr. Rev.* **14,** 269 (1993).
6. M. Pfahl, *Endocr. Rev.* **14,** 651 (1993).
7. H. Selye, *Endocrinology (Baltimore)* **30,** 437 (1942).
8. C. Frye and J. F. DeBold, *Brain Res.* **612,** 130 (1993).
9. R. Rupprecht, J. M. H. Reul, T. Trapp, B. van Steensel, C. Wetzel, K. Damm, W. Zicglgausberger, and F. Holsboer, *Neuron* **11,** 523 (1993).
10. B. S. McEwen, *TIPS* **112,** 141 (1991).
11. S. D. Erulkar and D. M. Wetzel, *J. Neurophysiol.* **61,** 1036 (1989).
12. J. M. Rendt, L. Toro, and E. Stefani, *Am. J. Physiol.* **262,** C293 (1992).
13. M. Boyle, N. J. MacLusky, F. Naftolin, and L. Kaczmarek, *Nature (London)* **330,** 373 (1987).
14. E. Levitan, L. Hemmick, N. Birnberg, and L. Kaczmarek, *Mol. Endocrinol.* **5,** 1903 (1991).
15. M. Wong and R. Moss, *J. Neurosci.* **12,** 3217 (1992).
16. N. G. Weiland, *Endocrinology (Baltimore)* **131,** 662 (1992).
17. N. G. Weiland, *Endocrinology (Baltimore)* **131,** 2697 (1992).
18. M. Schumacher, H. Coirini, and B. S. McEwen, *Brain Res.* **484,** 178 (1989).
19. R. Loy, G. L. Gerlach, and B. S. McEwen, *Dev. Brain Res.* **39,** 245 (1988).
20. C. S. Woolley and B. S. McEwen, *J. Neurosci.* in press (1994).
21. A. Korneyev, B. S. Pan, A. Polo, E. Romeo, A. Guidotti, and E. Costa, *J. Neurochem.* **61,** 1515 (1993).
22. C. Corpechot, J. Young, M. Calvel, C. Wehrey, J. Veltz, G. Touyer, M. Mouren, V. Prasad, C. Banner, J. Sjovall, E. Baulieu, and P. Robel, *Endocrinology (Baltimore)* **133,** 1003 (1993).
23. B. Schlinger and A. Arnold, *Proc. Natl. Acad. Sci. U.S.A.* **88,** 4191 (1991).
24. M. Birmingham, M. Sar, W. Stumpf, and E. Stumpf, *Cell. Mol. Neurobiol.* **13,** 373 (1993).
25. A. Miller, R. Spencer, M. Pulera, S. Kang, and B. S. McEwen, *Biol. Psychiatry* **32,** 850 (1992).
26. B. S. McEwen, L. Plapinger, C. Chaptal, J. Gerlach, and G. Wallach, *Brain Res.* **96,** 400 (1975).
27. G. Hammond, *Endocr. Rev.* **11,** 65 (1990).

28. W. Rosner, D. Hryb, M. S. Khan, A. Nakhla, and N. Romas, *J. Steroid Biochem. Mol. Biol.* **40,** 4 (1991).
29. S. Mellon and C. Deschepper, *Brain Res.* **629,** 283 (1993).
30. S. Paul and R. Purdy, *FASEB J.* **6,** 2311 (1992).
31. P. M. Stewart and C. Edwards, TEM 225 (1990).
32. R. E. Brinton and B. S. McEwen, *Neurosci. Res. Commun.* **2,** 37 (1987).
33. R. R. Sakai, V. Lakshmi, C. Monder, and B. S. McEwen, *J. Neuroendocrinol.* **4,** 101 (1992).
34. R. C. Melcangi, F. Celotti, P. Castano, and L. Martini, *Endocrinology (Baltimore)* **132,** 1252 (1993).
35. I. Lieberburg and B. S. McEwen, *Endocrinology (Baltimore)* **100,** 588 (1977).
36. B. McEwen, *Reprod. Physiol. IV* (R. Greep, ed.) **27,** 99 University Park Press, Baltimore, MD (1983).
37. M. Frankfurt, E. Gould, C. Woolley, and B. S. McEwen, *Neuroendocrinology* **51,** 530 (1990).
38. A. Segarra and B. S. McEwen, *Neuroendocrinology* **54,** 365 (1991).
39. M. Frankfurt and B. S. McEwen, *Neuroendocrinology* **54,** 653 (1991).
40. A. Matsumoto and A. Yasumasa, *Neuroendocrinology* **42,** 232 (1986).
41. B. Parsons, T. Rainbow, and B. S. McEwen, *Endocrinology (Baltimore)* **115,** 1412 (1984).
42. B. S. McEwen, *Semin. Neurosci.* **3,** 497 (1991).
43. E. Gould and B. S. McEwen, *Curr. Opinion Neurobiol.* **3,** 676 (1993).
44. R. Roof, *Brain Res.* **610,** 148 (1993).
45. E. Gould, C. Woolley, and B. S. McEwen, *Psychoneuroendocrinology* **16,** 67 (1991).
46. D. F. Sherry, L. F. Jacobs, and S. J. Gaulin, *TINS* **15,** 298 (1992).
47. C. Woolley and B. S. McEwen, *J. Comp. Neurol.* **336,** 293 (1993).
48. E. R. DeKloet, D. A. Voorhuis, Y. Boschma, and J. Elands, *Neuroendocrinology* **44,** 415 (1986).
49. M. Schumacher, H. Coirini, D. W. Pfaff, and B. S. McEwen, *Science* **250,** 691 (1990).
50. S. Mendelson and B. S. McEwen, *Neuroendocrinology* **55,** 444 (1992).
51. D. Chalmers, S. P. Kwak, A. Mansour, H. Akil, and S. Watson, *J. Neurosci.* **13,** 914 (1993).
52. P. W. Burnet, I. N. Mefford, C. C. Smith, P. W. Gold, and E. M. Sternberg, *J. Neurochem.* **59,** 1062 (1992).
53. C. D. Toran-Allerand, R. C. Miranda, W. Bentham, F. Sohrabji, T. J. Brown, R. B. Hochberg, and N. J. MacLusky, *Proc. Natl. Acad. Sci. U.S.A.* **89,** 4668 (1992).
54. A. Ferreira and A. Caceres, *J. Neurosci.* **11,** 392 (1991).
55. B. S. McEwen, H. Cameron, H. Chao, E. Gould, A. M. Magarinos, Y. Watanabe, and C. Woolley, *Cell Mol. Neurobiol.* **13,** 457 (1993).
56. P. Liang and A. Pardee, *Science* **257,** 967 (1992).
57. J. D. Blaustein, *Endocrinology (Baltimore)* **131,** 1336 (1992).
58. D. Morris, *J. Steroid Biochem. Mol. Biol.* **45,** 19, 25 (1993).
59. J. J. Reilly, D. B. Mamani, J. Schulkin, B. Slotnik, B. S. McEwen, and R. R. Sakai, *Soc. Neurosci. Abstr.* **17,** No. 239.6, p. 582 (1993).

60. J. Benavides, F. Bourdiol, A. Dubois, and B. Scatton, *Neurosci. Lett.* **125,** 219 (1991).
61. D. W. Pfaff, "Estrogens and Brain Function." Springer-Verlag, New York, 1980.
62. B. S. McEwen, K. Jones, and D. Pfaff, *Biol. Reprod.* **36,** 37 (1987).
63. B. S. McEwen, H. Coirini, A. Danielsson, M. Frankfurt, E. Gould, S. Mendelson, M. Schumacher, A. Segarra, and C. Woolley, *J. Steroid Biochem. Mol. Biol.* **40,** 1 (1991).
64. D. Tempel and S. Leibowitz, *Brain Res. Bull.* **23,** 553 (1989).
65. B. S. McEwen, R. Sakai, and R. Spencer, *in* Hormonally Induced Changes in Mind and Brain'' (J. Schulkin, ed.), pp. 157–189. Academic Press, San Diego, 1993.
66. R. R. Sakai, S. Nicolaidis, and A. Epstein, *Am. J. Physiol.* **251,** R762 (1986).
67. J. Schulkin, "Sodium Hunger." Cambridge Univ. Press, London, 1991.
68. L. Ma, B. S. McEwen, R. Sakai, and J. Schulkin, *Horm. Behav.* **27,** 40 (1993).
69. M. McCarthy, E. Schlenker, and D. Pfaff, *Endocrinology (Baltimore)* **133,** 433 (1993).
70. M. Pepin, F. Pothier, and N. Barden, *Nature (London)* **355,** 725 (1992).
71. M. Meaney, S. Bhatnagar, J. Diorio, S. Larocque, D. Francis, D. O'Donnell, N. Shanks, S. Sharma, J. Smythe, and V. Viau, *Cell. Mol. Neurobiol.* **13,** 321 (1993).
72. R. Sapolsky, "Stress, the Aging Brain and the Mechanisms of Neuron Death." MIT Press, Cambridge, Massachusetts, 1992.
73. B. S. McEwen, *Progr. Brain Res.* **93,** 365 (1992).

Index